Power Electronics Basics

OPERATING PRINCIPLES, DESIGN, FORMULAS, AND APPLICATIONS

Power Electronics Basics

OPERATING PRINCIPLES, DESIGN, FORMULAS, AND APPLICATIONS

Yuriy Rozanov • Sergey Ryvkin
Evgeny Chaplygin • Pavel Voronin

CRC Press
Taylor & Francis Group
Boca Raton London New York

CRC Press is an imprint of the
Taylor & Francis Group, an **informa** business

CRC Press
Taylor & Francis Group
6000 Broken Sound Parkway NW, Suite 300
Boca Raton, FL 33487-2742

Printed on acid-free paper
Version Date: 20150305

International Standard Book Number-13: 978-1-4822-9879-6 (Hardback)

Visit the Taylor & Francis Web site at
http://www.taylorandfrancis.com

and the CRC Press Web site at
http://www.crcpress.com

Contents

Preface

Efficiency of the use of electrical energy can be significantly improved by means of power electronics. The latter is the fastest growing field of electrical engineering. This book is devoted to the world of power electronics. It presents and explains the basics and the most important concepts of power electronics. It also gives us the fundamental knowledge for analysis and design in its area. Readers will find many practical examples that demonstrate the achievements and prospects of developing and using power electronics including mainstream such as renewable energy production, transferring, and distribution.

This book will be of interest to many readers including graduate students, industry professionals, researchers, and academics.

The authors express their gratitude to all who helped in the creation of this book.

Yuriy Rozanov
Sergey Ryvkin
Evgeny Chaplygin
Pavel Voronin

Authors

Yuriy Rozanov graduated from the Moscow Power Engineering Institute (MPEI) in 1962 and earned a diploma in electromechanical engineering. After graduation, he worked at plant "Searchlight" of electrical industry in various positions from engineer to deputy chief designer. In 1969, he earned his PhD. In 1987, he earned his PhD in Technical Sciences. In 1989, he joined the MPEI to head the Department of Electrical and Electronic Apparatus where currently he is a professor.

He is the author of 7 books and more than 160 articles and 24 patents. Under his leadership and supervision, postgraduate students earned master's degrees.

For exemplary services in the fields of education and science and technology, he was awarded the honorary title of Honored Scientist of the Russian Federation. He is the winner of the Government of the Russian Federation award in the field of science (2001) and education (2005).

He is the chairman of the Joint Chapter PEL/PES/IES/IAS of the Russian section of the IEEE (Institute of Electrical and Electronics Engineers). In 2012, he was awarded the title of Fellow. From 2010 onward, he has been the editor-in-chief of the journal *Russian Electrical Engineering*.

His areas of technical and scientific interests are power electronics and its applications.

Sergey Ryvkin graduated with high honors as an engineer from the Moscow Institute for Aviation Engineering (Technical University), after which he earned his PhD from the Institute of Control Sciences (USSR Academy of Science) in Moscow and was awarded a DSc from the Supreme Certifying Commission of Russian Ministry of Education and Science in Moscow. He is currently a professor at the Moscow Power Engineering Institute (Technical University) and a main researcher at the Trapeznikov Institute of Control Sciences from the Russian Academy of Sciences. His research interests include application of the sliding mode technique to control electrical drives and power systems and their parameter observation. Professor Ryvkin holds six patents and has published 2 monographs,

5 textbooks, and more than 130 technical papers in international journals and conference proceedings. He is a senior member of the IEEE, a full member of the Russian Academy of Electrotechnical Sciences and a full member of the Power Electronics and Motion Control Council (PEMC-C). He is a deputy editor-in-chief of the leading Russian electrical engineering journal *Electrotechnika* (English version *Russian Electrical Engineering*) and a member Board of Editors of the international journals *International Journal of Renewable Energy Research, Transactions on Electrical Engineering* and the *International Journal of Advances in Telecommunications, Electrotechnics, Signals and Systems (IJATES2)* and leading Russian electrical engineering journal *Elektrichestvo*.

Evgeny Chaplygin graduated from the Department of Radiotechnics, Moscow Power Engineering Institute (MPEI) in 1965. In 1974, he was awarded PhD on "Power Electronics." Since 1966, he has taught at the Department of Industrial Electronics, Moscow Energy Power Institute, as an associate professor since 1980 and as a professor since 2011. He supervised 11 students toward their master's degree.

He has published two books, the textbook *Industrial Electronics*, more than 100 papers, and has 70 patented inventions.

He is a member of the Academic Council of the Institute of Radio Electronics of MPEI, member of the editorial board of the journal *Russian Electrical Engineering*.

His areas of interest include modeling of power electronic devices, improved EMC converters and mains, increasing the quality of electricity by means of power electronics.

Pavel Voronin earned his certificate in electrical engineering from the Moscow Power Engineering Institute (MPEI), USSR in 1980, and a PhD in power electronics from the same institute in 1983.

From 1983 to 1985 he worked at MPEI and then he moved to the industrial electronics department, where he worked as an assistant, and since 1988 as an assistant professor.

He has published more then 70 papers, 2 books, and 31 patents in the field of power electronics.

His areas of interest include power converters, multilevel inverters, soft switching circuits, and computer simulations of power electronic devices.

chapter one

Basic concepts and terms in power electronics

1.1 Conversion of electrical energy: Classification of converters

Power electronics involves the conversion of electrical energy or the switching (on or off) of an electrical power circuit, with or without control of the electrical energy (IEC, 551-11-1). In the Russian literature, power electronics associated with the conversion of electrical energy is often referred to as converter engineering.

Electronic power conversion may be defined as the modification of one or more parameters of electrical energy by means of electronic devices, without significant power losses (IEC, 551-11-2).

A converter is a device for power conversion, with one or more switching components and also, if necessary, transformers, filters, and auxiliary devices (IEC, 551-12-01).

A converter for power-supply purposes is often called a *secondary power source*, in contrast to primary sources such as batteries, solar cells, and ac grids.

The main forms of power conversion are as follows (Kassakian et al., 1991; Mohan et al., 2003; Rozanov, 2007; Zinov'ev, 2012):

1. *Rectification* (conversion from ac to dc). The corresponding converter is known as a *rectifier* or an ac/dc converter for rectification. The energy source for most consumers is a single- or three-phase general-purpose ac grid. At the same time, dc voltage is required for control and communications devices and computers. It is also required for certain components of drives, electrical equipment, and optical equipment. In autonomous systems, the ac sources are rotary generators. Rectifiers are the most common power converters.
2. *Inversion* (conversion from dc to ac). The corresponding converter is known as an *inverter* or an ac/dc converter for inversion. The dc source of an inverter may be a battery, a solar cell, or a dc transmission line or else another converter.
3. *Conversion from ac to dc and vice versa*. The corresponding converter is known as an *ac/dc converter*. Such converters are able to change the

direction of the energy flux and may operate either in the rectifier mode, when the energy is transmitted from an ac grid to a dc circuit, or in the inverter mode, when the energy is transmitted from a dc circuit to an ac grid. In an electric drive, changing the direction of the energy flux ensures a recuperative braking of the machine.

4. *Conversion from ac to ac*. The corresponding converter is known as an *ac converter*. Three types exist:
   a. The *ac voltage converter*, with the same number of phases and the same voltage frequency at the input and output, changes the voltage amplitude (upward or downward) and/or improves the voltage by stabilizing the fundamental harmonic or by adjusting the harmonic composition.
   b. The *frequency converter* transforms an m_1-phase voltage of frequency f_1 to an m_2-phase voltage of frequency f_2. Many components of drives, electrical equipment, and optical equipment require alternating current whose frequency is variable or differs from the industrial frequency of 50 (60) Hz. Such components include ac frequency drives, inductive heaters, and power sources of optical equipment. Rotary generators in autonomous systems often produce voltage of unstable frequency, whose stabilization requires frequency converters.
   c. The *phase converter* transforms single-phase to three-phase voltage and vice versa. Conversion of single-phase to three-phase voltage permits power supply to three-phase components in the absence of a three-phase grid. The connection of high-power single-phase loads to a single phase of a three-phase grid imposes an asymmetric burden on the grid, and it is expedient to use the appropriate phase converter in that case.
5. *Conversion from dc to dc*. The corresponding converter is known as a *dc converter*. Such conversion improves the power of a dc source and matches the voltage of the source and consumers. It is most often employed for components whose power source is a low-voltage battery.
6. *Reactive power conversion*. The corresponding converter is known as a *reactive power converter*. Such conversion compensates the reactive power generated or consumed (Section 1.3). Converters of this type draw active power from the grid only for the compensation of losses.

These are the basic types of power conversion, but others also exist. For example, in technological systems, units for the generation of powerful single pulses are employed. As technology develops, the list of power-electronic components may grow.

As already noted, converters may change the direction of the energy flux. A *converter with one possible direction of power flow* transmits energy only in one direction: from the source to the load. Converters in which the direction of the energy flux may change are known as *reversible converters*. *Two-quadrant converters* may change the direction of the energy flux by changing the polarity of the voltage or current in the load circuit. *Four-quadrant converters* may change the direction of the energy flux by changing the direction of both the voltage and the current. *Multiquadrant converters* capable of changing the direction of the energy flux may be based not only on ac/dc converters, but also on ac converters of different types and dc converters.

In direct power conversion, there is no conversion to other types of electrical energy. In this case, we use *direct converters*. *Indirect converters* are also widely used, as illustrated in Figure 1.1. Figure 1.1a shows a frequency converter with an intermediate dc component consisting of a rectifier supplied from a grid of frequency f_1 and an inverter producing voltage of frequency f_2. Figure 1.1b shows a dc converter with a high-frequency intermediate ac component consisting of an inverter and a rectifier. The intermediate component includes a transformer. Figure 1.1c shows an indirect rectifier consisting of a rectifier Rc1, an inverter, and a rectifier Rc2, with

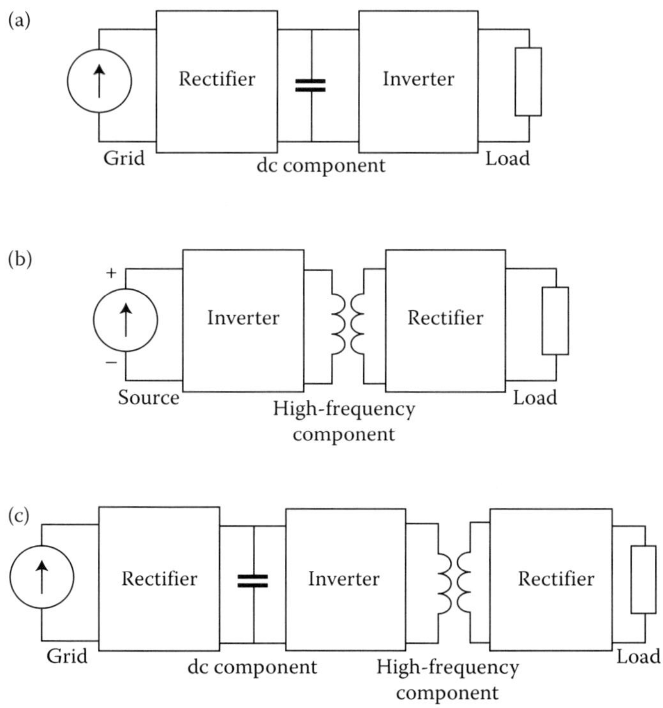

Figure 1.1 Functional diagrams of indirect converters: (a) frequency converter; (b) dc converter; and (c) rectifier.

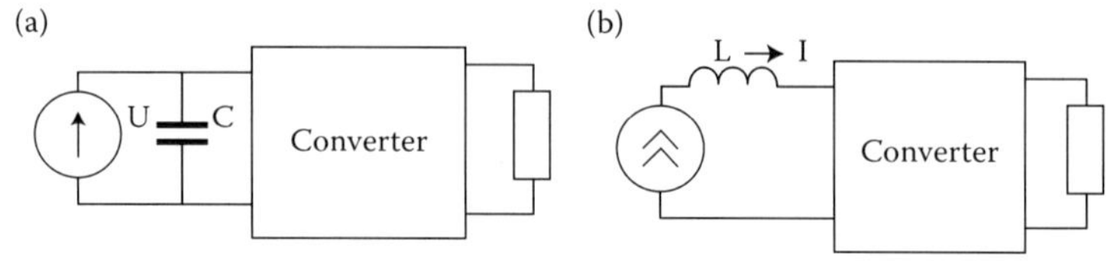

Figure 1.2 (a) Voltage converter and (b) current converter.

two intermediate components: a dc component between the rectifier Rc1 and the inverter and a high-frequency ac component between the inverter and rectifier Rc2.

Although indirect conversion increases the energy losses and reduces the efficiency, it is widely used, as it offers benefits such as better quality of the voltage and current at the converter output and/or input and reduced size and mass of the equipment (thanks to high-frequency transformation).

Depending on the properties of the power source, we may distinguish the voltage and current converters. At the input of a *voltage converter,* the properties of the power source U resemble those of an emf source. As a rule, a capacitor C is connected in parallel to the source (Figure 1.2a). At the input of a *current converter,* the properties of the power source I resemble those of a current source. A choke L is connected in series with the source (Figure 1.2b).

The converters consist of a power component and also a *control system,* in the form of information-processing components that generate pulses sent to the control electrodes of the power switches. The interface components of the control system are voltage and/or current sensors installed in the power component of the converter, in the load circuit, or in the power source.

Converters based on uncontrollable switches (diodes) are known as *noncontrollable converters.* Such converters have no control system.

1.2 *Output parameters and characteristics of converters*

Consumers of electric power impose numerous requirements on converters, often without adequate specificity. A switch-based converter generates voltage and current of complex harmonic composition. As an example, we show the output voltage of a controllable rectifier in Figure 1.3a and the output voltage of a voltage inverter formed by pulse-width modulation in Figure 1.3c; the corresponding spectra are shown in Figure 1.3b and d.

The useful component of the output voltage (or current) of a converter with a dc output is the constant component of the Fourier-series expansion in Figure 1.3b. The constant component of the spectrum (the harmonic with $k = 0$) corresponds to the mean over the repetition period T:

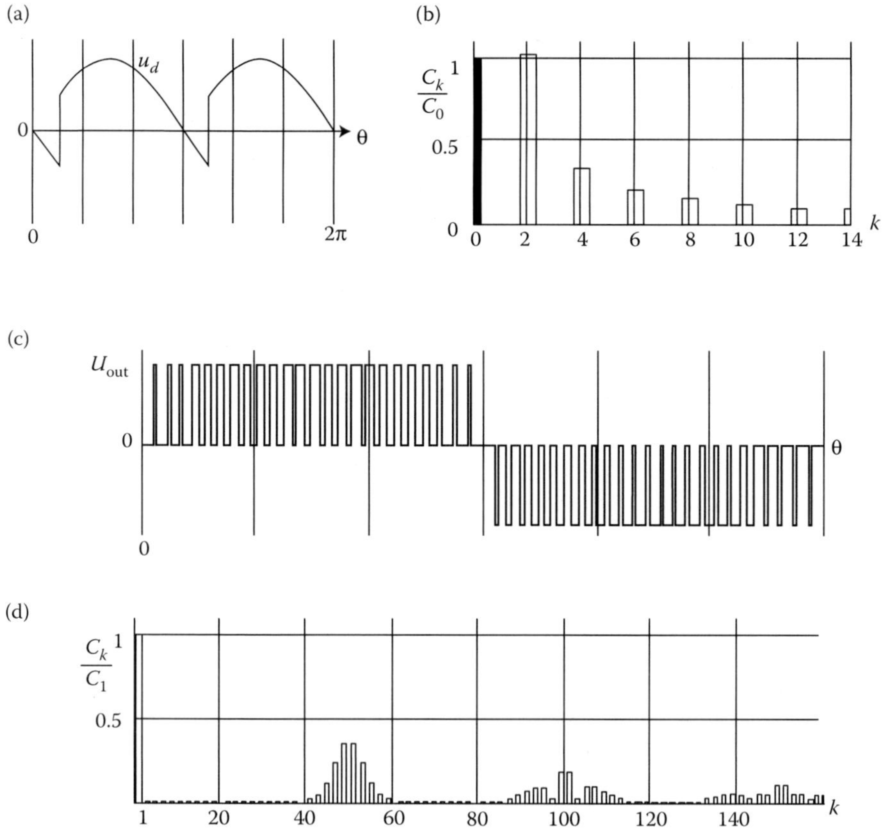

Figure 1.3 (a) Output voltage of a controllable rectifier (b) and its spectrum and (c) output voltage of an inverter formed by pulse-width modulation (d) and its spectrum.

$$U_d = \frac{1}{T}\int_0^T u_d(t)\mathrm{d}t = \frac{1}{\pi}\int_0^\pi u_d(\theta)\mathrm{d}\theta, \tag{1.1}$$

where $\theta = \omega t$. For the example shown in Figure 1.3a, the repetition period of the voltage is half the grid period.

The useful component of the output voltage or current of a converter with an ac output is the *fundamental* with the converter's output frequency (the harmonic with $k = 1$), determined in Fourier-series expansion of the voltage or current (Figure 1.3d).

For consumers whose operation is not critical to the harmonic composition of the voltage or current, the effective value is generally adopted as the useful component of the output voltage or current.

As is evident from Figure 1.3b and d, the spectra contain not only the useful component, but also other harmonics, which may sometimes be subharmonics. In accordance with consumer requirements, these harmonics may be suppressed by means of filters at the converter output (Section 1.5).

Spectra give the best idea of the quality of the output voltage or current. However, the spectral composition of a converter changes with its operating conditions. Therefore, we need generalized criteria for assessing the harmonic composition so as to identify the least favorable conditions and also to compare different methods of generating the output voltage.

The quality of alternating voltage or current may be assessed on the basis of the following characteristics:

1. The *total harmonic distortion* (THD), which is the ratio of the effective value of the higher harmonics to the fundamental of the periodic function

$$k_{\text{thd}} = \frac{\sqrt{\sum_{k \neq 1} C_k^2}}{C_1}. \tag{1.2}$$

For a sinusoidal function, the THD is zero. An increase in k_{thd} indicates deterioration in the harmonic composition. If we use a voltage converter, the THD of the output voltage is largely independent of the load; in contrast, the THD for the output current characterizes the converter—load system.

2. The *fundamental factor*, which is the ratio of the effective values of the fundamental and the periodic function

$$k_{\text{fu}} = \frac{C_1/\sqrt{2}}{\sqrt{\sum_{k=1}^{\infty} \left(C_k/\sqrt{2}\right)^2}} = \frac{C_1}{\sqrt{\sum_{k=1}^{\infty} C_k^2}}. \tag{1.3}$$

For a sinusoidal function, $k_{\text{fu}} = 1$.

In dc circuits, we consider not only the useful component, but also the variable component (the *ripple*). The current quality is assessed by means of the *dc ripple factor*, which is the ratio of half the difference of the maximum and minimum values of the pulsating current to its mean value. Various methods are used to assess voltage pulsation:

- From the effective value of the variable component
- From the difference in the maximum and minimum instantaneous values of the voltage
- From the amplitude of the lowest harmonic of the pulsation

The following parameters are significant: the stability of the fundamental and the possibility of its regulation, the maximum and minimum values of U_{out} and I_{out}, and the maximum active power in the load P_{out}.

For the power consumer, the *characteristic of the converter* is important: this curve shows the relation between the output voltage and the output current. In dc circuits, U_{out} and I_{out} are characterized by constant components; in ac circuits, they are characterized by the effective values of the fundamental with the output frequency. Typical converter characteristics are shown in Figure 1.4.

For the family of converter characteristics shown in Figure 1.4a, a variation in load power has no pronounced influence on the voltage. In other words, the converter characteristics resemble those of the emf source. Such characteristics are said to be hard. In this case, γ is the control parameter of the converter, specified by the control system. The slope of the converter characteristics corresponds to the active power losses in the power and often also to various unrelated factors. In the case of Figure 1.4b, the converter has the properties of a current source. Such converter characteristics are often used in electric drives and in technological equipment. The converter characteristics in Figure 1.4c have three sections. In section III, the load current is stabilized. With increasing voltage, the converter passes to section II, where the power is stabilized (as a rule, at the maximum permissible power of the converter or load). With further decrease in the load current, the converter reaches section I, where the output voltage is limited to permissible values.

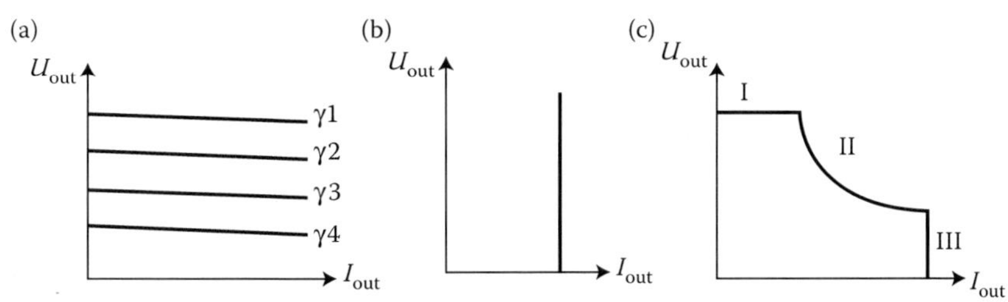

Figure 1.4 Typical converter characteristics: (a) hard characteristics (for converters with the properties of an emf source); (b) characteristic of a converter with the properties of a current source; and (c) characteristic of a converter with voltage (I), power (II), and current (III) stabilization.

The converter characteristic determined solely by the power equipment of the converter is known as the *natural characteristic*. It is recorded when there is no feedback with respect to the converter's output parameters. As a rule, voltage converters have a hard natural characteristic.

The converter characteristic obtained in the presence of modifying equipment—for example, in voltage, current, or power stabilization by the corresponding feedback loop—is known as a *forced characteristic*. Examples of forced characteristics are seen in Figure 1.4b and c.

1.3 Influence of converters on the grid

Power-electronics components based on semiconductor switches constitute a nonlinear load for the grid (Czarnecki, 1987; Emanuel, 1999; Kassakian et al., 1991; Mohan et al., 2003; Zinov'ev, 2012). Consider the operation of a converter with a single-phase sinusoidal grid. The characteristic electromagnetic processes are illustrated in Figure 1.5. Figure 1.5a shows the grid voltage u and the current i consumed by the converter. The instantaneous power, shown in Figure 1.5b, is $p(\theta) = u(\theta) \cdot i(\theta)$, where $\theta = \omega t$. When $p > 0$, energy is transmitted from the grid to the converter and then to the load. When $p < 0$, energy is returned from the load to the grid.

By definition, the *active power* consumed by the converter from the grid is

$$P = \frac{1}{T}\int_{0}^{T} p(t)\mathrm{d}t. \tag{1.4}$$

We divide the fundamental (first harmonic) i_1 of the current i (shown in Figure 1.5a) into two components: the *active component* i_{1a}, with the same phase as the grid voltage (Figure 1.5b), and the *reactive component* i_{1r}, with a phase lag of $\pi/2$ (Figure 1.5c). Then the current may be expressed in the form

$$i = i_1 + i_h = i_{1a} + i_{1r} + i_h, \tag{1.5}$$

where i_h is the sum of higher harmonics of the current i (Figure 1.5d).

The instantaneous power p may also be divided into three components

$$p(\theta) = u(\theta) \cdot [i_{1a}(\theta) + i_{1r}(\theta) + i_h(\theta)] = p_{1a}(\theta) + p_{1r}(\theta) + p_h(\theta). \tag{1.6}$$

We plot p_{1a}, p_{1r}, and p_h in Figure 1.5b through d, respectively.

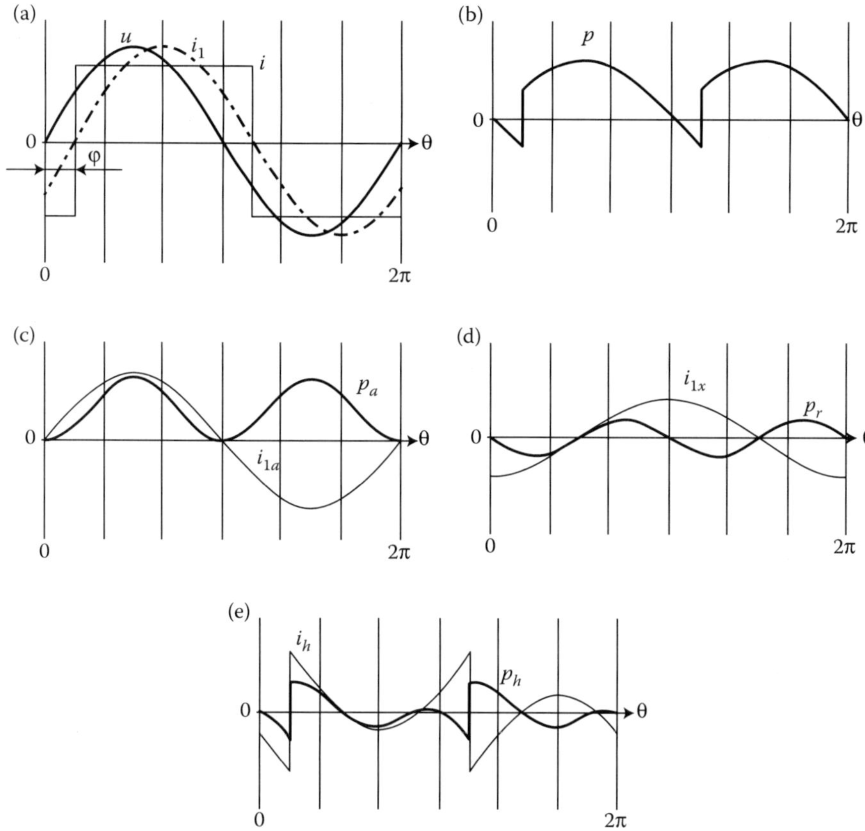

Figure 1.5 Influence of the converter on the grid: (a) grid voltage and current consumed by the converter; (b) instantaneous power; (c) active component of the current fundamental and the corresponding instantaneous power; (d) reactive component or the current fundamental and the corresponding instantaneous power; and (e) sum of the higher current components and the corresponding instantaneous power.

The instantaneous power $p_{1a}(\theta)$ (Figure 1.5b) includes a constant component. This means that the active power transmits the active component of the fundamental of current i to the load. Conversely, no constant component is seen in $p_{1r}(\theta)$ or $p_h(\theta)$. Therefore, the reactive current of the fundamental and higher harmonics is not involved in the transmission of active power and is responsible for the useless oscillation of the energy between the grid and the converter.

The total (apparent) power consumed by the converter from the grid is $S = UI$, where U and I are the effective values of the voltage u and current i, respectively. The total power in the single-phase sinusoidal grid includes the active power P, the reactive power of the fundamental harmonic Q, and the distortion power T:

$$S = \sqrt{P^2 + Q^2 + T^2}. \tag{1.7}$$

The *power factor* is the ratio of the active power to the total power

$$\chi = \frac{P}{S}. \tag{1.8}$$

In a single-phase sinusoidal grid, the power factor is

$$\chi = \nu \cdot \cos\varphi, \tag{1.9}$$

where ν is the fundamental factor of current I (the ratio of the effective values of the fundamental to the total current: $\nu = I_1/I$), which characterizes the nonsinusoidal form of current i and the distortion power T; φ is the phase shift of current i_1 relative to the grid voltage u (Figure 1.5a), characterizing the reactive power Q.

The active and reactive power and the distortion power are expressed in terms of the total power:

$$\begin{aligned} P &= UI_1 \cos\varphi = S \cdot \nu \cdot \cos\varphi, \\ Q &= UI_1 \sin\varphi = S \cdot \nu \cdot \sin\varphi, \\ T &= \sqrt{S^2 - P^2 - Q^2}. \end{aligned} \tag{1.10}$$

In a three-phase grid, the total power, active power, and reactive power are equal to the sums of the corresponding components for the phase components:

$$\begin{aligned} S &= U_A I_A + U_B I_B + U_C I_C = S_A + S_B + S_C, \\ P &= P_A + P_B + P_C, \\ Q &= Q_A + Q_B + Q_C. \end{aligned} \tag{1.11}$$

In a symmetric three-phase sinusoidal grid, Equation 1.9 applies.

In an asymmetric three-phase sinusoidal grid, an additional component appears: the asymmetry power N characterizing the energy transfer between the phases

$$S = \sqrt{P^2 + Q^2 + T^2 + N^2}. \tag{1.12}$$

With an asymmetric grid, the power is transmitted to the load only by the active component of the grid current's direct sequence.

When a converter is connected to a nonsinusoidal grid, the active power and reactive power are equal to the sum of the active and reactive power of all the harmonics of voltage *u*:

$$P = \sum_{k=1}^{\infty} P_k, \quad Q = \sum_{k=1}^{\infty} Q_k, \tag{1.13}$$

where *k* is the number of harmonics.

We cannot use Equation 1.9 for nonsinusoidal and three-phase asymmetric grids.

The appearance of inactive components in the grid current reduces the power factor and has undesirable technological and economic consequences.

1. In the system, not only the active, but also the reactive power must be kept in balance. If the reactive power is not taken in hand, the grid frequency will decline.
2. The inactive components of the total power perform no work. To transmit the same active power to the consumer, with a decrease in the power factor, the current must be increased. That necessarily increases expenditures on conductors, transformers, and switchgear.
3. In grids of limited power, which do not fully correspond to the properties of an emf source, the reactive current reduces the voltage at the consumer terminals, whereas the higher current harmonics produce higher harmonics in the grid voltage.
4. The heating of the insulation due to higher harmonics is considerably greater than for the fundamental frequency. That reduces the life of the insulation and may lead to accidents. In transformers and electrical machines, higher harmonics increase the losses in steel and copper components.
5. When the grid is nonsinusoidal, the operation of safety system is impaired; unnecessary shutdowns will occur. In addition, deterioration will be observed in the operation of computers and also communications and automation systems connected to the grid.

For many years, researchers in power electronics have been preoccupied with increasing the power factor of converters and reducing the grid distortion that they produce. Considerable progress has been made possible by the appearance of powerful and completely controllable semiconductor components and microprocessor controllers.

The ideal load for the grid is a device consuming sinusoidal current in phase with the fundamental of the grid voltage. In asymmetric three-phase grids, the ideal load is a device consuming sinusoidal symmetric current in phase with the direct sequence of the fundamental of the grid voltage.

The power factor of converters may be increased by two approaches considered in this book:

1. *Internal measures*: change in the converter's power circuit and/or control algorithm
2. *External measures*: the connection of filters and compensators to the converter (or other load)

1.4 Basic converter parameters

A converter consists of the power component and the control system (Figure 1.6a), each of which has a considerable influence on its basic parameters.

The *transfer factor of the converter* is the ratio of the voltage at the load to the voltage at the source. The voltage on the dc side is characterized by the mean, and the voltage on the ac side by the effective value of the fundamental, as a rule. The converter's maximum transfer factor is an important parameter.

The converter's *efficiency* is the ratio of the active power P_{out} transmitted to the load to the active power P_{in} consumed from the source

$$\eta = \frac{P_{out}}{P_{in}} = \frac{P_{out}}{P_{out} + P_{los}}, \tag{1.14}$$

where P_{los} is the power lost in the converter.

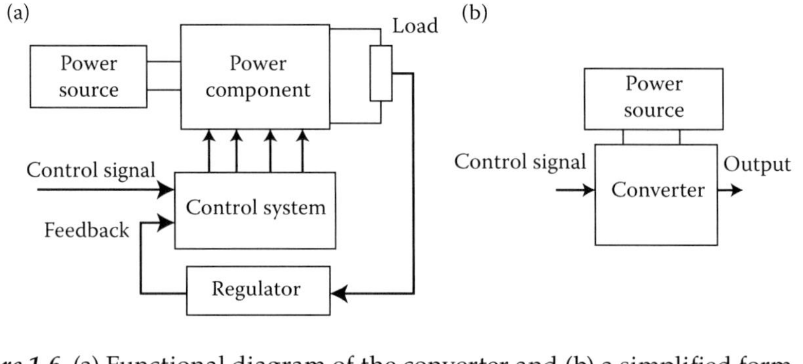

Figure 1.6 (a) Functional diagram of the converter and (b) a simplified form.

The losses in the converter consist of the losses in the semiconductor switches, the losses in the other power components (capacitors, chokes, transformers, etc.), and the power losses in the control system. The losses in the semiconductor switches depend on the number of switches in series through which the basic converter current passes. The proportion of the losses in the semiconductor switches increases significantly with a decrease in the working voltages in the converter—for example, when the converter is connected to low-voltage batteries—and also with an increase in the switching frequency (Chapter 2). The losses in the control system are only significant in low-power converters. The losses in the converter may be reduced not only by increasing the efficiency, but also by reducing the energy consumption in cooling.

The *functionality* of the converter is determined by its ability to ensure all the required electrical input and output parameters in all conditions (Sections 1.2 and 1.3).

The converter's *reliability* depends on the reliability of its components, the complexity of the system, the design of the components, the manufacturing technology, the cooling efficiency, and the safety measures employed. The control system plays a considerable role in ensuring reliable operation. In the event of faults in the control system or the anomalous development of dynamic processes (e.g., in unstable operation of the automatic control system), a fault may appear in the converter; in some cases, the converter, the load, or the source may become inoperative.

Depending on the purpose and application of the converter, different requirements may be imposed on its efficiency, reliability, cost, mass, and size. These requirements are often contradictory.

An analog or digital control signal (or signals) is sent to the input of the control system, specifying the basic electrical parameters in the load. Many specialists represent converters as amplifiers, to whose input a control signal is supplied, while the output is a voltage. Of course, this is a very approximate description, and the actual converter properties differ considerably from those of an amplifier.

The converter (including the control system) may also be characterized as an information transmitter. The converter's forced characteristic is consistent with this approach (Section 1.2).

The *control characteristic* is the dependence of the converter's basic output parameter (e.g., the effective value of the fundamental of the output voltage) on the control signal (e.g., the voltage). The control characteristic may be recorded with or without feedback. In some cases, the control characteristic may be the dependence of the output parameter on the modulated parameter of the power component (e.g., the switching delay α or the filling coefficient γ, which is the ratio of the time that the switch is on to the total switching period).

The dynamic parameters of the converter include the following:

- The maximum (ΔU_{max}) and minimum (ΔU_{min}) deviations of the output voltage in transient conditions
- The duration of the transient control process, ending when the system enters a steady state within the permissible range
- The margin of stability
- The degree of oscillation of the transient process

1.5 ac and dc filters

As noted in Section 1.2, secondary sources based on converters form an output voltage and current of complex harmonic composition. The higher harmonics are limited by means of filters based on reactive components: chokes and capacitors. Despite many types of converters, these filters are of similar structure.

In Figure 1.7, we show the standard filter structures used in power electronics.

The L-shaped LC filter (Figure 1.7a) is used in dc and ac filters. Figure 1.8 shows the filter's transmission factor and the modulus of its input impedance as a function of frequency, with different values of the load resistance: the rated load Z_{rat} (continuous curves) and $20Z_{rat}$ (dotted curves).

The frequency characteristics may be divided into three regions:

1. The transmission region, where the filter's transmission factor $K_f \approx 1$, whereas the input impedance is large
2. The suppression region, where $K_f \ll 1$, whereas the input impedance is large

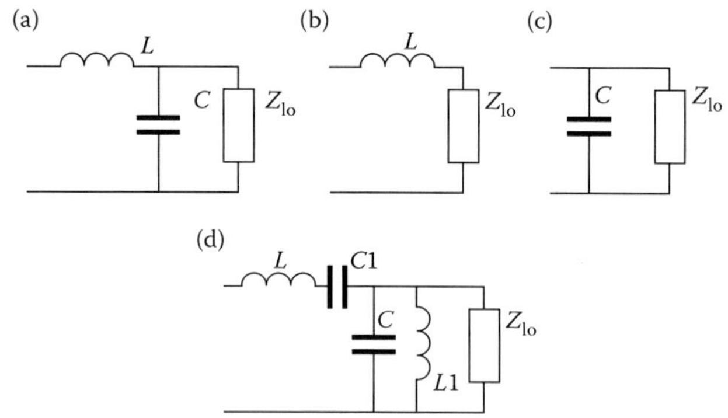

Figure 1.7 Standard filter structures: (a) LC-filter; (b) L-filter; (c) C-filter; and (d) LCLC-filter.

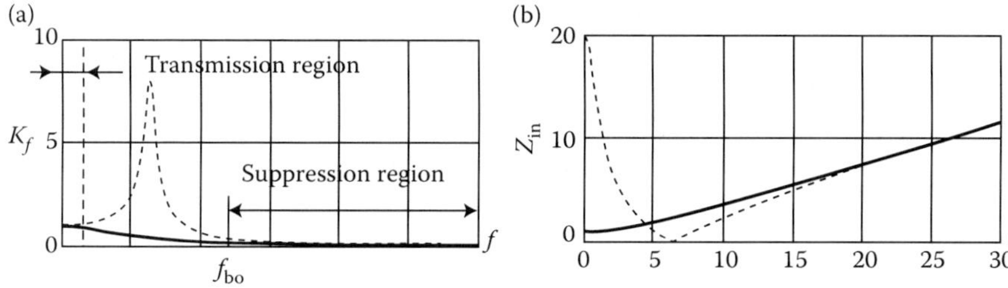

Figure 1.8 Frequency characteristics of an L-shaped LC filter: (a) Transmission factor and (b) modulus of its input impedance.

3. The intermediate region, where the transmission factor increases sharply at the filter's resonant frequency $\omega_{rez} = 1/\sqrt{LC}$ with large load resistance, whereas the input impedance vanishes

In the intermediate region, the frequency spectrum of the converter's output voltage should not contain harmonics, because it is not suppressed but rather amplified by the filter. The upper bound on the intermediate region is $f_{bo} \approx (2\text{–}3)f_{res}$.

The dc filter is intended to reduce (smooth out) the pulsations on the dc side. The smoothing factor S_f is the ratio of the voltage ripple factors at the filter input and output.

Consider the filters for voltage converters with a forced characteristic. When using an LC filter (Figure 1.7a), if we neglect the losses, the filter's transmission factor for the constant component is $K_f(0) = 1$. In the suppression region, the filter's transmission factor in the worst conditions (zero load) for single-phase rectifier is

$$K_f(f) = \frac{1}{4\pi^2 f^2 LC - 1}. \tag{1.15}$$

The smoothing factor is

$$S_f = \frac{1}{K_f(f_{bo})} = 4\pi^2 f_{bo}^2 LC - 1. \tag{1.16}$$

As follows from Figure 1.8, the L-shaped LC filter is effective over a broad range of load values (to zero load).

If we neglect the losses, the transmission factor of the inductive filter (Figure 1.7b) for the constant component is again $K_f(0) = 1$. With an active load R_{lo}, the smoothing factor (indice de pulsation = 2) is

$$S_f = \frac{\sqrt{R_{lo}^2 + 4\pi^2 f_{bo}^2 L^2}}{R_{lo}}. \quad (1.17)$$

The smoothing efficiency is sharply reduced with an increase in R_{lo}. Therefore, the inductive filter is only used for low-resistance loads with little variation.

When connected to a voltage converter, the capacitive filter (Figure 1.7c) either eliminates the converter's voltage-source properties or impermissibly overloads the converter switches with high current harmonics.

Conversely, in current converters, the capacitive filter is always used. The high-frequency components of the filter current are shunted directly through this filter, bypassing the load.

The ac filter is used to reduce the content of higher harmonics in the load voltage and current. The following requirements are imposed on ac filters for voltage converters:

1. The filter's transmission factor for the fundamental must be as close as possible to 1 (i.e., $K_{f1} \approx 1$) at all loads. The losses of the fundamental voltage harmonic in the filter must be minimal. That permits the retention of the hard characteristic and eliminates the need to increase the converter's supply voltage.
2. The converter's higher voltage harmonics must be suppressed enough to permit the functioning of the load circuits. The transmission factor in the suppression region is limited: $K_{f.h} \leq K_{req}$. Here K_{req} is the ratio of the required THD on the load side to the THD at the filter input.
3. Limits must be imposed on the reactive currents corresponding to the fundamental and the higher harmonics shunted through the filter, bypassing the load. This permits reduction in the currents through the converter's semiconductor switches and the currents through the converter at zero load.

With a load that varies widely, the LC filter is commonly used (Figure 1.7a).

To minimize the losses of the fundamental voltage harmonic in the filter, the choke inductance L must be limited. In Figure 1.9, we show the dependence of the filter's transmission factor for the fundamental voltage harmonic on its choke inductance. The inductance is presented in

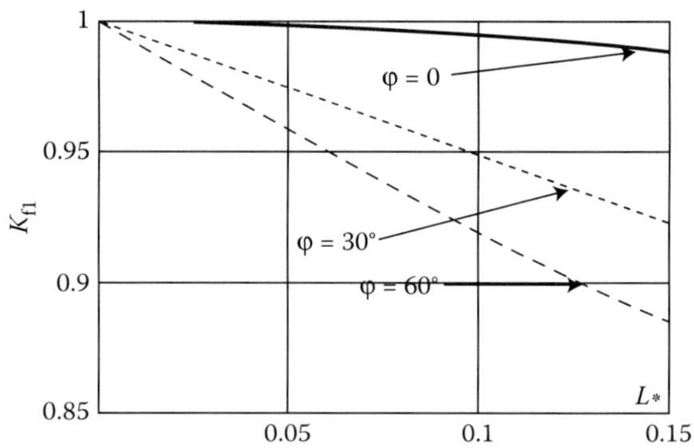

Figure 1.9 Dependence of the transmission factor of the LC filter for the fundamental voltage harmonic on its choke inductance, for different load angles.

the normalized form: $L_* = 2\,L\pi f_{out}/Z_{lo.min}$, where Z_{lo} and φ are the modulus of the impedance and phase angle of the load, for the fundamental, respectively.

The inductance selected from Figure 1.9 may be insufficient for limitation of the reactive current's higher harmonics, which are shunted through the filter, bypassing the load, as the modulus of the filter's input impedance for higher harmonics is $Z_{in.high.} = 2\,L\pi f$, where $f \geq f_{bo}$.

In that case, the inductance is selected so as to limit the reactive current. To ensure a hard characteristic, capacitor C1 is added to the filter (Figure 1.7d). Resonance in the L–C1 series circuit is observed at the fundamental. Therefore, neglecting the losses, we may assume that $K_{f1} = 1$.

The fundamental of the reactive current shunted through the filter, bypassing the load, is determined by capacitor C at the fundamental. If this reactive current is large, we must introduce an additional choke L1 in the filter (Figure 1.7d), so that the resonant frequency of the L1–C parallel circuit corresponds to the fundamental frequency of the converter, and the filter blocks the fundamental of the reactive current.

The filters corresponding to Figure 1.7d have significant deficiencies due to the oscillatory circuits with resonance at the fundamental.

1. They are significantly heavier, larger, and more expensive than filters corresponding to Figure 1.7a.
2. The filter may only function at constant output frequency.

At small f_{bo}/f_{fu}, we need to switch from the filter in Figure 1.7a to that in Figure 1.7d. The use of powerful semiconductor switches with high switching frequency permits expansion of the output-voltage range that is free of harmonics (e.g., see Figure 1.3d) and an increase in f_{bo}/f_{fu}. In that case, the use of an L-shaped LC filter meets all the requirements on the filter parameters, with minimal filter cost.

As a rule, ac output filters for current converters correspond to Figure 1.7c. The expansion of the output-current range that is free of harmonics reduces filter costs and improves its mass and size.

1.5.1 Dynamic processes in filters

The output filters of converters contain reactive elements in which energy is stored. The transient processes when the operating conditions are changed (at startup, adjustment of the load, modification of the supply voltage, etc.) are largely determined by the processes in the filter. When the load varies widely, LC filters are used in both dc and ac circuits.

Now consider the dynamic processes in an L-shaped LC filter (Figure 1.7a).

When current flows through choke L, the energy stored there is

$$W_L = \frac{Li^2}{2}. \tag{1.18}$$

With an abrupt decrease in the load current—for example, on switching off a power consumer—the energy at the choke is supplied to capacitor C, thereby increasing its energy

$$W_C = \frac{CU_C^2}{2} \tag{1.19}$$

and the load voltage. The output voltage begins to oscillate at the filter's resonant frequency ω_{res}, until the excess energy has been scattered in the load. The surges in the output voltage are proportional to $\sqrt{L/C}$ and may considerably exceed the steady output voltage.

The intensity of the output-voltage surges may be reduced by reducing L and increasing C, with constant resonant frequency of the filter. However, this is not always feasible, as it increases the currents through the converter switches or results in a converter characteristic that is no longer hard. To reduce the voltage surges in the load circuit, we may introduce nonlinear elements: varistors.

With an abrupt increase in the load power, the filter capacitor discharges its stored energy to the load. Consequently, the capacitor voltage falls, and there is a trough in the load voltage. The transient process continues until the capacitor's energy has been replenished from the converter's power source.

References

Czarnecki, L.S. 1987. What is wrong with the Budeanu concept of reactive and distortion power and why it should be abandoned. *IEEE Trans. Instrum. Meas.*, 3(3), 834–837.

Emanuel, A.E. 1999. Apparent power definitions for three-phase systems. *IEEE Trans. Power Deliv.*, 14(3), 767–771.

Kassakian, J.C., Schlecht, M.F., and Verghese, G.C. 1991. *Principles of Power Electronics*. Addison-Wesley.

Mohan, N., Underland, T.M., and Robbins, W.P. 2003. *Power Electronics—Converters, Applications and Design*, 3rd edn. John Wiley and Sons.

Rozanov, Ju.K. 2007. Power electronics: Tutorial for universities. (Silovaja jelektronika: Uchebnik dlja vuzov), Rozanov, Ju.K., Rjabchickij, M.V., Kvasnjuk, A.A. M.: Izdatel'skij dom MJeI (in Russian).

Zinov'ev, G.S. 2012. Bases of power electronics (Osnovy silovoj jelektroniki). Moskva: Izd-vo Jurajt (in Russian).

chapter two

Semiconductor power switches and passive components

2.1 Introduction

The limiting mean current of semiconductor power switches is 10 A or more. They may be divided into controllable components (transistors and thyristors) and noncontrollable components (diodes). The designation of power switches indicates the basic semiconductor material, the purpose, or the operating principle of the switch (Rozanov et al., 2007).

In terms of the basic semiconductor material, we distinguish between silicon (Si), germanium (Ge), and gallium-arsenide (GaAs) components. Most power switches are based on silicon. Recently, other semiconductors have also been employed: indium compounds (InP and InAs) and silicon carbide (SiC).

2.2 Power diodes

In terms of function, power diodes may be divided into the following types: rectifier diodes, avalanche rectifier diodes (voltage limiters), avalanche rectifier diodes with controllable breakdown (voltage stabilizers), and pulsed diodes.

2.2.1 Power diodes with p^+–n^-–n^+ structure

A power diode with a p–n junction has the semiconductor structure p^+–n^-–n^+ (Figure 2.1). The central n^- layer is the base. In terms of the electrophysical characteristics, this is the intrinsic silicon semiconductor, with a relatively low donor concentration: $N_d \approx (5–7) \times 10^{13}$ cm^{-3}. In contrast, the external positive layers of the diode structure are highly doped and accordingly are generally described as emitters (Yevseyev and Dermenzhi, 1981).

When a reverse voltage is applied to the diode, the external electric field is mainly concentrated within the high-resistance base. The presence of the highly doped emitter layer n^+ ensures an almost rectangular electric field strength in the base. Consequently, the working voltage of the power diode may be practically twice that of the standard p^+–n^- structure.

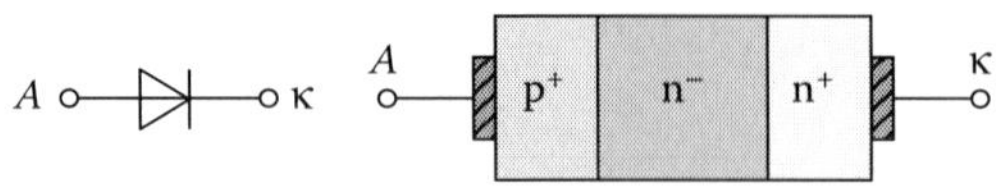

Figure 2.1 A power diode and its structure.

The maximum value of the reverse working voltage is determined by the avalanche breakdown voltage $U_{(BR)}$ and is selected with some margin of safety

$$U_{RRM} = k_0 U_{(BR)}, \tag{2.1}$$

where U_{RRM} is the maximum possible repeating reverse pulsed diode voltage and k_0 is the safety factor (usually 0.75).

The thickness W_0 of the space-charge layer, where the electric field is mainly concentrated when the diode is off, may be tens or hundreds of microns. Its limiting value is estimated as

$$W_0 \approx 0.52\sqrt{\rho_n U_{(BR)}}, \tag{2.2}$$

where ρ_n (Ω/cm) is the resistivity of silicon in the n^- base and $U_{(BR)}$ (V) is the avalanche breakdown voltage in the diode structure.

If the calculated value of W_0 is less than 150 μm, we select 150 μm as the thickness of the n^- base so as to ensure the required mechanical strength in the silicon plate.

When $W_0 > 150$ μm, we select the thickness W_{n^-} of the n^- base as equal to W_0, or even a little less than W_0 in view of the protective n^+ emitter layer limiting electric field propagation.

The total thickness of the initial silicon plate also includes the depth w of the p^+–n^- and n^+–n^- junctions

$$W_{Si} = W_{n^-} + w_{p^+n^-} + w_{n^+n^-}. \tag{2.3}$$

For power diodes, usually, $w_{p^+n^-} \approx 75$–125 μm and $w_{n^+n^-} \approx 30$–50 μm.

With a forward voltage at the diode, the electron–hole junction p^+–n^- acquires a forward bias, and hole injection from the highly doped p^+ layer to the diode base begins.

In filling the base, the injected holes will move toward the n^+–n^- junction. The initial doping of the n^+ emitter layer is specified so that most of the holes cannot overcome the potential barrier of the n^+–n^- junction. They begin to accumulate close to the n^+–n^- junction. Their positive charge is compensated by an influx of electrons from the n^+–n^- junction, which acquires a forward bias. Thus, double hole injection occurs in the p^+–n^-–n^+ structure of the power diode. The n^- layer (the base) is filled with electron–hole plasma from both sides.

The voltage drop at the n^+–n^- junction is somewhat less than that at the p^+–n^- junction. The total voltage at the diode's junction may be estimated as

$$U_{\mathrm{p-n}} \approx 2\varphi_T \ln\left(\frac{I_F}{I_S}\right), \tag{2.4}$$

where φ_T is the thermal potential (0.025 V at room temperature), I_F the forward diode current, and I_S the saturation current ($I_S \approx 10^{-6}$–10^{-12} for silicon junctions).

With high forward-current density in the n^- base, injection is vigorous. As the semiconductor structure is characterized by double injection, the electrical conductivity of the base increases in proportion to the current. Consequently, the forward voltage drop at the n^- base remains practically constant and does not depend on the diode current or the resistivity of the base. The voltage drop at the base in this case may be estimated as

$$U_{\mathrm{n}^-} \approx 1.5\varphi_T \exp\left(\frac{W_{\mathrm{n}^-}}{2.4L_p}\right), \tag{2.5}$$

where L_p is the diffusional length of the holes in the n^- base.
With an increase in the current density to 100–300 A/cm^2, the mutual scattering of electrons and holes begins to affect the forward voltage in the base. In that case, the concentration of charge carriers in the base reaches $(7–8) \times 10^{16}$ cm^{-3} or more; their mobility is inversely proportional to their concentration and the conductivity of the base no longer depends on the current. An additional component of the voltage drop proportional to the forward current I_F is then seen in the diode's base.

Taking account of all the relevant factors, we may write the forward voltage at the power diode in the form

$$U_F = 2\varphi_T \ln\left(\frac{I_F}{I_S}\right) + 1.5\varphi_T \exp\left(\frac{W_{\mathrm{n}^-}}{2.4L_p}\right) + \frac{W_{\mathrm{n}^-}}{16S} I_F, \tag{2.6}$$

where S is the active area of the diode structure.

In Equation 2.6, we must express W_{n^-} in cm and S in cm^2.

2.2.2 Schottky power diodes

The operational principle of the Schottky diode is based on the interaction between a metal and a depleted layer of a semiconductor, and the contact between which has rectifier properties in certain conditions. Schottky diodes are based on n^- silicon with electron conductivity. The highly doped n^+ substrate has a donor concentration of 5×10^{18}–5×10^{19} cm^{-3} and its thickness is 150–200 μm; this is determined by the thickness of the initial silicon plate. The presence of a highly doped substrate considerably reduces the resistance of the diode and ensures satisfactory ohmic contact with the metallized cathode layer. The active n^- base of the Schottky diode has a lower impurity concentration (3×10^{15} cm^{-3}); its thickness w_B is determined by the diode's working voltage and is in the range from a few microns to tens of microns. To minimize extreme avalanche breakdown and to increase the electric field strength in the base, the diode includes a system of guard rings with a p–n junction, whose depth is a few microns (Figure 2.2).

The voltage drop at the junction in the Schottky diode is less than that for a diode with a p–n junction, whereas the reverse currents are greater (Melyoshin, 2005).

The forward voltage in the Schottky diode consists of two main components: the voltage at the junction and the voltage at the active-region resistance in the n^- base of the diode

$$U_F = \varphi_T \ln\left(\frac{I_F}{I_S} + 1\right) + \frac{I_F \rho_n w_B}{S}. \tag{2.7}$$

With an increase in the maximum reverse voltage of the Schottky diode, the resistance of the n^- base increases, as the creation of higher reverse voltage requires a more extended weakly doped region with lower charge-carrier concentration. As a result, the resistance of the n^- base in high-voltage Schottky diodes is significantly increased. This is the

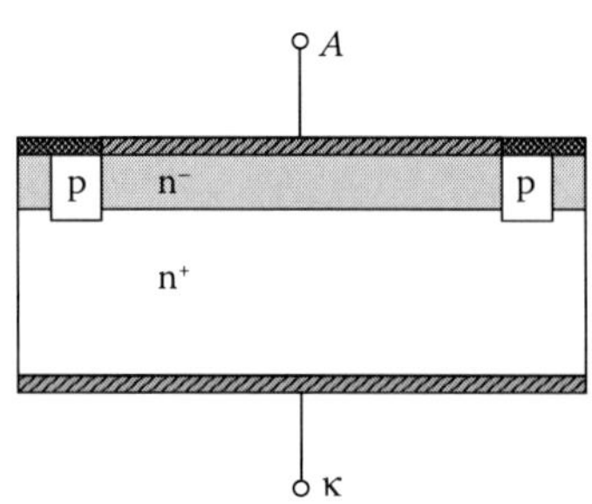

Figure 2.2 Structure of a Schottky diode.

main reason for the upper limit of 200–400 V on the working voltage of such diodes.

2.2.3 Pulsed diodes

Pulsed diodes are intended for pulsed and high-frequency operation; their transient processes are relatively brief. Two basic structures may be noted (Figure 2.3): with a p–n junction (diffused and epitaxial pulsed diodes) and with a metal–semiconductor junction (pulsed Schottky diode).

Pulsed Schottky diodes have the shortest switching time (a few nanoseconds) because there is no accumulation of secondary charge carriers. In the transient process occurring when the diode is switched off, there are practically no reverse current surges; the reverse recovery time is determined solely by the recharging time for the junction's barrier capacitance. However, the blocking voltage is not more than 200–400 V for silicon-based diodes. For pulsed Schottky diodes based on gallium arsenide (GaAs), the maximum reverse voltage is somewhat larger, up to 600 V. Currently, the largest blocking voltages are obtained for diodes based on SiC, up to 1700 V. In the absence of injection and modulation of the base layers, the working currents of the pulsed Schottky diodes are not more than tens of amperes.

Various parameters determine the switching inertia of pulsed diodes with a p–n junction. When the pulsed diode is switched on, the initial voltage surge is relatively large. The voltage surge is determined by the initial resistance of the diode's base layer r_{B0} and the forward-current amplitude I_F and may be a few tens or even hundreds of volts. As charge accumulates in the base, the forward voltage at the diode falls to a steady value (Figure 2.4).

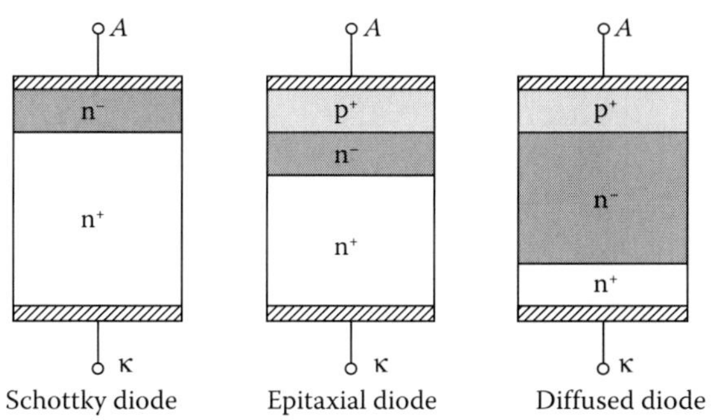

Figure 2.3 Main types of pulsed diodes.

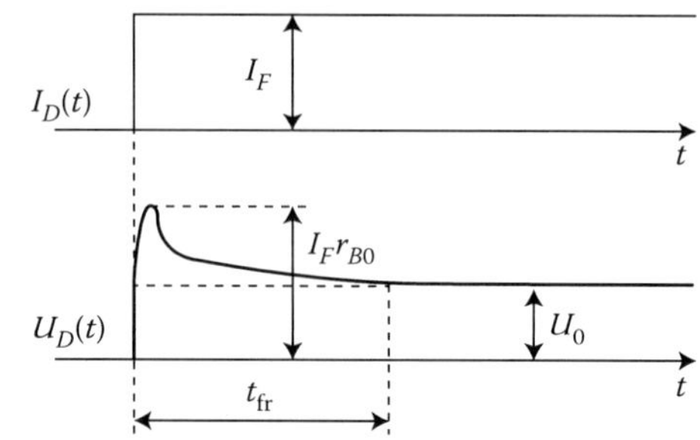

Figure 2.4 Voltage and current variations when a pulsed diode is switched on.

The time in which the forward voltage at the diode first rises and then falls to 110% of its steady value U_0 is known as the time t_{fr} to establish the diode's forward resistance.

When a pulsed diode with a p–n junction is switched off, the reverse current rises sharply and then falls to zero (Figure 2.5).

On account of the excess charge in the diode's base, the maximum reverse current may be several times the amplitude of the forward current. The surge in the reverse current may be characterized by a dynamic parameter: the reverse recovery charge Q_{rr} of the pulsed diode. Essentially, Q_{rr} is the integral of the reverse current at the diode. The time in which the reverse current begins to rise and then falls to 0.2 of the maximum value is known as the recovery time t_{rr} of the diode's reverse resistance (Voronin, 2005).

The maximum blocking voltage of epitaxial pulsed diodes with a relatively narrow base is 600–1200 V. As a rule, diodes with a reverse voltage

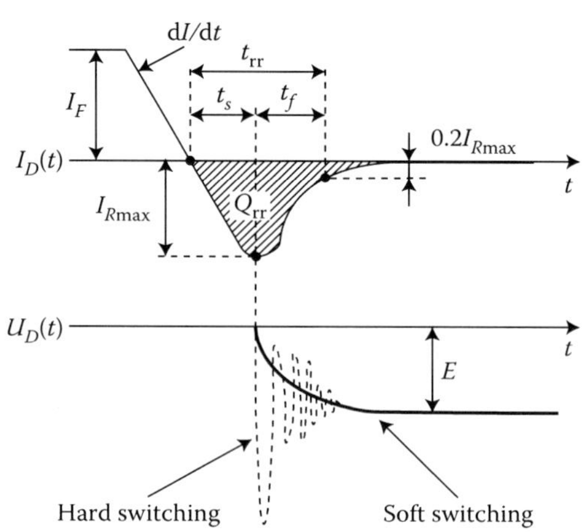

Figure 2.5 Voltage and current variations when a pulsed diode is switched off.

greater than 1200 V are produced by diffused technology. The maximum reverse blocking voltage in such diodes is around 6500 V. A typical value of the mean forward current for the semiconductor crystals (chips) of pulsed diodes with a p–n junction is 50, 100, or 150 A. By connecting individual chips in parallel, working currents up to thousands of amperes may be obtained.

The use of pulsed diodes in high-power systems imposes an additional requirement on the dynamics of reverse recovery, which must correspond to a specified softness characteristic. To that end, we introduce the softness coefficient S, which is the ratio of the components of the recovery time t_{rr} for the diode's reverse resistance: the time t_f for the reverse current to fall and the time t_s in which it rises is

$$S = \frac{t_f}{t_s}. \tag{2.8}$$

The voltage surge at the diode in recovery depends on the rate di/dt at which the forward current falls (usually regulated in the circuit) and on the parasitic installation inductance L_s. However, with sharper drop in the diode's reverse current, the voltage surge at the diode will be greater. To minimize the voltage surge, the softness coefficient must be greater than 1. As a rule, for diodes with soft recovery, the rate at which the reverse current falls is approximately equal to the rate at which it rises.

2.3 *Power bipolar transistors*

At the relatively high working voltages of power bipolar transistors (10^2–10^3 V), it is difficult to create an n^+–p^-–n^+ structure with a thin base. However, unless that is accomplished, the frequency properties of the transistor will be significantly impaired. In particular, there will be a significant reduction in the limiting amplification frequency f_T at which the current's amplification factor is reduced to 1. The limiting frequency f_T may be determined from the formula

$$f_T = |h_{21E}|f, \tag{2.9}$$

where $|h_{21E}|$ is the modulus of the current's amplification factor (presented in handbooks for each type of the transistor) and f is the frequency at which $|h_{21E}|$ is measured.

Therefore, the n^+–p–n^-–n^+ structure is used for power bipolar transistors. The lightly doped n^- layer is the high-resistance region of the collector (Figure 2.6).

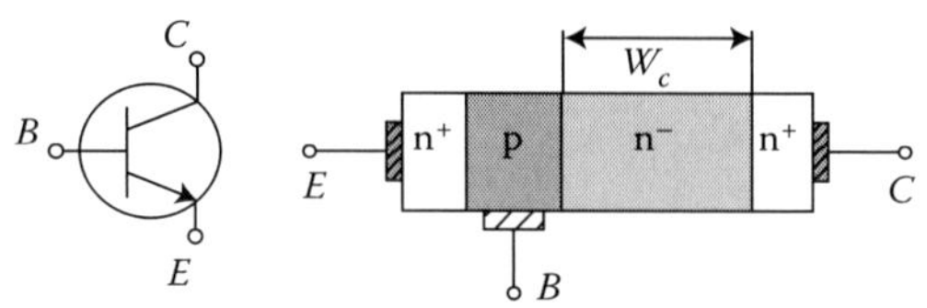

Figure 2.6 A power bipolar transistor and its structure.

Power transistors are primarily intended for operation, with a common emitter. Therefore, their working voltage is determined by the maximum permissible constant collector–emitter voltage U_{CEmax}. Knowing U_{CEmax}, we may calculate the avalanche breakdown voltage of the collector–emitter system

$$U_{\text{CE0}} = \kappa_0 U_{\text{CEmax}}, \tag{2.10}$$

where κ_0 is the safety margin ($\kappa_0 \geq 1$).

Having found U_{CE0}, we may determine the breakdown voltage of the collector–base junction

$$U_{\text{CB0}} = (1 + \beta_N)^{1/n} U_{\text{CB0}}, \tag{2.11}$$

where $n \approx 3$–6 for silicon transistors.

We may then calculate the thickness of the space-charge region, which mainly extends toward the high-resistance region of the collector

$$W_{\text{C0}} \approx 0.52\sqrt{\rho_n U_{\text{CB0}}}, \tag{2.12}$$

where ρ_n is the resistivity of the collector's n^- region.

Taking account of the constraint on the electric field of the highly doped n^+ layer, we select the required thickness W_{C} of the collector's high-resistance region as about half of W_{C0}.

At saturation, the emitter and collector junctions of the bipolar transistor acquire forward bias. The residual voltage at the transistor is

$$U_{\text{CE(sat)}} = \varphi_T \ln \frac{\alpha_I}{1 + I_C/(1+\beta_I)I_B} + (I_C + I_B)r_E + I_C r_C, \tag{2.13}$$

where β_I is the amplification factor in reverse switching of the transistor, $\alpha_I = \beta_I/(1 + \beta_I)$, r_E is the resistance of the emitter layer, and r is the resistance of the collector layer.

Depending on the resistance of the collector layer, the voltage $U_{CE(sat)}$ for silicon transistors is 1–2 V for low-voltage components and 4–5 V for high-voltage components.

From the limiting amplification frequency f_T, we may calculate the transition time τ_C of the charge carriers through the base and the lifetime τ_B of the charge carriers in the transistor, which determine its dynamic properties

$$\begin{cases} \tau_C = \dfrac{\alpha_N}{2\pi \cdot f_T}, \\ \tau_B = \beta_N \tau_C, \end{cases} \tag{2.14}$$

where α_N is the amplification factor of the current in a system with a common base and β_N is the amplification factor of the current in a system with a common emitter. Note that $\alpha_N = \beta_N/(1 + \beta_N)$.

In the amplifier mode, the bipolar transistor operates in the active region of its output characteristics. In the switching mode, which is more typical of power transistors, a control signal switches the component from the closed state (cutoff) to the open state (saturation) and back, so that it traverses the active region at the switching fronts. To guarantee saturation, the basic control current in the transfer is specified with a margin, depending on the transistor's saturation factor

$$N = \frac{I_B^+}{I_{\text{B.li}}}, \tag{2.15}$$

where I_B^+ is the forward base current, $I_{\text{B.li}} = I_{\text{lo}}/\beta_N$ is the limiting base current, and $I_{\text{lo}} = I_{\text{Csat}}$ is the switch's load current.

From the saturation factor, we may estimate the basic dynamic parameters of the transistor as follows.

1. The duration t_{on} of the switching front when the transistor is switched on is

$$t_{\text{on}} \approx \tau_0 \ln\left[\frac{N}{N-1}\right]. \tag{2.16}$$

2. The delay time in switching off the transistor, which may be known as the storage time t_s, is

$$t_s \approx \tau_0 \ln\left[\frac{N+1}{2}\right]. \tag{2.17}$$

3. The duration t_{off} of the switching front when the transistor is switched off is

$$t_{\text{off}} \approx \tau_0 \ln 2. \tag{2.18}$$

Here $\tau_0 = \tau_B + (1 + \beta_N)C_{\text{CB}}R_{\text{lo}}$ is the equivalent time constant when the influence of the collector–base capacitance of the transistor is taken into account and R_{lo} is the load resistance.

The presence of a high-resistance collector layer leads to an additional working region (the presaturation region) on the transistor's output characteristics (Figure 2.7). The boundary between the active region and the presaturation region is described by the equation

$$I_C = \frac{U_{\text{CE}}}{r_{C0}}, \tag{2.19}$$

where $r_{C0} = \rho_n W_C / S$ is the initial resistance of the collector's n^- region and S is the area of the transistor structure.

When the transistor's switching trajectory intersects with the presaturation region, the variation in the collector current slows significantly. In the transient switching process, we observe slow intervals, which are known as quasi-saturation intervals and quasi-storage intervals, respectively, when the transistor is being switched on and off. With an increase in resistivity of the collector's high-resistance region, these intervals become longer. When the resistance of the collector's n^- layer is modulated by the accumulated charge, the transistor's base region expands (the Kirk effect and the induced base effect). The current transfer coefficient β_N declines here, and the charge-carrier lifetime τ_B increases. That largely explains the slowing of the current fronts t_{on} and t_{off} and limits the speed of high-voltage bipolar transistors. For example, the

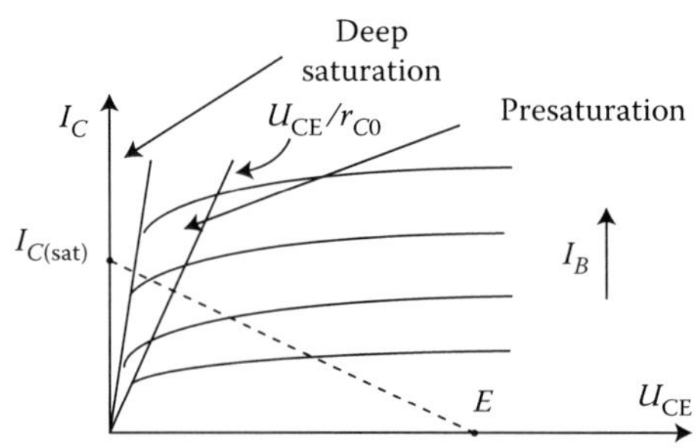

Figure 2.7 Volt–ampere characteristics of a power bipolar transistor.

switching time of a bipolar transistor calculated for a working voltage $U_{CEmax} = 500$ V may be 5–10 μs.

Another feature of power bipolar transistors is the displacement of the current associated with the longitudinal resistance of the p base. As the transistor's base current flows parallel to the emitter plane, the potential of the p base declines in the direction of the current. If the decrease in the potential exceeds $2\varphi_T$, the current density in the central part of the emitter will differ by practically an order of magnitude from that at its periphery. Correspondingly, the load current density will be localized within the transistor—at the edges of the emitter when the transistor is switched on and in the central part of the emitter when the transistor is switched off. Thus, only the part ΔX_E of the emitter is in the conducting state, where

$$\Delta X_E \approx 0.54 W_B \sqrt{\beta_N}, \tag{2.20}$$

where W_B is the width of the transistor's p base.

To reduce the influence of current displacement, the emitter is produced in annular, involute, and ridged forms, whose cross-section corresponds to ΔX_E.

The deficiencies of power bipolar transistors may be partially overcome by using composite designs.

To increase the amplification factor of the current, we may use composite transistors based on Darlington and Sziklai circuits (Figure 2.8). In the composite transistor, the current transfer factor is practically $\beta_{N1}\beta_{N2}$.

To increase the maximum permissible voltage and the cutoff speed, an avalanche design with emitter switching is used for the transistor (Figure 2.9). In that case, the cutoff speed is increased by a factor of β_N, whereas the limiting voltage is the breakdown voltage U_{CB0} of the collector–base junction, which is about 150% of the value for an ordinary common emitter transistor.

To improve the frequency characteristics of the transistor and to reduce the storage time t_s, unsaturated switches with nonlinear feedback may be used (Figure 2.10).

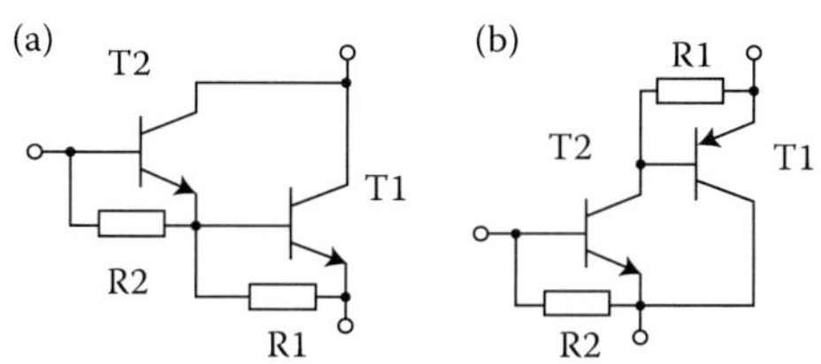

Figure 2.8 Composite (a) Darlington and (b) Sziklai circuits.

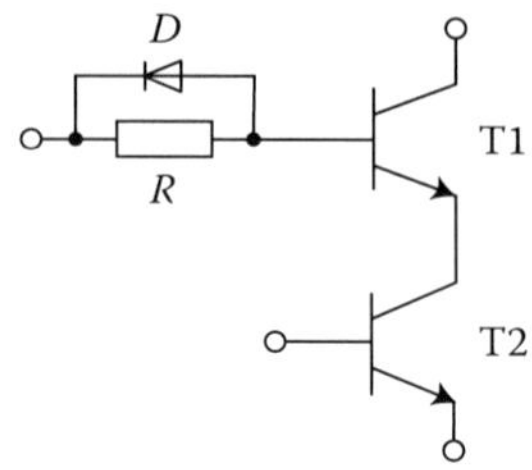

Figure 2.9 A cascode circuit of transistor switch.

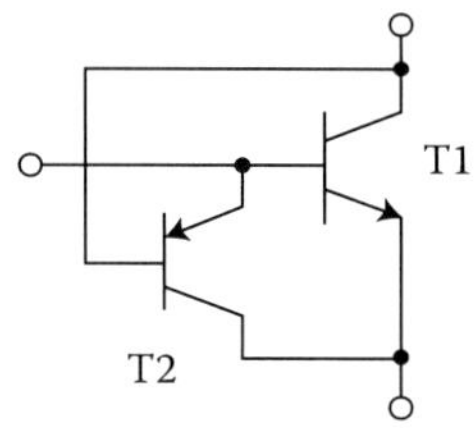

Figure 2.10 An unsaturated switch circuit.

2.4 Thyristors

2.4.1 Controllable semiconductor switches with p–n–p–n structure

Switches with p–n–p–n structure have two stable states. Their operation depends on internal feedback. The thyristor structure is based on four layers p–n–p–n and three p–n junctions: the anode junction, the central junction, and the cathode junction (Figure 2.11).

If we regard the thyristor as consisting of an n–p–n transistor and a p–n–p transistor, such that the collector of each transistor is connected to the base of the other (Figure 2.12), we may describe its volt–ampere characteristics in the form (Gentry et al., 1964)

$$I_A = \frac{\alpha_{NPN} I_g + I_{C0}}{1 - \alpha_{NPN} - \alpha_{PNP}}, \tag{2.21}$$

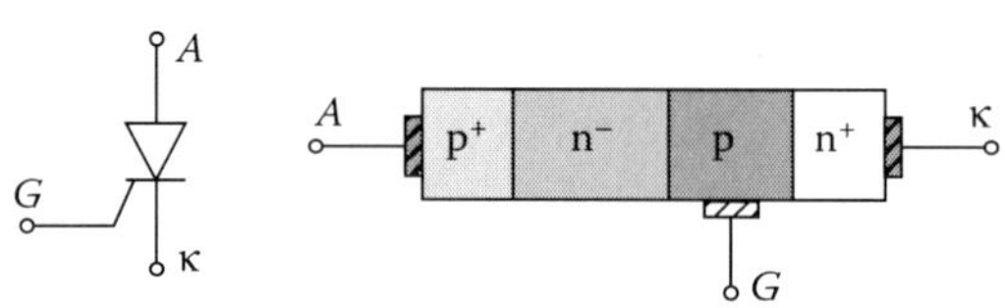

Figure 2.11 A thyristor and its structure.

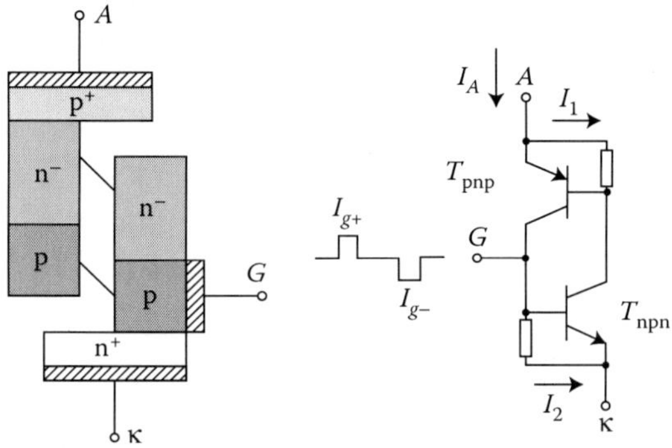

Figure 2.12 A composite thyristor structure.

where I_A is the thyristor's anode current, I_g the control current in the thyristor's p base, α_{NPN} the transfer factor of the current in the n–p–n transistor, α_{PNP} the transfer factor of the current in the p–n–p transistor, and I_{C0} the thermal current of the central junction.

When the thyristor is off, forward and reverse blocking is observed (Figure 2.13).

If a forward voltage is applied to the thyristor (with positive polarity at the anode), the central p–n junction of the structure is closed. The space-charge region begins to penetrate mainly into the most lightly doped n base of the thyristor, moving away from the central junction. If a reverse voltage is applied to the thyristor, the anode and cathode p–n junctions of the structure are closed. As the cathode junction adjacent to the control electrode is characterized by very poor closing properties (as a rule, corresponding to a voltage not greater than 15–20 V), the space-charge region penetrates into the *n* base of the thyristor, moving away from the anode junction. Thus, both forward and reverse blocking will depend on the dimensions and degree of alloying of the *n* base.

If the switch is off and the electric field strength in the space-charge region reaches the critical value, avalanche breakdown begins. The

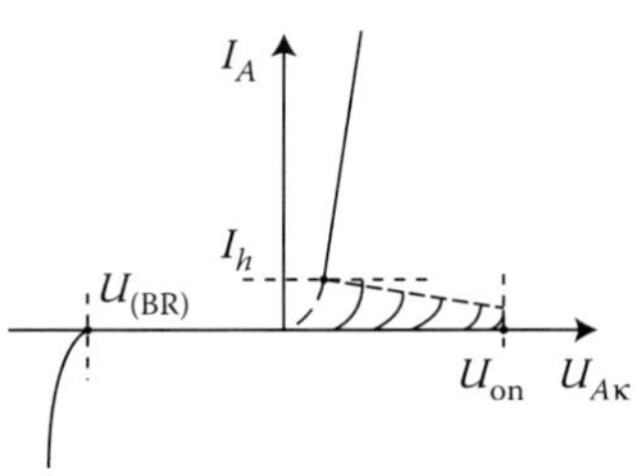

Figure 2.13 Volt–ampere characteristics of a thyristor.

limiting dimensions of the space-charge region (in micrometers) may be estimated from the formula

$$W_0 \approx 0.52\sqrt{\rho_n U_{(BR)}} \leq W_n, \tag{2.22}$$

where ρ_n (Ω/cm) is the resistivity of the silicon in the n base and $U_{(BR)}$ (V) is the avalanche breakdown voltage in the thyristor.

When a control current pulse is sent to the p base of the thyristor, positive feedback of the transistor structures begins, and the current transfer factors α_{NPN} and α_{PNP} begin to increase.

Using the differential form $\alpha^* = \Delta I_C/\Delta I_E$, we may obtain the condition for transition of the p–n–p–n structure to a stable conducting state

$$\alpha^*_{NPN} + \alpha^*_{PNP} \geq 1. \tag{2.23}$$

When the thyristor is on, it corresponds to a diode p^+–i–n^+ structure with a saturated base. Then, by analogy with a high-voltage diode, the forward voltage drop at the open thyristor may be calculated as (Yevseyev and Dermenzhi, 1981)

$$U_T = 2\varphi_T \ln\left(\frac{I_A}{I_S}\right) + 1.5\varphi_T \exp\left(\frac{W_n + W_p}{2L}\right) + \frac{W_n + W_p}{16S} I_A, \tag{2.24}$$

where W_n is the width of the n base, W_p the width of the p base, and L is the effective diffusional length of the charge carriers in the saturated bases of the thyristor ($L \approx 1.2L_p$).

The dynamic parameters of the thyristor are determined by the switching times of the p–n–p–n structure.

When the thyristor is switched on, we may identify four basic stages: the switching delay, regeneration, establishment of a steady state, and propagation of the on state over the whole area of the structure (Figure 2.14).

The switching delay depends on the amplitude of the control current supplied to the thyristor's p base. For powerful thyristors, as a rule, $I_{co} \geq 1$ A, and the delay time is relatively small (not more than 0.2 μs).

The regeneration stage is characterized by an avalanche increase in the thyristor current

$$I_A(t) \approx B_0 I_g \exp\left(\frac{t}{\tau_0}\right), \tag{2.25}$$

where $\tau_0 \approx 0.1$–0.2 μs is the time constant of the current rise and $B_0 \approx 0.4$–1.2 is the current amplification factor.

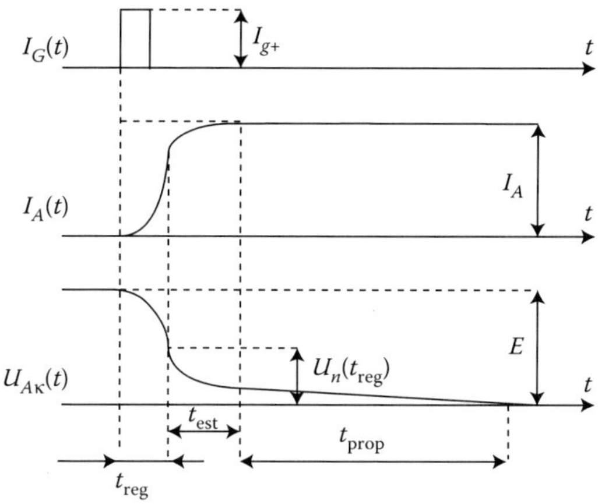

Figure 2.14 Transient process when the thyristor is switched on.

With an increase in the anode current, the thyristor voltage declines, whereas the voltage drop at the high-resistance n base increases. When the anode voltage and the voltage at the base are approximately equal, the avalanche growth in the anode current ends, and the establishment of a steady state begins. At that stage, the modulation of the n base's conductivity increases. The voltage drop at the thyristor's base layer declines more or less exponentially

$$U_{AC}(t) \approx U_T + [U_n(t_{reg}) - U_T]\exp\left(-\frac{t - t_{reg}}{\tau_e}\right), \tag{2.26}$$

where $U(t_{reg})$ is the amplitude of the voltage at the thyristor's n base at the end of regeneration and $\tau_e \approx 0.5–2.0$ μs is the time constant for the establishment of a steady state, which is approximately equal to the hole life in the n base.

The maximum value of $U_n(t_{reg})$ may be estimated as

$$U_n(t_{reg}) \approx W_n E_{cr}, \tag{2.27}$$

where $E_{cr} \approx 10^4$ V/cm is the critical electric field strength in the n base, at which saturation of the charge-carrier speed begins.

The area S_0 of the region that is initially switched on in the thyristor structure is not more than 0.1 cm^2, as a rule. The propagation of the on state over the structure is characterized by some rate approximately proportional to the anode current. To assess the change in the area of the on state over time, we use the formula

$$S(t) \approx \sqrt{kI_A t + S_0^2}, \tag{2.28}$$

where k is a constant of proportionality.

Then the change in the thyristor voltage at the propagation stage may be written in the form

$$U_{AC}(t) \approx U_{T0} + r_{T0} I_A \frac{S}{S(t)} = U_{T0} + k^* \sqrt{\frac{I_A}{t}}, \tag{2.29}$$

where U_{T0} and r_{T0} are the linear approximation parameters of the thyristor's volt–ampere characteristic and $k^* = r_{T0} S/\sqrt{k}$ is an empirical constant, in which $k^* \approx 0.2–0.3\ \mathrm{V\ A^{-1/2}\ \mu s^{1/2}}$.

Assuming that at the end of the propagation stage $U_{AC}(t_{prop}) = U_T = U_{T0} + r_{T0} I_A$, we obtain the total duration of propagation as

$$t_{prop} = \frac{(k^*/r_{T0})^2}{I_A}. \tag{2.30}$$

Different methods may be used to switch off the thyristor.

- Reduction in the forward anode current to the holding current I_h
- Reversal of the polarity of the anode voltage U_{AC}
- Supply of a reverse (negative) current I_g^- to the control electrode (gate)

A thyristor that is switched off in the anode circuit is known as a semiconductor-controlled rectifier. A thyristor that is switched off by reverse current in the control circuit is known as a gate turn-off (GTO) thyristor.

In all cases, essentially, the excess charge that accumulates as a result of the forward anode current is reduced to some minimal or critical value, at which the device is able to withstand the anode voltage and remains in the off state.

The time for the excess charge in the thyristor bases to fall to the critical value is sometimes known as the shutoff time t_q.

When the thyristor is switched off by reversing the polarity of the anode voltage, as a rule, the n–p–n transistor is the first to shut down, as its p base is more highly doped. Thereafter, the p–n–p transistor operates with a broken base, and the carriers that have accumulated in the n base may only disappear by recombination. The shutdown time may be calculated as

$$t_q = \tau_p \ln \frac{Q_0}{Q_{cr}}, \tag{2.31}$$

where τ_p is the carrier lifetime in the n base, Q_0 the initial stored charge, and Q_{cr} the critical charge.

The charge Q_0 is proportional to the forward anode current, whereas Q_{cr} is proportional to the holding current I_h, which is provided in handbooks. Then the shutdown time may be estimated as

$$t_q \approx \tau_p \ln \frac{I_A}{I_h}. \tag{2.32}$$

When a pulse of forward anode voltage (amplitude ΔU and speed dV/dt) is applied to a thyristor that is off, additional charge is supplied to the base regions through the barrier capacitance C_{CB} of the central junction. This is known as the dV/dt effect in the thyristor and is taken into account in calculating the shutdown time

$$t_q = \tau_p \ln \frac{Q_0}{Q_{cr} - C_{CB}\Delta U} \approx \tau_p \ln \frac{I_A}{I_h - C_{CB}\Delta U/\tau_p}. \tag{2.33}$$

2.4.2 *Power photothyristors*

A photothyristor is a unidirectional p–n–p–n current switch that is turned on by the action of light.

Two main designs are known: with an illuminated emitter and with an illuminated base (Yevseyev and Dermenzhi, 1981).

The photothyristor with an illuminated base is the main component in power electronics (Figure 2.15). This structure responds not to the intensity of illumination, but to the radiant flux. Consequently, the sensitivity to light may be increased without a significant increase in the gain of the component transistors. This, in turn, permits an increase in the critical rate of voltage growth at the photothyristor and reduces the temperature dependence of the switching voltage. The resistance of the photothyristor to fast growth in the anode current is ensured by including a regenerative control electrode in the structure. The auxiliary structure employed is a light-controlled p–n–p–n structure, whose anode current stimulates the basic structure in the p base.

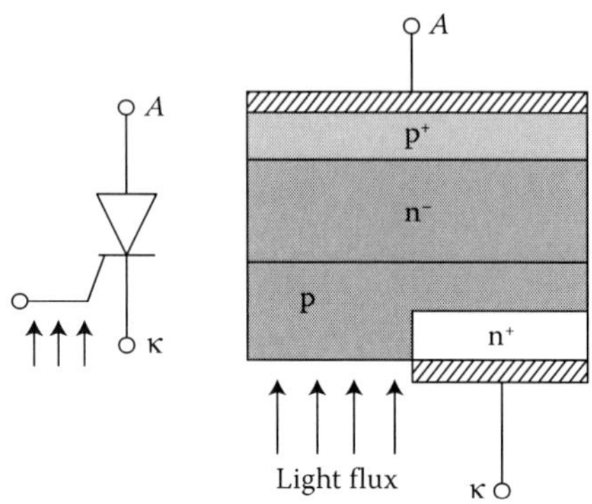

Figure 2.15 A photothyristor and its structure with an illuminated base.

The volt–ampere characteristic of the photothyristor with an illuminated base takes the form

$$I_A \approx \frac{\alpha_{NPN}\beta_{PNP}(I_{ph1} + I_{ph2})}{1 - \alpha_{NPN} - \alpha_{PNP}}, \tag{2.34}$$

where I_{ph1} is the photoflux through the anode junction and I_{ph2} is the photoflux through the central junction.

According to Equation 2.34, a photothyristor with an illuminated base corresponds to a p–n–p–n structure in which the collector of the p–n–p transistor in a common emitter circuit is supplied to the p base, whereas the photocurrent $I_{ph1} + I_{ph2}$ is supplied to the n base. The volt–ampere characteristic of the photothyristor with an illuminated base qualitatively resembles that of a current-controlled p–n–p–n structure.

In design terms, the photothyristor is analogous to a high-voltage tablet thyristor, but the photothyristor has an optical input rather than an electrical control output.

In the center of the photothyristor's cathode base, there is an optical window, which is close to the photosensitive region of the semiconducting structure.

The photothyristor is controlled by means of optical interface cable (Figure 2.16).

At one end, it is inserted into the slot of the tablet's cathode base; the other end is connected through an optical socket to a laser diode. The light pulse is supplied to the photosensitive region of the silicon structure from the laser diode through a fiber-optic cable, an adapter, and an optical window.

The photothyristor is controlled by an IR pulse. The control pulse is transmitted to the photosensitive region of the silicon structure through

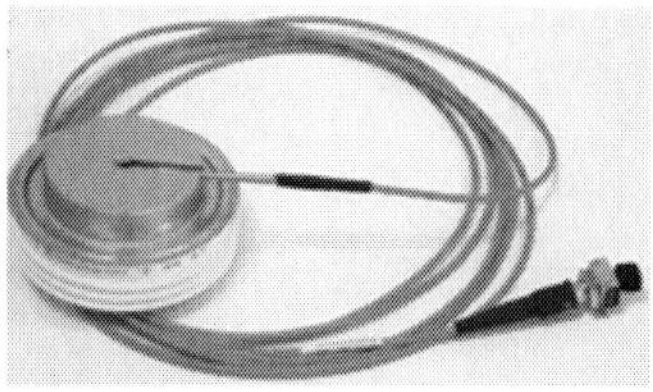

Figure 2.16 A photothyristor with optical cable.

a fiber-optic light guide. It is connected optically to the photothyristor housing and to a laser diode, for example, an Osram SPL-PL90 diode, with an optical wavelength of 0.88–0.98 μm. The length of the light guide is practically unlimited, as the damping of the control signal is very slight (about 1 dB/km). The laser diode converts the electric signal from the control driver into a light pulse that mimics the shape and length of the electrical pulse.

2.4.3 Symmetric thyristors

The properties of five-layer n–p–n–p–n structures are employed in symmetric thyristors, also known as triacs. The volt–ampere characteristics of these components include two sections of negative resistance on the forward and reverse branches, which permits ac control (Figure 2.17).

The structure of a symmetric n–p–n–p–n switch with shunted emitters is shown in Figure 2.18.

This component is called a diac. It consists of two p–n–p–n sections in parallel but in opposite directions. The diac structure becomes conducting on account of the application of a voltage exceeding the switching voltage to its electrodes or by a sharp increase in the applied voltage, with a steep front. On the basis of that structure, controllable switches may be created by adding gates to the base regions p1 and p2. However, this design is unpromising for powerful components because of the presence

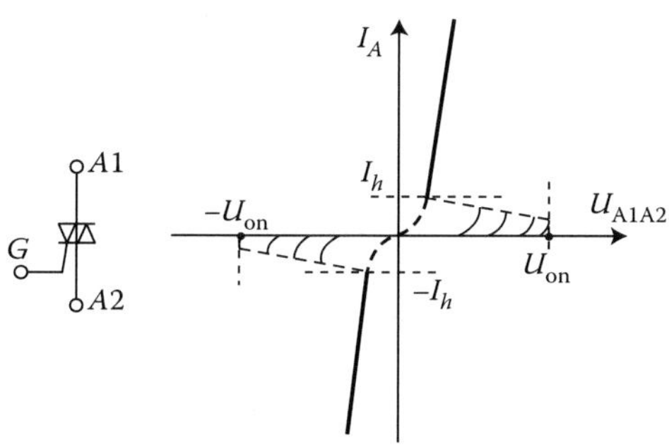

Figure 2.17 A triac (symmetric thyristor) and its volt–ampere characteristic.

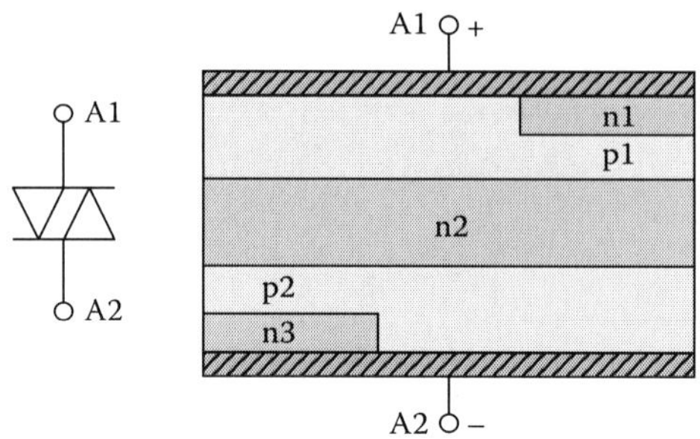

Figure 2.18 A diac and its structure.

of two gates, one of which must pass through the lower plane of the semiconducting plate.

In a regular thyristor, the gate is connected directly to the p base and is unable to inject charge carriers. This is known as an ohmic gate. Another thyristor design includes an additional p–n junction directly below the gate; this junction is able to inject carriers into the base. This is known as an injecting gate. Note that the control current in this structure flows from the gate.

Then, for an ordinary thyristor built on an n-type silicon plate, the base supports an anode (a p-type emitter). This is a forward-biased structure. If the base supports a cathode (an n-type emitter), we obtain a reverse-biased structure.

In the thyristor, forward current flows toward the plane with the gate. The incoming gate current is positive. Then we distinguish the following basic types of thyristor p–n–p–n structures (Yevseyev and Dermenzhi, 1981):

1. A forward-biased structure with an ohmic gate, which conducts forward current, with positive gate current
2. A reverse-biased structure with an injecting gate, which conducts reverse current, with negative gate current
3. A forward-biased structure with an injecting gate, which conducts forward current, with negative gate current
4. A reverse-biased structure with an ohmic gate, which conducts reverse current, with positive gate current

Symmetric controllable switches are assembled from different combinations of these structures.

For example, the TS161 triac combines elementary structures 1 and 4 on a single chip. The KU208 triac combines structures 1 and 2 on a single chip. The TS222 triac combines structures 1 and 3 on a single chip.

In the transient process when elementary triac structures with ohmic and injecting gates are switched on, two structures are turned on in succession: first, the auxiliary (control) structure and then, by means of the control structure's current, the basic triac structure. The practical model of the transient process when such double structures are turned on may be reduced to a single structure. In fact, when a double structure is turned on, stages of regeneration, establishment of a steady state, and propagation of the on region are observed, as in a regular thyristor. For example, the time to turn on the TS161 triac is about 20 μs, which corresponds to the time required to turn on a regular thyristor of the same power.

As the application of negative voltage to the gate switches on any triac, the structure may only be switched off by applying reverse voltage to the power anode and cathode. Note, however, that this is not the case for gates with positive control current. In that case, a negative control current may be supplied to the gate without switching off the triac.

The resistance of triacs to the d*V*/d*t* effect is characterized by two parameters. Here, we distinguish between the critical growth rate of the forward voltage with the same polarity as the forward current and the critical growth rate of the forward voltage with the opposite polarity to the forward current. Typical values of the critical rate d*V*/d*t* are 10–20 V/μs.

2.5 Switched thyristors

2.5.1 The GTO thyristor

Ordinary p–n–p–n structures can only be switched off by a negative current in the control circuit at relatively low load currents, when the sum of the current transfer factors of the component transistors is slightly more than 1.

An increase in the load current requires an increase in the switching current. Because of the finite longitudinal resistance of the base p layer, a reverse bias appears over some of the cathode junction. As the charge is removed from the p base by a negative current, we observe avalanche breakdown of the section of the cathode junction closest to the control electrode. As a result, the remainder of the cathode junction will be shunted, and a bypass route becomes available for the negative control current. The switching of the thyristor by the gate is terminated.

To ensure high switching current, we must increase the avalanche breakdown voltage of the cathode junction. At the same time, the longitudinal resistance of the p base must be reduced. However, these two requirements are mutually inconsistent. As a compromise, the longitudinal resistance of the p base is reduced by reducing the cathode's emitter band. For example, a GTO thyristor with a maximum cutoff current of 600 A has a

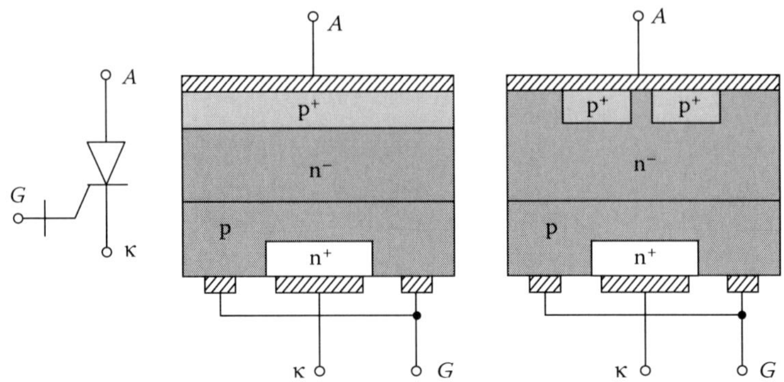

Figure 2.19 A GTO and its basic structures.

cathode structure consisting of 200 emitter bands (width 30 μm and length 4 mm), each of which is covered by regions of the branched gate.

There are two basic GTO structures: with reverse blocking (used mainly in current inverters) and without reverse blocking (used mainly in voltage inverters). Both structures are shown in Figure 2.19. In the structure without reverse blocking, the anode emitter junction is shunted.

The volt–ampere characteristic of the GTO is of the same form as that for the structure that may be switched from the basic circuit. If we substitute a negative control current in the formula for the volt–ampere characteristic, we may calculate the cutoff factor G, which is the ratio of the anode current to the switched current

$$G = \frac{I_A}{I_g^-} = \frac{\alpha_{NPN}}{\alpha_{NPN} + \alpha_{PNP} - 1}. \tag{2.35}$$

The cutoff factor is a structural parameter and allows us to determine the amplitude of the switching current required to shut down the thyristor with specified load current. To increase G, the current transfer factor of the p–n–p transistor is increased, while ensuring that the sum of the transistor transfer factors is slightly more than 1. For modern GTOs, $G = 4–8$.

In the transient cutoff process, we may distinguish three states: resorption of the saturation charge in the thyristor bases; regenerative drop in the anode current; and recovery of the resistance of the central junction (Figure 2.20).

In the first stage, under the action of the cutoff current, the concentration of charge carriers in the thyristor bases is reduced. As a result, the central junction is no longer saturated. Then, both the component transistors become active, and avalanche (regenerative) drop in the

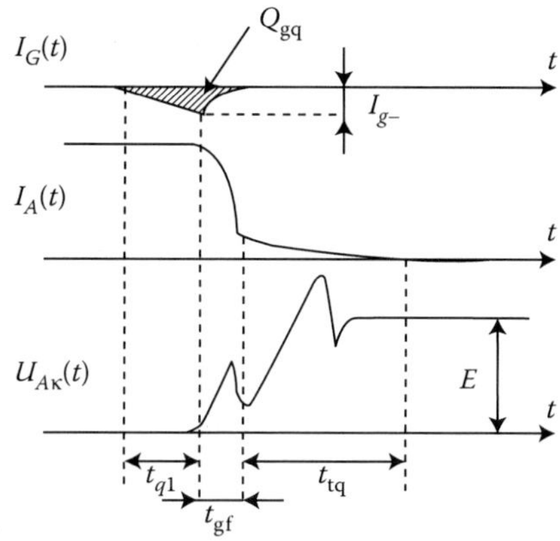

Figure 2.20 Transient process when the GTO is switched off.

thyristor's anode current begins on account of positive feedback. As negative control current develops only in one p base, the charge first falls to zero in that base. Positive feedback stops, and the n–p–n transistor switches to the cutoff mode. In the third stage, the resorption of the charge in the base of the p–n–p transistor occurs with a broken base—in other words, by recombination. As a result, the residual anode current in the structure slowly falls, with a time constant practically equal to the carrier lifetime.

Handbooks for GTO thyristors present three parameters corresponding to these stages: t_{gl}, the time delay with respect to the gate; I_{fg}, the time in which the anode current declines (with respect to the gate); and t_{tq}, the duration of the final cutoff stage.

2.5.2 *Gate-commutated thyristors (GCTs, ETOs, MTOs)*

The dynamic characteristics of GTOs are determined by their cutoff time, the main component of which is the time delay t_{gl} with respect to the gate.

For specified amplitude of the anode current, the cutoff charge Q_{gq} in the control circuit is practically independent of the growth rate of the negative control current (Figure 2.20).

The time t_{gl} may be determined from the formula

$$t_{gl} = \sqrt{2Q_{gq}\left(\frac{dI_g^-}{dt}\right)^{-1}}, \tag{2.36}$$

where dI_g^-/dt is the growth rate of the cutoff current.

To reduce t_{gl}, we may increase dI_g^-/dt. However, that reduces G. Some reduction in G is useful in terms of increased speed, improved dynamic distribution of the current between the structural elements in cutoff, and reduction in the dynamic losses. However, at very high dI_g^-/dt, the duration t_{tq} of the final stage begins to increase, along with the amplitude of the residual current. Therefore, by means of a choke in the control circuit, dI_g^-/dt is maintained within the range from 10–20 to 80–120 A/μs.

These problems are largely overcome by switching the load current from the gate of the GTO. Then, in shutdown, the anode current is transmitted to the thyristor's control circuit with $G \approx 1$. Several versions of this approach may be noted (Li et al., 2000).

1. Hard switching with connection of the negative voltage source to the control circuit (Figure 2.21). Such structures are known as gate-commutated turn-off (GCT) thyristors.
2. Emitter switching of the anode current by means of additional switches in the cathode and control circuits (Figure 2.22). Such structures are known as emitter turn-off (ETO) thyristors.
3. Switching of the anode current by shunting the control electrode using an additional switch (Figure 2.23). Such structures are known as metal oxide semiconductor (MOS) gate turn-off (MTO) thyristors.

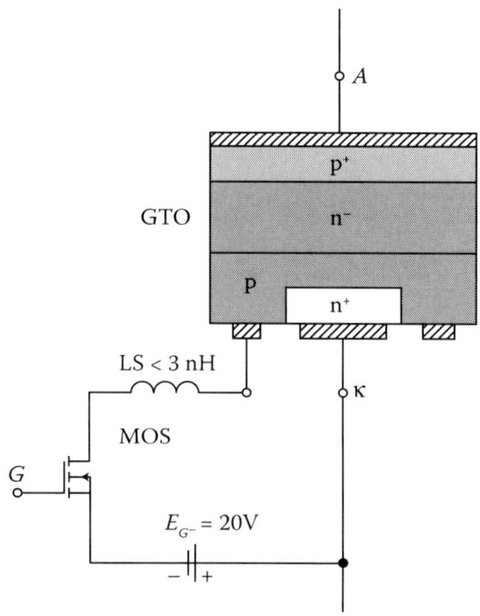

Figure 2.21 A GCT and its structure.

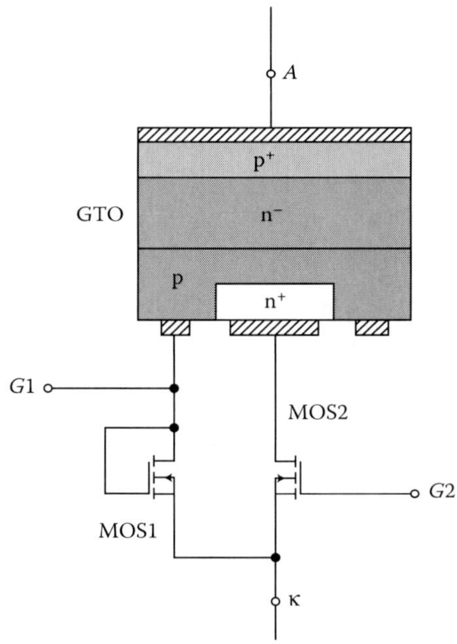

Figure 2.22 An ETO and its structure.

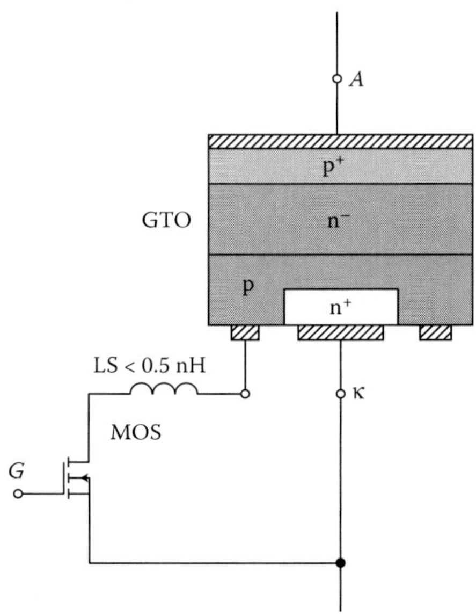

Figure 2.23 An MTO and its structure.

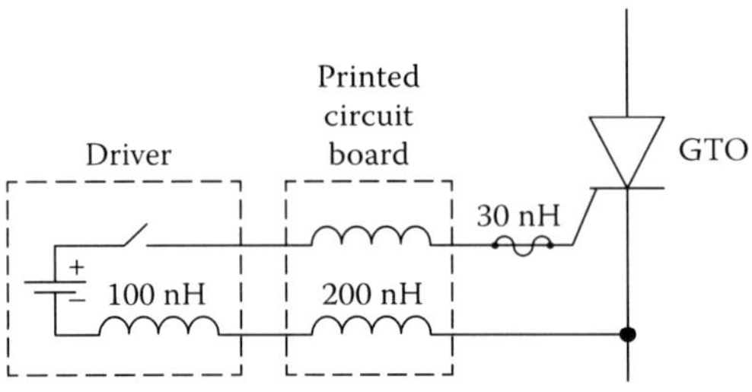

Figure 2.24 Parasitic inductance in the GTO control circuit.

2.5.3 The integrated GCT thyristor

One of the main problems in the control of gate-commutated thyristors is the distributed parasitic inductance L_S in the switching circuit. It consists of the inductance of the driver circuit (~100 nH), the supply buses (~200 nH), and the control electrode (~30 nH). At high switching speeds, with powerful anode currents (>10^3 A), the parasitic inductance L_S may lead to unacceptable voltage surges in the control circuit (Figure 2.24).

For safe switching of a powerful anode current in the thyristor's control circuit, a technique has been devised for connecting the contact areas of the housing to individual segments of the chip. To reduce the parasitic inductance of the supply buses, the control driver is integrated with the thyristor housing (Figure 2.25). This structure is known as an integrated GCT (IGCT) thyristor (Hidalgo, 2005).

2.5.4 The MOS-controlled thyristor

The MOS-controlled thyristor (MCT) represents a relatively new class of semiconductor power components. In terms of switched power and switched current density, these components match GTOs. The MCT is controlled from a circuit with an insulated gate. The basic module of the MCT is analogous to the semiconductor structure of a powerful insulated

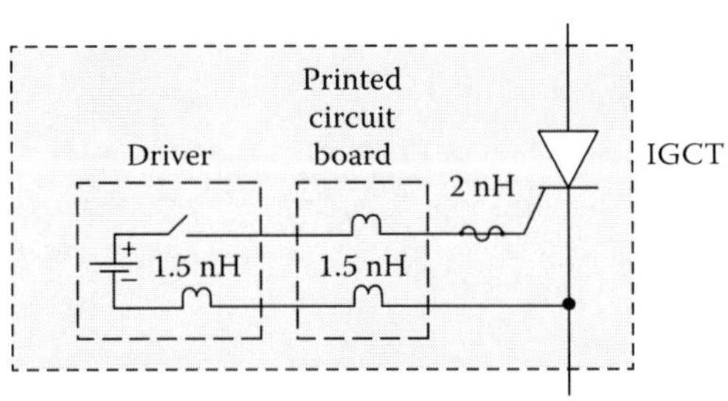

Figure 2.25 An IGCT and the parasitic inductance in its control circuit.

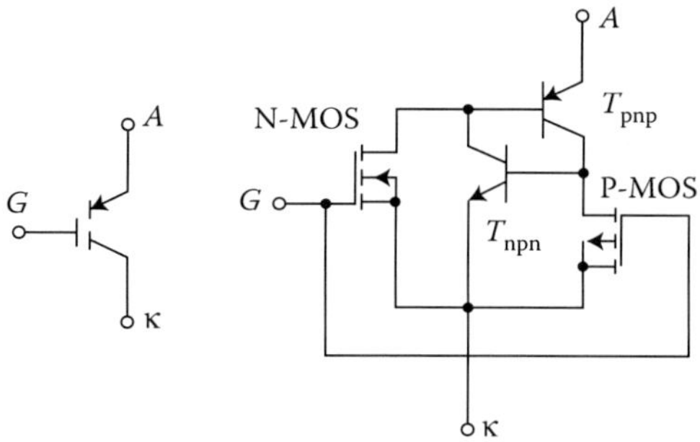

Figure 2.26 The MCT and its equivalent circuit.

gate bipolar transistor (IGBT), into which two control MOS transistors with opposite types of conductivity are integrated (Figure 2.26).

The N-channel control transistor ensures positive feedback in the thyristor's p–n–p–n structure on shutdown. The P-channel control transistor interrupts the feedback by shunting the emitter junction of the p–n–p–n structure through the cathode in thyristor shutdown.

The main benefits of MCTs are increased speed and relatively low power consumption from the control circuit. The voltage-blocking properties of the MCT are comparable with those of GTOs, but their voltage in the open state is lower, the impact current is greater, and the resistance to dI/dt is greater.

Numerous MCT designs have been developed: with P and N current channels; with symmetric and asymmetric blocking; with unidirectional and bidirectional gates; and with different switching methods (e.g., by means of a light flux). However, most of these proposals are at the experimental stage, in view of the complexity of the technology and the low yield of usable products.

2.6 Field transistors

2.6.1 Powerful short-channel MOS transistors

Transistors with metal–dielectric–semiconductor structure exist in two main forms: with a built-in current channel and with an induced current channel. For powerful transistors, the induced-channel design is employed. In the transistor, the gate electrode is separated from the semiconductor by a layer of dielectric, usually silicon dioxide (SiO_2). Accordingly, they are generally known as MOS transistors. The operation of such transistors is monitored on the basis of the gate voltage. An MOS transistor with an induced channel is nonconducting if there is no gate voltage. In order to produce current, the voltage applied to the gate must be of the same polarity as the main mobile carriers within the semiconductor. Charges of opposite sign are then induced in the semiconductor and reverse its type of conductivity; accordingly, a channel for current transmission is formed. Devices of N- and P-channel type are produced, although the former is preferable on account of the greater electron mobility (Figure 2.27).

Different types of MOS transistors are compared in terms of their design quality, which is determined by the ratio of the component's amplification factor to the length of the pulse front at the output (Oxner, 1982)

$$D = \frac{b}{2.2C_g}, \tag{2.37}$$

where b is the transistor gradient and C_g is the parasitic capacitance between the dielectric and the channel.

The gradient b is calculated as

$$b = \frac{\varepsilon\varepsilon_0\mu Z}{Ld}, \tag{2.38}$$

where ε is the dielectric permittivity of the oxide layer, ε_0 the absolute permittivity, μ the mobility of the primary charge carriers, Z the channel width, L the channel length, and d the thickness of the dielectric.

The capacitance between the gate and the channel is

$$C_g = \frac{\varepsilon\varepsilon_0 ZL}{d}. \tag{2.39}$$

Thus, the design quality of the MOS transistor is determined by the formula

$$D = \frac{\mu}{2.2L^2}. \tag{2.40}$$

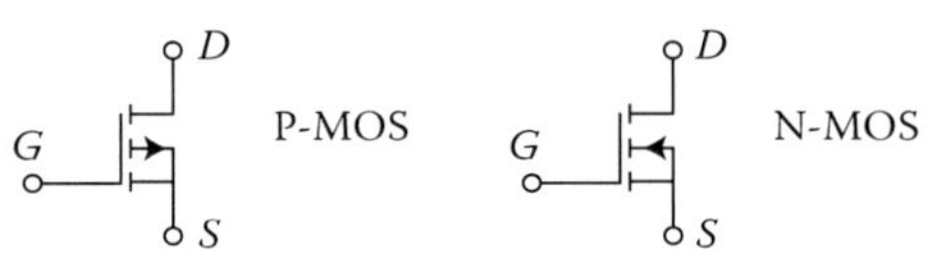

Figure 2.27 Powerful MOS transistors with an induced channel.

It follows that the primary method of increasing the design quality for MOS transistors is to reduce the channel length. To this end, we use horizontal and vertical multichannel structures with a short channel, created by double diffusion. The short channel ensures saturation of the carrier speed due to the strong electric field created by the drain voltage. As a rule, a structure with a horizontal cell is used for the manufacture of relatively low-voltage transistors (not more than 100 V), although the currents in such components may reach tens or even hundreds of amperes.

The drain of an N-channel power MOS transistor contains a lightly doped n^- region and a highly doped n^+ region (Figure 2.28). When the transistor is off, the space-charge region created by the drain's external voltage mainly travels toward the lightly doped n^- region.

Thus, the limiting voltage of the components is determined by the dimensions of the drift region and its resistivity. With an increase in length of the drain's drift region, the breakdown voltage increases. However, the auxiliary resistance between the drain and the source increases. The resistance R_D of the drift region may be estimated as

$$R_D = \kappa U^{n}_{(BR)DS}, \tag{2.41}$$

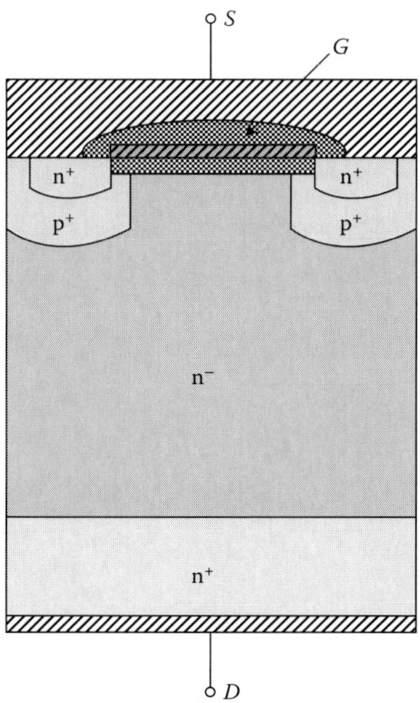

Figure 2.28 Vertical structure of a powerful short-channel MOS transistor.

where $\kappa = 8.3 \cdot 10^{-9}\ \Omega\ V^{1/n}$ is a constant of proportionality for silicon with a chip area $S = 1\ cm^2$; $n = 2.4–2.6$ is an exponent; and $U_{(BR)DS}$ is the voltage of avalanche drain–source breakdown.

When the MOS transistor is on, its voltage is relatively high on account of current transfer by primary carriers and the impossibility of modulation in the drain's drift region. The resistance $R_{DS(ON)}$ of the MOS transistor, when it is on, consists of several parasitic resistances of different layers in the transistor structure; the contribution of the drift region is the most important. For transistors with a horizontal structure, $R_{DS(ON)}$ is distributed approximately equally between the drift region and the channel. For higher-voltage transistors with a vertical structure, the drift region accounts for 95% of $R_{DS(ON)}$.

To reduce the resistivity, the number of parallel cells in the structure is increased. However, this increases the chip area and cost. Therefore, as a rule, the maximum working voltage of MOS transistors is 600–800 V.

For MOS transistors with a U-shaped gate channel, $R_{DS(ON)}$ is lower than that for the standard structure. This technology permits the use of smaller basic cells and a greater channel density in the crystal (Figure 2.29). Monolithic structure is more effective for high-current transistors.

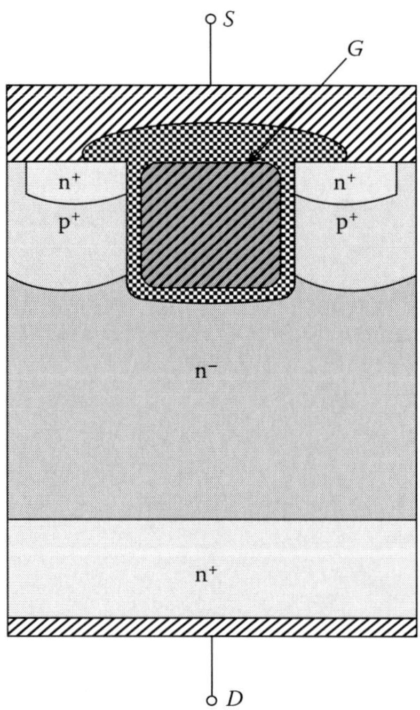

Figure 2.29 Structure of an MOS transistor with a U-shaped gate channel.

Finally, the relatively low-voltage components with a horizontal structure are practically ideal switches in terms of both speed and residual voltage when on. For powerful MOS transistors with a working voltage up to 100 V at load currents of 100–300 A, for example, $R_{DS(ON)}$ is not more than a few milliohms.

We should also note that, if part of the p region of a powerful MOS transistor is connected to a metallized layer of the drain so as to prevent the appearance of a parasitic bipolar n–p–n transistor, an internal diode appears in parallel to, but also in the direction opposite to, the current channel (Figure 2.30). In terms of its electrical parameters, this diode corresponds to a power transistor and may therefore be used in practice. Note that the cutoff time of this diode is around 100 ns, which is greater than that for most independent diodes.

The dynamic characteristics of the MOS transistor are determined by the recharging rate of the parasitic capacitances between the basic electrodes in the structure: C_{GS}, C_{GD}, and C_{DS}. Handbooks present the input C_{iss}, through C_{rss}, and output C_{oss} capacitances measured at external electrodes. These quantities are linearly related to the parasitic capacitances

$$\begin{cases} C_{iss} = C_{GS} + C_{GD}, \\ C_{rss} = C_{GD}, \\ C_{oss} = C_{DS} + C_{GD}. \end{cases} \tag{2.42}$$

It is of interest to estimate the time delays and the rise and fall times of the current and voltage fronts on switching, especially in the case of an inductive load, which is the most common in practice (Figure 2.31).

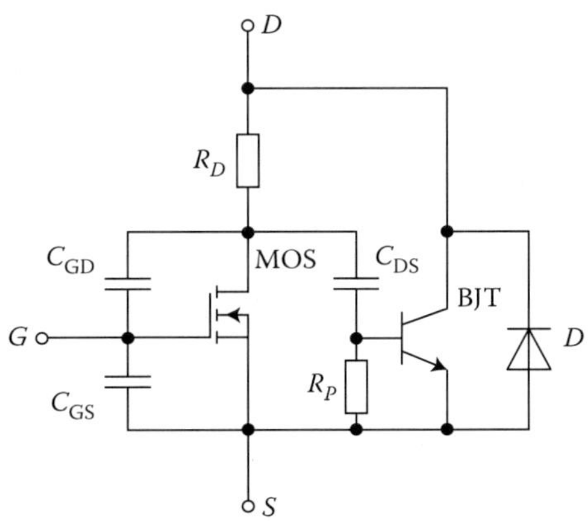

Figure 2.30 Equivalent circuit of a powerful MOS transistor.

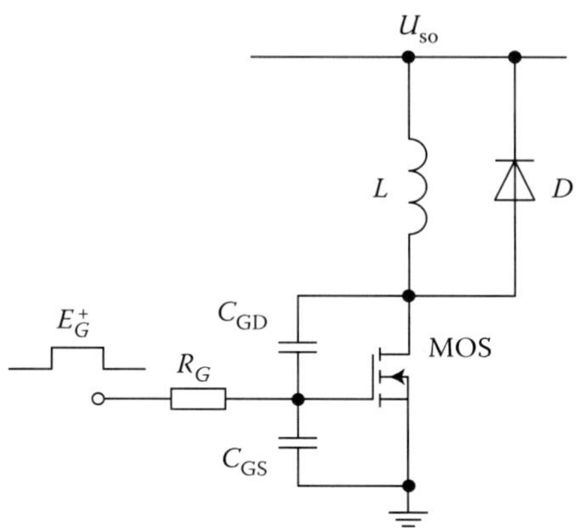

Figure 2.31 MOS transistor with an inductive load shunted by a diode.

When switching on the transistor, the delay $t_{d(ON)}$ depends on the threshold voltage V_{th} at which a conducting channel is induced in the structure

$$t_{d(ON)} = R_G C_{iss} \ln \frac{E_G^+}{E_G^+ - V_{th}}, \tag{2.43}$$

where R_G is the series resistance in the gate circuit and E_G^+ is the positive control voltage pulse.

When switching on the transistor with an inductive load, the time for the drain current to reach the load current I_{lo} may be calculated as

$$t_{rI} = \frac{I_{lo} C_{iss} R_G}{E_G^+ S_0}, \tag{2.44}$$

where S_0 is the slope of the transistor's transfer characteristic.

With a faster increase in the drain current (at a few A/ns), which is typical for powerful MOS transistors, a marked voltage drop is seen in the drain circuit, on account of the parasitic inductance L_S of the installation joints. The voltage drop may be estimated from the formula

$$\Delta U = L_S \frac{E_G^+ S_0}{R_G C_{iss}}. \tag{2.45}$$

In a system with an inductive load, an increase in the drain current leads to a simultaneous current surge in the antiphase diode. After the shutoff properties of the antiphase diode are restored, the voltage at the MOS transistor begins to fall. Due to influence of negative feedback via the gate-drain capacitance C_{GD} of transistor (the Miller effect), the input current in the gate circuit is practically completely compensated by the current recharging capacitance C_{GD}. In those conditions, the front of the voltage drop at the drain may be determined as follows:

$$t_{fU} \approx 0.8\frac{R_G C_{GD}(U_{so} - \Delta U)}{E_G^+ - (V_{th} + I_{lo}/S_0)}, \tag{2.46}$$

where U_{so} is the supply voltage in the transistor circuit.

When switching off the transistor, the delay $t_{d(OFF)}$ is determined by the discharge of the transistor's input capacitance to the critical voltage V_{cr} in the gate circuit, during which the channel cannot conduct the specified load current ($V_{cr} = V_{th} + I_{lo}/S_0$):

$$t_{d(OFF)} = R_G C_{iss} \ln\frac{E_G^+}{V_{cr}}. \tag{2.47}$$

When switching off the transistor with an inductive load, the drain voltage first rises to the supply voltage U_{so}. The time for such voltage increase in the transistor is

$$t_{rU} \approx \frac{U_{so} C_{OSS}}{I_{lo}}. \tag{2.48}$$

Then, the drain current falls. In the transistor, this process takes a time

$$t_{fI} \approx 2.2 R_G C_{iss}. \tag{2.49}$$

The basic oscillograms of the transient process in the switch circuit are shown in Figure 2.32.

2.6.2 *CoolMOS technology*

In Figure 2.33, we show the cross-section of a CoolMOS power transistor.

In the CoolMOS device, in contrast to a regular transistor, the p column penetrates deeply into the active region of the structure. That ensures greater doping of the drift layer, adjacent to the n column. The resistance of the CoolMOS transistor in the on state is several times less than that of a regular MOS transistor (Arendt et al., 2011). The doping of

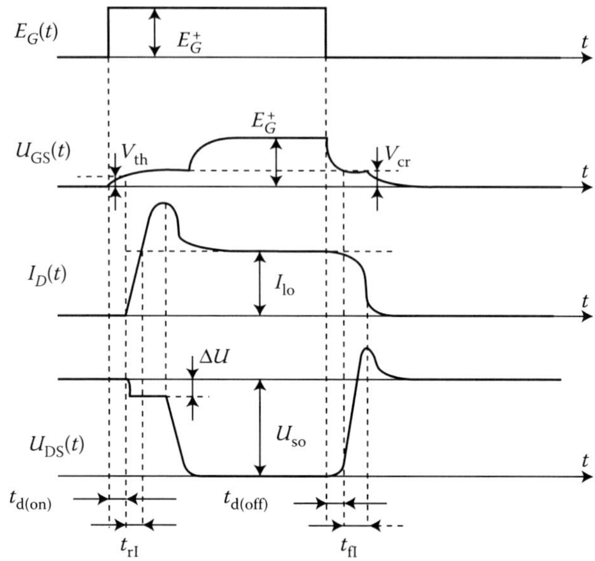

Figure 2.32 Oscillograms of the transient process in the MOS switch.

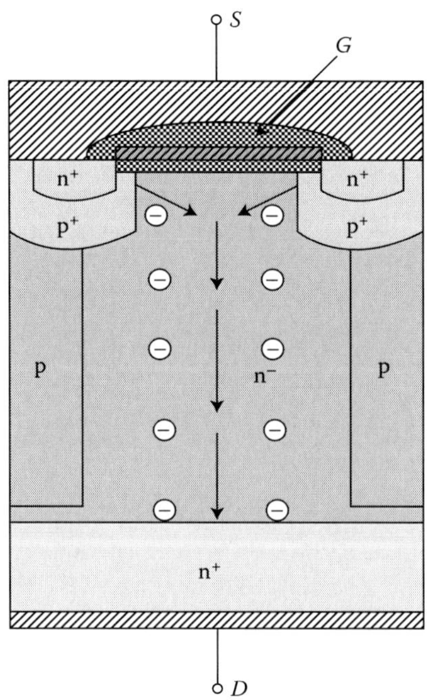

Figure 2.33 Structure of a CoolMOS transistor.

the p region is carefully monitored during the manufacture of the structure, as there must be absolutely no free carriers in that region when the transistor is off. The resistance of the drift layer in the CoolMOS transistor depends more or less linearly on the maximum permissible voltage $U_{(BR)DS}$:

$$R_D = \kappa U^n_{(BR)DS}. \tag{2.50}$$

Here $n \approx 1.17$ and $\kappa = 6.0 \times 10^{-6}\ \Omega\ V^{1/n}$ is the constant of proportionality with a chip area $S = 1\ cm^2$.

The electric field strength along the drift region of the CoolMOS transistor is practically rectangular. That increases the avalanche breakdown voltage. For mass-produced CoolMOS transistors, the maximum permissible sink–drain voltages are 600 and 900 V, respectively.

Another benefit of CoolMOS transistors is that the output capacitance C_{DS} is significantly nonlinear. With an increase in the drain voltage to tens of volts, the p region becomes depleted, with a marked decrease in C_{DS}. Therefore, at high working voltages, the output capacitance of the CoolMOS transistor is much less than that for regular transistors. In the voltage range 350–400 V, the energy stored in the output capacitance is reduced by about 50%, with marked reduction in the switching losses as the switching frequency increases.

2.6.3 Static induction transistors

The static induction transistor (SIT) is a field transistor with a control p–n junction and a built-in current channel. In contrast to a regular field transistor, the very short channel in the SIT does not permit saturation of the drain current, analogous to the vacuum triode tube. For this reason, the SIT is sometimes regarded as the solid-state version of the vacuum triode (Oxner, 1982).

The SIT is based on an n-type silicon plate of thickness 0.4–0.9 mm. The n^+ substrate is strongly doped with donor impurity (to concentrations around $5 \times 10^{19}\ cm^{-3}$). Then, an epitaxial n^- layer is grown on the substrate; this layer determines the structure's breakdown voltage. The subsequent technological operations form the drain region and the control gate (grid).

The SIT has two basic structures, with the grid at the surface and in the depth (Figure 2.34).

With the grid in the depth, we obtain a low-frequency component used in devices at the acoustic frequency range. High-frequency SITs are manufactured with a surface gate, which has direct ohmic contact with the metallization layer. This permits sharp reduction in the series resistance in the gate circuit.

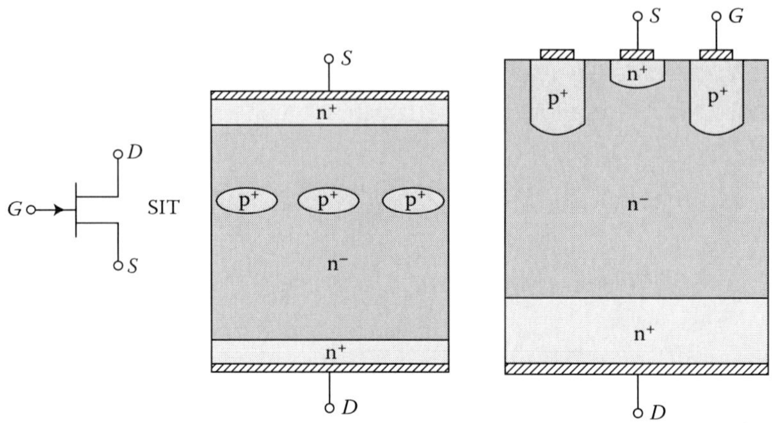

Figure 2.34 The SIT and its basic structures.

The SIT is able to withstand large voltages between the drain and the source. The SIT is reliably turned off by applying a voltage that is negative to the gate, relative to the drain. The gate voltage may be determined as

$$U_G = \frac{U_D}{\mu}, \tag{2.51}$$

where U_D is the drain voltage of the SIT when it is off and μ is the external voltage amplification factor (the blocking factor) of the SIT.

The required value of μ is ensured in the manufacture of the transistor. Its standard value is 10^2–10^3, depending on the voltage in the circuit.

The family of triode volt–ampere characteristics of the SIT (Figure 2.35) is described by the equation

$$I_D = I_0 \exp\left(-\frac{\eta(U_G - U_D/\mu^*)}{\varphi_T}\right), \tag{2.52}$$

where I_D is the drain current, the parameters I_0 and η are determined by the structure and the impurity profile of the SIT, respectively, μ* is the internal voltage amplification factor of the SIT (proportional to μ), U_G is the gate voltage, and U_D is the drain voltage.

With high-current density in the transistor's output circuit, we need to take account of the series resistance in the drain (R_D) and source (R_S). Then the volt–ampere characteristic takes the form

$$I_D = \frac{U_D}{(1+\mu^*)R_S + R_D} - \frac{U_G\mu^*}{(1+\mu^*)R_S + R_D}. \tag{2.53}$$

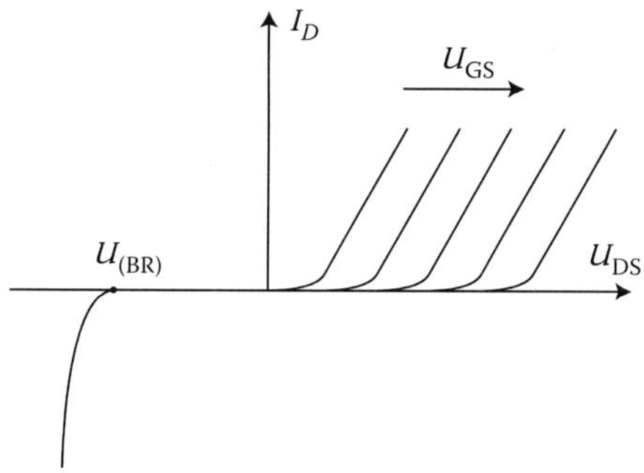

Figure 2.35 Output volt–ampere characteristics of an induction transistor.

On account of the relatively high resistance in the drift region and the increase in the forward voltage drop at the SIT, the drain current is not more than 10 A, as a rule.

At large load currents, the SIT switches to bipolar operation, with the application of positive voltage to the gate circuit. Then, a resistor whose magnitude specifies the amplitude of the forward control current I_G^+ appears in series with the gate. Hole injection from the controlling p^+ grid of the gate sharply reduces the resistance of the drift region and reduces the forward voltage drop at the SIT.

The effectiveness of the SIT in bipolar operation is characterized by the current transfer factor $B_0 = I_D/I_G^+$, which is in the order of 10^2, in contrast to a regular bipolar transistor.

To ensure SIT operation at currents of 10^2–10^3 A, the structure is modified: a silicon plate characterized by hole conductivity is used as the substrate (Figure 2.36).

This structure is switched on simply by removing the cutoff potential from the gate. Then the injection of charge carriers from the substrate to the drift region begins. This ensures effective modulation of the transistor's conductivity. An SIT of that structure is known as a field-controlled thyristor (FCT). Note that the FCT is switched in the absence of positive feedback. Consequently, the FCT is relatively resistant to pulsed noise (Voronin, 2005).

With injection from the control grid, as in bipolar SIT operation, or from the substrate (anode), as in the FCT, secondary charge carriers appear in the n^- base. This significantly changes the transient processes, especially during shutdown. During shutdown, the holes are drawn from the n^- base by a reverse-biased gate. Consequently, a current pulse that is relatively short but close in amplitude to the load current will act in the control circuit. In the FCT, we then observe the appearance of a parasitic p–n–p transistor whose emitter is the thyristor's drain (anode), whereas its collector

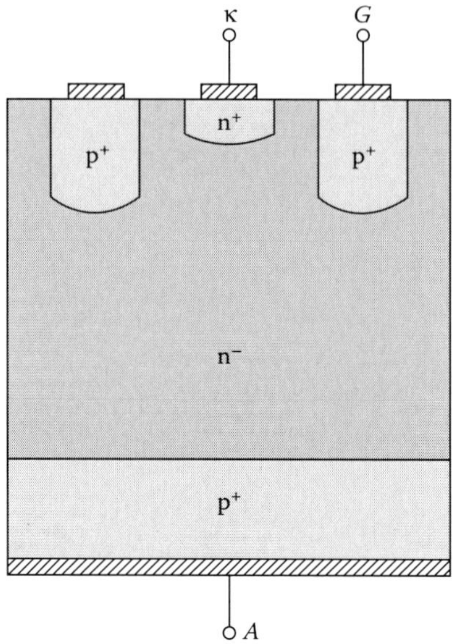

Figure 2.36 Structure of the FCT.

is the gate. Its base is rapidly shut off, with closure of the primary current channel. The FCT then shuts down analogously to a p–n–p transistor with a broken base, that is, the charge declines slowly by recombination. Then, a residual current appears in the thyristor's output and control circuits; its duration is approximately three times the charge-carrier lifetime.

Note that the powerful reverse-current flow appearing in the gate circuit on shutdown induces an additional potential, both at p^+ points of the grid and at the output resistance of the control circuit. This reduces the cutoff voltage (Figure 2.37).

With relatively large resistance, the negative voltage in the gate circuit may be insufficient to form a reliable cutoff potential in the FCT channel. As a result, it is closed, and the thyristor is in a secondary-breakdown region.

The widespread practical introduction of the SIT and the FCT is hindered because of manufacturing difficulties and because these switches are in conditions in which there is zero potential in the control circuit.

2.7 The IGBT

Field control and bipolar conductivity are integrated in the IGBT. Like the powerful MOS transistor, its structure includes a lightly alloyed drift n^- layer, whose dimensions determine the maximum permissible voltage at the component. At the surface of the n^- layer, a controlling MOS structure

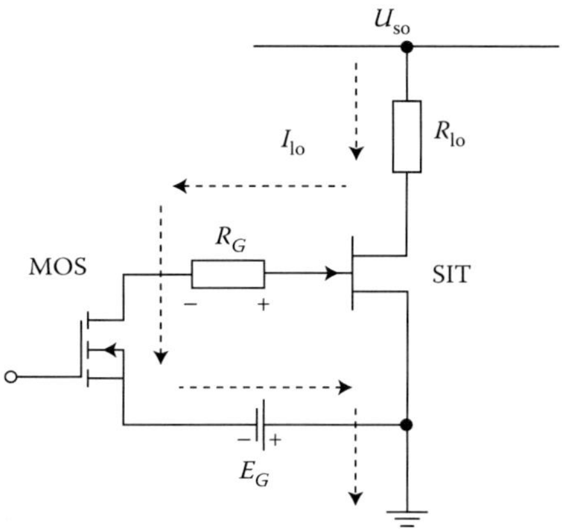

Figure 2.37 A switch based on an induction thyristor.

with an insulated gate is formed. However, the bottom of the drift layer is in contact with a highly doped p^+ layer, characterized by hole conductivity (Figure 2.38).

To switch off the IGBT, a positive voltage must be applied to the gate, so as to induce an n-type channel at the surface of the p base region under

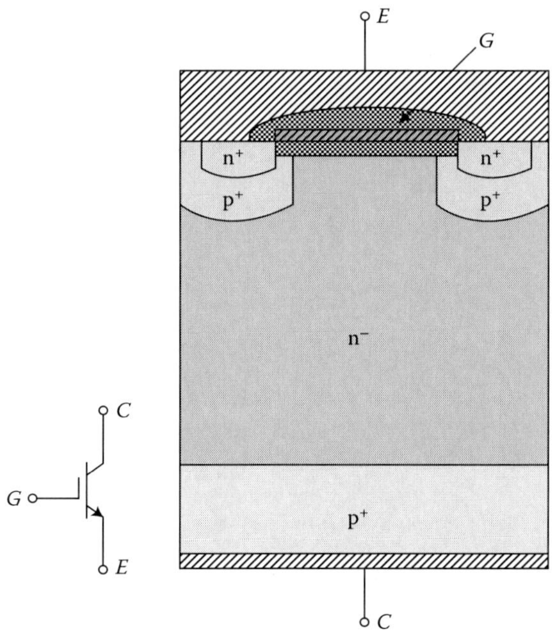

Figure 2.38 The IGBT and its structure.

the gate. Electric contact is formed between the n^+ emitter structure and the n^- drift region. The positive bias at the emitter of the p–n–p transistor in the IGBT structure results in the injection of secondary charge carriers from the highly doped p^+ layer to the drift region, with a consequent decrease in resistance. Thus, the monolithic IGBT structure combines control of the voltage at the high input resistance and the transfer of forward current with high permissible density.

The equivalent circuit of the IGBT is shown in Figure 2.39. It includes a controlling MOS transistor, a bipolar p–n–p transistor, and a field n-channel transistor with a controlling p–n junction. The latter blocks the external voltage applied to the IGBT when it is closed, thereby protecting the controlling MOS from high voltage.

The collector current of the IGBT has two components: the primary electron current, which passes through channels in both field transistors, and the hole component in the output circuit of the p–n–p transistor.

The voltage drop at the channel resistance produces negative bias in the collector–base circuit of the p–n–p transistor, which is thereby switched to active operation. This is largely responsible for the increased saturation voltage in the monolithic IGBT, as the reverse bias of the collector junction in the p–n–p transistor ensures intense hole capture from the transistor's base region, with a consequent increase in its resistance.

The saturation voltage of the IGBT consists of three components: the voltage drop at the base–emitter junction of the p–n–p transistor; the voltage drop at the channel of the controlling MOS transistor; and the ohmic

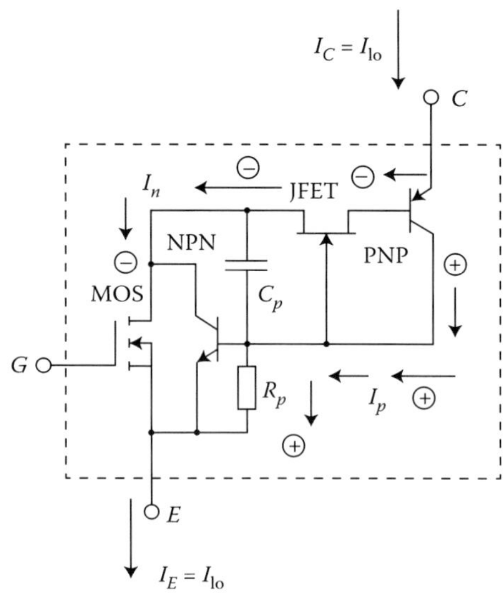

Figure 2.39 Equivalent circuit of the IGBT.

voltage drop at the n⁻ drift layer, which is modulated by charge carriers. Thus, we may write

$$U_{\mathrm{CE(sat)}} = \varphi_T \ln\frac{I_C}{I_S} + I_C(1-\alpha_{\mathrm{PNP}})r_{\mathrm{MOS}} + I_C r_{\mathrm{n}^-}, \tag{2.54}$$

where I_C is the collector current of the IGBT, I_S the reverse saturation current of the base–emitter junction, α_{PNP} the current transfer factor in the common-base circuit of the p–n–p transistor, r_{MOS} the resistance of the controlling MOS channel, and r_{n^-} the resistance of the modulated n^- base.

The hole flux captured by the collector junction of the n–p–n transistor traverses the longitudinal resistance of the section of the IGBT's p base directly under the n^+ emitter (Figure 2.40).

With current overload at the critical hole-current density, a forward bias is applied to the n^+–p junction at the longitudinal resistance, and the parasitic n–p–n transistor formed by layers of the n^+ emitter, the p base and the n^- drift layer begins to operate. As the parasitic n–p–n transistor and the basic p–n–p transistor form a trigger circuit, the IGBT becomes uncontrollable and malfunctions.

The critical IGBT collector current at which such triggering occurs is (Kuzmin et al., 1996)

$$I_{\mathrm{C(cr)}} = \frac{25\varphi_T S}{R_P^* \alpha_{\mathrm{PNP}} l(l/2+d)}, \tag{2.55}$$

where S is the area of the IGBT chip, R_P^* the resistivity of the p base, α_{PNP} the current transfer factor in the common base circuit of the transistor, l the width of the elementary n^+ emitter, and d the half-width of the elementary n^- base region below the gate.

The formula for the critical current permits the identification of the basic methods for stabilizing IGBT operation with current surges.

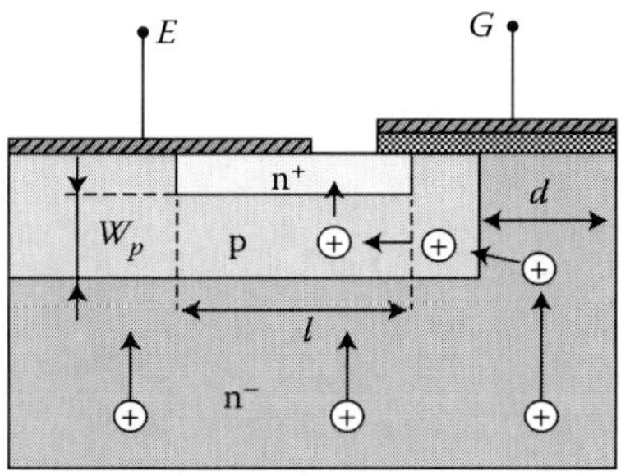

Figure 2.40 Region of the IGBT structure under the n^+ emitter.

- Reducing the overall resistance of the p base
- Reducing the strip width of the n^+ emitter
- Reducing the width of the region below the gate
- Reducing the amplification factor of the p–n–p structure

Better results may be obtained by combining these methods. However, the most effective method of increasing the IGBT's critical current is to reduce the strip width of the n^+ emitter.

The dynamic characteristics of the IGBT are assessed by the same parameters as for powerful MOS transistors. Manufacturers indicate the standard time delays and switching times, usually measured with an inductive load. The bipolar operation of the IGBT results in lower speed than for MOS transistors. When an IGBT is switched off, the controlling MOS channel is first closed, thereby shutting off the base output of the p–n–p transistor. Thus, the residual charge in the drift layer disappears solely as a result of relatively slow recombination. As a result, the plot of the switched current includes a long tail (Figure 2.41). The length of the tail is three to five times the carrier lifetime (from hundreds of nanoseconds to a few microseconds). The initial amplitude $I_T(0)$ of the residual current is proportional to the load current and depends nonlinearly on the voltage (Hefner and Blackburn, 1988). It may be estimated as

$$I_T(0) = \frac{I_{\text{lo}}}{1 + \left(W_n^2 / 2D\tau_B \right)}, \tag{2.56}$$

where I_{lo} is the load current of the switch, W_n the width of the n^- base with the specified supply voltage U_{so}, $D = 2000\ \mu\text{m}^2/\mu\text{s}$ is the ambipolar diffusion coefficient, and τ_B is the carrier life in the n^- base.

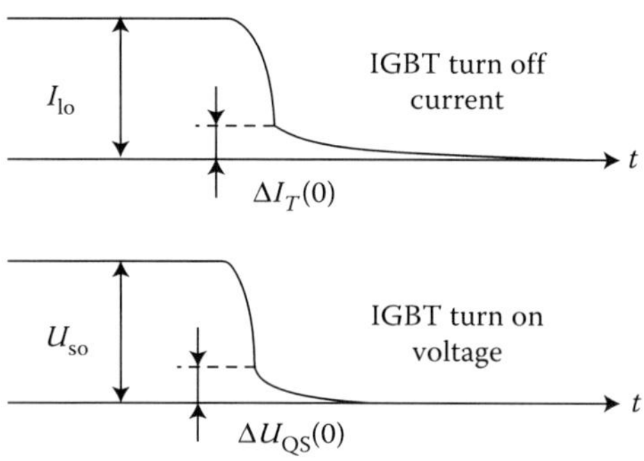

Figure 2.41 Tail current and dynamic saturation voltage of the IGBT.

The width of the n⁻ base at voltage U_{so} may be calculated from the formula

$$W_n = W_{n0} - \kappa\sqrt{U_{so}}, \tag{2.57}$$

where W_{n0} is the initial width of the drift region n⁻ at zero voltage and $\kappa = 2.56\ \mu\text{m V}^{-1/2}$ is a constant of proportionality.

In the transient process when an IGBT (especially a high-voltage IGBT) is switched on, the voltage curve (Figure 2.41) includes a slow stabilization stage (dynamic saturation). This stage may last a few hundred nanoseconds. When the IGBT is off, the space-charge region mainly extends over the lightly doped drift region. However, it also penetrates slightly into the p base. The barrier capacitor formed by the depleted region in the p base offers practically no paths for fast discharge when the structure is switched on, which also gives rise to a dynamic saturation stage. The initial amplitude $U_{QS}(0)$ of the dynamic saturation voltage hardly depends on the load current; it is determined by the external supply voltage. In fact, $U_{QS}(0)$ depends on the doping of the IGBT's base regions and may be 10–10² V, depending on the IGBT design and the supply voltage. It may be determined from the formula

$$U_{QS}(0) \approx \frac{N_D(U_{so})^{1.4}}{N_A + N_D}, \tag{2.58}$$

where N_D is the donor concentration in the n⁻ base, N_A the acceptor concentration in the p base, and U_{so} the supply voltage.

Estimation of the dynamic losses in the IGBT on switching (the switching losses) is of great practical interest. In the literature, we may find values of the dynamic losses when the IGBT is switched on (E_{ON}) and off (E_{OFF}), measured in specified electrical conditions. A relatively precise formula permits conversion of the handbook data to the dynamic losses W_{dyn} for actual loads

$$W_{dyn} = (E_{ON} + E_{OFF})\frac{I_{lo}}{I_0}\left(\frac{U_{so}}{U_0}\right)^n, \tag{2.59}$$

where I_{lo} is the actual load current of the IGBT, I_0 the handbook current (usually selected as the average IGBT current), U_{so} the supply voltage, U_0 the handbook voltage (usually selected as half of the maximum permissible IGBT voltage), and $n \approx 1.5$–1.8.

In estimating the dynamic parameters of the IGBT, we must identify the critical factors that determine the safe region. In the transient process when the IGBT is switched on, the critical factor is the growth rate of the collector current in switching a load current from an antiphase diode. The critical factor presented in handbooks is the minimum permissible resistance R_G connected in series with the IGBT's gate. We do not recommend this choice.

In the transient process when an IGBT is switched off, the critical factor is the parasitic installation inductance L_S, which is limited to 100 nH, as a rule.

The IGBTs in widest use are third-generation devices with a planar gate, manufactured by two basic technologies: the punch-through IGBT (PT-IGBT) and non-punch-through IGBT (NPT-IGBT). Fourth-generation devices use a vertical gate (the trench-gate IGBT), which permits maximum packing density of the components on the chip and reduces the static losses. Finally, fifth-generation devices [the soft punch-through IGBT (SPT-IGBT), field-stop IGBT (FS-IGBT), and carrier-stored trench-gate IGBT (CSTBT)] are characterized by optimized switching characteristics and conductivity.

2.7.1 Epitaxial (PT) and homogeneous (NPT) IGBT structures

The PT-IGBT is manufactured on a highly doped p^+ substrate with hole conductivity. A high-resistance drift layer of n^- type is then grown epitaxially on the substrate (Figure 2.42). Accordingly, this IGBT technology is also said to be epitaxial.

For a transistor with a working voltage of 600–1200 V, the drift layer measures 100–120 μm. Then a controlling MOS structure is formed on the n^- layer by a technology analogous to that used in the production of powerful MOS transistors. In epitaxial technology, the structure includes an additional n^+ electron layer between the drift layer and the substrate; this is known as the buffer layer. The n^+ layer is small (around 15 μm). The additional highly doped layer limits the electric field strength in the base of the IGBT when it is off, thereby increasing the resistance to avalanche breakdown. By adjusting the doping of the n^+ layer, the injection properties of the emitter in the p–n–p transistor within the IGBT may be regulated. Hence, the required charge-carrier concentration may be established in the base of the IGBT, and the carrier life may be regulated. By means of the buffer layer, it is also possible to modify the critical collector-current density at which the loss of controllability of the PT-IGBT is triggered. A deficiency of epitaxial technology is the temperature dependence of the residual current, which increases the energy losses when the transistor is switched off.

The NPT-IGBT (Figure 2.43) is manufactured using a uniform n^- substrate with electron conductivity (thickness 200 μm). A planar MOS gate is created on the upper side of the plate; on the opposite side, a p^+ emitter is formed by ionic doping. The equivalent circuit of the resulting structure is the same as for a PT-IGBT.

As the NPT-IGBT structure lacks the highly doped buffer layer, a relatively large n^- substrate is used to ensure high resistance to breakdown.

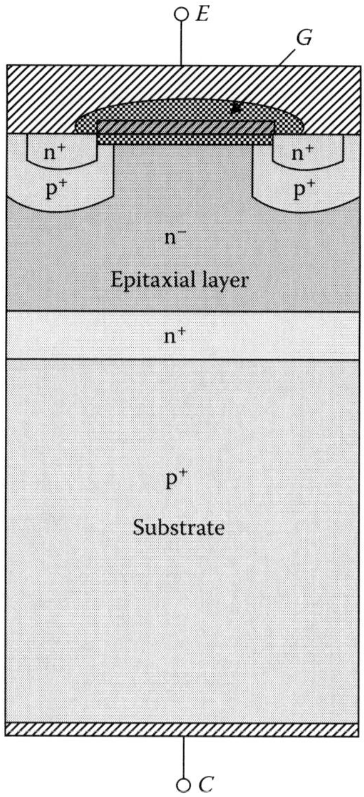

Figure 2.42 Structure of PT-IGBT.

This means that NPT-IGBT has a higher saturation voltage. However, the uniform structure is highly resistant to short circuits, has a positive temperature coefficient of the voltage at all load currents, and has a safe operating region that is rectangular. The residual current of the NPT-IGBT persists for a relatively long time (a few microseconds), but the amplitude of the tail is about half that for the PT-IGBT and it hardly depends on the temperature.

2.7.2 *Trench-gate IGBT*

To reduce the saturation voltage of the IGBT, designers have created a structure with a vertical gate in a trench (Figure 2.44). The depth of the gate and the width of the trench are a few micrometers.

The operating principle of the trench-gate IGBT is as follows. Some of the holes, injected from the p^+ region of the p–n–p transistor's emitter to its p collector, reach its boundary and are captured by the electric field of the reverse-biased junction, as in a regular IGBT with a planar gate. Other holes, injected from the p^+ region in the direction of the vertical gate's base, cannot be captured at once by the collector of the p–n–p transistor,

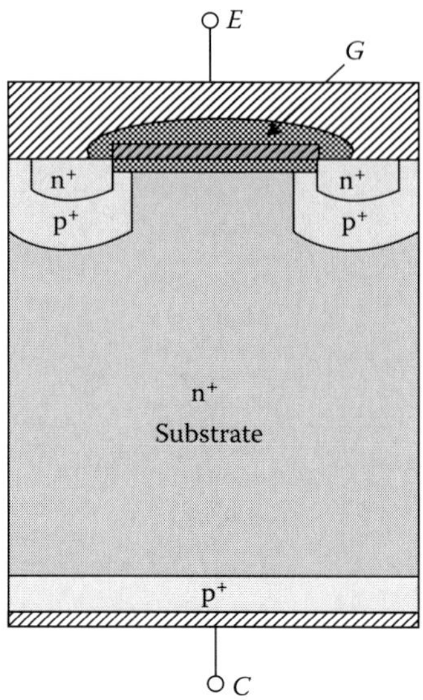

Figure 2.43 Structure of NPT-IGBT.

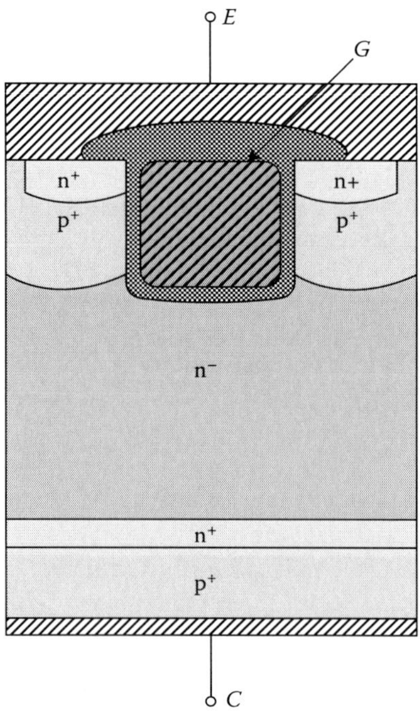

Figure 2.44 Structure of trench-gate IGBT.

and their charge accumulates at the section of the n^- base immediately beneath the gate. To compensate the positive charge of the holes, intensified electron injection begins from the emitter's n^+ region, and the electron concentration increases in the region beneath the gate. As a result, a structure equivalent to a p–i–n diode, with a characteristic charge-carrier distribution close to its emitter layers in bilateral injection, is formed in the corresponding section of the IGBT (about half of the total transistor base). Then the structure of the IGBT base with a vertical gate may be regarded as a combination of elementary p–i–n and p–n–p structures of identical width (Udrea and Amaratunga, 1997). As a result, the saturation voltage of the trench-gate IGBT is reduced to around 1.4–1.7 V, which is about 30–40% less than that in a standard IGBT.

A downside of the trench-gate IGBT is its relatively high input capacitance and the gradient for forward-current transfer. This requires the use of special control drivers and additional current-surge protection.

2.7.3 The trench-FS and SPT

On the basis of the PT-IGBT and NPT-IGBT structures, various new chips for power transistors have been developed. Note, in particular, the SPT (soft PT-IGBT) and trench-FS (trench-field stop) structures (Arendt et al., 2011). They both include a built-in buffer n^+ layer and employ smaller chip area (Figure 2.45).

The main parameters determining the frequency properties and application of the IGBT are the saturation voltage, the gate charge and circuit, and the switching losses.

SPT chips have optimized cutoff properties: a smooth and linear increase in the voltage at the transistor when it is off and limited duration

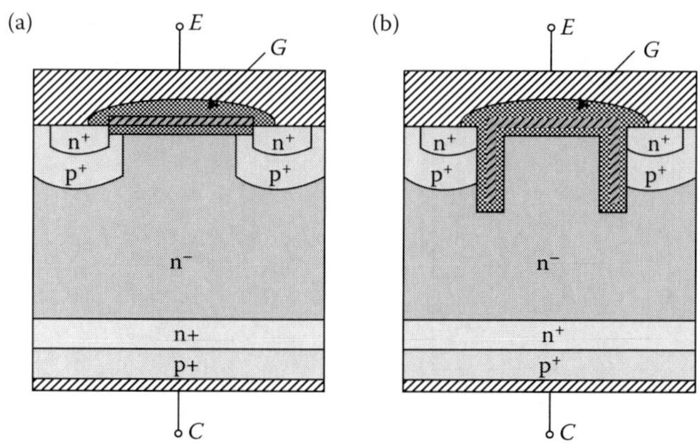

Figure 2.45 Structure of the (a) SPT IGBT and (b) trench-FS.

of the residual current. The switching losses are less for the SPT than for the standard IGBT.

The trench-FS chip has a deep vertical gate and a modified emitter structure, thereby ensuring optimal distribution of the carriers in the n⁻ substrate. As a result, the saturation voltage is 30% less than that for standard NPT-IGBT. The increased current density in the trench-FS permits significant reduction in the chip area, by almost 70%. However, the thermal resistance and charge in the gate circuit are higher than those in the standard design.

Both the SPT and trench-FS chips are characterized by elevated resistance to short circuits. The upper limit on the current is six times the rated value.

2.7.4 The CSTBT and SPT+

In the IGBT of regular structure, hole capture is most intense in the section of the base layer that is below the gate and is in direct contact with the reverse-biased collector junction p–n–p of the transistor. To increase the charge-carrier density in the base, an additional layer with electron conductivity is integrated into the structure of fifth-generation chips, between the p–n–p collector and the n^- base (Voronin, 2005). The doping of the additional *n* layer is greater than that in the base, which creates an additional potential barrier to the hole current. As a result, most of the holes cannot overcome the potential barrier and collect close to the junction, thereby reducing the resistance of the base and the transistor's saturation voltage. This technology is used in an IGBT with a planar gap (the SPT+; Figure 2.46) and an IGBT with a vertical gap (the carrier-stored trench-gate IGBT, CSTBT; Figure 2.47).

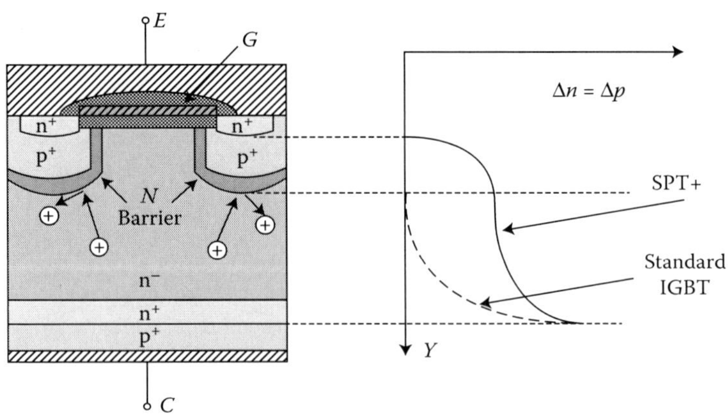

Figure 2.46 Structure of the SPT+.

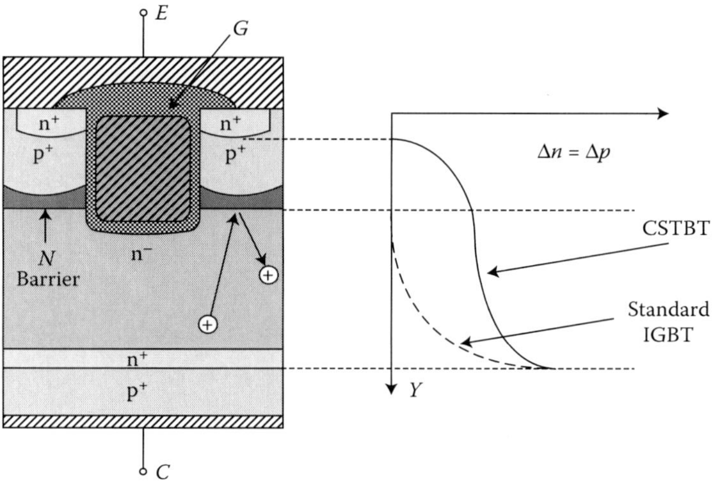

Figure 2.47 Structure of the CSTBT.

2.8 Switch modules

2.8.1 Topology of integrated power modules

We may identify the following basic types of integrated power modules.

1. A single switch and a switch with an opposed parallel diode (Figure 2.48)
2. Upper- and lower-level choppers (Figure 2.49)
3. Two semiconductor switches (Figure 2.50) in series, with an output at the midpoint (a half-bridge)
4. A single-phase bridge: four controllable switches or two controllable switches plus two diodes (Figure 2.51)
5. A three-phase bridge (Figure 2.52)
6. A three-phase bridge with an additional (braking) switch (Figure 2.53)

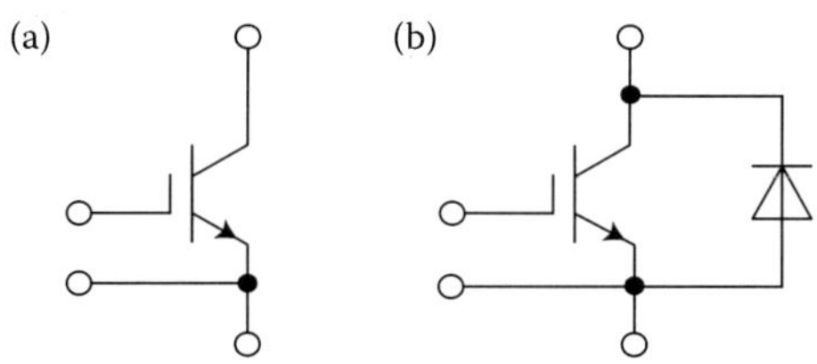

Figure 2.48 (a) Single switch and (b) switch with opposed parallel diode.

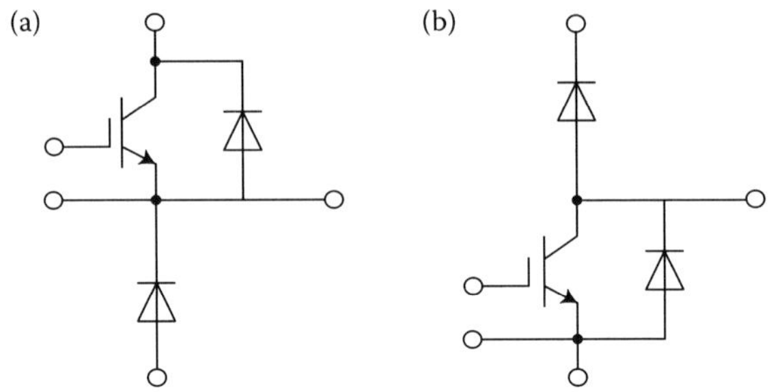

Figure 2.49 (a) Upper- and (b) lower-level choppers.

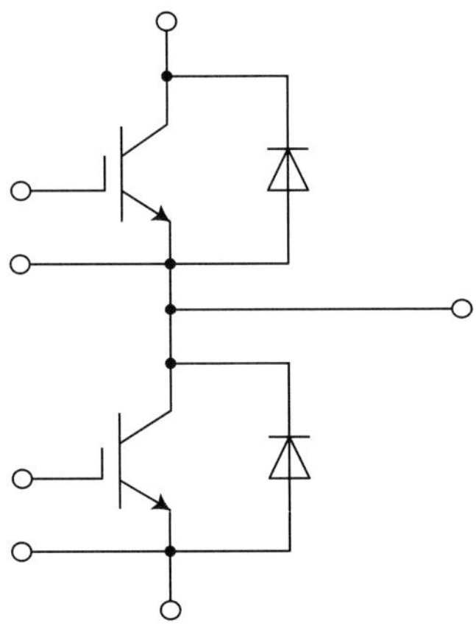

Figure 2.50 Two semiconductor switches with an output at the midpoint (a half-bridge).

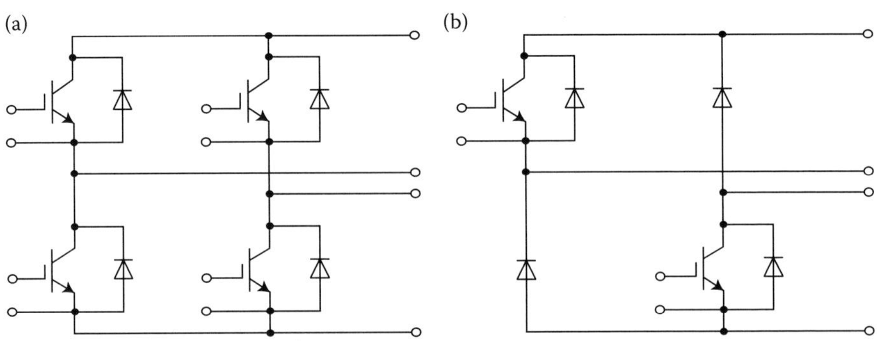

Figure 2.51 A single-phase bridge: (a) four or (b) two controllable switches.

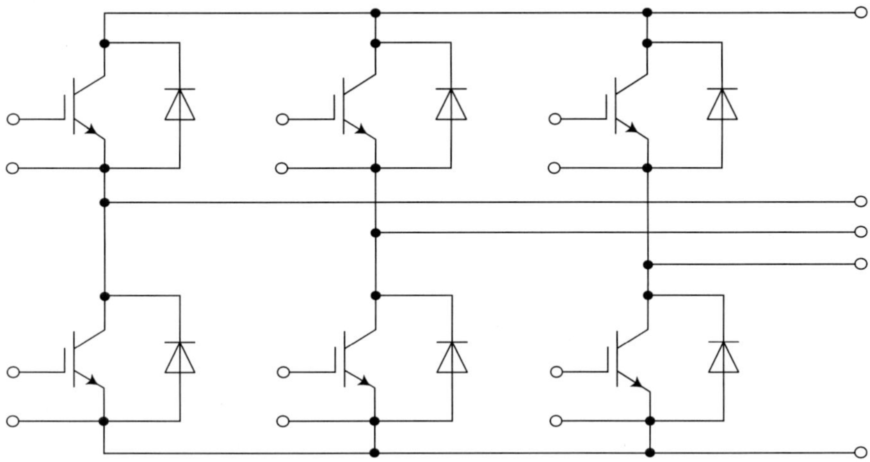

Figure 2.52 A three-phase bridge.

7. Specialized modules, for example, a three-level neutral-point-clamped inverter (Figure 2.54)
8. Integrated modules, for example, of topology B6U + B6I (Figure 2.55) or topology B2U + B6I (Figure 2.56)

2.8.2 Assembly of power modules

The following are the key technologies in the assembly of power modules (Figure 2.57).

- Soldering
- Ultrasonic welding
- Sealing with silicon gel

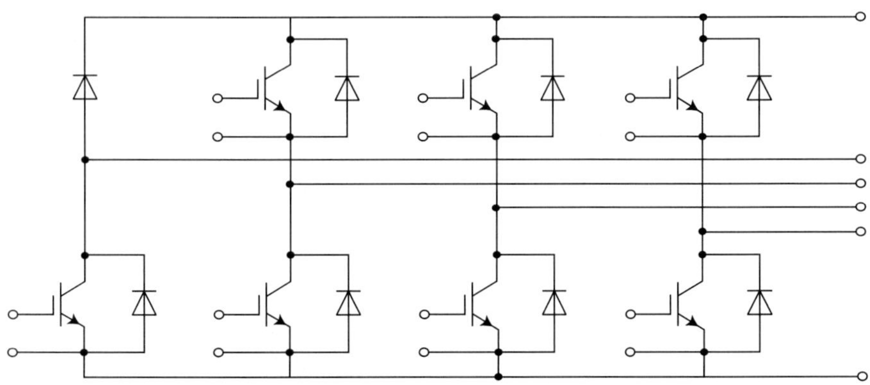

Figure 2.53 A three-phase bridge with an additional (braking) switch.

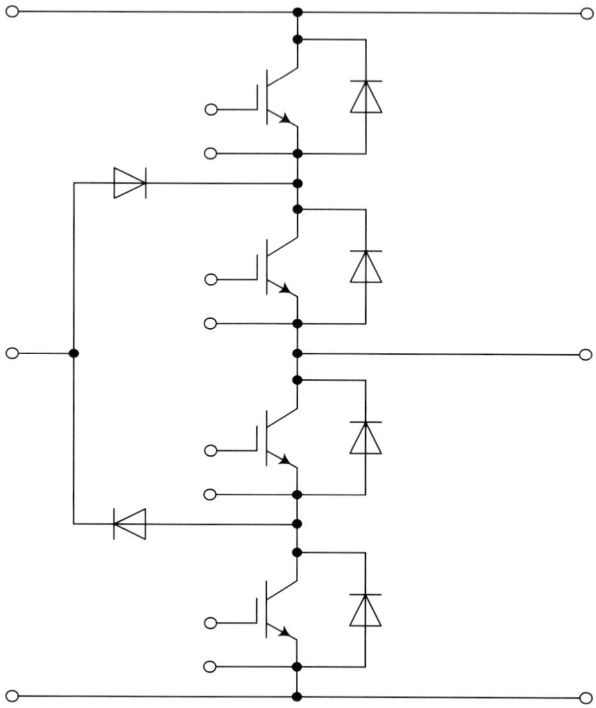

Figure 2.54 A three-level half-bridge with a fixed null point.

Soldering is used to mount chips on a DBC (direct bonded copper) ceramic substrate. The quality of the soldering processes determines the thermal characteristics of the module and its thermocyclic strength. A promising alternative to soldering is low-temperature sintering of the semiconductor and the substrate by means of superfine silver powder. This technology permits an increase in the module's temperature to

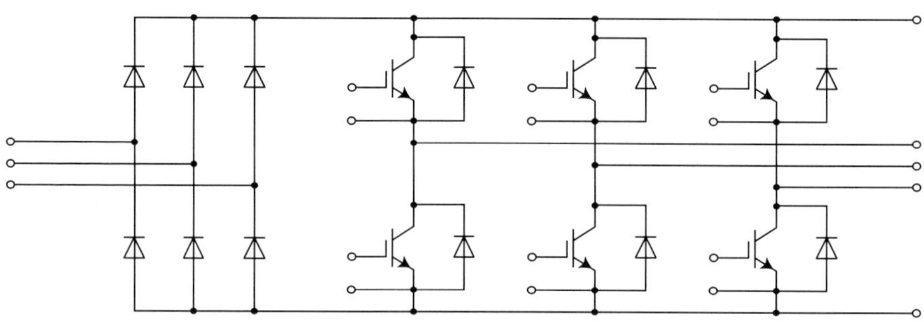

Figure 2.55 Integrated module: topology B6U + B6I.

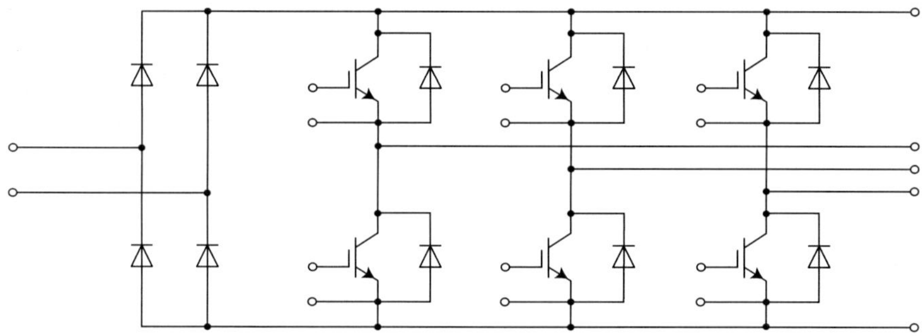

Figure 2.56 Integrated module: topology B2U + B6I.

300°C. Accordingly, low-temperature sintering is suitable for the installation of chips on an SiC base.

The buses of the DBC substrate, the chip outputs, the electrical connections, and the output terminals of the module housing must be designed to withstand the required current density. The critical point most sensitive to current overloads is the joint between the chip outputs and the metallized buses of the DBC substrate. This joint is subject to considerable mechanical stress due to thermal expansion and contraction. Therefore, it is important to ensure high-quality microwelded joints. The current density may be increased if the aluminum conductors are replaced by flat copper terminals welded to the DBC substrate. Such joints tolerate much higher current loads and are more resistant to thermocycling.

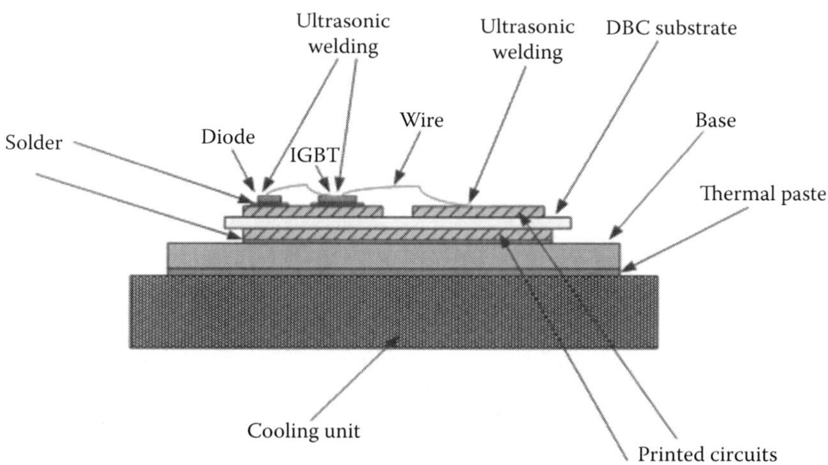

Figure 2.57 A soldered power module.

A promising approach is the use of spring contacts, so that the power terminals have sprung contact with the copper buses, and the signal outputs take the form of springs. New modules of this type dispense with the base, as they are built on the ceramic substrate, which rests directly on the radiator.

In improving power-module topology, there is great interest in reducing the thickness of the DBC substrate and in enlarging the copper plating. Recently, the thickness of the aluminum-oxide substrate was reduced from 0.63 to 0.38 mm. A new technology with zirconium additives reduces this value to 0.32 mm. Another promising approach is to use DBC substrates based on aluminum nitride or silicon nitride, which are characterized by better thermal conductivity and mechanical strength. An increase in thickness of the copper coating on the DBC substrate from 0.3 to 0.6 mm is now feasible.

In the final stage of power-module manufacture, the space within the housing is sealed with silicon gel. The use of this gel ensures sealing and electrical insulation and also improves heat propagation.

2.8.3 *Connecting the module to the power circuit*

In intellectual power modules (IPMs) for home electronics, the chips are mounted directly on a printed circuit board (PCB), along with components such as capacitors, filters, and connectors.

The power of such modules is not more than 100 W.

With an increase in the current density, the power module is mounted separately on control circuits. However, up to 100 A, direct contact of the module with the PCB is possible.

With an output current above 100 A, threaded contact of the module's power terminals with the buses or cables of the power circuit is required. In the old designs, the power outputs are at the top surface of the module. In new designs, such outputs are placed at the edges of the housing, so as to accommodate the driver in the upper part (Figure 2.58).

For modules with very high current density (output current above 600 A), special spring systems are used. For example, in the StakPak module,

(a) (b)

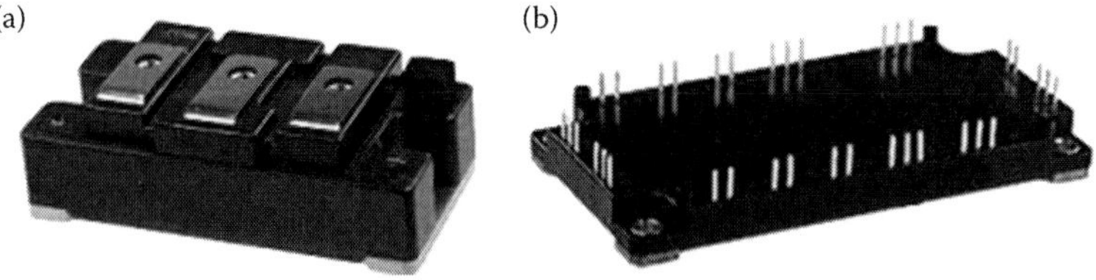

Figure 2.58 Power modules with threaded couplings (a) at the top surface of the module and (b) at the edges of the housing.

a group of spring units is combined in a single assembly (a stack). These switches have no electrical insulation. They are connected to the power circuit by means of their upper and lower surfaces, which serve as the power terminals.

2.9 Power assemblies

2.9.1 Integrated power modules

An integrated power module with an output power of 10 kW consists of a B6U rectifier and a B6I inverter mounted in a housing with a braking switch and current and temperature sensors (Figure 2.59).

2.9.2 Intellectual power module

The IPM represents a higher level of integration and includes an intellectual control driver.

Modern designs use a single-chip control driver for all the switches in the IPM. The driver is attached directly to the DBC substrate of the module.

2.9.3 Power assemblies of basic topology (1/2B2, B2, B6) with a dc element and a cooling system

Such power assemblies are equipped with a forced-cooling system and accordingly are mounted on an air- or liquid-cooled radiator. The power bus of the dc element is designed so as to be compact (minimum parasitic inductance due to installation) and reliable (limiting electrical

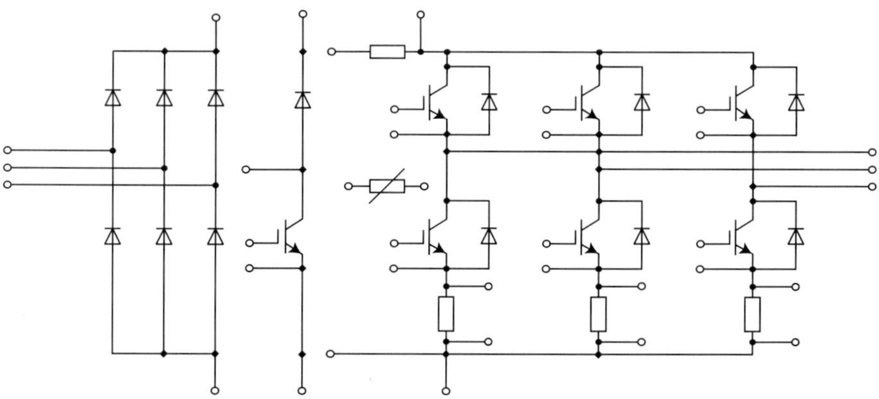

Figure 2.59 Circuit diagram of an integrated power assembly (output power up to 10 kW).

and thermal parameters). The dc element also includes an array of electrolytic and safety capacitors. The output power of such assemblies is 10–100 kW.

2.9.4 Power assemblies of B6U + B6I topology: Inverter platforms

An inverter platform provides a wide power range on the basis of a set of standard modules or cubes.

A single module takes the form of a cube whose faces measure around 40 cm. The module consists of a power cascade with a dc element; driver circuits responsible for control, safety, and monitoring; and a set of sensors generating the basic analog signals required for the formulation of control pulses. The output power of the inverter platform is 10–900 kW.

2.9.5 Power bipolar assemblies

Structurally, power bipolar assemblies are diode–thyristor systems of topology B6U + B6C, equipped with cooling radiators, internal electrical couplings based on copper buses, and safety circuits. In mechanical terms, we may distinguish two groups: those with modular power devices and those with disk power devices (power columns). As a rule, those with modular power devices are used at voltages up to a few kilovolts, whereas those with disk power devices are used at 10^2–10^3 kV. The output power of bipolar assemblies ranges from hundreds of kilowatts to a few megawatts.

2.10 Applications of power switches

Although the basic types of semiconductor power switches are far from ideal, there are no fundamental obstacles to the creation of highly efficient power components over a wide power range: from a few watts to a few megawatts.

To identify their applications, power switches are classified in terms of their switching frequency and power.

The total world market for power modules is about $2.5 billion and is growing rapidly. The total may be broken down as follows.

- Bipolar diodes and thyristors: ~12%
- IGBT: ~48%
- Integrated power modules (rectifier + inverter): ~8%
- IPM (consisting of a power avalanche with a control circuit and sensors): ~32%

More than half of the power modules produced (56%) are used in drives. Manufacturers of industrial drives are interested in product lines that cover a broad power range. Note that these products are based on a single design platform, which facilitates the production of standardized converters.

Electrical transportation is the second largest consumer of power modules (10% of the total). The key requirements for such applications are high reliability and long-term availability of the components from a range of suppliers.

Home electronics accounts for about 9% of the total. The most popular components are low-current IPMs, whose power modules are produced in single in-line (SIL) and dual in-line (DIL) packages.

Two other important applications are renewable energy and auto electronics. Wind power accounts for a relatively small share (5%) but is growing extremely rapidly (by 25% a year). The requirements on components for renewable-energy applications are similar to those in transportation: reliable long-term operation and the ability to withstand challenging environmental conditions.

Likewise, auto electronics currently accounts for a relatively small share (4%) but is growing rapidly (by 19% a year). The requirements on such components are very rigorous and specific, including a broad range of operating temperatures and high resistance to thermocycling.

2.11 *Cooling systems for semiconductor power devices*

At currents of a few amperes or more, semiconductor power switches are mounted on radiators rapidly dispersing heat to the surroundings: air or water (Gentry et al., 1964).

The heat flux p (W) from the radiator to the surroundings may be calculated as

$$p = hA\eta\Delta T, \tag{2.60}$$

where h is the radiator's heat-transfer coefficient (W/cm^2 °C), A the radiator's surface area (cm^2), η the efficiency of the radiator's vanes, and ΔT the temperature difference between the radiator surface and the surroundings (°C).

For radiators of specified structure, the heat-transfer coefficient h is the sum of the radiant component h_R and the convective component h_C:

$$h = h_R + h_C. \tag{2.61}$$

With a vertical vane and laminar air flow, the convective heat-transfer coefficient h_C (W/cm² °C) may be estimated as

$$h_C \approx 4.4 \times 10^{-4} \sqrt[4]{\frac{\Delta T}{L}}, \tag{2.62}$$

where L is the vertical dimension of the vane (cm).

If the air is driven by a fan, h_C may be determined as

$$h_C \approx 0.38 \times 10^{-2} \sqrt{\frac{V}{L}}, \tag{2.63}$$

where V is the linear free-flow velocity of the air (m/s) and L is the length of the vane along the air flow (cm).

The radiant heat-transfer coefficient h_R depends on the emissivity ε of the coolant, the ambient temperature T_A, and the temperature of the radiator (heat sink) T_S as follows:

$$h_R \approx 0.235 \times 10^{-10} \varepsilon \left(\frac{T_S + T_A}{2} + 273 \right)^3. \tag{2.64}$$

The thermal resistance between the radiator and the ambient air may be calculated as

$$R_{S-A} \approx \frac{1}{2(h_R + h_A)A\eta}. \tag{2.65}$$

2.11.1 Radiators for air cooling

With an increase in converter power, there is a growing demand for cooling equipment able to disperse considerable heat in a limited space. Attention focusses here on the material from which the radiator is made. The alloy employed must have the following characteristics (Mikitinets, 2007):

- High thermal conductivity
- Ease of machining
- Excellent corrosion resistance

As an example, Table 2.1 presents the basic characteristics of high-quality 6060 alloy (according to standard European terminology).

Table 2.1 Characteristics of Radiator Material (6060 Alloy)

Density (kg/dm³)	2.7
Resistivity (μΩ m)	0.031
Thermal conductivity (W/m °C)	209
Melting point (°C)	635
Elasticity coefficient (N/mm²)	69,000

The radiator structure must exhibit high machining precision. Table 2.2 presents the linear and angular precision of Tecnoal cooling radiators.

Cooling radiators may be divided into the following groups:

- Configurations with one or more zones for attaching the housings of standard switches (say, TO switches)
- Configurations that also serve as housings
- Configurations with great thermal inertia
- Flanged configurations
- Universal composite configurations

2.11.2 Radiators for liquid cooling

Forced air cooling is limited by the maximum possible air speed (15–20 m/s) and the low heat-transfer efficiency. The only option at that point is to increase the radiator surface, but that means increasing its mass and size.

By switching to liquid cooling, the heat-transfer coefficient may be increased to 0.1–0.7 W/cm² °C.

Figure 2.60 shows some practical radiators with liquid cooling.

The use of water has two series drawbacks: its relatively high freezing point and its low electrical strength. This prevents the use of water cooling at negative temperatures and in high-voltage equipment.

Table 2.2 Dimensional Precision of Tecnoal Cooling Radiators

Nonplanarity (% of dimension)	0.5
Linear precision (mm)	
for 100 × 150 mm² component	±1.2
for 150 × 200 mm² component	±1.5
for 200 × 250 mm² component	±1.8
for 250 × 300 mm² component	±2.1
Angular precision (°)	
(for angles greater than 20°)	±1

Figure 2.60 Liquid-cooled radiators.

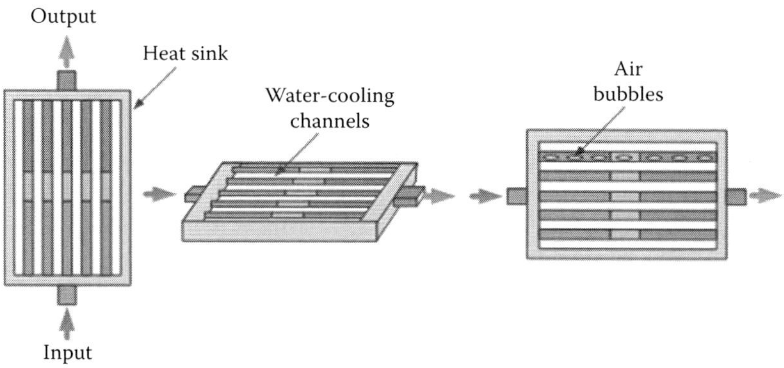

Figure 2.61 Possible configurations of liquid-cooled radiators.

At high heat flux density (around 20 W/cm^2), liquid cooling becomes ineffective; it is reduced to evaporative cooling.

Figure 2.61 shows possible configurations of radiators for liquid cooling. The best option is to use vertical cooling pipes with upward liquid flow. Another acceptable approach is a horizontal configuration with the cooling pipes in the same plane (Arendt et al., 2011). The use of horizontal pipes at different heights proves unsuccessful, as there is a risk that air bubbles will form in the upper channel.

2.12 Promising developments in power electronics

2.12.1 Power switches based on SiC

Semiconductor power switches based on SiC have the following main benefits (Lebedev and Sbruev, 2006):

- Because the band gap is larger for silicon (Si) and GaAs, the working temperature range is considerably greater (in theory, up to ~1000°C).
- As the critical field strength in SiC is almost an order of magnitude greater than that in silicon, the doping of a structure with fixed breakdown voltage may be increased by two orders of magnitude,

with a consequent decrease in the resistance and an increase in the unit power.

- The high critical field strength in SiC affords considerable radiation stability.
- Cooling is simplified on account of the high thermal conductivity of the material. (The value for polycrystalline SiC matches that for copper.)
- Faster charge-carrier saturation permits greater saturation currents in field transistors based on SiC.
- Because of the high Debye temperature, which determines the temperature at which elastic lattice vibrations appear, the thermal stability of SiC semiconductors is improved.
- SiC is versatile. Because SiC components contain a large substrate made of the same material as the semiconductor structure, while silicon dioxide (SiO_2) is also present, and both n- and p-type conductivity may be created in SiC, any type of semiconductor power device may be created.

The industrial production of semiconductor power devices based on SiC depends on the availability of high-quality substrates. Today, SiC substrates are in great demand. With each year, their quality improves, and their diameter increases. Today, diameters of about 100 mm are possible.

The mass production of power transistors based on SiC still lies in the future. The current stage may be appraised on the basis of the following data.

In 2002, specialists at Kansai Electric Power (Japan) reported the development of a 5.3 kV junction field-effect transistor (JFET) (drain current 3.3 A).

Research by SiCLAB (Rutgers University) and United Silicon Carbide has led to developments such as the following:

- The vertical junction field effect transistor, with a maximum permissible drain voltage of 1200 V and a drain current of 10 A; resistivity in the on state not more than 4 mΩ cm^2
- A metal oxide semiconductor field-effect transistor (MOSFET) with a maximum permissible drain voltage of 2400 V and drain current of 5 A; resistivity in the on state 13.5 mΩ cm^2
- A bipolar junction transistor (BJT) with a maximum permissible collector voltage of 1800 V and a collector current of 10 A; resistivity in the on state 4.7 mΩ cm^2

Semiconductor power devices developed by Cree (USA) include the following:

- A thyristor with a breakdown voltage of 5200 V, a forward current of 300 A, and a leakage current less than 100 μA
- Prototype Schottky diodes with a 5.6×5.6 mm working region at 1200 V/50 A and 600 V/100 A

- An SiC MOSFET (working area 3.8 × 3.8 mm), with a drain voltage of 1200 V and a drain current of 10 A and with $R_{DS(ON)} = 0.1\ \Omega$ at 150°C
- A 3 × 3 mm BJT, with a collector–emitter voltage of 1700 V, a saturation voltage of 1 V, and a collector current of 20 A
- A 4H-SiC p–i–n diode with a pulsed power of ~3 MW (mean current 20 A, maximum pulsed current more than 300 A, leakage current 300 μA), reverse voltage more than 9 kV, and 8.5 × 8.5 mm working region. The parameters of the p–i–n diode in the housing are 10 kV and 20 A

The benefits of semiconductor power devices based on SiC are clearly demonstrated by a recent joint project of Cree and Kansai Electric Power. A three-phase voltage inverter based entirely on SiC switches has been developed and successfully tested. This inverter (output power 110 kV A) fits in the space required for an inverter based on silicon technology with an output power of only 12 kV A.

2.12.2 Highly integrated power modules

To increase the reliability and thermocyclic strength of power switches and to also expand their operational temperature range, we may use fundamentally new technologies such as the following.

- Replacement of soldered joints with low-temperature sintering
- Replacement of conductor joints with reliable welded contacts
- The use of spring systems
- Greater integration, especially for high-power components

The new technologies permit the creation of new classes of power modules. Along with the use of promising new semiconductor materials, the development of highly integrated power modules is of great interest in high-temperature electronics (Arendt et al., 2011).

Recently, a construction of a three-phase electric drive (effective current 36 A and voltage 48 V) without standard power modules (Moser et al., 2006) is developed.

This drive is based on a housing-free design and is produced in 10 operations, using a minimum of materials. The creation of housing-free power circuits permits the development of electronic systems with a very high-power density and excellent mechanical characteristics. The lack of soldered joints ensures high thermocyclic strength, and the limiting temperature of semiconductor power chips may reach 200°C. Due to the compact design and integrated dc element, there is little electromagnetic radiation.

2.13 Control of semiconductor power switches

The control-pulse generator is the part of the converter's control system that forms a logical sequence of control pulses for power switches and then amplifies the pulses to the required power.

The main sources of noise in the control system are the converter's power component and, to some extent, the amplifier module of the control-pulse generator. The switching of large load currents creates powerful pulsed interference that penetrates into the electrical circuit of the control-pulse generator and may disrupt the operation of the information and logical components of the control system. Therefore, one of the main requirements in developing the power circuit is electrical uncoupling between the power and control components of the converter.

Such uncoupling is also necessary in relation to the control of higher-level power switches that are not directly connected to the system's common bus.

In terms of the approach to uncoupling, we may distinguish between the following types of control-pulse generators (Voronin, 2005):

- Transformer-based control-pulse generators with combined transmission of the electrical and informational components of the control signal.
- Control-pulse generators with uncoupling of the control signal's informational component and subsequent amplification of that signal to the required power.

In turn, the first group may be divided into control-pulse generators using a voltage transformer and those using a current transformer.

For control-pulse generators with separate transmission of the energy and information, the uncoupling in the information channel may be based on a high-frequency transformer or on optrons.

As a rule, control-pulse generators with combined energy and information transmission are used for current-controlled (charge-controlled) power switches such as power bipolar transistors and thyristors. The main benefits of such control-pulse generators are the lack of auxiliary energy sources for amplification of the control pulses and the high insulation voltage (as much as 6.5 kV).

However, the use of transformer-based control-pulse generators is associated with certain problems.

- With an increase in the switching frequency, the amplitude of the control signals begins to depend on the quality of the transmitted pulses.
- The minimum and maximum lengths of the control signals are limited by the characteristics of the transformer's magnetic core.

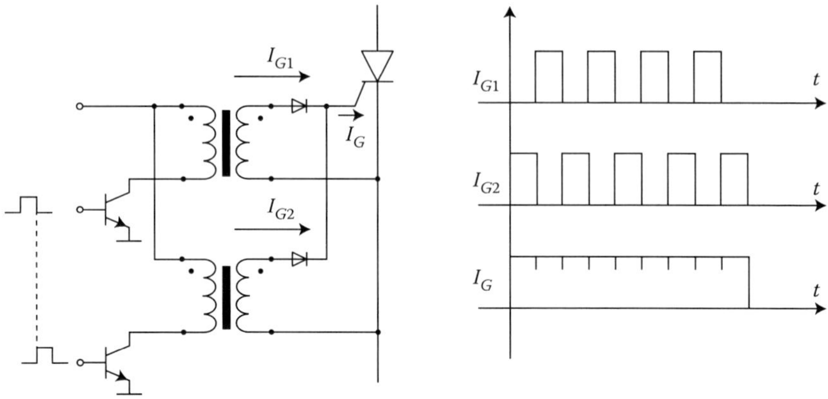

Figure 2.62 Formation of a prolonged control signal.

In power circuits based on high-inductance thyristors, packet transmission of antiphase pulses is used in the load circuit. This permits the production of relatively long (a few milliseconds) control signals (Figure 2.62).

Control-pulse generators with separate transmission of the energy and information contain three basic components.

- A circuit for uncoupling of the information channel
- An amplification circuit for the control pulses
- A power-supply circuit for the amplifier

In addition, such control-pulse generators may contain a circuit protecting the power switch from current and voltage surges and a circuit controlling the voltage in the amplifier's power source.

In control-pulse generators with separate transmission of the energy and information, uncoupling may be based on optrons or a pulsed transformer.

Deficiencies of optronic uncoupling include temperature instability of the parameters, a low current transfer factor (for a diode optron), and a large delay in control-signal transmission (for a transistor optron). The insulation in systems with optronic uncoupling cannot withstand voltages greater than 2.5 kV.

However, in contrast to transformer-based uncoupling, the optronic systems permit the transmission of a continuous control signal.

With broad temperature fluctuation, optrons are replaced by pulsed transformers, characterized by more stable parameters and stronger insulation. Packet transmission of high-frequency signals is used to eliminate the dependence of the information signal on the length (Figure 2.63).

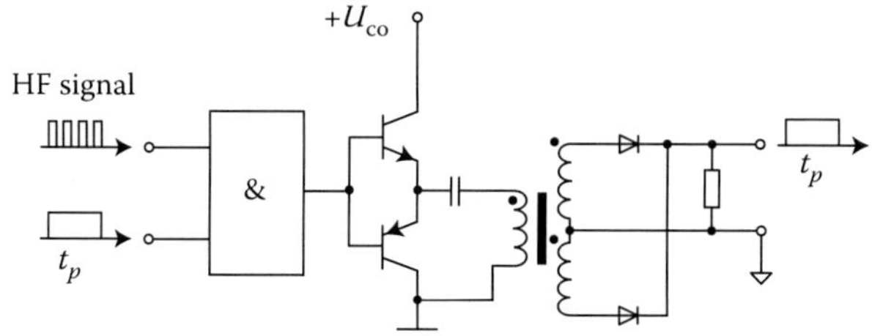

Figure 2.63 Transformer-based uncoupling of the information signal.

When they are produced in the form of separate integrated circuits, control-pulse generators with separate transmission of the energy and information are known as drivers. Industrial drivers are mainly produced for power switches, controlled by an insulated gate (MOS transistors or IGBT). There are drivers for the control of a single switch or a half-bridge or bridge assembly of switches.

As a rule, unipolar power supply (+10 or +15 V) is used for the drivers in MOS transistors. For more powerful IGBT switches, bipolar supply is used (+15 and –10 or ±15 V). This improves the switching speed and the protection against pulsed interference.

In standard drivers, the pulsed currents used are 6, 12, 15, 35, 50, and 65 A. With an increase in the switching frequency, the mean driver current rises significantly. The minimum necessary mean current is chosen in accordance with the power and input capacitance of the switch.

Calculation of the required average driver current I_a is based on the dynamic characteristic of the switch's gate circuit, which is the dependence of the switch's input voltage on the charge in the gate circuit (Figure 2.64).

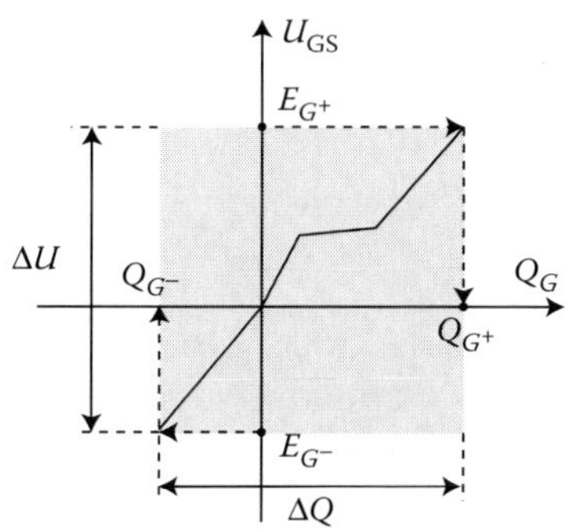

Figure 2.64 Dynamic characteristics of a power switch's gate circuit.

First, we calculate the driver energy W required for switching of the transistor with specified gate-circuit characteristic

$$W = \Delta Q \Delta U, \tag{2.66}$$

where ΔQ is the range of the charge in the switch's gate circuit and ΔU is the range of the switch's input voltage at specified driver supply voltage.

With a bipolar supply voltage, the range of the switch's input voltage is calculated as

$$\Delta U = E_{G^+} + |E_{G^-}|, \tag{2.67}$$

where E_{G^+} is the supply voltage of positive polarity and E_{G^-} is the supply voltage of negative polarity.

Then, the mean driver power may be determined from the formula

$$P = W \cdot f_k = \Delta Q \cdot \Delta U \cdot f_k, \tag{2.68}$$

where f_k is the switching frequency.

The average driver power depends on the average current consumed

$$P = I_a \Delta U. \tag{2.69}$$

Then the required average driver current is

$$I_a = \frac{P}{\Delta U} = \Delta Q \cdot f_k. \tag{2.70}$$

Thus, the selection of a particular driver depends on the characteristic of the power switch's gate circuit and its switching frequency.

2.14 Passive components

2.14.1 Introduction

By *passive components* in electrical circuits, we mean transformers, reactors, capacitors, resistors, varistors, and other elements whose operation does not require additional power sources, internal or external. In power electronics, we speak of passive components in contrast to semiconductor devices, which directly control the power flux and are active components. At the same time, passive components form the fundamental framework that permits power conversion. They are present in practically all power electronic devices, in which they perform functions such as the following:

- Transformers match the voltage levels and ensure the galvanic uncoupling of circuits.
- Reactors are the basic components of filters, switching circuits, and intermediate energy stores.
- Capacitors are used in ac and dc filters and also as intermediate energy stores.

In power electronic devices, passive components operate under the action of nonsinusoidal high-frequency currents and voltages. This significantly complicates their choice in the development of power electronic devices. In addition, their choice is often critical to the success of the design. The great range of functions performed by passive components clearly indicates their importance in power electronics.

2.14.2 Electromagnetic components

2.14.2.1 Basic characteristics of ferromagnetic materials

The dependence of the magnetic induction B on the magnetic field strength H is different for different ferromagnetic materials. Options include the basic magnetization curve, a limiting static hysteresis loop, and a dynamic hysteresis loop. Less common dependences take account of the specific features of the magnetization or demagnetization process, for instance, the dynamic magnetization curve.

The basic magnetization curve is the locus of the vertices of particular steady hysteresis loops. For magnetically soft materials, which are predominantly used in transformers and reactors, this curve is practically identical to the initial magnetization curve obtained in the first magnetization of completely demagnetized material. Magnetization curves are described as follows:

$$B = \mu_a H \quad \text{or} \quad B = \mu_0(H + M), \tag{2.71}$$

where μ_a is the absolute magnetic permeability, in which

$$\mu_a = \mu_0\mu_r, \tag{2.72}$$

and $\mu_0 = 4\pi \times 10^{-7}$ H/m is the magnetic constant (the permeability of vacuum). In addition, μ_r is the relative magnetic permeability, characterizing the increase in induction in the ferromagnet on account of its magnetization M.

The section where the magnetization curve rises modestly with an increase in the field strength H corresponds to saturation of the ferromagnet, which occurs at the saturation magnetization M_s and saturation magnetic induction B_s (Figure 2.65). With further increase in the field strength, the variation in the induction is practically linear: $dB/dH = \mu_a$.

Depending on the conditions in which the permeability is determined—in a steady or variable magnetic field—it is known as the static or dynamic magnetic permeability. The static and dynamic magnetic permeabilities differ in magnitude, on account of the influence of eddy currents, magnetic viscosity, and resonant phenomena on the magnetization processes in alternating fields.

The concept of dynamic magnetic permeability is directly related to the definition of the dynamic induction L_d, which relates the variation in flux linkage $\Delta\Psi$ and current Δi in the circuit of the electromagnetic component

$$L_d = \frac{\Delta\Psi}{\Delta i}. \tag{2.73}$$

The magnetization curve is nonlinear, and therefore μ_a varies with change in H. The dependence of B on H in the case of small variations at a specific point of the magnetization curve may be characterized by means of the differential magnetic permeability μ_d:

$$\mu_d = \left(\frac{dB}{dH}\right)_{H_a \times B_a}, \tag{2.74}$$

where H_a and B_a are the coordinates of point a, where the value of μ_d is determined (Figure 2.65).

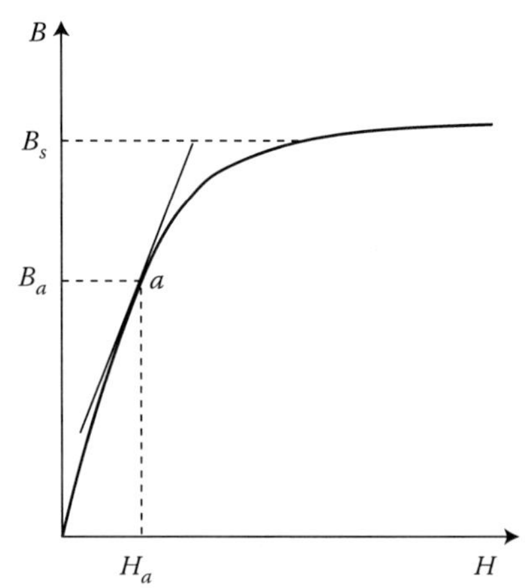

Figure 2.65 Generalized magnetization curve.

The properties of a ferromagnet may be more completely characterized using hysteresis loops. The limiting static hysteresis loop determines the dependence of B on H within sections of the complete magnetization and demagnetization cycle with slow variation in the external field strength—in other words, effectively at constant current ($dH/dt = 0$). In Figure 2.66, we show the limiting static hysteresis loop. Its characteristics are as follows.

- The maximum values B_m and H_m, which exceed the values corresponding to the saturation induction B_s
- The residual induction B_r
- The coercive force H_c

The sections of the hysteresis loop from $+B_m$ to $-H_c$ and from $-B_m$ to $+H_c$ correspond to demagnetization, whereas the sections from $-H_c$ to $-B_m$ and from $+H_c$ to $+B_m$ correspond to magnetization. With a decrease in the magnetic field strength to zero, the ferromagnet will have residual induction $+B_r$ or $-B_r$, depending on the polarity of the initial B_m values.

The rectangularity factor permits approximate assessment of the form of the static hysteresis loop

$$K_{\text{rec}} = \frac{B_r}{B_m}. \tag{2.75}$$

The area of the hysteresis loop determines the energy losses in the ferromagnet on remagnetization (the hysteresis losses).

The electromagnetic components in power electronic devices usually operate under the action of high-frequency alternating voltages and currents. Therefore, they are described by dynamic hysteresis loops,

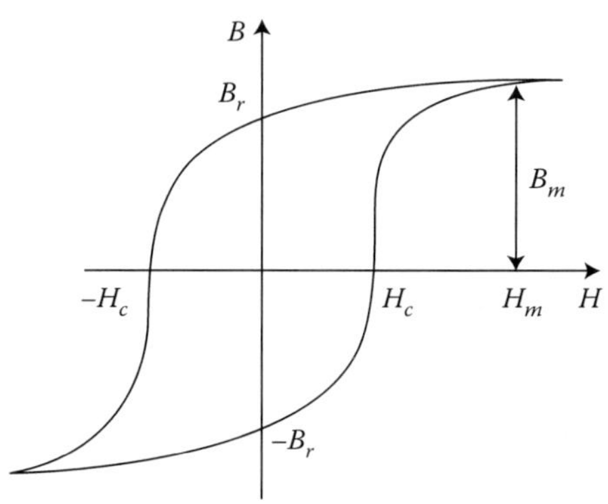

Figure 2.66 A general form of hysteresis loop.

which relate B and H when $dH/dt \gg 0$. With an increase in the frequency, dynamic hysteresis loops differ considerably from static hysteresis loops. The area of the dynamic hysteresis loop increases with an increase in the magnetization frequency, that is, the energy losses in the ferromagnet increase. In addition, the steep sections of the dynamic hysteresis loop become more shallow (Figure 2.67). These physical processes may be attributed to magnetic viscosity, that is, delay in the orientation of the domains as a function of the magnetic field strength. In addition, high-frequency electromagnetic fields produce eddy currents in the ferromagnet, which hinder remagnetization. The losses due to eddy currents and magnetic viscosity are said to be dynamic. Besides the properties of the ferromagnet, other significant factors affect the form of the dynamic hysteresis loop. For example, the dynamic hysteresis loops recorded with remagnetization by means of a current source will differ markedly from those recorded for the same material by means of a voltage source. Other relevant factors include the form of the currents and voltages present and the design of the magnetic system.

At low frequencies (between 50 Hz and 5 kHz), various magnetically soft metals are employed, such as electrical-engineering steels with added silicon (with low coercive force $H_c < 4$ A/m) or alloys of iron with nickel (permalloy). These alloys are characterized by high relative magnetic permeability and low coercive force. Therefore, the losses on remagnetization are small, which is especially important in high-frequency systems.

Above 5 kHz, semiconducting ferrites or magnetic dielectrics are employed. Ferrites are made from powder compounds of iron oxides with

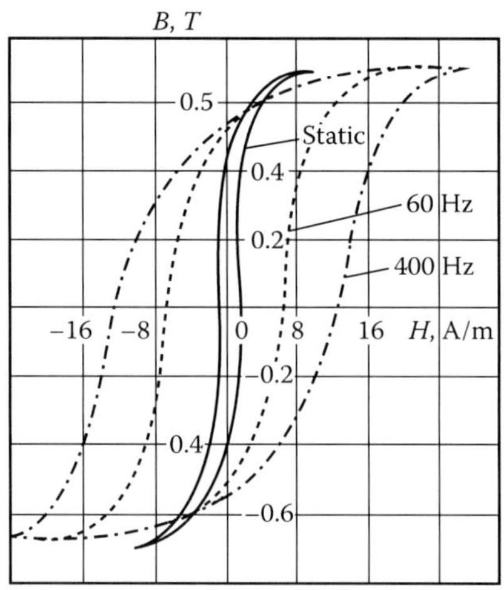

Figure 2.67 Hysteresis loops recorded with a sinusoidal voltage source.

zinc, manganese, and other metals, by a ceramic technology. Ferrites are categorized as semiconductors. Therefore, their bulk electrical resistivity is many orders of magnitude greater than that of steel and alloys. This high resistance significantly reduces the eddy currents and associated energy losses. On account of their low energy losses due to eddy currents and the ability to produce magnetic cores of different shapes, ferrite magnetic systems are widely used in power electronics. At present, manganese–zinc (MnZn) ferrite is widely used in magnets with high saturation induction and magnetic permeability and low energy losses in the range from 300 kHz to 1 MHz. Above 1 MHz, nickel–zinc (NiZn) ferrites are recommended. However, their use is limited by their temperature sensitivity and saturation characteristics (Melyoshin, 2005).

In developing magnetic systems for reactors, in which low inductance is often required at high current, the characteristics employed are different than those for the magnetic systems for transformers. In addition, when using a reactor as a filter in dc circuits, its magnetic system is subject to bias magnetization. In that case, constant inductance must be maintained with wide variation in the current.

Traditionally, this problem is solved by the manufacture of magnetic systems with one or more air gaps. This reduces the reactor's inductance and also reduces its dependence on the current in the winding. However, this method has a number of deficiencies. In particular, magnetic scattering fluxes appear close to the gaps, and the electromagnetic compatibility is generally impaired.

At present, magnetic dielectrics with low (required) magnetic permeability are under development. They are based on composites of powder structure, in which magnetic materials are combined with dielectrics by means of special binders. The resulting material is characterized by low magnetic permeability because the air gap is effectively distributed over the whole length of the magnetic system. The high resistivity of magnetic dielectrics practically eliminates eddy currents. One such material is Alsifer, an alloy of aluminum, silicon, and iron. Low μ_r values (up to about a hundred) permit its effective use in the manufacture of magnetic systems for filter reactors, with practically linear dependence of the induction over a wide range of magnetic field strengths.

Other new magnetic materials of note are amorphous magnetically soft alloys, which lack crystalline structure. Such alloys are based on iron, boron, and silicon, with the addition of various components, such as chromium, to improve their properties. Amorphous alloys have good magnetic, mechanical, and anticorrosion properties. In particular, they ensure low energy losses. In the mass production of electromagnetic components, this permits considerable economic savings in terms of the consumption of metal and energy.

2.14.2.2 *Influence of high frequencies and nonsinusoidal voltage on the operation of transformers and reactors*

Losses in the magnetic system. Various physical processes are responsible for the losses in the magnetic system, which may be determined, in the general case, as the sum of the losses due to factors such as hysteresis, eddy currents, and magnetic viscosity. Precise calculation of individual components is often more complex than the calculation of the total losses on the basis of experimental data obtained with a sinusoidal field. For example, the following formula was presented for the specific losses P_{sp} (W/cm^3) in a magnetic system (Rozanov et al., 2007):

$$P_{sp} = \left(\frac{f}{f^*}\right)^{\alpha}\left(\frac{B_m}{B_m^*}\right)^{\beta} = A_0 f^{\alpha} B_m^{\beta}, \tag{2.76}$$

where f is the working frequency, f^* the baseline frequency (1000 Hz), B_m the maximum induction, B_m^* the baseline induction (1 T), and A_0, α, and β the empirical constants.

For the materials used in magnetic systems, $\alpha > 1$, and according to Equation 2.76, the losses in the system increase with increasing working frequency. In steel magnetic systems, eddy-current losses predominate at high frequencies; in ferrite systems, hysteresis losses dominate. This difference may be taken into account by assigning different values to the constants. In particular, the following simple formula may be used for the losses:

$$P_{sp} = Af^{3/2}B_m^2, \tag{2.77}$$

where the empirical constant A takes account of factors such as the types of losses in different materials (Rozanov et al., 2007).

Under the action of nonsinusoidal periodic voltages, the losses in the magnetic systems are greater than those for sinusoidal voltage at the same fundamental frequency. This is due to the presence of high-frequency components in the nonsinusoidal voltage.

The influence of higher harmonics on the losses in the magnetic system may be taken into account by summing the losses for each harmonic

$$P_{sp} = \sum_{n=1}^{\infty} P_n, \tag{2.78}$$

where P_n denotes the power losses for the nth component. For practical purposes, only the most significant higher harmonics need be taken into account.

If the voltage at the transformer contains a constant component, bias magnetization will occur, with shift in the working induction in the magnetic system. As an example, consider the processes in the magnetic core of a pulsed transformer with a unipolar voltage. Suppose that the period of the voltage is greater than the duration of the transient processes in the transformer, and its scattering inductance and the active resistance of the windings are zero. In Figure 2.68a, the voltage generator is shown as an ideal dc voltage source E, which is periodically connected to the primary winding of transformer T by switch S. When switch S is on, voltage E is applied to the primary winding (with N_1 turns). This is equivalent to the action of a voltage pulse with amplitude E and length t_p. The induction in the transformer's magnetic core begins to vary. Under the given assumptions, we show the simplified equivalent circuit of the transformer in Figure 2.68b; the transformer is replaced by a nonlinear impedance z_μ with magnetization current i_μ and the load by the resistance reduced to the primary winding $R'_{\text{lo}} = R_{\text{lo}} \cdot N_1/N_2$. In Figure 2.68b, we also show time diagrams of the voltage u'_2 at the secondary winding, reduced to the primary winding and for the induction in the magnetic system in a transient process when the core is completely demagnetized initially. Under the action of voltage E, the mean value of the induction at time $t = t_p$ is

$$\Delta B_{\text{me}} = \frac{E \cdot t_p}{N_1 S_M}, \tag{2.79}$$

where S_m is the cross-section of the magnetic core. In Figure 2.68c, the variation in the induction under the action of the first voltage pulse corresponds to motion over the initial magnetization curve from point O to point A_1.

When switch S is turned off, demagnetization of the magnetic system begins. Under the given assumptions, the demagnetization current will fall in the circuit formed by impedance z_μ and load R'_{lo}. Assuming that the switch is off for a time that permits the current i_μ to fall to zero, we may say that the induction B varies over the demagnetization cycle from point A_1 to point O_1 before the switch is again turned on. Then, when switch S is turned on, the magnetization process begins again, but from point O_1. At constant E and t_p, in accordance with Equation 2.79, ΔB_{me} will also remain constant. As a result of periodic pulsed stimulation, the initial and final values of the induction will be displaced to point O_k, corresponding to point A_k in Figure 2.68c. Further pulses will result in remagnetization of the magnetic system over the cycle from point O_k to point A_k and back. In steady conditions,

$$\Delta B_{\text{me}} = B_{\text{A}k} - B_{\text{O}k}, \tag{2.80}$$

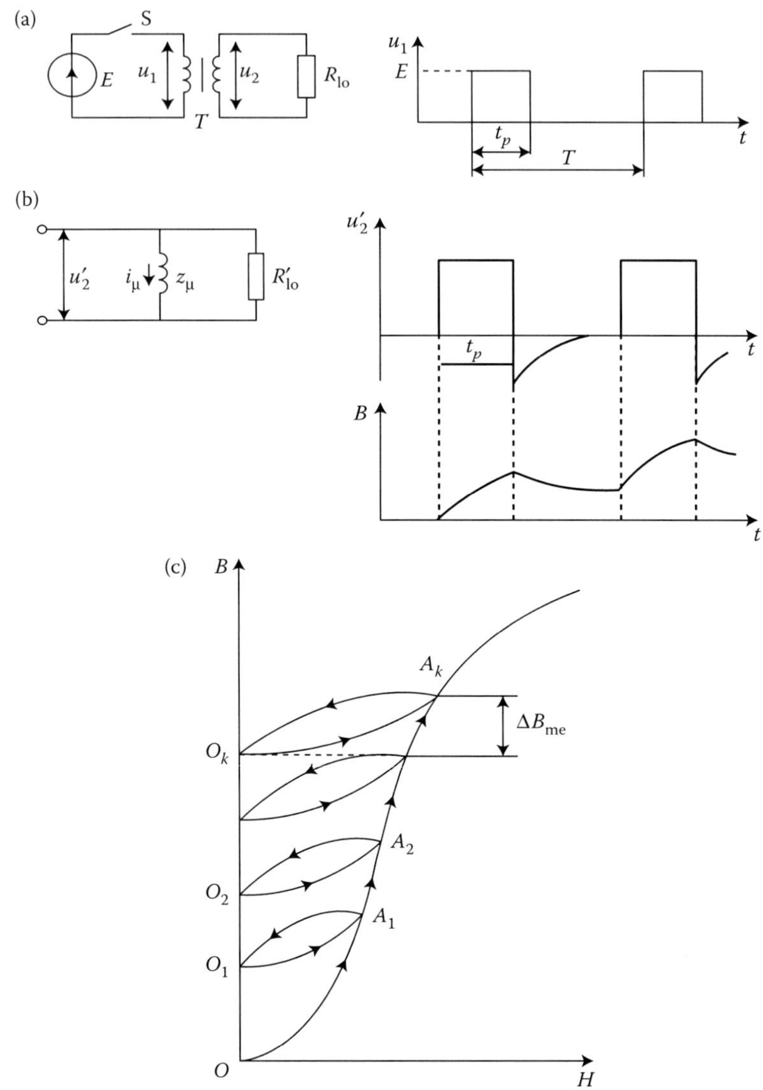

Figure 2.68 Magnetization processes in a pulsed transformer: (a) circuit diagram and voltage diagram for the primary winding; (b) equivalent circuit diagrams of the voltage and induction; and (c) magnetization diagram of the magnetic system.

where ΔB_{me} is the induction in Equation 2.79 and B_{Ak} and B_{Ok} are the induction values in the magnetic system at the end and beginning of the next pulse, respectively.

When the constant components of the magnetizing currents (bias currents) exceed the variable component, the cycle is shifted further to the right of the vertical axis (Figure 2.68c). With an increase in the bias currents, the cycle is shifted to a section of the magnetization curve that is less steep—the saturation zone, where the magnetic permeability is lower in dynamic operation. Accordingly, with an increase in the bias magnetization, the dynamic inductance declines. Note that the bias magnetization depends significantly on the operation of the circuit containing the transformer or reactor—in particular, the internal impedance of the pulsed voltage source connected to the primary transformer winding.

Losses in windings. High-frequency voltages and currents, including nonsinusoidal voltages and currents, produce additional energy losses not only in the magnetic systems, but also in the windings of transformers and reactors. These losses are mainly due to the skin effect in conductors, in which the current is displaced to the surface under the action of the electromagnetic fields. As a result, the active impedance in alternating current is greater than the resistance R_0 in direct current. This increase in impedance is due to the decrease in effective cross-section of the wire. In the skin effect, the current is displaced radially. Current displacement also occurs under the action of the electromagnetic fields produced by adjacent conductors. In that case, the redistribution of the current depends on the design of the windings and their configuration in the magnetic system. The additional losses in the winding with alternating current are taken into account by means of the coefficient K_{add}:

$$K_{add} = \frac{R_{\sim}}{R_0}. \tag{2.81}$$

The value of K_{add} must be calculated for each specific transformer design, with allowance for the current or voltage frequency.

With nonsinusoidal voltages and currents, the additional losses due to each harmonic are determined on the basis of Fourier expansion. These losses may be taken into account by the equivalent value of K_{add}:

$$K_{add} = \frac{\sum_{n=1}^{\infty} I_n^2 K_{addn}}{I^2}, \tag{2.82}$$

where K_{addn} takes account of the losses at the frequency of the nth harmonic and I and I_n are the effective values of the total current and its harmonics, respectively.

It is difficult to determine K_{addn} as it depends on many factors—notably, the cross-sectional area, the design of the windings, and their configuration in the magnetic system. For example, the skin effect is characterized by the coefficient δ, which determines the penetration depth of the current in the conductor, that is, the distance over which the current density declines by a factor of *e* from its maximum value at the surface (sometimes known as the depth of the skin effect).

The coefficient δ is highly frequency-dependent. For copper at 100°C, δ = 8.9 mm at 50 Hz, 0.89 mm at 5 kHz, and 0.089 mm at 500 kHz. With an increase in the working frequency, an increase in the winding's rated current requires special measures to limit the skin effect. The most common approach is to use special multicore conductors (litz wires). Litz wire consists of a large number of small-diameter conductors, which are insulated from one another. The conductors of each pair are wound together, so as to prevent the formation of the magnetic flux produced by the currents of short-circuited pairs. Then all the pairs are combined so that the winding produced has two external terminals.

Another method of reducing the skin effect in conductors that carry large currents is to use a thin copper strip with an insulated surface.

If the wire diameter $d \ll \delta$, the skin effect will be slight. However, in transformers and reactors operating at high frequencies, the mutual influence of the conductors will change the current density in the turns of the winding, with a consequent increase in the power losses. In that case, the calculation of the actual losses is a complex field problem.

A simple qualitative approach here is to consider a two-dimensional problem and to take account of the symmetry of the winding configuration in the magnetic system (Rozanov et al., 2007).

In Figure 2.69, we show the distribution of the magnetomotive force in the layers of single-section reactor windings. With an increase in the magnetomotive force, the magnetic field strength increases. There is also a quadratic increase in the additional losses within the winding layers. The losses are greatest in the surface layer, in which the magnetic field strength is a maximum. To reduce these losses in high-frequency transformers, the windings are divided into sections; this reduces the magnetic field strength in the outermost layers of the winding (Rozanov et al., 2007). In Figure 2.69b, we show an example in which the secondary winding is divided into two sections and the primary winding into three sections. This reduces the magnetic magnetomotive force and magnetic field strength by a factor of 4.

An increase in the frequency boosts the influence of parasitic parameters in the transformers and reactors, for example, the scattering inductances and the interturn and interwinding capacitances. In Figure 2.70, we show the equivalent circuit of a transformer with the scattering inductances of the primary L_{S1} and secondary L_{S2} windings, the capacitance C_{12} between the windings, and the input and output capacitances (C_1 and C_2). Obviously, an increase in frequency of the input voltage will be accompanied by significant distortion of the output voltage, depending on the circuit parameters. This, in turn, will impair the operation of the power

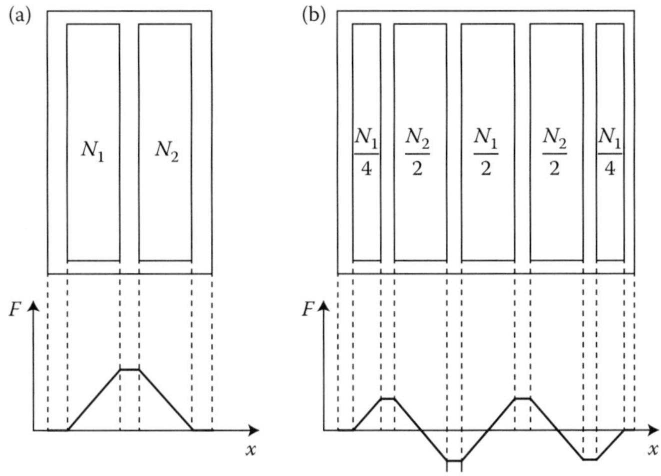

Figure 2.69 Distribution of the magnetomotive force over the layers of transformer windings: (a) transformer without separation of the windings into sections and (b) transformer with a two-section secondary winding and a three-section primary winding.

electronic device and its energy characteristics, including the efficiency and energy density. In some cases, the parasitic parameters may be put to good use. For example, the scattering inductance at high frequencies may serve as a current limiter in the event of short circuits in the load circuit.

With an increase in the working frequency, it is difficult to ensure electromagnetic compatibility of transformers and reactors with other circuit elements. Moreover, the design of electronic devices is complicated, especially for reactors. In power electronic devices, reactors perform various functions such as filtration, energy storage, the formation of switching currents for thyristors, and reactive-power compensation. Most such reactors

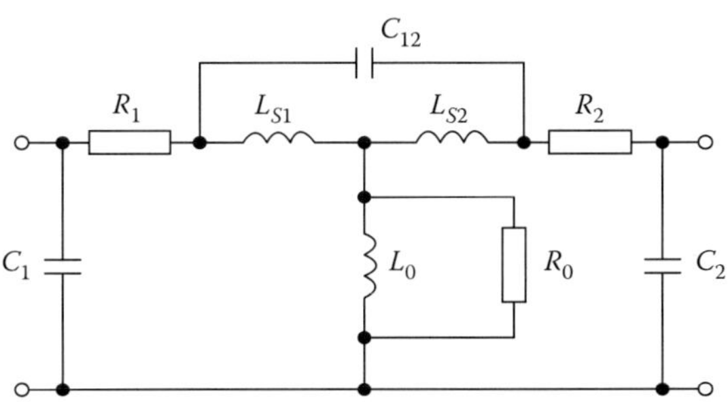

Figure 2.70 Equivalent circuit of a transformer with parasitic inductance and capacitance.

are characterized by small inductance values that do not vary greatly with broad variation in the magnetizing currents. In other words, the inductance must be practically constant. To that end, the magnetic system may be manufactured from a material with low magnetic permeability over a broad range of magnetic field strengths, such as Alsifer. At very low inductance, air reactors without a core may be used. However, to reduce the electromagnetic field produced by such reactors, the magnetic flux must be enclosed as much as possible within the reactor. A toroidal design with uniform distribution of the turns over the reactor circumference is suitable.

New technologies have been appearing lately to improve the high-frequency performance of transformers and reactors. The main trend is the use of flat windings printed on the magnetic system (Rozanov et al., 2007). This design offers numerous benefits: simple sectioning of the windings, decrease in parasitic capacitance, and minimization of other negative effects. In addition, they are highly compatible with hybrid and modular manufacture of power electronic devices.

2.14.3 *Capacitors: Basic definitions and characteristics*

A *capacitor* is able to store and deliver large quantities of electrical energy. The usual design consists of conducting elements (such as metal plates) separated by a dielectric. Under the action of an electric field, coupled electric charges (electrons, ions, and larger charged particles) move within the dielectric in accordance with the direction of the electrical field strength vector. This results in induced polarization of the dielectric, in which the centers of gravity of the positive and negative charges move in opposite directions.

In Figure 2.71, we show the simplified charge distribution in a plane capacitor connected to an external source of voltage *U*. The charge

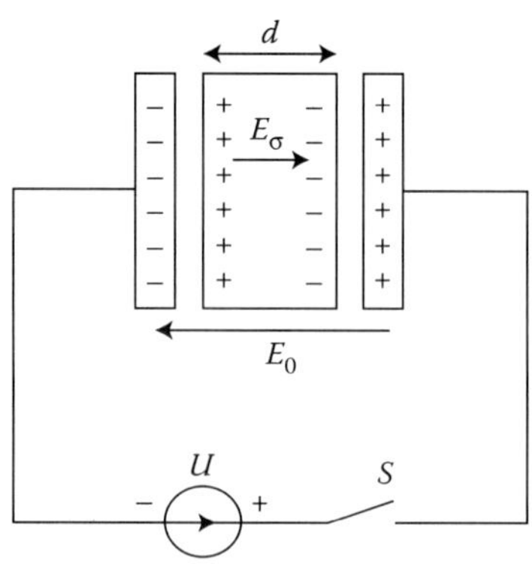

Figure 2.71 Simplified charge distribution in a plane capacitor.

appearing at the surface of the dielectric creates an electric field of strength E_σ, opposed to the external field $E_0 = U/d$. This reduces the field within the dielectric. The overall capacitance may be expressed in the form

$$C = \varepsilon_r C_0 = \frac{\varepsilon_r \varepsilon_0 S}{d}, \tag{2.83}$$

where S is the surface area of each plate, d is the distance between the plates, the constant $\varepsilon_0 = 8.857 \times 10^{-12}$ F/m characterizes the dielectric properties of free space (vacuum), and ε_r is the relative permittivity of the dielectric ($\varepsilon_r > 1$), which characterizes the increase in the charge, the capacitance, and the stored energy in comparison with the case in which is there no dielectric (in vacuum).

Different dielectric materials are used, depending on the required characteristics of the capacitor, its intended application, its production, and other factors. We may distinguish nonpolar dielectrics (with electrically neutral molecules) from polar and ionic dielectrics and ferroelectrics (Ermuratsky, 1982). Depending on the type of dielectric, ε_r may vary from one or two to 10^4–10^5.

In power circuits with constant and pulsating voltage, electrolytic capacitors are widely used. A common design consists of oxidized aluminum foil (the plate), dielectric, and unoxidized aluminum foil with a fibrous covering (the opposing plate). The capacitor has terminals for connection to a voltage of appropriate polarity. The cathodic terminal (–) is connected to the aluminum frame, whereas the anodic terminal (+), in the form of an individual lobe, is isolated from the housing and connected to the oxidized plate.

The basic capacitor parameters are the capacitance, the tangent of the loss angle, the leakage current, and the resistance of the insulation. In addition, the permissible voltage in different conditions, the reactive power, or the permissible stored energy is noted, depending on the type of capacitor.

With sinusoidal voltage, the tangent of the loss angle is defined as the ratio of the active P and reactive Q power

$$\tan\delta = \frac{P}{Q}. \tag{2.84}$$

It may also be expressed in terms of the parameters of the simplest equivalent circuits (Figure 2.72)

$$\tan\delta = \omega C_e R_e = \frac{1}{\omega C_e' R_e'}, \tag{2.85}$$

where ω is the angular frequency of the applied voltage, C_e and R_e the capacitance and resistance according to the equivalent circuit in Figure 2.72a, and C_e' and R_e' the capacitance and resistance according to the equivalent circuit in Figure 2.72b.

Note that, in the general case, the parameters of the equivalent circuits in Figure 2.72 depend on the frequency. Hence, according to Equation 2.85, tan δ is also frequency-dependent. In addition, more complete equivalent circuits include the inductance of the electrical terminals, their active impedance, and other parameters (Rozanov et al., 2007).

In selecting the type of capacitor, we need to take account of the operating conditions, the form and frequency of the currents and voltages, the available space, the cooling conditions, the total working life, the reliability, and many other factors. The age of the capacitor must also be taken into account, as the characteristics may vary considerably over time. For example, the capacitance of some capacitors may vary by 30% over time. Likewise, over time, we see considerable variation in tan δ and the resistance of the capacitor's insulation, which determines the leakage current.

In power electronics, capacitors operate in very diverse and specific conditions. In terms of operating conditions, it is expedient, in general, to distinguish between nonpolar ac capacitors and filter capacitors for dc circuits with little pulsation.

1. For nonpolar ac capacitors, alternating and pulsed voltages of different forms are present. The presence of a constant voltage component comparable with the amplitude of pulsation is possible. Such capacitors do not have terminals of different polarity; in other words, the polarity of the voltage applied is not critical.
2. Unipolar capacitors, such as electrolytic capacitors with oxide dielectric, are commonly used as filters for dc circuits with little pulsation. In that case, the capacitance is high, as is the stored energy density per unit volume. For such capacitors, operation with alternating voltage at the plates is impermissible.

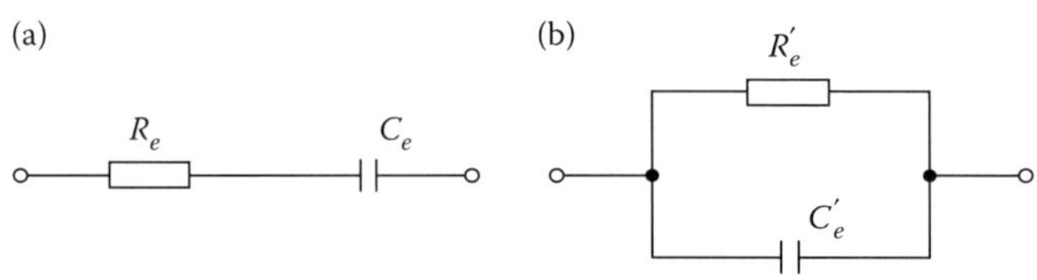

Figure 2.72 Simplified equivalent circuits of capacitors: (a) series circuit and (b) parallel circuit.

2.14.3.1 *Influence of the form and frequency of the voltage on capacitor operation*

The basic functions of ac capacitors in power electronic devices are as follows:

- Reactive-power compensation at the fundamental frequency of the ac voltage
- Energy storage for forced thyristor commutation
- Formation of the switching trajectory for electronic switches
- Filtration of the higher voltage and current harmonics in ac power circuits.

In reactive-power compensators and regulators, capacitors usually operate with high-frequency sinusoidal voltages. In that case, they conform to the conventional operating rules for power electronic systems. In some reactive-power compensators, periodic switching produces higher current harmonics. In that case, the higher current harmonics must be taken into account in capacitor design and selection.

In thyristor commutation, as a rule, switching capacitors operate with fast recharging from one polarity to the other. As a result, pulsed currents with relatively sharp pulse fronts act on the capacitors. The voltage will then be approximately trapezoidal. Capacitors used for the formation of the switching trajectory are usually less powerful than switching capacitors. However, they are generally able to operate at higher frequencies, corresponding to the spectral composition of the switching voltages. In addition, their basic parameters must be relatively frequency-independent. In particular, their design should minimize the inductance, which degrades the transient processes that occur when switches are turned off.

Capacitors in harmonic filters must also withstand nonsinusoidal currents, whose spectral composition must be taken into account in capacitor selection.

Nonsinusoidal currents and voltages result in greater power losses and also change many important capacitor parameters. We know that, with sinusoidal voltage, the losses in the capacitor are proportional to the tangent of the loss angle in the dielectric. In many calculations, constant $\tan \delta$ is assumed, although it depends on the operating conditions and especially on the frequency of the applied voltage. The frequency dependence of $\tan \delta$ must be taken into account in selecting capacitors for use at nonsinusoidal voltages. The frequency dependence of $\tan \delta$ stated in the technical specifications provides the basis for assessment of the additional power losses in the presence of high-frequency voltages. Typically, for nonpolar capacitors, there is little change in $\tan \delta$ at 50–1000 Hz, but approximately 10-fold increase in the range 1000–10,000 Hz. Temperature variation has less influence on $\tan \delta$ for such capacitors. Overall, precise assessment of the high-frequency losses in capacitors is a challenge even with sinusoidal voltage.

It is even more difficult to assess the losses in a capacitor with nonsinusoidal currents and voltages. The most general approach, which is very approximate, is based on frequency analysis of the voltage or current. In such calculations, the power losses in the capacitor due to each harmonic of the applied voltage are summed

$$P_C = C\omega \sum_{n=1}^{\infty} nU_n^2 \tan\delta_n, \tag{2.86}$$

where n is the number of the voltage harmonic, ω the angular frequency of the first voltage harmonic, U_n the effective voltage of the nth harmonic, and tan δ_n the tangent of the loss angle for the nth harmonic.

By harmonic analysis—for instance, on the basis of Fourier transformation—we may determine the most significant harmonics in the nonsinusoidal voltage and estimate the power losses from Equation 2.86. Analogous methods may be used with specified nonsinusoidal current.

With an increase in the active power losses, the permissible effective voltage at the capacitor must be reduced with an increase in the frequency. An increase in the effective values of the high-frequency currents creates the risk of failure of the contacts and other structural elements of the capacitor. That also entails reducing the permissible effective voltage at the capacitor with an increase in the frequency. In Figure 2.73, we show a typical dependence of the permissible effective value of a sinusoidal voltage at an ac capacitor (Ermuratsky, 1982).

Depending on the frequency and form of the voltage, different factors may be prioritized in selecting the required capacitor. For example, with a trapezoidal capacitor voltage at low frequencies and short fronts, the amplitude of the pulsed current is the critical factor. In contrast, with high frequencies of a sinusoidal voltage (above 1 kHz), the additional power

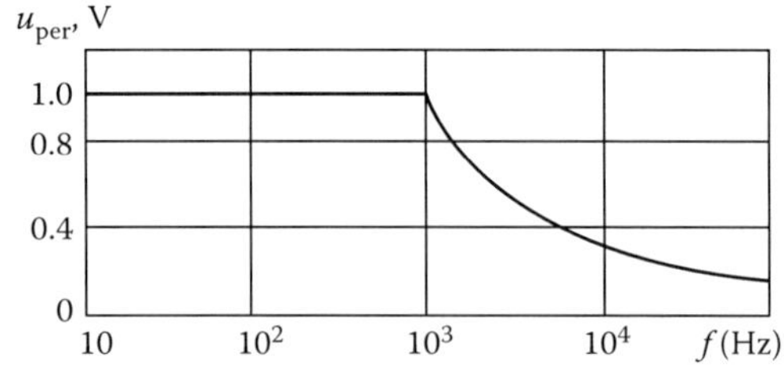

Figure 2.73 Frequency dependence of the permissible effective voltage at the capacitor.

losses are critical. Another important consideration in capacitor selection is its short-term electrical strength, which is taken into account in setting standards for the rated voltage. The permissible effective voltage at the capacitor may also be selected so as to limit the power of the discharge, by restricting the maximum temperature with constant losses.

As the reactive power of an ac capacitor depends directly on the frequency, its unit parameters (the ratio of the reactive power to the volume, mass, or other parameters) will also depend on the frequency. In Figure 2.74, we show the frequency dependence of the unit reactive power for some Russian ac capacitors. It is evident that, for each specific capacitor, there is an optimal frequency of the applied voltage, such that its volume is a minimum.

Electrolytic capacitors are the main components of dc filters. In operation, they are constantly subject to both dc and ac voltage components. Usually, the basic parameters that appear in the technical specifications for electrolytic capacitors include not only the capacitance, but also the rated constant component and the permissible variable component in a sinusoidal voltage of frequency $f = 50$ Hz. However, at higher frequencies, other factors that reduce the conductivity of the capacitor and hence also its filtering capacity must be taken into account (Rozanov et al., 2007). Thus, with sinusoidal voltage, the filtering capacity is determined by the total impedance Z_C of the capacitor. The corresponding equivalent circuit is shown in Figure 2.75a, in which C_d is the capacitance due to the dielectric, r_d and r_e are the active impedances corresponding to the losses in the dielectric and the electrolyte, respectively, and L_e is the equivalent inductance of the sections and terminals. According to the equivalent circuit, at frequency f,

$$Z_C = \sqrt{r_S^2 + \left(\frac{1}{2\pi \cdot f \cdot C_e}\right)^2}, \quad r_S = r_d + r_e, \quad C_e = \frac{C_d}{1 - (f/f_0)^2}, \tag{2.87}$$

where $f_0 = 1/2\pi\sqrt{L_e C_d}$.

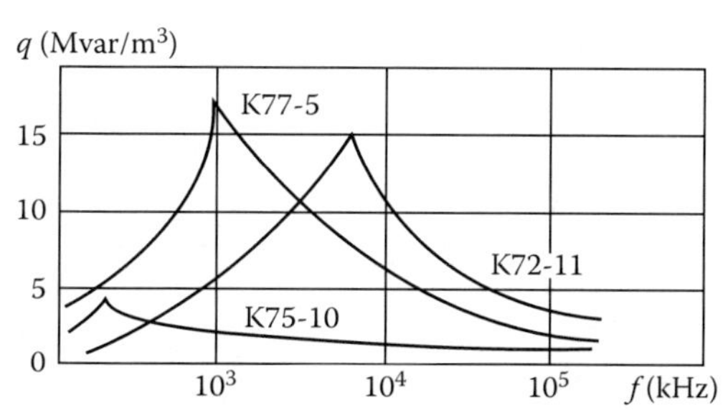

Figure 2.74 Frequency dependence of the capacitors' unit reactive power.

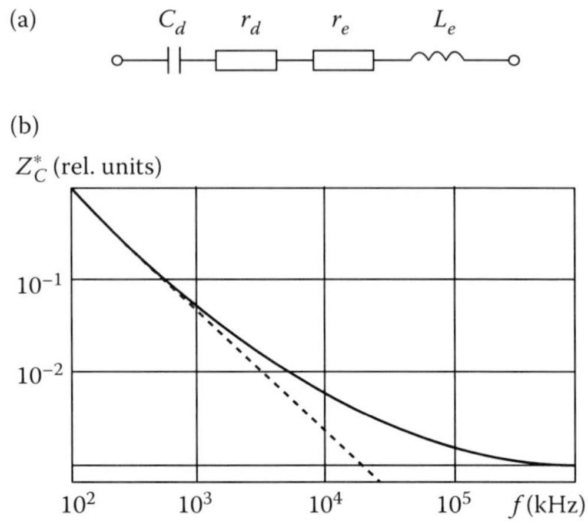

Figure 2.75 (a) Equivalent circuit of an electrolytic capacitor and (b) frequency dependence of the total impedance for a K50-20 capacitor.

In the calculations, we must take account of the influence of various factors on the equivalent circuit's parameters. The capacitance C_d depends on the type of capacitor, its parameters, and the frequency. The inductance L_e is a stable quantity. The tangent of the loss angle and other parameters depend on the frequency, time, and temperature. In addition, the parameters exhibit technological spreads, which are usually random. To take the influence of these factors into account, we assess and compare the unit parameters of the capacitors at high frequencies on the basis of their effective capacitance

$$C_{ef} = \frac{1}{2\pi \cdot f \cdot Z_c}. \tag{2.88}$$

In Figure 2.75b, as an example, we show the frequency dependence of the total impedance Z_C^* (in relative units) for a K50-20 capacitor at an ambient temperature of 25°C. The dashed line shows the frequency characteristic of an ideal capacitor ($L_e = r_e = 0$).

We see that the filtering capacity of the K50-20 capacitor begins to decline above 10 kHz. Above 20 kHz, its use is inexpedient. Instead, capacitors with an organic or ceramic dielectric must be employed.

If the variable component of the current passing through the capacitor is nonsinusoidal, the effectiveness of filtration again changes. For example, at large di/dt, the alternating voltage at the capacitor terminals due to the inductance L_e increases and may considerably exceed the alternating voltage at C_d.

Table 2.3 Unit Parameters of Some Capacitors

Type of capacitor	Energy density (J/kg) 2001	Energy density (J/kg) 2011	Unit mass (kg/kW) 2001	Unit mass (kg/kW) 2011	Frequency of alternating voltage component (Hz)
Polymer film	0.40	20.00	5.0	2×10^3	More than 100
Ceramic	0.01	5.00	10.0	10×10^3	More than 100×10^3
Electrolytic	0.20	2.00	0.2	10×10^3	More than 100
Mica	0.01	0.05	5.0	5×10^3	More than 1×10^6

With nonsinusoidal voltage pulsations at the capacitors, their filtering capacity and sustainable load will depend on the spectral composition of the pulsations. Therefore, for some types of oxide–electrolyte capacitors, the technical specifications mandate not only the frequency dependences already described, but also nomograms from which the permissible amplitude of a specific nonsinusoidal voltage—say, a trapezoidal voltage—may be determined as a function of the frequency.

For preliminary estimates at the design stage, it is sufficient to confine attention to the dominant harmonics in the voltage pulsations at the capacitor and to employ the superposition principle. The results must be refined experimentally. In particular, the effective currents should be measured (using thermoammeters), as well as the temperature of the capacitor housing and the ambient temperature.

Capacitors are fundamental to power electronics. Therefore, major electrical-engineering firms expend great resources in improving their performance. Table 2.3 presents the unit parameters of some capacitors, to illustrate the state of the art (Rozanov et al., 2007).

References

Arendt, W., Ulrich, N., Werner, T., and Reimann, T. 2011. *Application Manual Power Semiconductors*. Germany: SEMIKRON International GmbH.

Ermuratsky, V.V. 1982. *Handbook of Electric Capacitors*. Shtiintsa (in Russian).

Gentry, F.E., Gutzwiller, F.W., Holonyak, N.J., and Von Zastrov, E.E. 1964. *Semiconductor Controlled Rectifiers: Principles and Applications of p–n–p–n Devices*. Englewood Cliffs, NJ: Prentice-Hall.

Hefner, A. and Blackburn, D. 1988. An analytical model for steady-state and transient characteristics of the power insulated-gate bipolar transistor. *Solid-State Electron.*, 31(10), 1513–1532.

Hidalgo, S.A. 2005. Characterization of 3.3 kV IGCTs for medium power applications. Laboratoire d'Electrotechnique et d'Electronique Industrielle de l'ENSEEIHT.

Kuzmin, V., Jurkov, S., and Pomortseva, L. 1996. Analysis and modeling of static characteristics of IGBT. *Radio Eng. Electron.*, 41(7), 870–875 (in Russian).

Lebedev, A. and Sbruev, S. 2006. SiC—Electronics. Past, present, future. *Electron. Sci. Technol. Business*, 5, 28–41 (in Russian).
Li, Y., Huang, A., and Motto, K. 2000. A novel approach for realizing hard-driven gate-turn-off thyristor. IEEE PESC, pp. 87–91.
Melyoshin, V.I. 2005. Transistor converter equipment. *Technosphere* (in Russian).
Mikitinets, A. 2007. Tecnoal heat sinks. *Modern Electron.*, 8, 20–22 (in Russian).
Moser, H., Bittner, R., and Beckedahl, P. 2006. High reliability, integrated inverter module (IIM) for hybrid and battery vehicles. Proc. VDE EMA, Aschaffenburg. Oxner, E.S. 1982. *Power FETs and their Applications*. Englewood Cliffs, NJ: Prentice-Hall.
Rozanov, Yu.K., Ryabchitsky, M.V., and Kvasnyuk, A.A. 2007. *Power Electronics*. Publishing House MPEI (in Russian).
Udrea, F. and Amaratunga, G. 1997. An on-state analytical model for the trench insulated-gate bipolar transistor (TIGBT). *Solid-State Electron.*, 41(8), 1111–1118.
Voronin, P.A. 2005. *Power Semiconductors*. Dodeka-XXI (in Russian).
Yevseyev, Yu.A. and Dermenzhi, P.G. 1981. *Power Semiconductor Devices*. *Energoatomizdat* (in Russian).

chapter three

Control of power electronic devices

3.1 Mathematical models

3.1.1 One-dimensional and multidimensional models

In the general case, any power electronic device may be regarded as a control plant (Figure 3.1). One characteristic of such components is that they have two inputs.

1. A power input, to which supply voltages e_k ($k = 1,\ldots,K$) are supplied. These voltages must be converted into output voltage or current signals in accordance with the particular control problem. For the sake of simplicity, they may be expressed in the form of the supply voltage vector $(E)^{\mathrm{T}} = |e_1\ e_2 \ldots e_{K-1}\ e_K|$. Superscript T denotes transposition; in reality, this vector is a column one.
2. A control input, to which independent controls u_m ($m = 1,\ldots,M$) are supplied. They control the transformation of the supply voltages e_k. They may also be represented as a control vector $(U)^T = |u_1\ u_2 \ldots u_{M-1}\ u_M|$.

Another characteristic is that independent external disturbances f_l ($l = 1,\ldots, L$) act on the power electronic device. These may be expressed as an external perturbation vector $(F)^{\mathrm{T}} = |f_1\ f_2 \ldots f_{L-1}\ f_L|$.

Finally, the outputs or output variables are the electrical variables y_m, which are controlled by means of the control. These may be expressed as the output variable vector $(Y)^{\mathrm{T}} = |y_1\ y_2 \ldots y_{M-1}\ y_M|$.

If the number of independent control signals is greater than 1 and so is the number of output variables, the control plant is said to be multidimensional: it is a multi-input, multi-output (MIMO) one. Another important class consists of one-dimensional or scalar control plants: single-input, single-output (SISO) ones. Numerous special methods for the analysis and control design have been devoted to this class. They will be considered below.

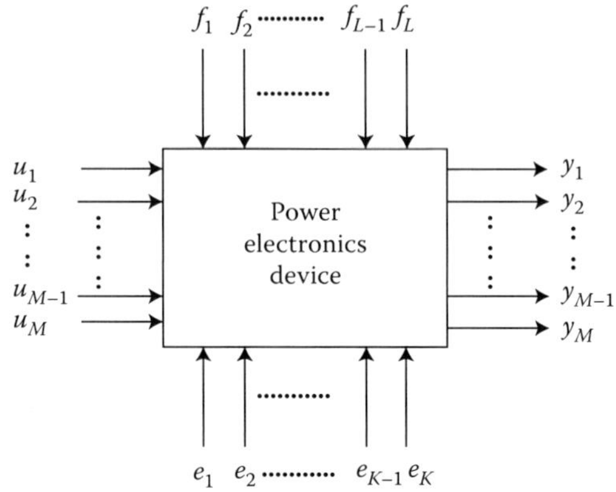

Figure 3.1 Generalized model of a power electronic device.

3.1.2 *Linear and nonlinear systems—Linearization*

From the viewpoint of mathematical description, a fundamental feature of power electronic devices is the presence of power electronic switches, which switch the electrical circuits in the devices in response to control. The controls are generated by the controller or under the action of the electromagnetic processes in the components.

A power electronic switch is a completely controllable semiconductor device. It has a control electrode to which a signal is sent for transition from the conducting to the nonconducting state or vice versa (Mohan et al., 2003; Rozanov et al., 2007). Examples include the following:

- The bipolar plane transistor
- The field-effect MOS transistor
- The insulated gate bipolar transistor
- The gate turn-off thyristor

In terms of the control of power electronic devices, these completely controllable semiconductor modules may be regarded, without loss of generality, as noninertial switches, with zero resistance in the conducting state and infinite resistance in the closed state, which are capable of infinitely large switching frequencies. Their idealized relay characteristic is shown in Figure 3.2.

Therefore, in terms of mathematical description, a power electronic device is a nonlinear control plant with a variable or switchable structure (Figure 3.3).

For each specific power electronic device, the properties of the individual structures are determined by the states of the power switches.

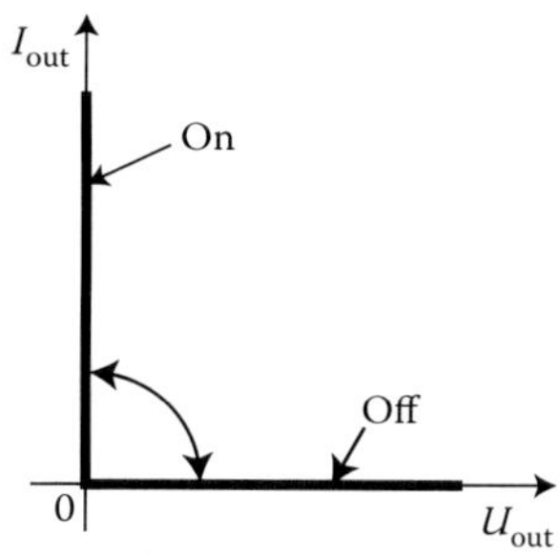

Figure 3.2 Idealized output characteristic of power switch: I_{out}, current at switch and U_{out}, voltage drop at switch (collector–emitter voltage drop).

Thus, analysis of the device's operation and design of the corresponding controls call for a mathematical model that reflects its most important properties. This model must be based on the laws of electrical engineering and must take into account the specific topology of the power electronic device. The complexity of the model is determined by the set of properties that must be taken into account.

In the first stage, only the basic properties of the components in the power electronic device are taken into account. In other words, the active and reactive components are assumed to be ideal. In that case, the description of the processes in the power electronic device is based on the following physical laws:

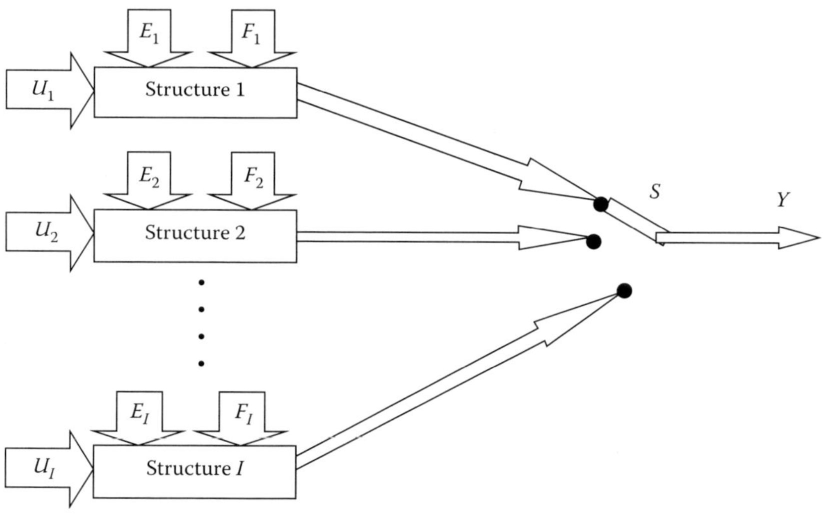

Figure 3.3 Generalized model of a power electronic device as a variable structure system (S-structure selector).

1. Ohm's law, which determines the relation between the resistance R of a conductor, the voltage u_R applied to it, and the current i_R that flows through it:

$$u_R = R \cdot i_R. \tag{3.1}$$

2. The law of capacitor charging (and discharging), which determines the relation between the capacitance C, the voltage applied u_C, and the current i_C:

$$u_C = \frac{1}{C}\int i_C \, dt. \tag{3.2}$$

3. The self-induction law, which determines the relation between the inductance L, the corresponding current i_L, and the self-induction electromotive force (emf) e_L that appears

$$e_L = -u_L = -L\frac{di_L}{dt}. \tag{3.3}$$

4. Kirchhoff's laws (rules), which determine the relation between the currents and voltages in the sections of any electrical circuit. These are of particular importance in electrical engineering on account of their universality: they may be applied to many circuit theory problems.
   4.1 *Kirchhoff's first law* (*the current law*). The algebraic sum of the currents at any node of any circuit is zero. A node is defined here as the point at which three or more conductors meet. (The incoming and outgoing currents are assigned opposite signs.)

$$\sum_{n=1}^{N} i_n = 0, \quad N \geq 3. \tag{3.4}$$

   4.2 *Kirchhoff's second law* (*the voltage law*). The algebraic sum of the voltage drops in any closed loop of the circuit is equal to the algebraic sum of the emfs acting along this loop. If there is no emf in the loop, the total voltage drop is zero. A loop is defined here as a closed secondary circuit. Each conductor may be a component of several loops

$$\sum_{k=1}^{K} e_k = \sum_{m=1}^{M} u_m = \sum_{p=1}^{P} u_{Rp} + \sum_{q=1}^{Q} u_{Lq} + \sum_{z=1}^{Z} u_{Cz}, \tag{3.5}$$

5. Switching laws, which determine the relation between the currents and voltages in the electrical circuit before and after switching.
   5.1 *The first switching law* (*the absence of a current discontinuity in an inductance*): the current $i_L(+0)$ passing through the inductance L immediately after switching is equal to the current $i_L(-0)$ passing through the inductance L immediately before switching, as the self-induction law rules out instantaneous change of the current in the inductance

$$i_L(+0) = i_L(-0). \tag{3.6}$$

   5.2. *The second switching law* (*the absence of a voltage discontinuity at a capacitance*): the voltage $u_C(+0)$ at capacitance C immediately after switching is equal to the voltage $u_C(-0)$ immediately before switching, as a voltage jump at the capacitance is impossible

$$u_C(+0) = u_C(-0). \tag{3.7}$$

6. Faraday's law of electromagnetic induction, which is fundamental to electrodynamics (describing the operation of transformers, chokes, and assorted electrical motors and generators). For any closed induction loop, the emf e is equal to the rate of change in the magnetic flux Ψ passing through this loop

$$\frac{d\Psi}{dt} = -e. \tag{3.8}$$

7. The mutual inductance law, which determines the relation between the current i_1 in inductance L and the magnetic flux Ψ_{21} passing through a second inductance that is created by i_1:

$$\Psi_{21} = M_{21} i_1, \tag{3.9}$$

   where M_{21} is the mutual inductance.

Linearization is widely employed to obtain a linear description of control plants. In this approach, the actual continuous nonlinear equation $y = \varphi(x)$ that describes an element of the system is replaced by a similar linear equation. The following linearization methods are employed.

1. *Analytical linearization*: the small-deviation method, based on Taylor expansion

$$y = y_0 + \left.\frac{dy}{dx}\right|_{\substack{x=x_0\\y=y_0}} \Delta x + \left.\frac{1}{2}\frac{d^2y}{dx^2}\right|_{\substack{x=x_0\\y=y_0}} (\Delta x)^2 + \cdots. \tag{3.10}$$

In Equation 3.10, (x_0, y_0) are the coordinates of the point selected for expansion, where $y = \varphi(x)$, $\Delta x = x - x_0$ is a small deviation from the point of expansion, x is the current value of the variable, and the derivatives of the function $y = \varphi(x)$ are calculated for the point of expansion.

Thus, the nonlinear equation $y = \varphi(x)$ may be replaced by a linear equation in the vicinity of the point of expansion, that is, by the linearized equation

$$y_{\text{lin}} = y_0 + k\Delta x \quad \text{or} \quad \Delta y = (y_{\text{lin}} - y_0) = k\Delta x. \tag{3.11}$$

The latter is the equation in terms of deviations.

For a continuous dependence of the output variable on several variables, the Taylor expansion is written in terms of partial derivatives with respect to different variables at the point of expansion. In other words, the nonlinear function $y = \varphi(x_1, x_2, \ldots, x_n)$ may be replaced at the point of expansion by the linear function

$$\Delta y = k_1\Delta x_1 + k_2\Delta x_2 + \cdots + k_n\Delta x_n, \tag{3.12}$$

where

$$k_i = \left.\frac{\partial y}{\partial x_i}\right|_{\substack{x_1 = x_{10} \\ x_2 = x_{20} \\ \vdots \\ x_n = x_{n0}}}$$

2. Statistical linearization for nonanalytic equations (e.g., on the basis of statistical data) is based on the least-squares method.

By using these methods, the linear approximation of statistical data is a line that has a minimum of the sum of squares of the error for all values of the variable x^i, for which values of the nonlinear function $\varphi(x^i)$ exist (here $i = 1, \ldots I$, where I is the number of measurements)

$$\begin{gathered} \Delta^i = \varphi(x^i) - y_{\text{lin}}(x^i) \\ \sum_{i=1}^{I} (\Delta^i)^2 \to \min. \end{gathered} \tag{3.13}$$

The result takes the form

$$y_{\text{lin}} = ax + b. \tag{3.14}$$

The problem is solved by using the LINEST function in Excel software. The result includes not only the calculated values of coefficient a and the free term b, but also statistical data such as the standard error, the F statistic, the degrees of freedom, the regression sum, and the residual sum, which permit assessment of reliability and quality of the model.

Using Equation 3.14, we may assess the relative error resulting from the replacement of the nonlinear characteristic by a linear characteristic for each measurement

$$\delta(x^i) = \left|\frac{\varphi(x^i) - y_{\text{lin}}(x^i)}{\varphi(x^i)}\right|. \tag{3.15}$$

3.1.3 Differential and matrix equations—Switching function

As already noted, in terms of mathematical description, a power electronic device must be regarded as a nonlinear control plant with a variable or switchable structure. Each structure defined by a particular combination of switch states is linear, on account of the electrical engineering laws employed.

In linear description, each variable and its derivatives appear in an equation offering first-order description of the properties of the device's structure. The variables employed usually include the circuit currents and voltages. The processes in the circuit are described by linear integro-differential equations.

If one variable is enough to describe the electrical processes, such control plants are said to be one-dimensional and are described by an nth-order nonhomogeneous linear differential equation (where n is the largest derivative employed)

$$a_n^i \frac{\mathrm{d}^n x}{\mathrm{d}t^n} + a_{n-1}^i \frac{\mathrm{d}^{n-1} x}{\mathrm{d}t^{n-1}} + \cdots + a_0^i x = f^i(t) + b_e^i e^i. \tag{3.16}$$

Here i is the number of the structure ($i = 1,\ldots, I$), x the independent variable, $a_0^i, a_1^i,\ldots,a_n^i$ the coefficients of the equation describing the behavior of x, and b_e^i the coefficient of the supply voltage for structure i. Superscript i indicates that the corresponding coefficient or equation describes structure i.

We may also write Equation 3.16 in the Cauchy form, that is, as a system of first-order equations

$$
\begin{aligned}
&\frac{dx}{dt} = x_1, \\
&\frac{dx_1}{dt} = x_2, \\
&\vdots \\
&\frac{dx_{n-2}}{dt} = x_{n-1}, \\
&\frac{dx_{n-1}}{dt} = -\frac{a_{n-1}^i}{a_n^i}x_{n-1} - \cdots - \frac{a_1^i}{a_n^i}x_1 - \frac{a_0^i}{a_n^i}x + \frac{1}{a_n^i}f^i(t) + \frac{b_e^i}{a_n^i}e^i.
\end{aligned}
\tag{3.17}
$$

We may write Equation 3.17 in the vector-matrix form by means of the state vector $X^T = |x\ x_1 \ldots x_{n-2}\ x_{n-1}|$, the supply voltage vector $(E^i)^T = |0\ 0 \ldots 0\ e^i|$, the external-disturbance vector $(F^i)^T = |0\ 0 \ldots 0\ f^i|$, and the state matrix A^i

$$
\frac{dX}{dt} = A^iX + \frac{1}{a_n^i}F^i + \frac{b_e^i}{a_n^i}E^i,
$$

where

$$
A^i = \begin{Vmatrix}
1 & 0 & \cdots & 0 & 0 \\
0 & 1 & \cdots & 0 & 0 \\
\vdots & \vdots & \vdots & \vdots & \vdots \\
0 & 0 & 0 & 1 & 0 \\
-\frac{a_0^i}{a_n^i} & -\frac{a_1^i}{a_n^i} & \cdots & -\frac{a_{n-2}^i}{a_n^i} & -\frac{a_{n-1}^i}{a_n^i}
\end{Vmatrix}. \tag{3.18}
$$

The space in which the one-dimensional system is described is known as the derivative space.

In such systems, the state variable is usually the output variable: $y = x$.

If the behavior of the structure is described by means of several state variables, the system is said to be multidimensional. Its dimensionality is determined by the number of independent variables required. Such structures are usually described by vector-matrix equations (Kwakernaak and Sivan, 1972)

$$
\frac{dX}{dt} = A^iX + B^iE^i + D^iF^i, \tag{3.19}
$$

$$
Y = CX, \tag{3.20}
$$

where X is the state vector, consisting of independent variables that describe the behavior of the control plant, E^i is the supply voltage vector, consisting of independent supply voltages, $(E^i)^T = |e_1^i\ e_2^i \ldots e_{K-1}^i\ e_K^i|$, F^i is the external disturbance vector, $(F^i)^T = |f_1^i\ f_2^i \ldots f_{L-1}^i\ f_L^i|$, Y^i is the output variable vector, $(Y^i)^T = |y_1^i\ y_2^i \ldots y_{M-1}^i\ y_M^i|$, and A^i, B^i, C, and D^i are matrices characterizing the features of the structure of the control plant. In general, the number of independent variables may be different from the number of output variables. In other words, the matrix C may be rectangular. We call Equation 3.18 the equation of state and Equation 3.20 the output variable equation.

To obtain the mathematical model for the power electronic device as a whole, rather than for each individual structure, we use switching functions. In that approach, the power switch (or a combination of switches) is described by a threshold or step function or by a sign function (Figure 3.4).

The argument α of the switching function, which determines the conditions in which the switch is turned on and off, may be a function of the time. In that case, the threshold function is

$$\Psi_m = \begin{cases} 1, & \text{if } t \in (0,t_k),(t_{k+1},t_{k+2}),\ldots, \\ 0, & \text{if } t \in (t_k,t_{k+1}),\ (t_{k+2},t_{k+3}),\ldots, \end{cases} \tag{3.21}$$

and the sign function is

$$\Psi_m = \begin{cases} 1, & \text{if } t \in (0,t_k),(t_{k+1},t_{k+2}),\ldots, \\ -1, & \text{if } t \in (t_k,t_{k+1}),(t_{k+2},t_{k+3}),\ldots, \end{cases} \tag{3.22}$$

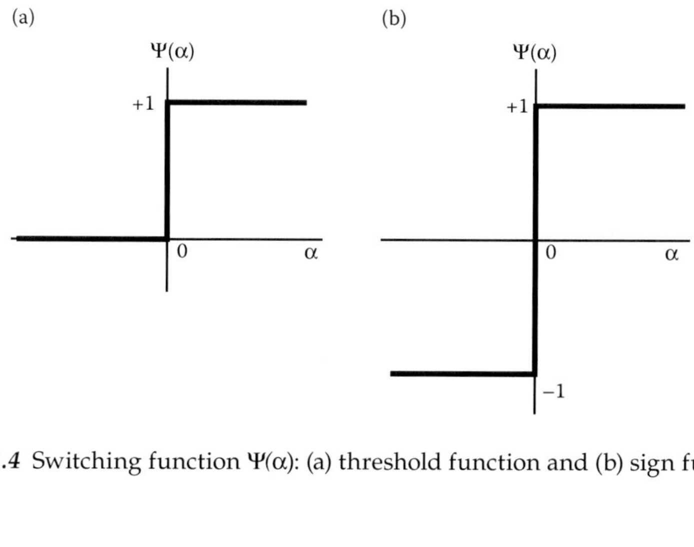

Figure 3.4 Switching function Ψ(α): (a) threshold function and (b) sign function.

where 1 corresponds to an on state and 0 or –1 to an off state. Alternatively, α may be a function of the control signal u_m. In that case

$$\Psi_m = \begin{cases} 1, & \text{if } u_m \geq A, \\ 0(-1), & \text{if } u_m \leq A, \end{cases} \tag{3.23}$$

or

$$\Psi_m = \begin{cases} 1, & \text{if } \operatorname{sgn}(u_m) > 0, \\ 0(-1), & \text{if } \operatorname{sgn}(u_m) < 0. \end{cases} \tag{3.24}$$

Note that the topology of the power electronic device will determine whether the switch state 0 or –1 will be used. In physical terms, the threshold switching function corresponds to unipolar switching of a dc supply voltage, when the power supply is connected or not. In bipolar switching, when different polarities of the supply voltage are connected, we use the sign switching function (e.g., relay control). The following algebraic relation between these two types of switching functions may be established:

$$\Psi_{1p} = \frac{(1 + \Psi_{2p})}{2} \quad \text{or } \Psi_{2p} = 2\Psi_{1p} - 1. \tag{3.25}$$

Formally, the transformation using the switching function may be written as the product of the initial function and a switching function

$$f_{\text{out}}(t) = \Psi(\alpha) f_{\text{in}}(t), \tag{3.26}$$

where $f_{\text{in}}(t)$ is the function to which the switching function is applied and $f_{\text{out}}(t)$ is the result of applying the switching function.

In that case, the mathematical model of the power electronic device takes the form of a differential equation with variable coefficients, external disturbances, and supply voltages, which change abruptly as a function of the combination of controls

$$a_n(\Psi)\frac{d^n x}{dt^n} + a_{n-1}(\Psi)\frac{d^{n-1} x}{dt^{n-1}} + \cdots + a_0(\Psi)x = \Psi_f f(\Psi) + b_e(\Psi)e. \tag{3.27}$$

Depending on which coefficients or parameters affect the switching function, we may distinguish between different types of systems, which are studied by different methods. For example, if only the coefficient of the supply voltage depends on the switching function and the left-hand side of Equation 3.27 is linear with coefficients that are constant and have constant signs, this is a relay system. The same remark applies to Equation 3.19, but the number of combinations of switch states is greater. The number of switch combinations is 2^g, where g is the number of switches. In that case

$$\frac{\mathrm{d}X}{\mathrm{d}t} = A(\Psi)X + B(\Psi)E + D(\Psi)F. \tag{3.28}$$

Besides equations in terms of absolute variables, two other options are possible: equations in terms of deviations and equations in terms of relative variables.

3.1.3.1 Equations in terms of deviations

As already noted, nonlinear equations may be linearized by considering the behavior of the system in the vicinity of some point. This significantly simplifies the analysis.

Usually, the points chosen in linearization are static states of the system in which all the derivatives are zero. Alternatively, a specified value of the output variable may be adopted. This is the most common choice in systems with switchable structure, in which, by definition, self-oscillation is a concern.

In the case of a one-dimensional system, if the point chosen is a specified value of x, denoted by x_z, then we may write the initial equation for each structure, Equation 3.16, in terms of the deviation or discrepancy Δx ($\Delta x = x_z - x$):

$$\begin{aligned} a_n^i \frac{\mathrm{d}^n \Delta x}{\mathrm{d}t^n} + a_{n-1}^i \frac{\mathrm{d}^{n-1} \Delta x}{\mathrm{d}t^{n-1}} + \cdots + a_0^i \Delta x &= -f^i(t) - b_e^i e^i + A^i, \\ A^i = a_n^i \frac{\mathrm{d}^n x_z}{\mathrm{d}t^n} + a_{n-1}^i \frac{\mathrm{d}^{n-1} x_z}{\mathrm{d}t^{n-1}} + \cdots + a_0^i x_z &= \text{const}. \end{aligned} \tag{3.29}$$

Likewise, for the system as a whole, described by Equation 3.27, we may write

$$\begin{aligned} a_n(\Psi)\frac{\mathrm{d}^n \Delta x}{\mathrm{d}t^n} + a_{n-1}(\Psi)\frac{\mathrm{d}^{n-1} \Delta x}{\mathrm{d}t^{n-1}} + \cdots + a_0(\Psi)\Delta x &= -\Psi_f f(\Psi) - b_e(\Psi)e + A, \\ A = a_n(\Psi)\frac{\mathrm{d}^n x_z}{\mathrm{d}t^n} + a_{n-1}(\Psi)\frac{\mathrm{d}^{n-1} x_z}{\mathrm{d}t^{n-1}} + \cdots + a_0(\Psi)x_z &= \text{const}. \end{aligned} \tag{3.30}$$

Analogously, we may write the matrix-vector forms in Equations 3.19 and 3.28 in terms of deviations.

3.1.3.2 Per-unit equations

In terms of relative variables (the per-unit system), the values of variables such as voltage, current, resistance, or power are expressed with respect to a specific baseline value (adopted as the unit). The conversion formula is

$$X_{\text{pu}} = \frac{X}{X_r}, \tag{3.31}$$

where X is the value of the physical quantity (parameter, variable, etc.) in the initial system of units (usually the SI system) and X_r is the base (baseline value) in the same system of units, adopted as the unit of measurement in the relative variable.

In the per-unit system, the baseline values usually adopted are the power, voltage, current, impedance, and admittance (complex conductivity); only two of these are independent. Thus, we may employ different per-unit systems, depending on the problem to be solved and on personal preference. The rated values are generally adopted as baseline values. It is conventional to use the same symbol pu (or sometimes p.u.) and to determine whether the voltage, current, or other parameters are intended on the basis of the context.

The calculation results may be converted back to dimensional values (volts, amperes, ohms, watts, etc.) by inverse use of Equation 3.31.

For example, in some SimulinkBlocksets/SimPowerSystems modules, the parameters are specified in per-unit form. The main electrical baseline values selected in the SimPowerSystem software are P_r, the baseline power, as the rated active power of the device (P_{rat}), and U_r, the baseline voltage, as the rated effective supply voltage of the device (U_{rat}). All the other electrical baseline values are obtained from these two on the basis of the electrical engineering laws. For example, the baseline current is

$$I_r = \frac{P_r}{U_r}, \tag{3.32}$$

whereas the baseline resistance is

$$R_r = \frac{U_r^2}{P_r}. \tag{3.33}$$

For ac circuits, the baseline frequency f_r must be specified; as a rule, the rated frequency f_{rat} of the supply voltage is chosen.

Switching to the per-unit system does not change the form of the equations, but only the numerical coefficients. The main benefits of such conversion are as follows:

1. Comparison of parameter values in different operating conditions is simplified. For example, if the voltage of some section of an electrical circuit is 2 pu, we understand that the voltage is twice the rated supply voltage.
2. There is little change in the total resistance with change in the power of the device and its supply voltage. Therefore, in the absence of precise parameter values for the specific device, we may use mean per-unit values, which are available in handbooks.
3. The calculations are simplified because the coefficients and variables are of lowest order.

3.1.4 *Two-dimensional mathematical description of a three-phase circuit*

Coupled three-phase circuits—that is, three-phase circuits without a null lead—are widely used on account of their operational benefits. Such a three-phase system (Figure 3.5) contains three identical and interconnected single-phase sinusoidal voltage sources, which differ only in that the sinusoid of each one is shifted by a third of a period (the angle $2\pi/3$) with respect to the next

$$
\begin{aligned}
e_A &= E_m \sin \omega t, \\
e_B &= E_m \sin\left(\omega t - \frac{2\pi}{3}\right), \\
e_C &= E_m \sin\left(\omega t + \frac{2\pi}{3}\right),
\end{aligned}
\tag{3.34}
$$

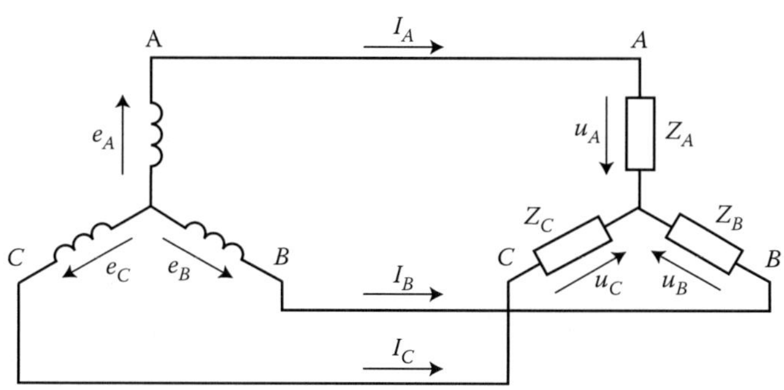

Figure 3.5 A coupled three-phase circuit.

where E_m is the amplitude of the supply voltage and ω is the frequency of the supply voltage. These sources operate with a symmetric three-phase load $Z_A = Z_B = Z_C$. In this case, Kirchhoff's first law in Equation 3.4 is always observed in the circuit. In other words, the phase-current vectors are related as follows:

$$\vec{I}_A + \vec{I}_B + \vec{I}_C = 0. \tag{3.35}$$

In terms of mathematical description, this means that, besides the three differential equations describing the electrical processes in the three phases, we have the algebraic relation between the phase currents (Equation 3.35). Consequently, one phase current is a function of the other two and may be excluded from consideration. Thus, the behavior of the three-phase system may be analyzed on the basis of a two-dimensional description. This significantly simplifies the analysis of the behavior of the power electronic device and the design of an appropriate control. The selection of the two-phase coordinate system depends on the conditions of the problem and the characteristics of the power electronic device. We now consider some well-known two-phase coordinate systems for the analysis of three-phase systems.

3.1.4.1 Motionless Cartesian coordinate system (α, β)

Like the initial three-phase coordinate system (*A*, *B*, *C*), the new system is motionless; its α axis runs along the *A*-axis of the three-phase coordinate system (Figure 3.6). In that case, we may use the following matrix transformation for direct coordinate conversion of vector *X*, that is, to express *X* in a motionless two-phase coordinate system, with allowance for the $2\pi/3$ phase difference of the axes of the three-phase system

$$\begin{vmatrix} X_\alpha \\ X_\beta \end{vmatrix} = k_{\text{pr}} \begin{vmatrix} 1 & -1/2 & -1/2 \\ 0 & 1/\sqrt{3} & -1/\sqrt{3} \end{vmatrix} \begin{vmatrix} x_A \\ x_B \\ x_C \end{vmatrix}. \tag{3.36}$$

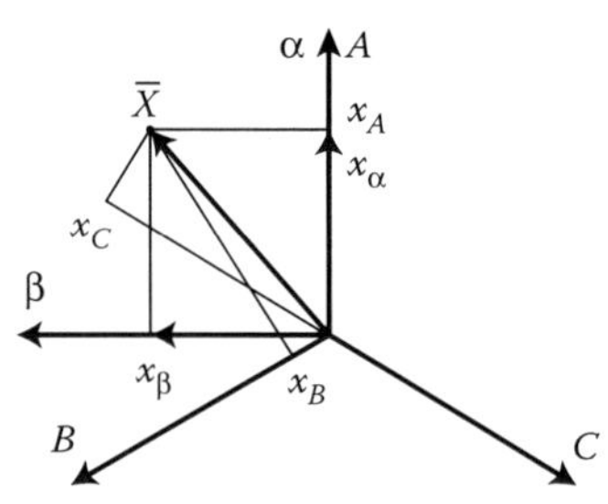

Figure 3.6 Projections of the vector *X* in the three-phase coordinate system (*A*, *B*, *C*) and in the motionless coordinate system (α, β).

Here k_{pr} is a constant of proportionality, selected on the basis that the power is invariant (i.e., the power is the same in the three- and two-phase systems). Therefore, k_{pr} will depend on the baseline values adopted for the coordinates and the modulus of the generalized vector. For example, $k_{pr} = 2/3$ to ensure equal amplitudes of the phase voltages in the three- and two-phase systems, but $k_{pr} = \sqrt{2/3}$ when describing a three-phase inverter system in terms of the effective line voltages.

For conversion of the coordinates from the two-phase to the three-phase system, we use the following formula, which may be regarded as the inverse of Equation 3.36:

$$\begin{vmatrix} x_A \\ x_B \\ x_C \end{vmatrix} = \begin{vmatrix} 1 & 0 \\ -1/2 & 1/\sqrt{3} \\ -1/2 & -1/\sqrt{3} \end{vmatrix} \begin{vmatrix} x_\alpha \\ x_\beta \end{vmatrix}. \quad (3.37)$$

3.1.4.2 *Rotating Cartesian coordinate system (d, q)*

In many cases, it is expedient to use a rotating Cartesian coordinate system in the analysis of the processes within the power electronic component and the design of appropriate control laws. The rate of rotation Ω of the system is chosen equal to that of one of the device's variables. For example, for electrical circuits, it is expedient to use the synchronous grid frequency. The behavior of synchronous machines (especially salient-pole machines) is often described using the Park equations, which are written in a coordinate system that rotates at the speed of the motor's driveshaft (Leonhard, 2001). In that case, the electromagnetic processes in the synchronous machine are described by differential equations with constant coefficients, rather than periodic coefficients; this simplifies the analysis. For induction motors, we use a coordinate system rotating at the speed of the rotor flux (Leonhard, 2001). This simplifies the formula for the motor's electromagnetic torque, which takes the form of the product of two variables.

In most cases, the axis from which the rotation of the coordinate system (d, q) is measured is the A-axis of the three-phase motionless coordinate system or, equivalently, the α-axis of the motionless Cartesian coordinate system (α, β) (Figure 3.7).

Conversion of vector $\vec{X}$ from the coordinate system (α, β) to the system (d, q) may be based on the formula

$$\begin{vmatrix} x_d \\ x_q \end{vmatrix} = \begin{vmatrix} \cos\vartheta & \sin\vartheta \\ -\sin\vartheta & \cos\vartheta \end{vmatrix} \begin{vmatrix} x_\alpha \\ x_\beta \end{vmatrix}, \quad (3.38)$$

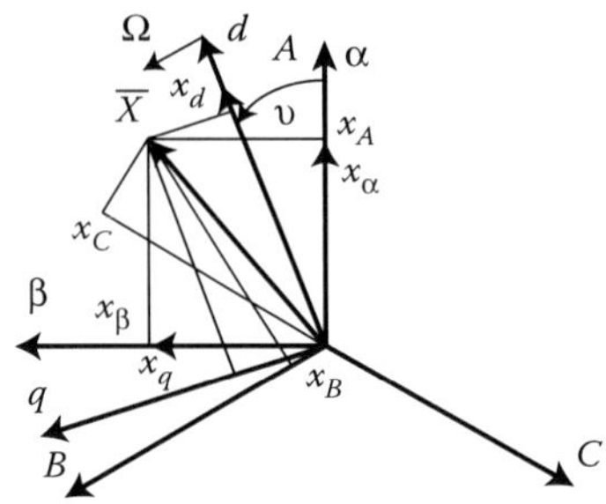

Figure 3.7 Projections of the vector X in the three-phase coordinate system (A, B, C) and in the moving coordinate system (d, q).

where ϑ is the angle of the rotating coordinate system relative to the motionless coordinate system

$$\vartheta = \int_0^t \Omega \, d\tau.$$

Here τ is the integration time.

Inverse conversion is based on the formula

$$\begin{vmatrix} x_\alpha \\ x_\beta \end{vmatrix} = \begin{vmatrix} \cos\vartheta & -\sin\vartheta \\ \sin\vartheta & \cos\vartheta \end{vmatrix} \begin{vmatrix} x_d \\ x_q \end{vmatrix}. \tag{3.39}$$

3.1.4.3 *Converting the instantaneous power of a three-phase system to the power of a two-phase system*

To control the components of the instantaneous power in the three-phase system, the signals of the three-phase system may be converted into the two-phase system.

The theory of such transformation, known as p–q theory, was outlined in Akagi et al. (2007). It has been used to develop control systems for reactive-power compensators and active filters. According to p–q theory, we may introduce the concepts of real and imaginary instantaneous powers in α–β coordinates

$$\begin{aligned} p(t) &= u_\alpha(t) \cdot i_\alpha(t) + u_\beta(t) \cdot i_\alpha(t), \\ q(t) &= -u_\alpha(t) \cdot i_\beta(t) + u_\beta(t) \cdot i_\alpha(t). \end{aligned} \tag{3.40}$$

The real component of the instantaneous power in Equation 3.40 corresponds to the active instantaneous power in the traditional sense. At the same time, the imaginary instantaneous power in Equation 3.40 does not exactly correspond to the traditional concept of the reactive instantaneous power.

If we express the currents in α–β coordinates as a function of the components of the instantaneous power in Equation 3.40, we obtain

$$\begin{bmatrix} i_\alpha \\ i_\beta \end{bmatrix} = \frac{1}{\left(u_\alpha^2 + u_\beta^2\right)} \cdot \left\{ \begin{bmatrix} u_\alpha & u_\beta \\ u_\beta & -u_\alpha \end{bmatrix} \cdot \begin{bmatrix} p \\ 0 \end{bmatrix} + \begin{bmatrix} u_\alpha & u_\beta \\ u_\beta & -u_\alpha \end{bmatrix} \cdot \begin{bmatrix} 0 \\ q \end{bmatrix} \right\} = \begin{bmatrix} i_{\alpha p} \\ i_{\beta p} \end{bmatrix} + \begin{bmatrix} i_{\alpha q} \\ i_{\beta q} \end{bmatrix}, \tag{3.41}$$

where $i_{\alpha p}$, $i_{\alpha q}$, $i_{\beta p}$, and $i_{\beta q}$ are the current components determining the real and imaginary instantaneous power

$$i_{\alpha p} = \frac{u_\alpha p}{\left(u_\alpha^2 + u_\beta^2\right)}, \quad i_{\alpha q} = \frac{u_\beta q}{\left(u_\alpha^2 + u_\beta^2\right)}, \quad i_{\beta p} = \frac{u_\beta q}{\left(u_\alpha^2 + u_\beta^2\right)}, \quad i_{\beta q} = \frac{-u_\alpha q}{\left(u_\alpha^2 + u_\beta^2\right)}. \tag{3.42}$$

According to p–q theory, the real and imaginary power may be expressed as the sum of constant and variable components

$$\begin{aligned} p &= \bar{p} + \tilde{p}, \\ q &= \bar{q} + \tilde{q}, \end{aligned} \tag{3.43}$$

where $\bar{p}$ and $\bar{q}$ are the constant components of the instantaneous power p and q, corresponding to the active and reactive power at the fundamental frequency, and $\tilde{p}$ and $\tilde{q}$ are the variable components of p and q due to the higher harmonics.

Thus, in compensating the reactive power at the fundamental and higher current harmonics, the control signal must take account of the components $\bar{q}$, $\tilde{p}$, and $\tilde{q}$ of the instantaneous power.

3.1.5 Laplace transformation and transfer function

Structural analysis usually reduces to finding the current or voltage variation in the branches of the equivalent circuit—that is, to finding the solution of Equation 3.16 for a one-dimensional control plant or of Equations 3.19 and 3.20 for a multidimensional one (Section 3.2.1).

If the structure is described by linear differential equations with constant coefficient, an effective method of solution is to convert the initial time functions into functions of a Laplace variable by Laplace transformation (Doetsch, 1974). The main benefit of such transformation is that the differentiation and integration of initial functions are replaced by algebraic operations on their Laplace transforms. In other words, we need to only solve simple algebraic equations, rather than complex differential equations. Inverse Laplace transformation to the time domain entails dividing the algebraic solution into the sum of simple fractions, finding their solutions from Table 3.1 for inverse Laplace transformation, and then superimposing the linear differential equations obtained.

The Laplace variable is denoted by s. The direct Laplace transform of the variable $x(t)$ takes the form

$$X(s) = L[x(t)] = \int_0^\infty x(t)e^{-st}\,\mathrm{d}t, \tag{3.44}$$

where L denotes direct Laplace transformation.

In this case, Equation 3.16, which describes the behavior of a one-dimensional linear system in Laplace space, takes the form

$$a_n^i s^n X + a_{n-1}^i s^{n-1} X + \cdots + a_0^i X = F^i(s) + b_e^i e^i(s) + a_n^i \sum_{k=0}^{n-1} \frac{\mathrm{d}^k x(0)}{\mathrm{d}t^k} + a_{n-1}^i \sum_{k=0}^{n-2} \frac{\mathrm{d}^k x(0)}{\mathrm{d}t^k} + \cdots + a_1^i x(0). \tag{3.45}$$

The solution of a linear differential equation depends continuously on the initial conditions. Accordingly, for the sake of simplicity, we adopt zero initial conditions when using Laplace transformation. In other words, we assume that the term on the right-hand side of Equation 3.35, which

Table 3.1 Common Laplace Transformations

Description	Function	Laplace transformation
Linearity	$Af(t)$ (A = const) $f_1(t) + f_2(t)$	$AF(s)$ $F_1(s) + F_2(s)$
Similarity	$f(at)$	$\frac{1}{a}F\left(\frac{s}{a}\right)$
Delay	$f(t - \tau_0)$	$e^{-\tau_0 s}F(s)$
Displacement	$e^{-\lambda t}f(t)$	$F(s + \lambda)$
Differentiation	$\frac{\mathrm{d}^n x(t)}{\mathrm{d}t^n}$	$s^n X(s) - \sum_{k=0}^{n-1} \frac{\mathrm{d}^k x(0)}{\mathrm{d}t^k}$
Integration	$\underbrace{\int\int\int\ldots\int}_{n} y(t)\,\mathrm{d}t^n$	$\frac{Y(s)}{s^n}$
Constant	A = const	$\frac{A}{s}$

depends on the initial conditions, is zero. Nonzero initial conditions are taken into account on inverse conversion to the time domain.

A control plant in Laplace space is most often described by means of a transfer function, which is the ratio of the Laplace transforms of the control plant's output and input variables by zero initial conditions (Kwakernaak and Sivan, 1972). The transfer function is usually denoted by W, with a subscript whose first and second digits characterize the output and input variables, respectively. In view of the linearity of Equation 3.16, which describes a one-dimensional system, we may apply the superposition principle. Correspondingly, the result of several disturbances of the linear system is equal to the sum of their individual results. In the present case, the supply voltage and an external disturbance act on the system. Those are the input variables. Thus, for our control plant, we may write two transfer functions.

1. For the supply voltage

$$W_{xe}^{i}(s) = \frac{X(s)}{E^{i}(s)} = \frac{b_e}{a_n^i s^n + a_{n-1}^i s^{n-1} + \cdots + a_0^i}. \tag{3.46}$$

2. For the external disturbance

$$W_{xf}^{i}(s) = \frac{X(s)}{F^{i}(s)} = \frac{1}{a_n^i s^n + a_{n-1}^i s^{n-1} + \cdots + a_0^i}. \tag{3.47}$$

We may write Equation 3.45 in the structural form shown in Figure 3.8.

This result may be expressed as a matrix transfer function. If we introduce the input variable vector $(H^i)^{\mathrm{T}} = (E^i(s)F^i(s))$, the vector transfer function takes the form

$$W_{xh}^{i}(s) = \frac{X(s)}{H^{i}(s)} = \begin{vmatrix} \dfrac{b_e}{a_n^i s^n + a_{n-1}^i s^{n-1} + \cdots + a_0^i} & 0 \\ 0 & \dfrac{1}{a_n^i s^n + a_{n-1}^i s^{n-1} + \cdots + a_0^i} \end{vmatrix}. \tag{3.48}$$

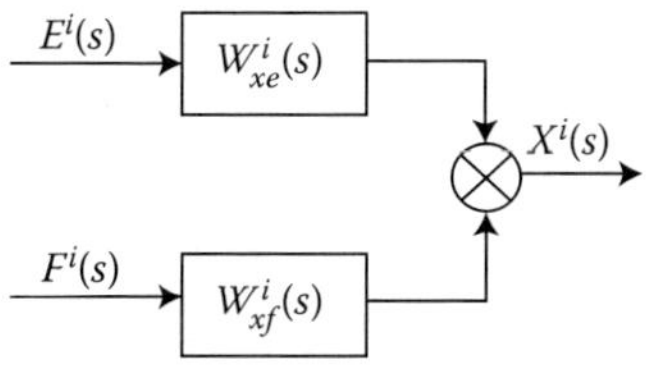

Figure 3.8 Structure of a one-dimensional control plant.

For a multidimensional system, Laplace transformation is applied to the components of the state vectors, the output variables, the supply voltage, and the external disturbances

$$sX(s) = A^iX(s) + x(0) + B^iE^i(s) + D^iF^i(s), \tag{3.49}$$

$$Y(s) = CX(s), \tag{3.50}$$

where $X(s)$ is the Laplace transform of the state vector X, consisting of the Laplace transforms of the independent variables describing the behavior of the control plant, $x(0)$ is the vector of initial conditions for the state vector X, $E^i(s)$ is the Laplace transform of the supply voltage vector, which consists of the Laplace transforms of the independent supply voltages, $(E^i(s))^{\mathrm{T}} = |\, u_1^i(s)\; u_2^i(s) \ldots u_{K-1}^i(s)\; u_K^i(s)|$, $F^i(s)$ is the Laplace transform of the external disturbance vector, $(F^i(s))^{\mathrm{T}} = |f_1^i(s)\; f_2^i(s) \ldots f_{L-1}^i(s)\; f_L^i(s)|$, and $Y^i(s)$ is the Laplace transform of the output variable vector, $(Y^i(s))^{\mathrm{T}} = |\, y_1^i(s)\; y_2^i(s) \ldots y_{M-1}^i(s)\; y_M^i(s)|$. Note that the number of independent variables determines the dimensionality of the control plant.

As we see, we already have algebraic matrix equations. The solution of Equation 3.49 with respect to $X(s)$, the Laplace transform of the state vector X, is

$$X(s) = (sI - A^i)^{-1}x(0) + (sI - A^i)^{-1}B^iE^i(s) + (sI - A^i)^{-1}D^iF^i(s), \tag{3.51}$$

$$I = \begin{vmatrix} 1 & \ldots & 0 \\ 0 & \ldots & 0 \\ 0 & \ldots & 1 \end{vmatrix},$$

where I is the unit matrix and $(sI - A^i)^{-1}$ is the inverse of matrix $(sI - A^i)$, so that $(sI - A^i)^{-1}(sI - A^i) = I$.

In view of the linearity of Equation 3.51, which describes the behavior of the system, we may apply the superposition principle. Correspondingly, the result of several disturbances of the linear system is equal to the sum of their individual results. If we assume that the initial condition vector is zero, we may write two vector transfer functions.

1. For the supply voltage

$$H_{ye}(s) = C(sI - A^i)^{-1}B^i, \tag{3.52}$$

$$Y(s) = H_{ye}^i(s)E^i(s). \tag{3.53}$$

2. For the external disturbance

$$H_{yf}(s) = C(sI - A^i)^{-1}D^i, \tag{3.54}$$

$$Y(s) = H^i_{yf}(s)F^i(s). \tag{3.55}$$

Each element h_{ij} of the matrix transfer function, where j is the number of rows in the matrix and i is the number of columns, is the transfer function from component j to output component i.

3.1.6 *Pulse modulation*

A pulse (or pulsed signal) is a brief change in a physical quantity, such as a voltage, current, or electromagnetic flux (Figure 3.9). On the assumption that the rise and fall of the physical quantity are brief in comparison with the pulse duration, the basic parameters determining the properties of the pulse are the amplitude A (the maximum deviation from a specified level) and the duration t_{imp}. Usually, we observe a sequence of pulses over time, characterized by the sampling period T of pulse repetition. Thus, the pulsed signal is characterized by three parameters: t_{imp}, A, and T. Instead of the sampling period, we may use the sampling frequency; instead of the pulse duration, we may use the duty factor, defined as the ratio of the pulse duration to the sampling period

$$\gamma = \frac{t_{\text{imp}}}{T}. \tag{3.56}$$

We may distinguish between two types of pulses.

1. Unipolar pulses, with deviation from the zero value in only one direction (with a single polarity)

$$A_{1p} = A \times \Psi_{1p} = \begin{cases} A, & \text{if } t \in (0, t_{\text{imp}}), \\ 0, & \text{if } t \in (t_{\text{imp}}, T). \end{cases} \tag{3.57}$$

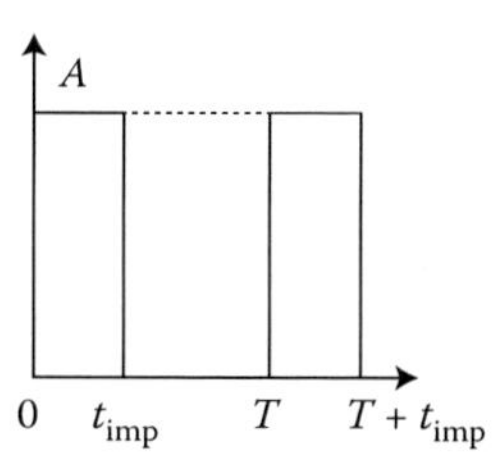

Figure 3.9 Pulse sequence.

2. Bipolar pulses, with deviation from the zero value in two directions (with two polarities)

$$A_{2p} = L \times \Psi_{2p} = \begin{cases} L, & \text{if } t \in (0, t_{\text{imp}}), \\ -L, & \text{if } t \in (t_{\text{imp}}, T), \end{cases} \tag{3.58}$$

where L is the amplitude of the bipolar pulse.

Obviously, taking account of the relation between the switching functions in Equation 3.25, with fixed amplitude of the pulse, we may write

$$A = \text{const}, \quad A_{1p} = A, \quad \left|A_{2p}\right| = \frac{A}{2}. \tag{3.59}$$

The algebraic relation between the pulses takes the form

$$A_{1p} = \frac{A}{2} + \frac{A_{2p}}{2} \quad \text{or } A_{2p} = 2A_{1p} - A. \tag{3.60}$$

Modification of a specific parameter of the pulse sequence over time in a particular manner is known as modulation. The time dependence is known as the modulating function or modulation law. We may identify three simple or elementary modulations, in which only one of the parameters is modulated (Table 3.2).

Other modulations used in power electronics are combinations of several above-mentioned simple ones. For example, relay control produces pulse-width-time modulation that is a combination of pulse-width modulation and time one.

In addition, we may define two kinds of modulation on the basis of how the parameter changes.

- In modulation of the first kind, the parameter changes in accordance with the current value of the modulating function.
- In modulation of the second kind, the parameter changes in accordance with the values of the modulating function at fixed times separated by an interval equal to the sampling period.

Table 3.2 Types of Elementary Modulation

Type of modulation	Amplitude (A)	Duration (t_{imp})	Period (T)
Amplitude modulation	var	const	const
Pulse-width modulation	const	var	const
Frequency or time modulation	const	const	var

The circuit component that produces modulation is known as a pulsed element. Its basic parameters are the pulse duration t_{imp} and the sampling period T (or equivalently the sampling frequency). These parameters are constant for a pulsed element of amplitude type and variable in all other cases.

A power electronic device includes a pulsed element (power switch) and a continuous structure. If the structure is linear, as is usually the case, and the pulsed element is of amplitude type, we obtain a linear pulsed structure. In the case of a pulsed element of pulse-width or frequency type, we have a nonlinear pulsed system, even if the continuous structure is linear.

3.1.7 Difference equations

For linear pulsed systems with modulation of the second kind, when the parameters vary in accordance with the values of the modulating function at fixed times separated by the sampling period, the concept of lattice functions has been introduced.

The values of the lattice function are calculated at discrete times, determined by the sampling period T. The lattice function consists of a set of discrete values of the continuous function $x(t)$ at times $t = kT$, where k is an integer (Figure 3.10) and is written in the form $x[kT]$ or, more briefly, as $x[k]$, as the sampling period T is constant. The lattice function is related to the modulating function as follows:

$$x[k] = x(t)\sum_{k=0}^{\infty} \delta(t - kT), \tag{3.61}$$

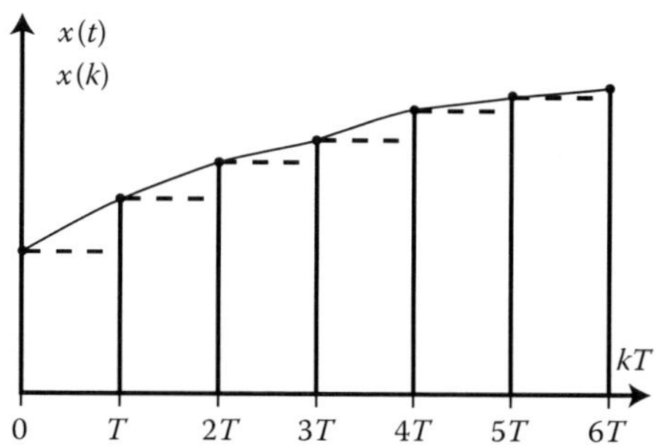

Figure 3.10 Continuous function and its lattice function.

where $\delta(t - kT)$ is a δ pulse

$$\delta(t - kT) = \begin{cases} 1, & \text{if } t = kT, \\ 0, & \text{if } t \neq kT. \end{cases} \tag{3.62}$$

The power electronic device may be regarded as a linear pulsed system, whose behavior is described by a difference equation. This equation is obtained from the differential equation of the power electronic device by substituting a ratio of infinitesimal quantities for the derivative

$$\frac{dx(kT)}{dt} = \lim_{T\to 0}\frac{\Delta x[kT]}{T}, \quad \Delta x[kT] = x[(k+1)T] - x[kT], \tag{3.63}$$

where $x[kT]$ are discrete values of the continuous function $x(t)$ and Δ^i is a finite-difference operator of ith order, analogous to the differential operator. Then the finite-difference operator takes the form

$$\begin{aligned} \Delta^0 x[k] &= x[k] \\ \Delta^1 x[k] &= x[(k+1)] - x[k] \\ \Delta^2 x[k] &= \Delta x[(k+1)] - \Delta x[k] \\ &\vdots \\ \Delta^i x[k] &= \Delta^{i-1} x[(k+1)] - \Delta^{i-1} x[k]. \end{aligned} \tag{3.64}$$

A finite difference of any order may be expressed in terms of the discrete values of the continuous function as follows:

$$\Delta^n x[k] = \sum_{i=0}^{n} C_n^i x[(n+k-i)](-1)^i, \tag{3.65}$$

where C_n^i are the coefficients of Newton's binomial formula (Appendix 3.1)

$$C_n^i = \frac{n!}{i!(n-i)!}.$$

Thus, a continuous linear equation 3.16 may be written as a linear nonhomogeneous finite-difference equation

$$a_n \Delta^n x[k] + a_{n-1} T \Delta^{n-1} x[k] + \cdots + a_0 T^n x[k] = T^n \{f[k] + b_e e[k]\}. \tag{3.66}$$

To solve this equation, we need to find the initial conditions of the function and its finite differences up to order (n – 1) (inclusive).

Another form of the finite-difference equation is expressed in terms of discrete values of the variable on the basis of Equation 3.63:

$$a_n^* x[n+k] + a_{n-1}^* x[n+k-1] + \cdots + a_0^* x[k] = f^i[k] + b_e^i e^i[k]. \tag{3.67}$$

The coefficients in Equations 3.66 and 3.67 are related by Equation 3.65. Equation 3.67 is a recurrence relation, in which each successive value of the discrete function is represented in terms of its preceding value.

3.1.8 *Discrete Laplace transformation (Z-transformation)*

To solve difference equations, we use discrete Laplace transformation, D-transformation, and Z-transformation (sometimes known as Laurent transformation) (Doetsch, 1974). These transformations are related as follows:

$$L[x(t)] \frac{s}{(1-e^{-s})} = D[x(t)] = Z[x(t)]_{z=e^{sT}}, \tag{3.68}$$

where $x(t)$ is the initial function, $L[x(t)]$ denotes direct Laplace transformation, $D[x(t)]$ denotes direct D-transformation, and $Z[x(t)]$ denotes direct Z-transformation with the introduction of the new variable $z = e^{sT}$. The basic advantage of this transformation is that the difference equations are converted into simple algebraic equations. Inverse transformation to the time domain involves dividing the algebraic solution obtained into the sum of simple fractions and finding their solution from the table of the inverse Z-transformation on the basis of the superposition of linear equations (Table 3.3).

The direct Z-transform of the variable $x(t)$ takes the following form.

- For a continuous function

$$X(z) = Z[x(t)] = \sum_{k=0}^{\infty} x(t)\delta(t-kT)z^{-k}. \tag{3.69}$$

Table 3.3 Common Z-Transformations

Description	Initial function	Z-transform
Linearity	$f[k] = \sum_{\nu=0}^{N} C_\nu f_v[k]$	$F(z) = \sum_{\nu=0}^{N} C_\nu F_v(z)$
Delay	$f[k-m], m > 0$	$z^{-m}\left\{F(z) + \sum_{(n-m)=1}^{m} f(-n+m)z^{n-m}\right\}$
Advance	$f[k+m], m > 0$	$z^{m}\left\{F(z) - \sum_{(n+m)=0}^{m-1} f(n+m)z^{n+m}\}\right\}$
Formulation of differences	$\Delta x[k]$ $\Delta^2 x[k]$ $\vdots$ $\Delta^n x[k]$	$(z-1)X(z) - zx(0)$ $(z-1)^2 X(z) - z(z-1)x(0) - z\Delta x(0)$ $\vdots$ $(z-1)^n X(z) - z\sum_{\nu=0}^{n-1}(z-1)^{n-1-\nu}\Delta^\nu x(0)$
Summation		
Incomplete sum	$\sum_{\nu=0}^{k-1} x[k]$	$\frac{1}{z-1}X(z)$
Complete sum	$\sum_{\nu=0}^{k} x[k]$	$\frac{z}{z-1}X(z)$

- For a lattice function

$$X(z) = Z[x[k]] = \sum_{k=0}^{\infty} x[k]z^{-k}. \tag{3.70}$$

The delay formula is simplified if the lattice function with a negative argument is zero. The advance formula is simplified if the lattice function is zero up to $n = (m - 1)$. Analogous to Laplace space, we may use the concept of a pulsed transfer function, which is the ratio of the Z-transform of the control plant's output variable to the Z-transform of its input variable, with zero initial conditions. The transfer function is usually denoted by W. In its subscript, the first digit denotes the output value, whereas the second digit denotes the input value. As Equation 3.16, which describes the behavior of a one-dimensional system, is linear, we may apply the superposition principle. Accordingly, the result of the action of several disturbances on a linear system is equal to the sum of the results of each perturbation.

3.2 Analysis of the electrical processes in power electronic devices

3.2.1 Analytical solution of differential equations

As already noted, each structure of the power electronic device is described by linear differential equations. Therefore, a precise description of the electrical processes in the device may be obtained by solution of the differential equations describing the behavior of this structure.

The solution of Equation 3.16 is an n-times differentiable function $x(t)$, satisfying the equation at all points of its region of definition. Usually, there is a whole set of such functions, and the selection of a single one is based on an additional stipulation, in the form of a set of initial conditions

$$x(t_0) = x_0, \frac{\mathrm{d}x(t_0)}{\mathrm{d}t} = \frac{\mathrm{d}x_0}{\mathrm{d}t}, \ldots, \frac{\mathrm{d}^{n-1}x(t_0)}{\mathrm{d}t^{n-1}} = \frac{\mathrm{d}^{n-1}x_0}{\mathrm{d}t^{n-1}}, \tag{3.71}$$

where t_0 is a fixed time—usually the starting time ($t_0 = 0$) and $x_0, \mathrm{d}x_0/\mathrm{d}t, \ldots, \mathrm{d}^{n-1}x_0/\mathrm{d}t^{n-1}$ are, respectively, fixed values of function x and all its derivatives up to the (n – 1)th order (inclusive). The combination of Equation 3.16 and the initial conditions in Equation 3.71 is known as the initial problem or Cauchy problem

$$\begin{cases} a_n^i \dfrac{\mathrm{d}^n x}{\mathrm{d}t^n} + a_{n-1}^i \dfrac{\mathrm{d}^{n-1}x}{\mathrm{d}t^{n-1}} + \cdots + a_0^i x = f^i(t) + b_e^i e^i, \\ x(t_0) = x_0, \dfrac{\mathrm{d}x(t_0)}{\mathrm{d}t} = \dfrac{\mathrm{d}x_0}{\mathrm{d}t}, \ldots, \dfrac{\mathrm{d}^{n-1}x(t_0)}{\mathrm{d}t^{n-1}} = \dfrac{\mathrm{d}^{n-1}x_0}{\mathrm{d}t^{n-1}}. \end{cases} \tag{3.72}$$

It has a single solution.

The solution of the differential equation consists of two parts.

1. The general solution $x_{es}(t)$ of the homogeneous differential equation—that is, the linear differential equation whose right side is zero, which describes the intrinsic properties of the power electronic device.
2. The particular solution $x_{fs}(t)$, which is determined by the properties of the right side and describes the behavior of the power electronic device due to the applied disturbances.

Thus

$$x(t) = x_{es}(t) + x_{fs}(t). \tag{3.73}$$

For linear equations of first and second order, methods of solution have been developed. The solution is undertaken in two stages: first the homogeneous equation and then the nonhomogeneous equation (Appendix 3.2).

3.2.1.1 *Solution of differential equations by Laplace transformation*

As already noted, Laplace transformation is an effective means of solving linear differential equations with constant coefficients. The basic advantages of this method are as follows:

1. Replacement of the operations of differentiation and integration by algebraic operations on the Laplace transforms; in other words, algebraic equations are solved, rather than differential equations.
2. Immediate determination of the differential equation's complete solution.
3. Inverse transformation to the time domain by means of the superposition principle and the corresponding tables.

The solution of Equation 3.45 in Laplace space takes the form

$$X(s) = \frac{1}{a_n^i s^n + a_{n-1}^i s^{n-1} + \cdots + a_0^i} \times \left[F^i(s) + b_e^i e^i(s) + a_n^i \sum_{k=0}^{n-1} \frac{\mathrm{d}^k x(0)}{\mathrm{d}t^k} + a_{n-1}^i \sum_{k=0}^{n-2} \frac{\mathrm{d}^k x(0)}{\mathrm{d}t^k} + \cdots + a_1^i x(0) \right]. \tag{3.74}$$

The sequence

$$a_n^i s^n + a_{n-1}^i s^{n-1} + \cdots + a_0^i \tag{3.75}$$

is known as the characteristic polynomial and characterizes the free (intrinsic) motion of the control plant. If the characteristic polynomial (Equation 3.75) is set equal to zero, we obtain the characteristic equation of the control plant

$$a_n^i s^n + a_{n-1}^i s^{n-1} + \cdots + a_0^i = 0. \tag{3.76}$$

The solutions of this equation (its roots) are said to be poles of the system and determine the intrinsic motion of the control plant. The number of roots is equal to the order of the equation. The roots may be real, imaginary, or complex. Several identical roots (multiple roots) may exist.

The roots may be obtained in an explicit form for equations of up to fourth order (inclusive). Beginning with the fifth order, we use a general formula expressing the roots in terms of the coefficients by means of radicals. Approximate methods may be used to find the roots.

After finding the roots $s_1, s_2, \ldots, s_n$, the characteristic polynomial (Equation 3.75) is rewritten as the product of factors. In other words, it is expanded in terms of the roots of the equation. The solution of Equation 3.74 is written as the sum of regular simple fractions

$$X(s) = \left[\sum_{i=1}^{n} \frac{d_i}{s - s_i}\right] \times \left[F^i(s) + b_e^i e^i(s) + a_n^i \sum_{k=0}^{n-1} \frac{\mathrm{d}^k x(0)}{\mathrm{d}t^k} + a_{n-1}^i \sum_{k=0}^{n-2} \frac{\mathrm{d}^k x(0)}{\mathrm{d}t^k} + \cdots + a_1^i x(0)\right], \tag{3.77}$$

where the coefficients d_1, d_2, and d_3 are determined from the condition

$$\frac{1}{a_n^i s^n + a_{n-1}^i s^{n-1} + \cdots + a_0^i} = \sum_{i=1}^{n} \frac{d_i}{s - s_i}.$$

Inverse transformation to the time domain is based on the superposition principle. Tables are used for inverse transformation of each term in Equation 3.77, and then convolution of the transformed terms is employed.

3.2.2 Fitting method

As already noted, the power electronic device is a system with variable structure, and each possible structure is described by linear differential equations. Therefore, precise description of the electrical processes in the power electronic device entails description of the processes in each possible structure and, hence, solution of the corresponding differential equations. We must also know the time at which each structure is switched on and the switching sequence of the structures.

This approach is known as the fitting method. When using this method, the solution of the differential equation describing the variation in the electrical variables is found for each structure, with unknown initial conditions. The different structures are combined on the basis of the switching rules. On transition from one structure to another, the initial conditions for the state variable of the power electronic device are assumed to be the final values in the previous structure. Thus, at the boundaries, the structures are stitched together by means of the continuous variable (the current at the inductance, the voltage at the capacitance, the mechanical coordinates of the electromechanical device, etc.).

The self-oscillation condition in the power electronic device may be determined on the basis that the initial conditions of the repeating structures must be the same. When the repeating structure is switched on a second time, its initial conditions are written in terms of the initial conditions for the first time, with allowance for the appearance of another structure.

3.2.3 Phase trajectories and the point-transformation method

The phase-plane method permits clear analysis of one-dimensional, second-order systems. In this approach, the behavior of the system is analyzed in the space corresponding to the state variable and its derivative. In that case, the time variable is eliminated and, instead of Equation 3.17, the system is described in the form

$$x_2 = f(x_1). \tag{3.78}$$

Each point of the phase plane, defined by the state variable x_1 and its time derivative x_2, corresponds to a single state of the system and is known as the phase point, image point, or representative point. The set of changes in state of the system as it functions is known as its phase trajectory. With fixed system parameters, only a single-phase trajectory passes through each point of the phase plane (except for singular points). Arrows on the phase trajectories show the displacement of the representative point over time. The complete set of phase trajectories constitutes the phase portrait. It provides information on all possible combinations of system parameters and on types of possible motion. The phase portrait is expedient for analysis of the motion in systems in which self-oscillation is possible.

When a second-order equation is written in terms of deviations Δx_1 and Δx_2 from the equilibrium point as a function of the position of its roots λ_1 and λ_2 on the complex plane, different types of system behaviors are possible (Figure 3.11).

a. Nondamping oscillatory motion (purely imaginary roots $\lambda_1 = j\omega$ and $\lambda_2 = -j\omega$)
b. Damping oscillatory motion (complex roots in the left half-plane $\lambda_1 = -\delta + j\omega$ and $\lambda_2 = -\delta - j\omega$)
c. Increasing oscillatory motion (complex roots in the right half-plane $\lambda_1 = \delta + j\omega$ and $\lambda_2 = \delta - j\omega$)

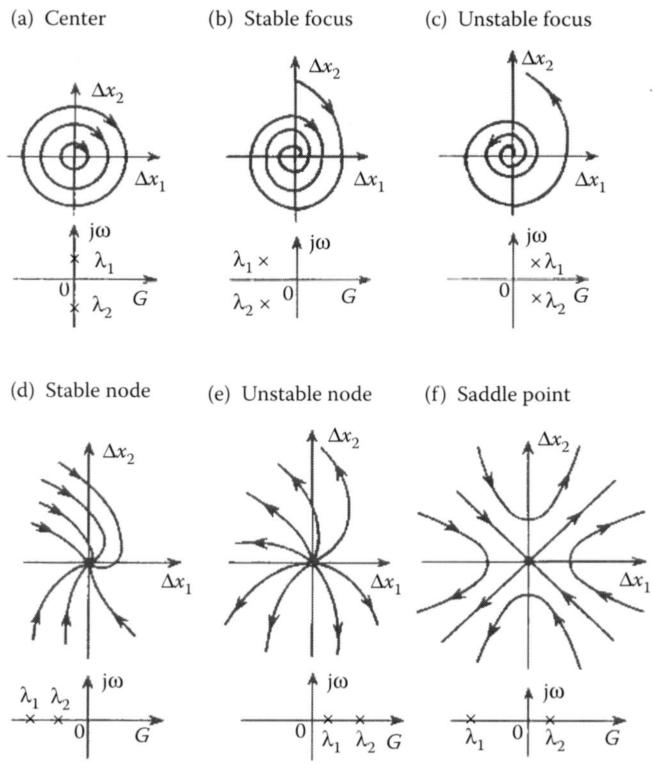

Figure 3.11 Phase trajectories of a second-order differential equation.

d. Damping aperiodic motion (two real roots in the left half-plane $\lambda_1 < 0$ and $\lambda_2 < 0$)
e. Increasing aperiodic motion (two positive real roots $\lambda_1 > 0$ and $\lambda_2 > 0$)
f. Nondamping motion (two real roots, one positive $\lambda_2 > 0$ and one negative $\lambda_1 < 0$)

In Figure 3.11a through f, we show the phase trajectories in the vicinity of the equilibrium points in all these cases and the corresponding position of the roots.

3.2.4 *The primary-component method*

The primary-component method is based on harmonic or spectral analysis and takes account of the filtering properties of the load.

Any periodic function *f*(*t*) with period *T* that satisfies the Dirichlet conditions—in other words, that has a finite number of discontinuities and continuous derivatives between them—may be represented as an infinite trigonometric Fourier series

$$f(t) = \frac{A_0}{2} + \sum_{k=1}^{\infty}\left[a_k \cos(k\omega_1 t) + b_k \cos(k\omega_1 t)\right], \tag{3.79}$$

or

$$f(t) = \frac{A_0}{2} + \sum_{k=1}^{\infty} A_k \cos(k\omega_1 t + \varphi_k), \tag{3.80}$$

where $\omega_1 = 2\pi/T$ is the repetition frequency (or the frequency of the first harmonic), k is the number of the harmonics, and a_k and b_k are the Fourier coefficients

$$\begin{aligned} a_k &= \frac{2}{T}\int_0^T f(t)[\cos(k\omega_1 t)]\,dt, \\ b_k &= \frac{2}{T}\int_0^T f(t)[\sin(k\omega_1 t)]\,dt. \end{aligned} \tag{3.81}$$

The amplitude A_k and phase φ_k of harmonic k are determined as follows:

$$\begin{aligned} A_k &= \sqrt{a_k^2 + b_k^2}, \\ \varphi_k &= -\mathrm{arctg}\left(\frac{b_k}{a_k}\right). \end{aligned} \tag{3.82}$$

Harmonic analysis is understood to mean Fourier-series expansion of $f(t)$ on the basis of Equation 3.79, with calculation of the Fourier coefficients a_k and b_k in Equation 3.81. Spectral analysis is Fourier-series expansion based on Equation 3.79, with the determination of the amplitude A_k and phase φ_k of the harmonics (cosinusoids) in Equation 3.82. The spectrum of the function $f(t)$ is its set of harmonics forming the Fourier series. The spectrum of the function includes the amplitude spectrum—that is, the set of amplitudes $A_k = A[k]$—and the phase spectrum $\varphi_k = \varphi[k]$, corresponding to the frequency spectrum of the Fourier-series expansion, $\omega_k = \omega[k] = k\omega_1$, $k = 0,1, \ldots, \infty$. Thus, any periodic signal may be regarded as a set of harmonic signals of different amplitudes and phases, with a fixed frequency spectrum.

In view of its physical properties, the load to which the periodic function is supplied acts as a filter. In other words, it only transmits a limited frequency spectrum. For example, if high-frequency pulsed voltage is supplied to a dc motor, the inductance of the armature circuit and the moment of inertia form a low-frequency filter for this voltage. Therefore, the motor speed will depend only on the mean component of this voltage, and the speed pulsation due to the high-frequency spectral components may be disregarded.

Thus, in view of the filtering properties of the power electronic device and the load, we may identify a dominant component in the Fourier-series expansion of the spectrum of the periodic discontinuous function. This is known as the primary, smooth, or useful component. All the other components are regarded as noise that impairs the operation of the consuming systems (their mechanical, energy, and other characteristics).

Usually, in view of the properties of the power electronic device, the primary component is assumed to be the constant component in converters with a dc output and the first voltage harmonic in converters with an ac output.

By identifying the primary component and by confining our attention to the corresponding electric process, we may significantly simplify the analysis of the power electronic device, which may be regarded as a linear continuous system described by equations of the following form.

- In the scalar case

$$a_n \frac{d^n x}{dt^n} + a_{n-1}\frac{d^{n-1}x}{dt^{n-1}} + \cdots + a_0 x = f(t) + u, \tag{3.83}$$

where x is an independent variable, $a_0, a_1, \ldots, a_n$ are the coefficients, $f(t)$ is the perturbation to which the power electronic device is subjected, and u is the control of the power electronic device.

- In the vector case

$$\frac{dX}{dt} = AX + BU + DF, \tag{3.84}$$

$$Y = CX, \tag{3.85}$$

where X is the state vector, which consists of independent variables describing the behavior of the control plant (the number of such variables determines the dimensionality of the object), U is the control vector, which contains all the independent control signals of the power electronic device, $(U)^T = |u_1\ u_2 \ldots u_{K-1}\ u_K|$, F is the external disturbance vector, $(F)^T = |f_1\ f_2 \ldots f_{L-1}\ f_L|$, Y is the output-variable vector, $(Y)^T = |y_1\ y_2 \ldots y_{M-1}\ y_M|$, and A, B, C, and D are the matrices characterizing the given control plant. In the general case, the number of independent variables may be different from the number of output variables. In other words, the matrix C may be rectangular. We call Equation 3.84 the equation of state and Equation 3.85 the output-variable equation.

3.2.5 Stability

In the analysis and design of power electronic devices, stability is a primary concern. A power electronic device is said to be stable if, after leaving the equilibrium (rest) state under the action of an external disturbance, it returns to that state when the disturbance is removed. If it does not return to the equilibrium state when the disturbance is removed, it is unstable. Normal functioning of the system requires that it is stable. Otherwise, even small changes in the initial conditions will produce large errors.

The Lyapunov definition of stability is the classical form (Kwakernaak and Sivan, 1972). The solution $x_0(t)$ of a nonlinear differential equation

$$\frac{dx}{dt} = f[x(t), t] \tag{3.86}$$

is stable in the Lyapunov sense if, for any t and any $\varepsilon > 0$, there exists a value of $\delta(\varepsilon, t) > 0$ (depending on ε and possibly on t_0), such that if the initial deviation of the vector from the rated value is $\|x(t_0) - x_0(t_0)\| \leq \delta$, the subsequent deviation $\|x(t) - x_0(t)\| \leq \varepsilon$ for all $t \geq t_0$. This definition employs the vector norm: $\|x\| = \sqrt{\sum_{i=1}^{n} x_i^2}$.

For linear systems, the necessary and sufficient condition for stability is that the roots of the characteristic equation be on the left half-plane of the complex plane; in other words, they must have a negative real component. Numerous algebraic and frequency methods permit determination of the system's stability without finding the roots of the characteristic equation. Those methods may be used in analyzing the device's behavior if it is described in terms of the primary component. For example, the Routh–Hurwitz algebraic criterion is as follows. The necessary and sufficient condition for a linear steady dynamic system to be stable is that all the n diagonal minors of the Hurwitz determinant be positive, where n is the order of the characteristic equation.

The Hurwitz determinant is formulated from the coefficients of the characteristic equation as follows.

1. All the coefficients of the characteristic equation are ranked along the primary diagonal from left to right, beginning with the coefficient of the term of order $(n - 1)$ and ending with the free term.

2. From each element of the diagram, the corresponding column of the determinant is arranged, so that the subscripts decline from top to bottom.
3. The coefficients with below-zero subscripts or large n are replaced by zero.

The diagonal minor is the determinant obtained from the Hurwitz determinant by eliminating the row and column corresponding to the given diagonal element.

For nonlinear systems, the stability is determined by means of the second Lyapunov method, which is a sufficient stability condition. According to that criterion, the system is stable if it is possible to select a sign-definite Lyapunov function that includes all of the system's state variables and whose derivative is sign-definite or sign-constant but with the opposite sign.

A sign-definite function is a function that tends to zero only at the coordinate origin. A sign-constant function is a function that vanishes not only at the coordinate origin, but also at all other points of space. If the derivative is sign-definite, asymptotic convergence occurs. Usually, the Lyapunov function is selected as a quadratic function, which is, by definition, a sign-definite positive function. Its derivative is then analyzed: if the system is to be stable, the derivative must be sign-definite or sign-constant and negative.

3.3 Control methods

3.3.1 Control problems and principles

The purpose of a power electronic device is to convert the input electrical energy into output energy with specified characteristics, corresponding to the needs of the overall system. The actual state of the power electronic device is characterized by one or more control variables.

As noted in Section 3.1, we may distinguish between multidimensional (MIMO) and one-dimensional (SISO) systems.

There are three fundamental control principles (Figure 3.12).

1. Open control
2. Control based on disturbances (the compensation principle)
3. Closed control (the feedback principle, control based on deviations)

In open control, the control at the input of the power electronic device is calculated on the basis of the technological requirements and complete information regarding not only the power electronic device, but also all the disturbances present. The controlled variable is not monitored. The advantage of this method is its simplicity. A significant disadvantage is the need for complete information regarding the power electronic device

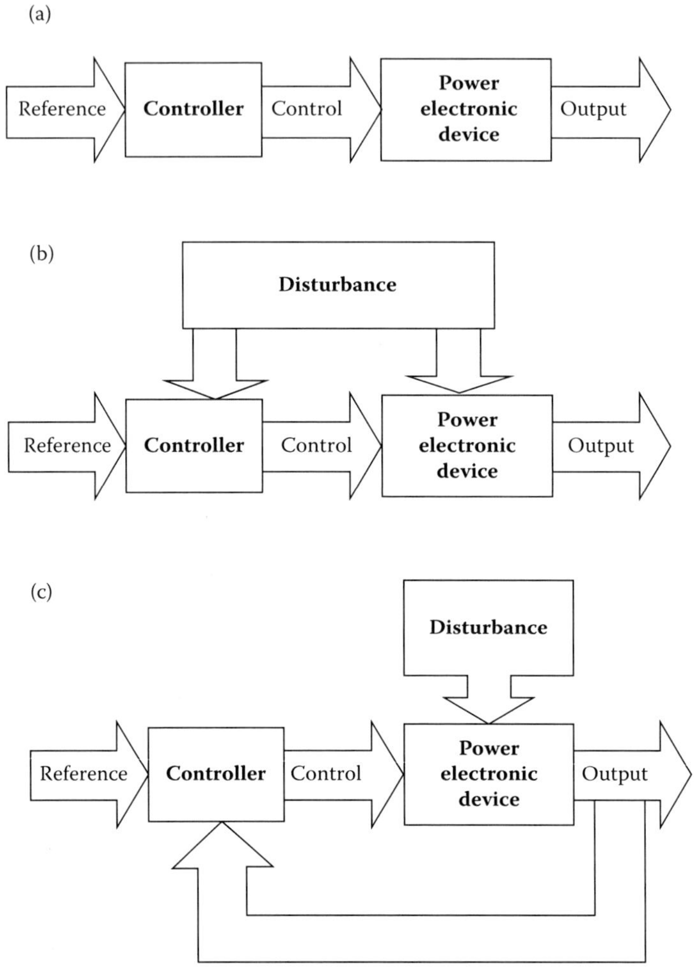

Figure 3.12 Control principles: (a) open control; (b) control based on the disturbances (the compensation principle); and (c) closed control (control based on deviations).

and all the disturbances present, as inaccuracy will lead to error in the controlled variable.

To eliminate the influence of disturbances on the control process, the compensation principle is employed. In that case, as well as a control designed to meet the technological requirements with specified disturbance or no disturbance, we also supply a control based on information regarding the current disturbance and intended to compensate for that disturbance. The compensation principle is effective if it is possible to obtain information regarding the disturbances.

Closed control is based on the comparison of the specified value of the controlled variable with the actual value in the presence of unknown disturbances. The discrepancy is used to select the control for the power electronic device. In other words, the control takes account not only of the specification, but also of the actual state of the power electronic device and the disturbances. Therefore, this is the most widely employed control principle. It permits successful control where the disturbances and the parameters of the control plant are indeterminate. The automatic systems based on closed control are known as automatic control systems.

The last two control principles are sometimes used together. This constitutes composite control, offering the highest precision.

In the most general case, the reference is an arbitrary function of the time, which is determined by an external control unit (controller). The processing of that reference is a tracking process. Particular cases of tracking are as follows:

1. Stabilization (maintenance) of the control value at a specified constant level
2. Program control, which involves adjustment of the specification according to a predetermined schedule

Precise and fast processing of the reference is required, with arbitrary external disturbances.

In multidimensional systems, with several independent controls, we are dealing with the control of several controlled variables.

3.3.2 *Structure of control system*

We may distinguish between several approaches to the design of automatic control systems, based in different mathematical models of the power electronic device.

- Single-loop control (Figure 3.13a)
- Two-step single-loop control (Figure 3.13b)
- Cascade (subordinate) control (Figure 3.13c)

In the first case, we use a precise model of the power electronic device, taking account of the discontinuous switch characteristic. Control design entails solving a nonlinear problem, whose complexity is due to the significant nonlinearity of the power electronic device. The switching frequency and the time at which different power switches are on will be generated automatically in the closed loop, as a secondary element in the overall control of the power electronic device. Such dynamic systems are characterized by high speed and low sensitivity to variation in the parameters and external disturbances. Regrettably, in this case, the automatically

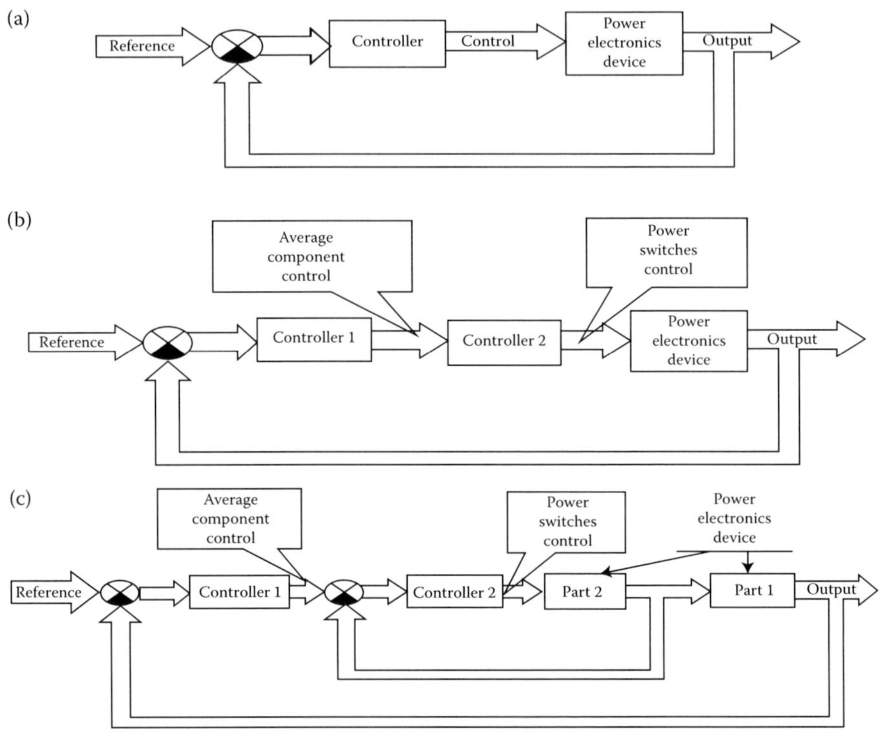

Figure 3.13 Structures of control systems for power electronic devices: (a) single-loop control; (b) two-step single-loop control; and (c) cascade (subordinate) control. CS, control system.

generated switching frequency of the power switches is not constant but depends on the initial conditions. This increases the switching losses of the power switches.

In the second case, we use a model of the power electronic device with a mean component and solve two independent problems: control of the power electronic device as a linear object and high-frequency control of the power switches. Control system 1 is synthesized within the framework of linear control theory, on the assumption that control system 2 creates the necessary high-frequency sequence of voltage pulses, whose mean component corresponds to the primary control problem, and the high-frequency voltage components are filtered out on account of the low-frequency properties of the load.

The control of the power switches is formulated independently. Open control is chosen, with modulation. The modulation law and the switching law, determining the frequency and the duty factor of the switches, are specified externally. However, in that case, as already noted, the automatic compensation of the external disturbances and fluctuations in the internal parameters rely on the control loop for the power electronic device.

Finally, cascade control is based on subdivision of the initial control problem in terms of the speed of the processes in the system. Each of the resulting control problems is then solved independently. As in the preceding case, we employ the natural division into fast and slow processes in the power electronic device; again, the natural filtering properties of the power electronic device and the load are taken into account. The slow external loop includes control system 1 and power electronic device 1 and is described by the mean component, as the fast internal loop, consisting of control system 2 and power electronic device 2, instantaneously processes its input signal; in other words, its transfer function is one. The internal loop is responsible for control of the switches on the basis of feedback. Correspondingly, it is relatively fast and insensitive to variation in the parameters and external disturbances. Note that, as in the first case, the switching frequency of the power switches is generated automatically and depends on the initial conditions. This may increase the switching losses of the power switches and the mechanical noise in electromechanical energy converters.

3.3.3 *Linear control methods*

As already noted, if we adopt the model of the power electronic device in terms of the mean, analysis of the device and control design may be based on linear automatic control theory. This theory is well developed, especially for one-dimensional systems (Kwakernaak and Sivan, 1972).

3.3.3.1 *Controller design for one-dimensional control systems*

As noted in Section 3.2.4, the first requirement in any automatic control system is stability: after removal of disturbances, the system must return to the specified state or nearby steady conditions. This will occur if the intrinsic motion of the system is damping. A necessary and sufficient condition for stability of the system is that all the roots of its characteristic equation be in the left half-plane of the complex plane.

The introduction of feedback changes the characteristic equation of the automatic control system, which includes the power electronic device and a controller.

The following controllers are widely used.

- The proportional controller (P controller), in which the control is proportional to the discrepancy

$$u = K(x_z - x), \tag{3.87}$$

where K is a constant of proportionality.

- The proportional–integral controller (PI controller), in which the control is the sum of a value proportional to the discrepancy and a value corresponding to the integral of the discrepancy

$$u = K(x_z - x) + \int (x_z - x)\mathrm{d}t. \tag{3.88}$$

- The proportional–integral–derivative controller (PID controller), in which the control is the sum of a value proportional to the discrepancy, a value corresponding to the integral of the discrepancy, and a value corresponding to the differential of the discrepancy

$$u = K_p(x_z - x) + K_i \int (x_z - x)\mathrm{d}t + \frac{\mathrm{d}(x_z - x)}{\mathrm{d}t}. \tag{3.89}$$

The equation of an automatic control system including the power electronic device and a P controller is then based on Equations 3.83 and 3.87:

$$a_n \frac{\mathrm{d}^n x}{\mathrm{d}t^n} + a_{n-1} \frac{\mathrm{d}^{n-1} x}{\mathrm{d}t^{n-1}} + \cdots + (a_0 + K)x = f(t) + Kx_z, \tag{3.90}$$

and its stability is evaluated on the basis of the characteristic equation

$$a_n r^n + a_{n-1} r^{n-1} + \cdots + (a_0 + K) = 0. \tag{3.91}$$

We may use Equation 3.91 to identify stability regions if the constant of proportionality K is selected as the variable parameter. This is known as D-division. For two parameters—for example, in the case of a PI controller—we obtain a Vyshegradskii diagram.

The selection of the controller coefficients will determine the position of the roots of the characteristic equation and hence the system's stability and transient. Selection of the desired root configuration is known as modal control.

3.3.3.2 *Controller design for multidimensional control systems*

In contrast to one-dimensional systems, multidimensional systems are characterized by several independent variables and several independent controls; in general, the number of independent variables will not be the same as the number of independent controls. Therefore, the control design in such systems is more complex and includes several stages. The first is to determine whether the system is controllable or may be stabilized.

A controllable system may be moved from any initial state X_0 to an arbitrary state X_1 in finite time by piecewise-continuous control signal U.

For systems with constant parameters, the necessary and sufficient condition for controllability is nondegeneracy of the controllability matrix

$$P = \begin{vmatrix} B & AB & A^2B & \dots & A^{n-1}B \end{vmatrix}, \quad \det P \neq 0. \tag{3.92}$$

When the available controls do not ensure controllability, the next step is to investigate whether the system may be stabilized. A system that may be stabilized is understood to mean a linear dynamic system, in which two vectors X^* and X^{**} may be formed from the components of the state vector X: $X^{\mathrm{T}} = |X^* \quad X^{**}|$. The state vector X^* may be moved from any initial state $X^*(0)$ to an arbitrary state $X^*(t)$ in finite time by a piecewise-continuous control signal U present in the system, while state vector X^{**} is stable. For the sake of clarity, we use the nondegenerate linear transformation of the state vector

$$X' = T^{-1}X \tag{3.93}$$

to convert such dynamic systems into the canonical form of controllability

$$\frac{\mathrm{d}X'}{\mathrm{d}t} = \begin{vmatrix} A'_{11} & A'_{12} \\ 0 & A'_{22} \end{vmatrix} X' + \begin{vmatrix} B'_1 \\ 0 \end{vmatrix} U + D'F, \tag{3.94}$$

where

$$\begin{vmatrix} A'_{11} & A'_{12} \\ 0 & A'_{22} \end{vmatrix} = T^{-1}A, \quad \begin{vmatrix} B'_1 \\ 0 \end{vmatrix} = T^{-1}B, \quad D' = T^{-1}D.$$

In that case, the second vector equation describes the motion of the state-vector components that are independent of the control. Those components must be stable if the system as a whole is to be stable. The first vector equation of motion of the state-vector components must be controllable. For these components, a controller may be designed to ensure that their actual values match their specified values. The stable values of the state-vector components in the second equation appear in the first equation as an external disturbance.

There may be several means by which a particular system may be stabilized, as the linear-transformation matrix satisfying Equation 3.93 is selected somewhat arbitrarily.

For a controllable or stabilized linear system, linear feedback

$$U = -KX, \tag{3.95}$$

such that the closed system that is stable may always be selected. Here K is the feedback matrix.

This may be explained, in that the stability of the system is determined by the characteristic numbers of the matrix $A - BK$, which depend on the selection of the elements in matrix K. In other words, the characteristic numbers may be in arbitrary positions on the complex plane (given that the roots are complex-conjugate).

As already noted, a system that may be stabilized is broken down into two subsystems. The first is uncontrollable but stable; the second is controllable, and the desired root configuration may be ensured by negative feedback.

Thus, in a system that may be stabilized, only some of the roots may be positioned arbitrarily. The remaining roots must lie in the left half-plane of the complex plane, in view of the system's properties.

3.3.4 Relay control

As noted in Section 3.1.3, if the switching function applies only to the input voltage and its variation is determined by the state of the system's control variables—as in Equations 3.23 and 3.24, for example—these power electronic devices are of relay type. At present, methods of analysis and design are best developed for one-dimensional second-order power electronic devices (Tsypkin, 1984). In particular, the phase-plane method permits clear representation of the processes in the system. This method is based on mathematical description of the power electronic device in terms of deviations from the rated value, as in Equation 3.30. In the Cauchy form

$$\begin{aligned} \frac{\mathrm{d}\Delta x}{\mathrm{d}t} &= x_1, \\ \frac{\mathrm{d}x_1}{\mathrm{d}t} &= -\frac{a_1}{a_2}x_1 - \frac{a_0}{a_2}\Delta x - \frac{1}{a_2}f(t) - \Psi\frac{b_e}{a_2}e + A, \end{aligned} \tag{3.96}$$

where

$$A = \frac{\mathrm{d}^2 x_z}{\mathrm{d}t^2} + \frac{a_1}{a_2}\frac{\mathrm{d}x_z}{\mathrm{d}t} + \frac{a_0}{a_2}x_z.$$

To plot phase trajectories of the system in the Cauchy form, we eliminate the time. From the first relation in Equation 3.96, we find that $\mathrm{d}t = \mathrm{d}\Delta x/x_1$. Substituting this result into the second relation, we obtain

$$x_1\,\mathrm{d}x_1 = \left(-\frac{a_1}{a_2}x_1 - \frac{a_0}{a_2}\Delta x - \frac{1}{a_2}f(t) - \Psi\frac{b_e}{a_2}e + A\right)\mathrm{d}\Delta x. \tag{3.97}$$

If the structure of the equation permits variable separation and integration of the left sides of the equation—for example, in the case in which $a_1 = 0$, $a_0 = 0$, $f(t) = \text{const}$, and $A = \text{const}$—we may obtain the phase-trajectory equation for each state of the switching function

$$x_1 = 2\sqrt{\left(-\frac{1}{a_2}f - \Psi\frac{b_e}{a_2}e + A\right)\Delta x} + C. \tag{3.98}$$

Here C is a constant of integration.

When integration is difficult, the phase trajectories may be approximately plotted without integration by the isoclines method. At each point of an isocline, the slope of the phase trajectories (integral curves) has the same constant value $dx_1/d\Delta x = k = \text{const}$. If we specify different values of the slope, we may construct a dense grid of isoclines, each of which includes small segments with specified slope k. Then, beginning at the initial point in phase space, we plot a line intersecting each isocline at the angle specified by the directional field. The resulting curve will be an approximate representation (sketch) of the phase trajectory.

The set of all phase trajectories is known as the phase portrait of the system.

The phase trajectories in different states of the switching function are spliced at the switching line in the space whose dimensions are the error and its derivative.

As the relay control system has a switching function, the self-oscillation parameters in the relay system will depend on the state of that function. The self-oscillation parameters may be determined from the parameters of the steady limiting cycle. The self-oscillation amplitude corresponds to the deviation of the limiting cycle from zero at the error axis.

The switching condition of the phase function in phase space is a zero value of the switching function

$$Z(\Delta x, x_1) = 0, \tag{3.99}$$

where Δx is the error and x_1 is its derivative.

In most cases, the switching function is a linear combination of the error and its derivative. Therefore, we may speak of the switching line—that is, the line at which the switching function changes state

$$Z(\Delta x, x_1) = x_1 + C\Delta x = 0, \tag{3.100}$$

where C is the slope of the switching line.

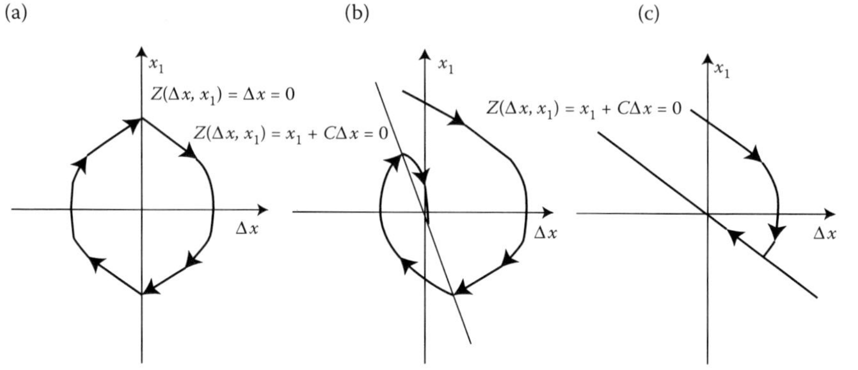

Figure 3.14 Phase trajectories of a relay system when the switching function depends on (a) the error and (b) the error and its derivative in the general case and (c) in the case with sliding motion.

In classical relay systems, $C = 0$ in the simplest case, and control may be based solely on the error (Figure 3.14a).

When a term depending on the error's derivative is introduced in the switching function, the parameters of the limiting cycle are changed. On the phase plane, the switching line is converted from a vertical line coinciding with the axis to an inclined line (slope C) passing through the coordinate origin.

In certain conditions, the phase trajectories of relay systems are directed toward the switching line from both sides. In that case, the point describing the behavior of the system moves along the switching line to the coordinate origin. This is known as sliding mode. In terms of the physical processes in the system, self-oscillation occurs at infinitesimally small frequency.

Then, with scalar control, the necessary and sufficient condition for sliding mode, which is the stability condition in switching-function space, takes the form

$$\lim_{Z \to +0} \frac{\mathrm{d}Z}{\mathrm{d}t} < 0, \lim_{Z \to -0} \frac{\mathrm{d}Z}{\mathrm{d}t} > 0. \tag{3.101}$$

Note that the sliding motion of the system is invariant with respect to external disturbance and change in dynamic properties of the control plant, as described by Equation 3.100, in which these factors do not appear. The sliding motion is an aperiodic transient process with time constant $1/C$. The selection conditions for the supply voltage e and the switching law of the switching function Ψ_{2p} depend on the conditions for satisfying Equation 3.101 in the system

$$e \geq \left| \frac{-a_2 C\Delta x - a_1 x_1 - a_0 \Delta x - f(t) + a_2 A}{b_e} \right|, \tag{3.102}$$

$$\Psi_m = \begin{cases} 1, & \text{if } \operatorname{sgn}(Z) > 0, \\ -1, & \text{if } \operatorname{sgn}(Z) < 0. \end{cases} \tag{3.103}$$

Equation 3.102 defines a region from which the system reaches the switching line and remains there, in the state space corresponding to Equation 3.99. When $Z = 0$, it determines a class of permissible disturbances and references that may be reproduced without dynamic tracking error. Equation 3.103 describes the operational algorithm for a relay control.

For one-dimensional first-order systems, sliding motion occurs at the equilibrium point

$$\frac{d\Delta x}{dt} = -\frac{a_0}{a_1}\Delta x - \frac{1}{a_1} f(t) - \Psi \frac{b_e}{a_1} e + A, \tag{3.104}$$

where

$$A = \frac{dx_{1z}}{dt} + \frac{a_0}{a_1} x_{1z}.$$

As already noted, the actual value of the control variable is equal to the specified value in self-oscillation.

In contrast to electronic power switches, mechanical switches—in particular, relays—are characterized by a limit on the switching number. Therefore, high-frequency self-oscillation is not an option in that case. Instead, hysteresis-based control is employed: the ideal relay is replaced by a relay with hysteresis. This permits reduction of the switching number per unit time. In that case, the controlled variable will be within a range that depends on the width of the hysteresis and will be equal to the required value, on average.

3.3.5 Sliding-mode control

In multidimensional relay systems with several relay elements, the control design is a complex problem, with no analytical solution in most cases. On account of the presence of discontinuous (relay) control in such systems, it seems natural to use the theory of sliding-mode systems (Utkin et al., 2009; Ryvkin and Palomar Lever, 2011). In particular, as shown for the example of one-dimensional relay systems in Section 3.3.4, the nonlinear dynamic systems with such control have attractive properties such as high quality of control and insensitivity to external disturbances and parameter variation of the control plant.

In terms of control, a multidimensional power electronic device is a nonlinear dynamic system with linear introduction of the control $u(t)$, which is discontinuous on account of operation of the power switches

$$\frac{\mathrm{d}x(t)}{\mathrm{d}t} = f(x,t) + B(x,t)u(t), \tag{3.105}$$

where $x(t)$ is the state vector, $x(t) \in R^n$; $f(x, t)$ is the column vector of the object of control; $f(x, t) \in R^n$; $u(t) \in R^m$; and $B(x, t)$ is a matrix of dimensionality $n \times m$. The components of the control vector $u(t)$ take one of the two vectors, depending on the state of the system, characterized by the state of the switching function $S_i(x)$:

$$u_i(x,t) = \begin{cases} u_i^+(x,t), & \text{if } S_i(x,t) > 0, \\ u_i^-(x,t), & \text{if } S_i(x,t) < 0. \end{cases} \tag{3.106}$$

The sliding motion in the discontinuous system corresponding to Equations 3.105 and 3.106 must ensure solution of the control problem—that is, zero error of the controlled variables $Z(t) = z_z(t) - z(t)$ and $z(x) \in R^l$. Here $z(x, t)$ is the controlled variable vector and $z_z(x,t)$ is the reference controlled variable vector. In other words, the existence of sliding motion is equivalent to stable motion of the initial system described by Equations 3.105 and 3.106, relative to the coordinate origin in the l-dimensional error space of the controlled variables. Using the terminology of stability theory for nonlinear systems, we may speak of the conditions of sliding motion in the large and small (global and local sliding motions, respectively). Stability in the small is equivalent to the existence of sliding motion over a surface of discontinuity, which corresponds to zero error of the components of the controlled variable vector $Z_j(t) = 0, j = \overline{1, l}$ as discontinuous control is reversed at this surface, or over the intersection of several surfaces of discontinuity. Stability in the large is not only associated with sliding motion over the surface of discontinuity or the intersection of surfaces of discontinuity, but also defines the incidence condition—that is, the condition guaranteeing that the mapping point will move from an arbitrary initial position to the surface of discontinuity or intersection of surfaces of discontinuity.

The existence of sliding motion is usually investigated by determining the Lyapunov stability (Section 3.2.5). In that case, we consider equations corresponding to the projection of Equations 3.105 and 3.106 onto the error subspace of the controlled variables

$$\frac{dZ}{dt} = \frac{dz_z}{dt} - (Gf + Du), \tag{3.107}$$

where G is an $l \times n$ matrix, whose rows are the gradient vectors of the functions $\Delta z_j(x)$, and $D = GB$.

In scalar control, the necessary and sufficient condition for sliding motion over a single surface of discontinuity $Z(x) = 0$ in Equation 3.99 is that the deviation from the surface of discontinuity and its rate of change must have opposite signs.

In vector control, there is no universal condition for sliding motion over the intersection of surfaces of discontinuity. Most known conditions for multidimensional sliding motion are expressed in terms of sliding with relation to the manifold $Z(x) = 0$, and the stability problem is solved by Lyapunov's second method. The corresponding sufficient conditions are written in terms of the D matrix of Equation 3.107.

3.3.5.1 *Sufficient conditions for the existence of sliding motion*

In a system corresponding to Equations 3.105 and 3.106, sliding motion over the manifold $Z(x) = 0$ exists if one of the following conditions is satisfied.

1. A matrix D with a primary diagonal ($|d_{aa}| > \sum_{b=1, a\neq b}^{m} |d_{ab}|$; d_{ab}, $a = \overline{1,m}, b = \overline{1,m}$, is an element of matrix D) and discontinuous control is selected in the form

$$u_i(x,t) = \begin{cases} -M_i(x,t), & \text{if } Z_i d_{ii} > 0, \\ M_i(x,t), & \text{if } Z_i d_{ii} < 0 \end{cases} \tag{3.108}$$

   with amplitude

$$M_i(x,t) > \frac{\left(|q_i| + \sum_{j=1, j\neq i}^{m} |d_{ij}|\right)}{|d_{ii}|}, \tag{3.109}$$

   where $q_i(x,t)$ are elements of the vector Gf. In that case, sliding motion occurs at each of the surfaces $Z_i(x) = 0$. In other words, the

m-dimensional sliding motion is broken down into m one-dimensional motions.

2. There is a control hierarchy in the system, so that the multidimensional problem is reduced to m one-dimensional problems, which are solved successively. In that case, one component of the control vector—say u_1—ensures sliding motion over the surface $Z_1(x) = 0$, regardless of the other components. After sliding motion over the surface $Z_1(x) = 0$ appears, control u_2 ensures sliding motion over the surfaces $Z_1(x) = 0$ and $Z_2(x) = 0$, regardless of the other control components and so on. In this approach, the sufficient conditions for multidimensional sliding motion are obtained on the basis of the analogous conditions for the scalar case in Equation 3.99:

$$\text{grad } Z_{k+1} b_k^{k+1} u_{k+1}^{+} < \min_{u_{k+2},\ldots,u_m} \left[-\text{grad } Z_{k+1} f^k - \sum_{j=2}^{m-k} \text{grad } Z_{k+1} b_k^{k+j} u_{k+j} \right],$$

$$\text{grad } Z_{k+1} b_k^{k+1} u_{k+1}^{-} > \max_{u_{k+2},\ldots,u_m} \left[-\text{grad } Z_{k+1} f^k - \sum_{j=2}^{m-k} \text{grad } Z_{k+1} b_k^{k+j} u_{k+j} \right], \tag{3.110}$$

where k is the number of discontinuity surfaces with sliding motion, $0 \le k \le m-1$, f^k is an n-dimensional vector, B_k is an $n \times (m-k)$ matrix with columns $b_k^{k+1}, \ldots, b_k^m$, and f^k and B_k are the elements of the differential equation describing the initial dynamic system in Equations 3.105 and 3.106 with sliding motion over the intersection of k surfaces.

Obviously, the sufficient condition of sliding motion obtained by means of a control hierarchy is the same as the sufficient condition obtained on the basis of the second Lyapunov method when the matrix D is diagonal.

The equivalent-control method is used to describe the motion of the system corresponding to Equations 3.105 and 3.106, with sliding motion over all or part of the manifold $Z(x) = 0$, which is necessary when using a control hierarchy (Utkin et al., 2009). It has been established that sliding motion may be described by the equivalent continuous control u_{eq}. This control ensures that the time derivative of vector $Z(x)$ is zero on the system's trajectories

$$u_{\text{eq}} = (D)^{-1} G f. \tag{3.111}$$

The equivalent control u_{eq} is substituted in the initial equation 3.105, which describes the sliding motion of the system

$$\frac{\mathrm{d}x(t)}{\mathrm{d}t} = f(x,t) + B(D)^{-1}Gf. \tag{3.112}$$

With the specified structure of the system, Equation 3.112 depends only on the elements of matrix G. Hence, by changing the position of the discontinuity surfaces in the system's state space, we may modify the sliding motion. Design of the desired sliding motion is a problem of lower order than our initial problem, because the sliding motion is described not only by Equation 3.112 but also by m algebraic equations of the discontinuity surfaces $Z(x) = 0$. This permits reduction by m in the order of Equation 3.112.

Thus, in the general case, the design of motion in systems with discontinuous control may be broken down into three problems.

1. The design of sliding motion
2. The existence of sliding motion
3. Incidence at the sliding manifold

The first problem is solved by selecting the switching functions that ensure the desired sliding motion. Classical methods from automatic control theory may be used here, as the right side of the differential equations is continuous.

The second and third problems are complicated because sufficient conditions have been stated for the existence of sliding motion. When using the first sufficient condition, matrix D in Equation 3.107 must be reduced to a special form. This is accomplished not by selecting the matrix G so as to ensure solution of the first problem, but by using the matrix $R_u(x,t)$ of linear nondegenerate transformation of the control vector

$$u^* = R_u(x,t)u, \tag{3.113}$$

where u^* is the new control vector or the matrix $R_S(x,t)$ of linear nondegenerate transformation of the discontinuity surfaces

$$Z^* = R_Z(x,t)Z, \tag{3.114}$$

where Z^* is the vector of new discontinuity surfaces. This approach is made possible by the invariance of the sliding equation to these transformations. Design requires the selection of the desired discontinuity surfaces and the subsequent transformation of these surfaces or the control vector corresponding to sufficient conditions for the existence of sliding motion or incidence at the sliding manifold.

Note that, if D in Equation 3.107 is reduced to a diagonal form, the transformations in Equations 3.113 and 3.114 will lead to different results.

When using linear nondegenerate transformation of the control vector, the corresponding matrix $R_u(x,t)$ is selected in the form D, and the transformed equation may be written as follows:

$$\frac{dZ}{dt} = Gf + u^*. \tag{3.115}$$

In that case, depending on the points x for which they are satisfied, the sufficient conditions ensure existence of the sliding motion or even incidence at the sliding manifold. In other words, they ensure sliding motion in the small or in the large.

When using linear nondegenerate transformation of the discontinuity surfaces, the corresponding matrix $R_Z(x,t)$ is selected in the form $(D)^{-1}$, and the transformed equation may be written as follows:

$$\frac{dZ^*}{dt} = (D)^{-1}Gf + u + \frac{d(D)^{-1}}{dt}DZ^*. \tag{3.116}$$

In contrast to the preceding case, we only obtain the condition of existence of sliding motion, as the last term on the right-hand side is absent only in the presence of sliding motion. In this approach, judgments regarding the incidence at the sliding manifold require additional information regarding the term $d(D)^{-1}/dt^{DZ^*}$.

3.3.6 Digital control

In terms of control system design, digital technology has introduced systems with quantization of magnitudes and time. In contrast to continuous signals, the signals to and from digital controllers take discrete values at discrete times (Kwakernaak and Sivan, 1972; Isermann, 1981). Quantization with respect to the time is a periodic process, with period T. If no allowance is made for the characteristics of digital control systems, the use of analog algorithms leads to reduced static precision of the system, the appearance of an oscillatory component (with amplitude proportional to the period), and beats.

The design of digital control systems is significantly different from the analysis and design of continuous systems, for the following reasons. First, the design of digital control systems is based on difference equations, which replace the differential equations describing continuous systems. Secondly, within the period T, the control may be particularized in terms of the rate of the processes, and the equation may be simplified because of the quasiconstant variables within this period. Thirdly, the memory (of capacity m) in the system stores the previous values of the state vector and control vector, which may be used in control. Fourth, the resulting digital algorithms are generally implemented on microcontrollers, with constraints on the length of the computation cycle and specific computational resources.

For computations in a microprocessor controller, the basic variable is the length of the computation cycle, which is closely related to the measurement times and computational capacity of the microprocessor. In what follows, we assume synchronous control and solution of the control problem within a single cycle (corresponding to the quantization period T). In other words, we assume sufficient computational capabilities to form the control command for the executive device within a single cycle. Note that, in principle, the solution of the control problem in this case cannot take less than two computational periods. In the first step $[k, k + 1]$, on the basis of the available information regarding the variables and specifications, the digital controller formulates the control commands for the executive device, such that solution of the control problem is possible at the end of the next step $[k + 1, k + 2]$. In the second step, these control commands are sent to the executive device. This delay, if significant for the given system, must be appropriately compensated. The time diagram of the calculations is shown in Figure 3.15.

In view of the foregoing, the design of high-quality digital control systems for power electronic devices requires the development of special design methods that actively utilize the characteristics of digital systems.

Thus, the first step in the design of digital control is to obtain a precise analytical difference model of the control plant based on the existing differential equations. We make the following assumptions here.

1. Attention focusses on the period T, which, in the case of pulse-width modulation, is the pulse-width modulation period.

Implementation of the control for step $[k - 1, k]$.	Implementation of the control for step $[k, k + 1]$.	***Implementation of the control for step*** $[k + 1, k + 2]$.
Computation of the control for step $[k, k + 1]$.	***Computation of the control for step*** $[k + 1, k + 2]$.	Computation of the control for step $[k + 2, k + 3]$.
Receiving the reference for step $[k + 1, k + 2]$.	Receiving the reference for step $[k + 2, k + 3]$.	Receiving the reference for step $[k + 3, k + 4]$.
Measuring the actual values of the variables.	Measuring the actual values of the variables.	Measuring the actual values of the variables.
Auxiliary computations.	Auxiliary computations.	Auxiliary computations.

Figure 3.15 Time diagram of digital control.

2. The period T is small in comparison with the time constant of the control plant.
3. The rates of the mechanical processes, magnetic processes (typical time constants 10–100 ms), and electrical switching processes (typical time constants 10–100 μs) are significantly different. Accordingly, some variables may be regarded as quasiconstant.

This approach allows complex control problems to be broken down into simpler problems.

The solution of the control problem entails solving algebraic difference equations, with the specification of conditions for as many steps as are required.

Note that the core of the digital controllers used for power electronic devices is a digital signal processor. The high frequency of such processors (up to 200 MHz) ensures a total cycle length of 25 μs (40 kHz). In that case, the use of digital controllers in the control system for a power electronic device is practically the same as that of analog controllers for continuous systems.

3.3.7 Predict control

At present, the control of dynamic objects by means of predictive models (predict control) rests on a formalized approach to the analysis and control design systems, based on mathematical methods of optimization (Linder et al., 2010; Rodriguez and Cortes, 2012).

The main benefits of this approach are the relative simplicity of creating feedback and excellent adaptation. Adaptation permits the control of multidimensional and multiple connected control plants with complex structures (including nonlinearity); real-time optimization of the processes with constraints on the control and controlled variables; and the accommodation of indeterminacy in the references of the control plant and disturbances. In addition, we may take account of transport delays, change in the control plant characteristics in the course of the process, and failures in the measurement sensors.

For a dynamic control plant, predict control is based on feedback and employs the following step-by-step procedure.

Program control is optimized using the mathematical model of the control plant (the predictive model), whose initial conditions correspond to the current state of the control plant. The goal of optimization is to reduce the error between the controlled variables of the predictive model and their corresponding references in some finite time period (the predictive horizon). In optimization, all the constraints on the control and controlled variables are taken into account, in accordance with the selected quality functional.

In the computation step, which takes up a fixed short segment of the predictive horizon, optimal control is implemented, and the actual state of the object at the end of the step is measured (or derived from the measured variables).

The predictive horizon is shifted one step forward, and this procedure is repeated.

Note the following points.

1. Systems of nonlinear differential equations may be used as the predictive model.
2. Constraints on the control and the components of the state vector may be taken into account.
3. The functional characterizing the quality of control is minimized in real-time.
4. Predict control enquires direct measurement or estimation of the state of the control plant.
5. In general, the predicted behavior of the dynamic control plant will differ from its actual motion.
6. For real-time operation, optimization must be rapid within the permissible delay.
7. Direct implementation of this control algorithm does not guarantee stability of the control plant, which must be ensured by special methods.

3.3.8 *Artificial intelligence in power electronics*

The control methods discussed so far are based on a precise mathematical model of the control plant. When it is difficult or impossible to derive this model, we may resort to artificial intelligence. Artificial intelligence employs information technology—in particular, the following methods.

- Fuzzy logic (Bose, 2000)
- Neural networks (Bose, 2000, 2007)
- Genetic algorithms (Bose, 2000, 2007)

3.3.8.1 *Fuzzy logic*

Fuzzy logic (the theory of fuzzy sets) is a generalization of the classical logic and set theory. In classical Boolean algebra, the characteristic function (indicator, indicator function) χ_A of a variable (the analog of the switching function in power electronics) is binary (Figure 3.16a)

$$\chi_A = \begin{cases} 1, & \text{if } x \in A, \\ 0, & \text{if } x \notin A, \end{cases} \tag{3.117}$$

where A is the set of values of variable x, $a \le x \le c$.

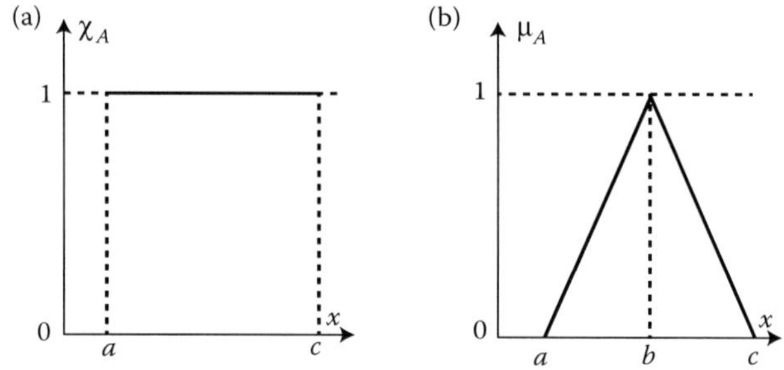

Figure 3.16 (a) Characteristic function and (b) membership function of the variable x.

Transformation by means of the characteristic function may be written in the following form:

$$x_{\text{out}}(t) = \chi_A x_{\text{in}}(t), \tag{3.118}$$

where $x_{\text{in}}(t)$ is the variable to which the characteristic function is applied and $x_{\text{out}}(t)$ is the result of applying the characteristic function.

In fuzzy logic, the characteristic function, sometimes known as the membership function and usually denoted by μ_A, may take any value in the interval [0, …, 1]. This allows us to take account of the fuzziness of the estimates, often expressed in terms of letters.

The membership function is formulated by expert assessment. More than 10 standard curves are used for the specification of the membership function; for example, triangular, trapezoidal, and Gaussian functions. The simplest triangular function is most commonly used (Figure 3.16b). It is specified by the analytical formula

$$\mu_A = \begin{cases} 0, & \text{if } x \le a, \\ (x-a)/(b-a), & \text{if } a \le x \le b, \\ (c-x)/(c-b), & \text{if } b \le x \le c, \\ 0, & \text{if } c \le x, \end{cases} \tag{3.119}$$

where $[a, c]$ is the range of x, and b is the maximum possible value of x.

Operations analogous to those in Boolean algebra may be applied to the membership function.

- Summation (OR) $\mu_{A \vee B} = \max\left[\mu_A, \mu_B\right]$
- Multiplication (AND) $\mu_{A \wedge B} = \min\left[\mu_A, \mu_B\right]$
- Negation (NOT) $\bar{\mu}_A = 1 - \mu_A$

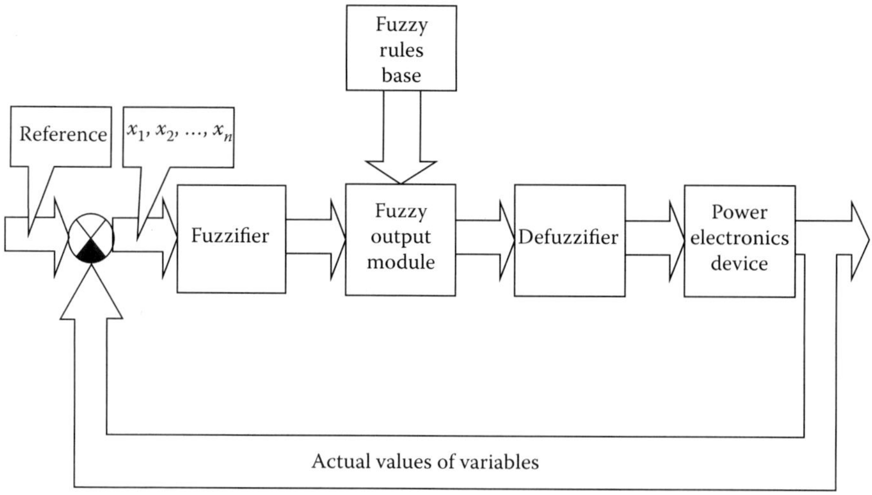

Figure 3.17 Structure of the fuzzy control system for a power electronic device.

Control using fuzzy logic is based on feedback (Figure 3.17). The control design includes three basic steps.

1. *Fuzzification*: the selection of a set of input signals for the control systems (usually the control errors), determination of the number of linguistic variables and the corresponding ranges of the input variables, and the conversion of each range to fuzzy format, which entails the selection of a membership function $\mu(x)$ for each range (Figure 3.18).

   The notation for linguistic variables in the theory of fuzzy sets takes a conventional form, with two letters. Each letter characterizes

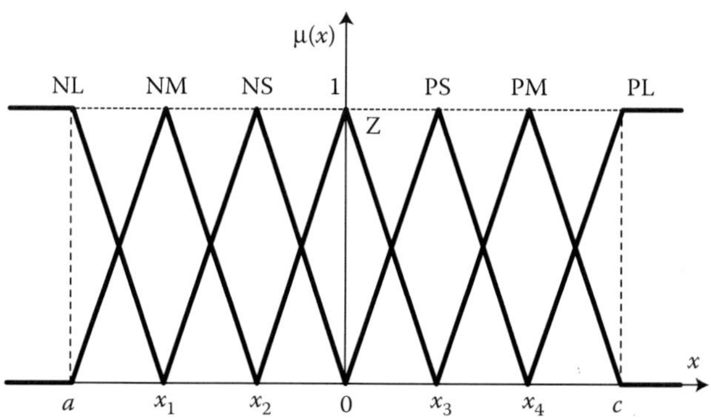

Figure 3.18 Division of the variation of x into ranges with triangular affiliation functions: $[-\infty,x_1]$, the NL range; $[a,x_2]$, the NM range; $[x_1,0]$, the NS range; $[x_2,x_3]$, the Z range; $[0,x_4]$, the PS range; $[x_3,c]$, the PN range; and $[x_4,\infty]$, the PL range.

a property of the range. The first denotes the sign of the range: N, negative; Z, zero; P, positive. The second characterizes the extent of the range: S, small; M, medium; L, large. Thus, NL is a negative large range; NM a negative medium range; and PL a positive large range. There may be any number of such ranges. However, larger numbers require experts formulating the rules for all combinations of input variables to have considerably greater experience.

2. *Formulation of the control in fuzzy-logic terms*: to this end, we generate a framework of fuzzy rules based on the classical IF–THEN operator, which designates an operation that only occurs on satisfaction of some conditions. These rules are formulated on the basis of expert assessments and establish a correspondence between the ranges of control variables and the ranges of control: for example, if $x_1 \in$ Z and $x_2 \in$ NS, then $u \in$ NS. For two input variables, these rules may be reduced to a tabular form (Table 3.4). In each cell of the table, the corresponding range of the control is recorded.

   Then, on the basis of information regarding the error in regulating the variable in accordance with these rules, the membership function of the control signal is truncated. The value of x at any time belongs simultaneously to several linguistic variables and will have a particular value of the membership function for each one. Within the framework of each rule, with each combination of input variables, the maximum value of the membership function is calculated in accordance with the conversion rules for logical variables. In other words, the membership function of the control is converted from a triangle into a trapezium. For two measurable variables, this approach is clearly illustrated in Figure 3.19. One combination of numerical values $x_1 = d$ and $x_2 = f$ corresponds to four combinations of linguistic variables, that is, to four fuzzy rules.

   a. Rule 1: if $x_1 \in$ Z and $x_2 \in$ NS, then $u \in$ NS.
   b. Rule 2: if $x_1 \in$ Z and $x_2 \in$ Z, then $u \in$ Z.

Table 3.4 Table of Fuzzy Rules

x_1 \ x_2	NL	NM	NS	Z	PS	PM	PL
NL	NL	NL	NL	NL	NM	NS	Z
NM	NL	NL	NL	NM	NS	Z	PS
NS	NL	NL	NM	NS	Z	PS	PM
Z	NL	NM	NS	Z	PS	PM	PL
PS	NM	NS	Z	PS	PM	PL	PL
PM	NS	Z	PS	PM	PL	PL	PL
PL	Z	PS	PM	PL	PL	PL	PL

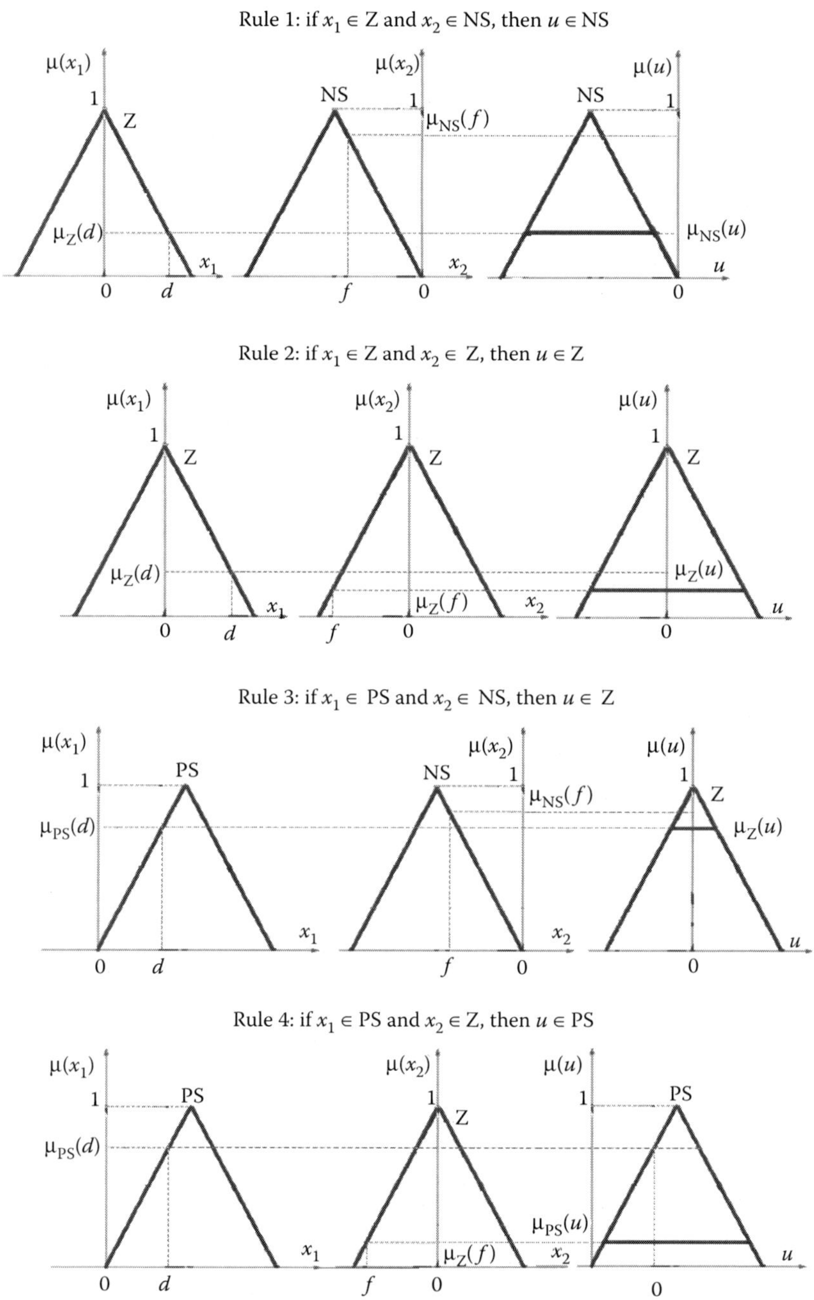

Figure 3.19 Determining the membership function of the control for each fuzzy rule.

c. Rule 3: if $x_1 \in$ PS and $x_2 \in$ NS, then $u \in$ Z.
d. Rule 4: if $x_1 \in$ PS and $x_2 \in$ Z, then $u \in$ PS.

For each of these fuzzy rules, we calculate the truncated membership function. As each combination is possible—in other words, the OR operation applies—the membership function for each combination of numerical values of x_1 and x_2 is the sum of the truncated membership functions for each fuzzy rule. The final operation in this step is to obtain the range of membership functions of the control for the given combination of x_1 and x_2 (Figure 3.20).

3. *Defuzzification*: conversion of the fuzzy-logic variable to a physical control for the power electronic device. Various defuzzification methods are possible. The most common is the mean-center or centroid method. The geometric interpretation of this method is that the physical control selected is the control corresponding to the center of gravity of the membership function range

$$u(d,f) = \frac{\int_{u_{\max}}^{u_{\min}} u\mu(u)\,du}{\int_{u_{\max}}^{u_{\min}} \mu(u)\,du}, \tag{3.120}$$

where $u_{\min}$ and $u_{\max}$ are the boundaries of the affiliation function with the specified values of x_1 and x_2.

3.3.8.2 *Neural networks*

In this approach, a controller that mimics a biological neural network is designed. An artificial neural network is based on an artificial neuron, that is, a simplified model of a natural neuron (Figure 3.21).

Inputs $(x_0, x_1, \ldots, x_n)$ are sent to the input of the neuron. Here x_0 is the displacement signal and $x_1, \ldots, x_n$ provide the information required for

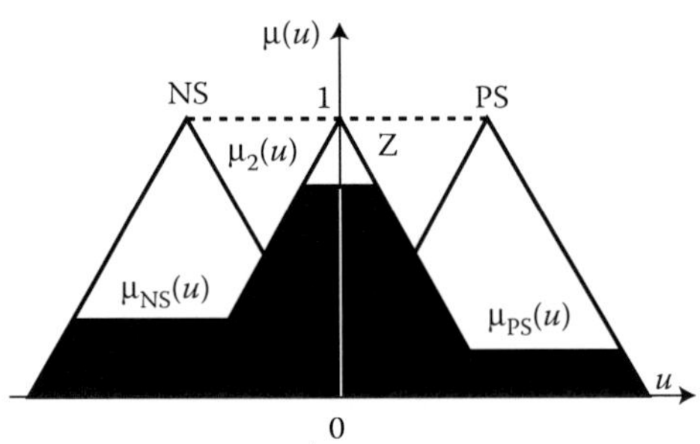

Figure 3.20 Range of the membership function for the control.

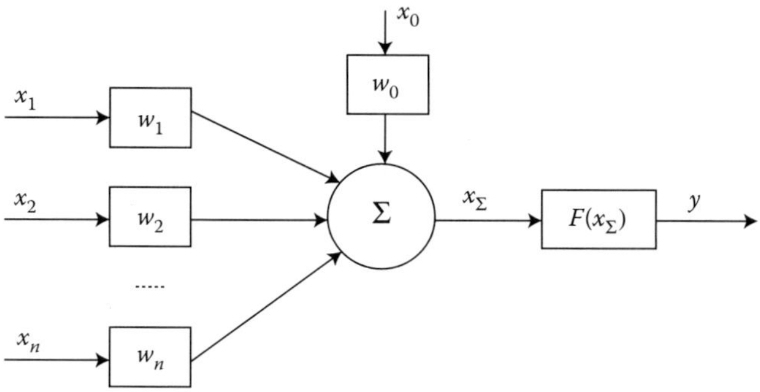

Figure 3.21 An artificial neutron.

the solution of the control problem. These signals, with corresponding weighting factors w_0, w_1, …, w_n, are summed and sent to the input of the module, calculating the transfer function (also known as the activation function or triggering function). This transfer function, which may be linear or nonlinear, completely characterizes the properties of the neuron. In Figure 3.22, we show the most common transfer functions.

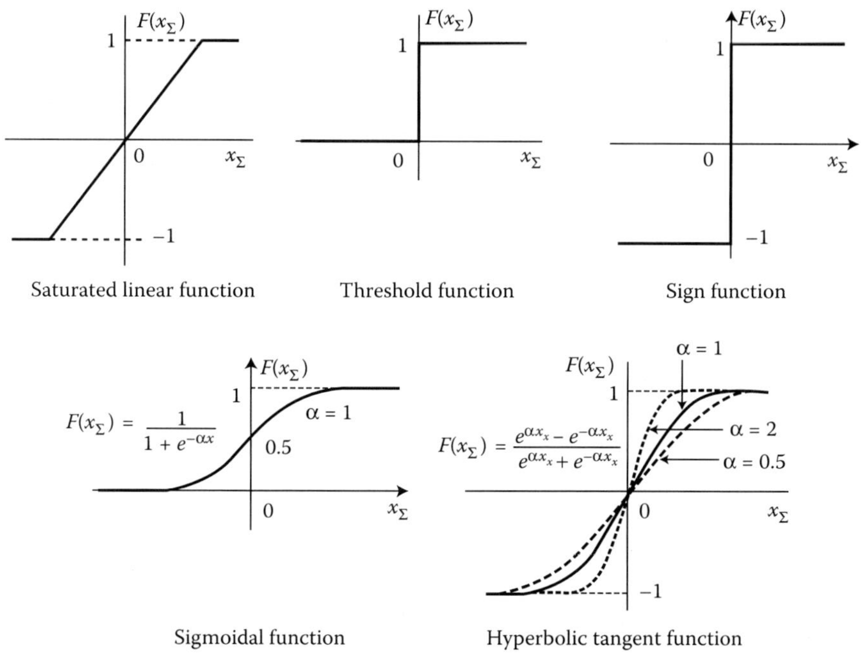

Figure 3.22 Transfer function of a neuron.

The output signal of the module calculating the transfer function is also the neuron's output signal. Thus, in mathematical terms, the neuron is some nonlinear function of a linear combination of input variables

$$y = F\left(\sum_{i=0}^{n} x_i\right), \tag{3.121}$$

where i is the number of the neuron's input variable.

By connecting the outputs of some neurons to the inputs of others, we obtain neural networks. Many such networks exist such as the perceptron, the adaptive neural network, the perceptron with dynamic links, and a network of radial basis functions. At present, about 90% of the neural networks employed are feedforward networks. In power electronics and electric drives, wide use is made of the multilayer (usually three-layer) perceptron with dynamic links.

Combinations of neurons to which the same combination of input signals is sent are known as layers of the neural network. An exception is the first layer, to each neuron of which one input signal is supplied. To eliminate the influence of the units of measurements, all the input signals are written as relative variables (Section 3.1.3). All the subsequent transformations also employ relative variables. The network's output variables are converted back into physical units. The input neurons usually have a linear transfer function, and the weighing factor of the input signal is one.

Each neuron in the second layer (the hidden layer) receives signals from all the input neurons and the displacement signal. The transfer function of these neurons is a hyperbolic tangent in the case of bipolar input signals and a sigmoid function in the case of a unipolar signal. The third layer (the output layer) contains neurons with a linear transfer function. The number of neurons in each layer except the first is established experimentally. In Figure 3.23, we show a 3–5–2 three-layer neural network. This network contains three input neurons, five neurons in the hidden layer, and two output neurons.

Before use, the neural network must be tuned up. This process, known as learning, involves sending successive values of the input vector from a representative training set to the input of the neural network. This set contains P pairs of values of the input vector and the corresponding output vector of the real control plant, for which the neural model is being designed. The weighing factors $w_0, w_1, \ldots, w_n$ of the neurons in the hidden and output layers are varied until the set of output-vector values is close to the requirements. We may assess the quality of training in two ways.

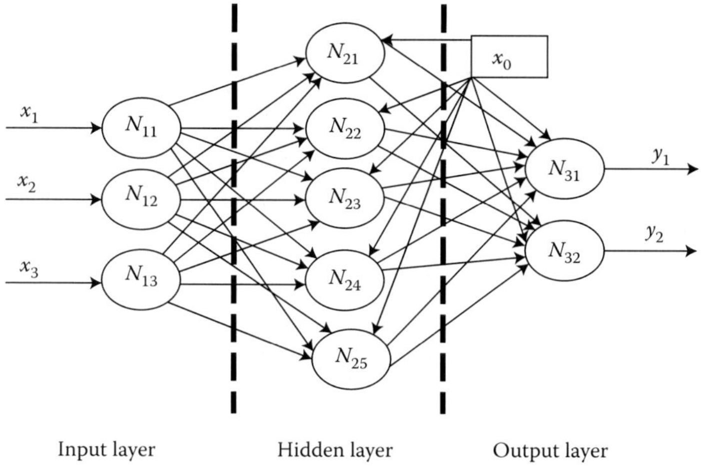

Figure 3.23 A 3–5–2 three-layer neural network.

1. On the basis of the least squares of the output-vector error

$$\sum_{p=1}^{P}\sum_{i=1}^{Q}(y_{iz} - y_i)^2 \rightarrow \min, \tag{3.122}$$

   where p is the number of the set of input-vector values, i the number of the output neuron, Q the number of output neurons, y_i the output value of output neuron i, and y_{zi} the desired output value of output neuron i with the selected set of input-vector values.
2. On the basis of the mean-square error

$$\frac{\left[\sum_{p=1}^{P}\sum_{i=1}^{Q}(y_{iz} - y_i)^2\right]}{Q} \rightarrow \min^{\cdot} \tag{3.123}$$

The Levenberg–Marquardt method, which ensures rapid convergence, most often determines the weighing factors.

With specified structure of the neural network, training ends on satisfying Equation 3.122 or 3.123. If we specify the square of the error in Equation 3.122 or the mean-square error in Equation 3.123, training also includes selection of the network structure—in other words, an increase in the number of neurons in the hidden layer, as the number of neurons in the input and output layers is determined by the physical properties of the object. This process continues until the calculated error is less than a specified value.

At the end of that process, training is regarded as complete. The network parameters are fixed, and the network is prepared for use in the control system. As the training set is representative, the neural network is able, after training, to adequately describe the behavior of the real control plant.

In this way, we obtain a static neural model of the object of control, which may be used for the analysis of the power electronic device's behavior and design of the control algorithms if the processes in the device are relatively slow.

Where we need to obtain a neural model for a dynamic system—in other words, where we need to take account of the control plant's dynamic properties—we use recurrent neural networks in which the input signals consist of information not only regarding the current input variables but also regarding the input and output variables in previous steps.

Such a neural controller is trained directly on the basis of the input and output data of the control plant, by means of an inverse neural model. In the limiting case, this model may serve as the controller, when the transfer function of the power electronic device with the neural controller is one and the output variable matches its specification. A significant deficiency of this approach is that identifying and adjusting the controller involves a complex multiparametric search for an extremum.

3.3.8.3 Genetic algorithms

Genetic algorithms may be used to find the global optimum required in the training of neural networks. In genetic algorithms, natural mechanisms are used for the recombination of genetic information, so as to ensure adaptation within a population. The algorithms represent stochastic heuristic optimization methods based on evolution by natural selection and operate with a set of individuals (a population). Each individual is a possible solution of the given problem and is assessed by a measure of its fitness, in terms of how well it corresponds to the solution of the problem. In nature, this is equivalent to assessing how effective the organism is in the competition for resources. The fittest individuals are able to reproduce, by crossbreeding with other individuals in the population. This generates new individuals, with a new combination of characteristics inherited from the parents. The least-adapted individuals are least likely to reproduce, so that their properties will gradually disappear from the population in the course of evolution. There will sometimes be mutations or spontaneous changes in the genes.

Thus, from generation to generation, good characteristics spread throughout the entire population. The interbreeding of the fittest individuals means that the most promising sections of search space are investigated. Ultimately, the population will converge to the optimal solution of the problem. The advantage of the genetic algorithm is that it fines approximately optimal solutions in a relatively short time.

Thus, the genetic algorithm used in searching for a global minimum in the case of a neural network, say, includes the following components:

- The initial population (a set of initial solutions)
- A set of operators (rules for the generation of new solutions based on the previous population)
- A target function assessing the fitness of the solutions (also known as the fitness function)

The initial population contains several individuals with particular chromosomes. Each gene in the chromosome conveys information regarding the value of a specific attribute of the object; for instance, the weighting factor of a specific neuron. All subsequent operations of the genetic algorithm occur at the level of the genotype. Accordingly, we may dispense with information regarding the internal structure of the object. This accounts for the wide use of this method.

The standard set of operators for all types of genetic algorithms consists of selection (reproduction), crossover (crossbreeding), and mutation.

Selection. In selection, we choose chromosomes on the basis of their fitness functions. There are at least two popular types of selection operator.

1. In roulette-wheel selection, the individuals are chosen on the basis of n spins of the roulette wheel. The wheel contains one sector for each member of the population. The size of sector i, corresponding to individual i, is proportional to the relative value of the fitness function for that individual

$$P_{\text{sel}}(i) = \frac{f(i)}{\sum_{i=1}^{N} f(i)}, \tag{3.124}$$

   where $f(i)$ is the value of the target function for individual i. The members of the population with the highest fitness will be selected with greater probability than those with low fitness.
2. In tournament selection, n individuals are selected on the basis of n tournaments. Each tournament includes k members of the population and selects the best participant. Most commonly, $k = 2$.

Crossover. In crossover, two (or more) chromosomes in the population exchange segments. We may distinguish between single-point selection and multipoint selection. In single-point selection, a division point is randomly selected. The division point is a section between adjacent genes of the chromosome. Both parental structures are broken into two segments

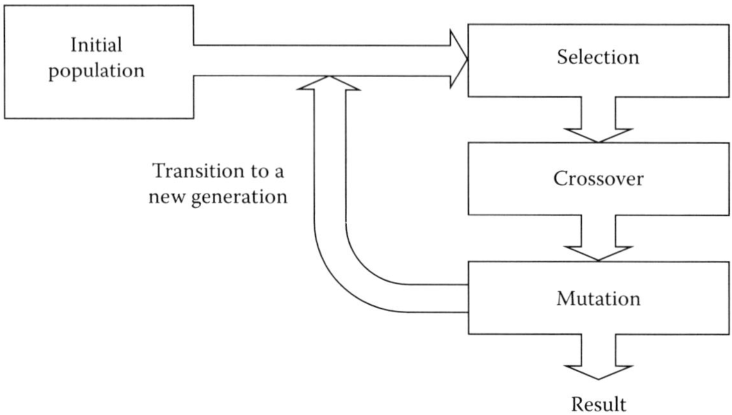

Figure 3.24 A genetic algorithm.

at this point. Then the corresponding segments of the different parents are spliced together, and two new genotypes are produced.

Mutation. A mutation is a random change in part of the chromosome.

The genetic algorithm operates iteratively. The process continues for a specified number of generations or until some other termination condition is satisfied. Each generation of the genetic algorithm includes selection, crossover, and mutation (Figure 3.24).

Appendix 3A: Newton's binomial formula

Newton's binomial formula permits expansion of the sum of two variables as terms of integral non-negative order.

The coefficients of Newton's binomial formula are obtained from Pascal's triangle

$$\begin{array}{ccccccccccccc}
 & & & & & 1 & & & & & & & i=0 \\
 & & & & 1 & & 1 & & & & & & i=1 \\
 & & & 1 & & 2 & & 1 & & & & & i=2 \\
 & & 1 & & 3 & & 3 & & 1 & & & & i=3 \\
 & 1 & & 4 & & 6 & & 4 & & 1 & & & i=4 \\
 & \swarrow\searrow & + & \swarrow\searrow & + & \swarrow\searrow & + & \swarrow\searrow & + & \swarrow\searrow & & & \\
1 & & 5 & & 10 & & 10 & & 5 & & 1 & & i=5
\end{array}$$

Each row of the triangle corresponds to a specific order i of the polynomial; the values in the rows correspond to the coefficients in the expansion. The triangle is constructed from the top down. In other words, from the polynomial of zero order at the top, each subsequent row corresponds to an increase in the order by 1. The arrows show which operations are performed. Specifically, each number is added to each adjacent term.

Appendix 3B: Solution of differential equations

1. **First-order equation**
   Consider the equation

$$a_1 \frac{\mathrm{d}x}{\mathrm{d}t} + a_0 x = f^i(t) + b_e^i e^i. \tag{B.1}$$

The corresponding homogeneous equation is

$$a_1 \frac{\mathrm{d}x_{\mathrm{es}}}{\mathrm{d}t} + a_0 x_{\mathrm{es}} = 0. \tag{B.2}$$

Separating the variables, we obtain

$$a_1 \frac{\mathrm{d}x_{\mathrm{es}}}{x_{\mathrm{es}}} = a_0\, \mathrm{d}t. \tag{B.3}$$

If we integrate the left and right sides, we find that

$$\begin{aligned} a_1 \ln x_{\mathrm{es}} &= -a_0 t + C^*, \\ x_{\mathrm{es}} &= C\, \mathrm{e}^{-(a_0/a_1)t}, \end{aligned} \tag{B.4}$$

where C is a constant of integration.
The complete solution is sought in the form

$$x = C(x)\, \mathrm{e}^{-(a_0/a_1)t}. \tag{B.5}$$

Then

$$\frac{\mathrm{d}x}{\mathrm{d}t} = \frac{\mathrm{d}C(x)}{\mathrm{d}t} \mathrm{e}^{-(a_0/a_1)t} - C(x)\left(\frac{a_0}{a_1}\right)\mathrm{e}^{-(a_0/a_1)t}. \tag{B.6}$$

After substitution in Equation B.1, we obtain

$$\begin{aligned} a_1\left[\frac{\mathrm{d}C(x)}{\mathrm{d}t}\mathrm{e}^{-(a_0/a_1)t} - C(x)\left(\frac{a_0}{a_1}\right)\mathrm{e}^{-(a_0/a_1)t}\right] + a_0 C(x)\mathrm{e}^{-(a_0/a_1)t} &= f^i(t) + b_e^i e^i, \\ \mathrm{d}C(x) &= \frac{1}{a_1}\left[f^i(t) + b_e^i e^i\right]\mathrm{e}^{(a_0/a_1)t}\, \mathrm{d}t. \end{aligned} \tag{B.7}$$

Hence

$$C(x) = \frac{1}{a_1}\int\left[f^i(t) + b_e^i e^i\right]\mathrm{e}^{(a_0/a_1)t}\,\mathrm{d}t + C_1. \tag{B.8}$$

Then the general solution is

$$x(t) = \frac{1}{a_1}\left[\int\left[f^i(t) + b_e^i e^i\right]\mathrm{e}^{(a_0/a_1)t}\mathrm{d}t + C_1\right]\mathrm{e}^{-(a_0/a_1)t}, \tag{B.9}$$

where C_1 is a constant of integration determined from the initial condition x_0

$$C_1 = x_0(t) - \frac{1}{a_1}\left\{\int\left[f^i(t) + b_e^i e^i\right]\mathrm{e}^{(a_0/a_1)t}\,\mathrm{d}t\right\}_{t=0}. \tag{B.10}$$

2. **Second-order equation**
   Consider the equation

$$a_2\frac{\mathrm{d}^2x}{\mathrm{d}t^2} + a_1\frac{\mathrm{d}x}{\mathrm{d}t} + a_0x = f^i(t) + b_e^i e^i. \tag{B.11}$$

The corresponding homogeneous equation is

$$a_2\frac{\mathrm{d}^2x_{\mathrm{es}}}{\mathrm{d}t^2} + a_1\frac{\mathrm{d}x_{\mathrm{es}}}{\mathrm{d}t} + a_0x_{\mathrm{es}} = 0. \tag{B.12}$$

We write the solution in the form

$$x_{\mathrm{es}}(t) = \mathrm{e}^{rt}. \tag{B.13}$$

Then, we may write Equation B.11 as follows:

$$e^{rt}\left(a_2r^2 + a_1r + a_0\right) = 0. \tag{B.14}$$

As the function $e^{rt} \neq 0$, it will be the solution of a differential equation if r is the root of Equation B.14, which is known as the characteristic equation.

In considering the roots, we may distinguish between three cases.

1. The roots are real and unequal when $D = a_1^2 - 4a_0a_2 > 0,\ r_1, r_2$. The solution is sought in the form

$$x_{es}(t) = C_1\, e^{r_1 t} + C_2\, e^{r_2 t} \tag{B.15}$$

2. The roots are real and equal when $D = a_1^2 - 4a_0a_2 = 0,\ r_1, r_2$. The solution is sought in the form

$$x_{es}(t) = (C_1 + C_2 t)e^{r_1 t}. \tag{B.16}$$

3. The roots are complex conjugate when $D = a_1^2 - 4a_0a_2 < 0$, $r_1 = \alpha + j\beta$ and $r_2 = \alpha - j\beta$. The solution is sought in the form

$$x_{es}(t) = (C_1\cos\beta t + C_2\sin\beta t)e^{\alpha t}. \tag{B.17}$$

In deriving this real solution, we take into account that each term of a complex solution is also a solution, according to the Euler formula

$$e^{(\alpha + j\beta)t} = e^{\alpha t}(\cos\beta t + j\sin\beta t). \tag{B.18}$$

The complete solution is sought in the form given earlier. We find the coefficients $C_1(t)$ and $C_2(t)$ by means of Lagrangian multipliers. In their selection, we employ the following conditions.

- The expression for the derivative dx/dt is the same as in the case of constant coefficients

$$\frac{dC_1}{dt}x_1 + \frac{dC_2}{dt}x_2 = 0, \tag{B.19}$$

  where x_1 and x_2 are the components of the solution.
- The complete solution chosen is the solution of the differential equation

$$\frac{dC_1}{dt}\frac{dx_1}{dt} + \frac{dC_2}{dt}\frac{dx_2}{dt} = f^i(t) + b_e^i e^i. \tag{B.20}$$

The solutions of Equations B.19 and B.20 are derivatives of the coefficients $C_1(t)$ and $C_2(t)$. These coefficients are determined by integration. The existing constants of integration are determined from the initial values x_0 and dx_0/dt.

References

Akagi, H., Watanabe, E.H., and Aredes, M. *2007. Instantaneous Power Theory and Applications to Power Conditioning*, 379 pp. Hoboken, NJ: John Wiley & Sons Inc.

Bose, B.K. 2000. Fuzzy logic and neural network. *IEEE Industrial Application Magazine*, May/June 2000, 57–63.

Bose, B.K. 2007. Neural network applications in power electronics and motor drives—An introduction and perspective. *IEEE Trans. Ind. Electron.*, 54(1), 14–33.

Doetsch, G. 1974. *Introduction to the Theory and Application of the Laplace Transformation*, 326 p. Berlin, Heidelberg: Springer.

Isermann, R. 1981. *Digital Control Systems*, 566 pp. Berlin: Springer-Verlag.

Kwakernaak, H. and Sivan, R. 1972. *Linear Optimal Control Systems*, 608 pp. New York: John Wiley & Sons Inc.

Leonhard, W. 2001. *Control of Electrical Drives*, 460 pp. Berlin: Springer.

Linder, A., Kanchan, R., Stolze, P., and Kennel, R. 2010. *Model-based Predictive Control of Electric Drives*, 270 pp. Göttingen: Cuvillier Verlag.

Mohan, N., Underland, T.M., and Robbins, W.P. 2003. *Power Electronics: Converters, Applications and Design*, 3rd edn, 824 pp. New York: John Wiley & Sons Inc.

Rodriguez, J. and Cortes, P. 2012. *Predictive Control of Power Converters and Electrical Drives*, 244 pp. Chichester, UK: Wiley-IEEE Press.

Rozanov, Ju.K., Rjabchickij, M.V., Kvasnjuk, A.A. 2007. *Power electronics: Textbook for universities*. 632 p. Moscow: Publishing hous MJeI (in Russian).

Ryvkin, S. and Palomar Lever, E. 2011. *Sliding Mode Control for Synchronous Electric Drives*, 208 pp. Leiden: CRC Press Inc.

Tsypkin, Ya. 1984. *Relay Control Systems*, 530 pp. Cambridge: Cambridge University Press.

Utkin, V., Shi, J., and Gulder, J. 2009. *Sliding Mode Control in Electro-Mechanical Systems*, 2nd Edn, 503 pp. Boca Raton: CRC Press.

chapter four

Line-commutated converters

4.1 Introduction

The operating principle of a converter depends on the type of power switches and commutation methods employed. We consider two groups of power electronic switches.

- Incompletely controlled switches
- Self-commutated (full controlled) switches

The first group includes diodes, whose controllability is limited because they are switched under the action of forward voltage, and silicon-controlled rectifiers (thyristors). The second group contains all electronic switches that are turned on and off by currents or voltages arriving at their control input.

A fundamental difference between the two groups is the commutation method. In an electronic converter, commutation is understood to mean the transfer of current from one or more simultaneously conducting switches to other switches during a finite interval in which all the switches being turned on and off are in the conducting state. For the power electronic switches in the first group, commutation is possible by means of an ac voltage such as a grid voltage. For single-throw thyristors, such commutation is said to be natural if the conducting thyristors are turned off as a result of polarity reversal of an external supply voltage. Therefore, converters with power switches from the first group are sometimes known as line-commutated converters (International Electrical-Engineering Dictionary, 1998, IEC, BO050-551). They correspond to the operational principles of many converter circuits and hence will be the focus of attention in the present chapter (Rozanov, 1992; Rozanov et al., 2007). They are also sometimes classified on the basis of less significant characteristics such as the following:

- The rated power (low, moderate, high, etc.)
- The working voltage and current (low-voltage, high-voltage, low-current, high-current, etc.)
- The frequency of the input or output voltage (low-frequency, high-frequency, etc.)
- The number of phases (single-phase, three-phase, multiphase, etc.)

- The modular design principle (multicell, multilevel, etc.)
- The method of thyristor commutation (with capacitor commutation, commutation by an LC circuit, commutation under the action of load resonances, etc.)
- The presence of resonant circuits to reduce switching losses (quasi-resonant dc converters, etc.)
- The control method (in terms of the input or the output, modification of the switch control algorithm, etc.)

In practice, other aspects of converter operation are sometimes used for classification purposes. However, they usually lack clear definitions and are not mentioned in the corresponding standards.

4.2 Rectifiers

4.2.1 The rectification principle

In electronic power rectification, the ability of power electronic switches to conduct unidirectional current is employed to convert alternating current to direct current without significant energy losses. The specifics of the rectification process will depend on factors such as the following:

- The type of switch and its control method
- The dc load
- The characteristics of the ac source

In considering the rectification principle, we make the following assumptions:

1. A sinusoidal voltage source with stable frequency is connected to the dc side.
2. The switches employed are diodes VD or thyristors VS with ideal characteristics.
3. The load consists of specific point components.
4. There are no additional losses in the rectification circuit.

For a more detailed study of the factors affecting the rectification process, we consider the simplest possible circuit, with a single switch; this is known as a half-wave circuit (Figure 4.1a). The switch employed is a diode VD or thyristor VS. If the thyristor is turned on at moments when the firing angle $\alpha = 0$ (Figure 4.1a), the processes in the circuit will correspond to those observed when a diode is turned on. The following loads are considered.

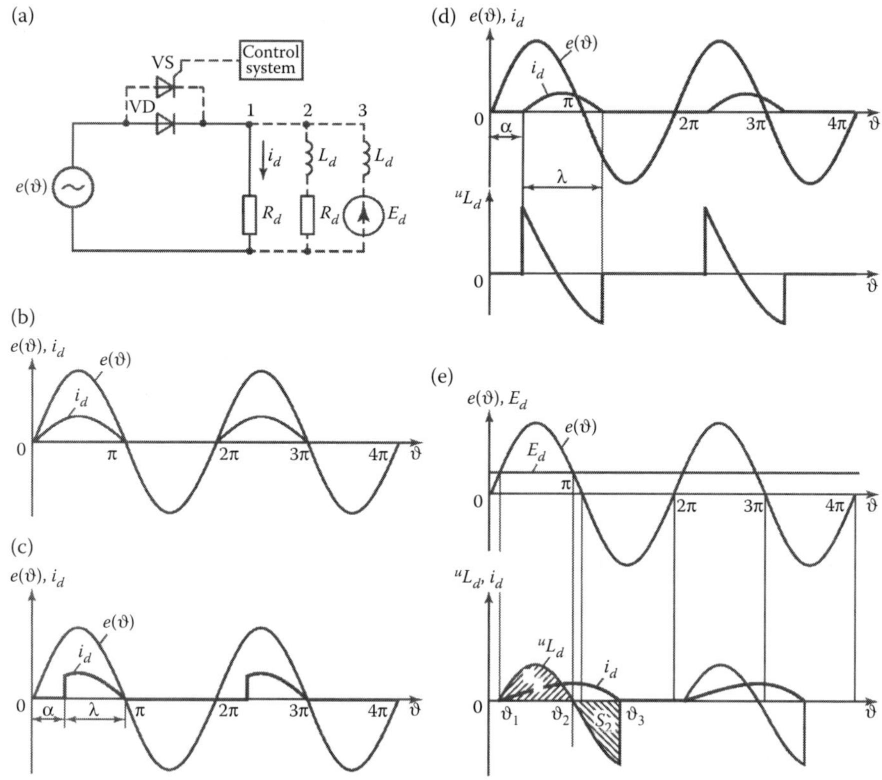

Figure 4.1 (a) A single-phase half-wave rectifier, voltage and current waveforms with an active load when (b) $\alpha = 0$ and (c) $\alpha = \pi/6$, with a resistive–inductive load when (d) $\alpha = \pi/6$, and (e) with a counter-emf load.

- An active load with resistance R_d (branch 1 in Figure 4.1a)
- An active–inductive load with resistance R_d and inductance L_d (branch 2 in Figure 4.1a)
- An opposing dc voltage source E_d with inductance L_d (counter-emf load; branch 3 in Figure 4.1a)

4.2.1.1 Circuit with active load

Here and in what follows, the time diagrams are plotted in terms of the angle $\vartheta = \omega t$, where ω is the angular frequency of the ac source. Thus, the input voltage is $e(\vartheta) = E_m \sin \vartheta$. In the circuit with diode VD, the current i_d begins to flow as soon as the forward voltage is applied. In other words, it conducts from the time $\vartheta = 0$ to $\vartheta = \pi$, when the voltage is zero and the diode is turned off. Negative voltage is applied to diode VD in the next half-period, and it is nonconducting. The current in load R_d reproduces the input voltage during the conducting interval from 0 to π. At time $\vartheta = 2\pi$, the cycle repeats (Figure 4.1b).

When diode VD is replaced by thyristor VS, current flow begins when the control pulse is supplied to the thyristor gate from the control system (CS). The delay after time $\vartheta = 0$ will depend on the firing angle α (Figure 4.1c). The thyristor will be switched off when voltage $e(\vartheta)$ (and hence current i_d) falls to zero, that is, when $\vartheta = \pi$. As a result, current i_d will flow for a shorter time than with a diode, specifically, for time $\lambda = \pi - \alpha$.

In this interval, the current reproduces the form of the voltage $e(\vartheta)$. As a result, periodic unidirectional currents i_d appear at load R_d, which indicates rectification. In other words, a constant component of current I_d will appear in the load R_d when voltage $e(\vartheta)$ is supplied by an ac source.

4.2.1.2 Circuit with resistive–inductive load

The character of the load on the dc side has considerable influence on the rectification process. For example, if a thyristor is turned on at time $\vartheta = \alpha$ in a load containing not only resistor R_d, but also a reactor with inductance L_d (branch 2 in Figure 4.1a), the current i_d will be determined by the equation

$$E_m \sin\vartheta = i_d R_d + L_d \frac{di_d}{dt}, \tag{4.1}$$

which is derived from the equivalent circuit in the presence of thyristor VS. With zero current in the inductance L_d when the thyristor is turned on, the solution of Equation 4.1 takes the form

$$i_d(\vartheta) = \frac{E_m}{\sqrt{R_d^2 + (\omega L_d)^2}}\left(\sin(\vartheta - \varphi) - \sin(\alpha - \varphi)\cdot e^{(-\vartheta+\alpha)/\tau\omega}\right), \tag{4.2}$$

where

$$\varphi = \operatorname{arctg}\frac{\omega L_d}{R_d} \quad \text{and } \tau = \frac{L_d}{R_d}.$$

Figure 4.1d shows the input voltage $e(\vartheta)$ and current i_d when $\alpha = \pi/3$. It is evident that the current i_d continues to flow through the thyristor after voltage $e(\vartheta)$ passes through zero. This is possible because the energy stored in the inductance L_d during the first half-period maintains the current i_d after the voltage reverses the sign until the time $\vartheta = \alpha + \lambda - \pi$, when current i_d is again zero.

4.2.1.3 Counter-emf load

A load in the form of a dc emf with polarity opposite to the switch may also be of practical interests. Such rectification circuits are used, for instance, in battery chargers and in systems for power recuperation from a dc source to an ac grid.

In some operating conditions, a large filter capacitor at the rectifier output may be regarded as a counter-emf source.

Branch 3 in Figure 4.1a corresponds to a half-wave rectifier circuit with diode VD and counter-emf E_d. At time $\vartheta = \vartheta_1$ (Figure 4.1e), the source voltage $e(\vartheta)$ exceeds the counter-emf E_d. Hence, forward voltage is applied to diode VD and it begins to conduct current i_d in the opposite direction to E_d. Under the assumptions already noted, connecting a source with voltage $e(\vartheta)$ to the counter-emf E_d will result in infinite growth in current i_d. To prevent this, a reactor with inductance L_d is introduced in the dc circuit. In that case, the current i_d will be

$$i_d(\vartheta) = \frac{1}{\omega L_d} \int_{\vartheta_1}^{\vartheta} (e(\vartheta) - E_d)\, d\vartheta. \tag{4.3}$$

The period corresponding to flow of current i_d may be divided into two parts an increase in i_d in the interval ϑ_1–ϑ_2 and a decrease in i_d in the interval ϑ_2–ϑ_3. At time ϑ_2, $e(\vartheta)$ is again equal to the counter-emf E_d. The second interval corresponds to voltage regions of opposite polarity at inductance L_d. The integral areas S_1 and S_2 of these regions (shaded in Figure 4.1e) are equal. That corresponds to balance of the stored and consumed energy in inductance L_d

$$\int_{\vartheta_1}^{\vartheta_2} u_L(\vartheta) d\vartheta + \int_{\vartheta_2}^{\vartheta_3} u_L(\vartheta) d\vartheta = 0. \tag{4.4}$$

When $\vartheta = \vartheta_2$, the current i_d is maximum. At specified values of $e(\vartheta)$ and E_d, the replacement of diode VD by a controllable thyristor permits the regulation of i_d by adjusting the firing angle α. This angle corresponds to the switching delay of the thyristor relative to the time ϑ_1 at which a forward voltage is applied to the thyristor.

4.2.2 Basic rectification circuits

We will consider idealized rectification circuits, under the following assumptions:

1. The semiconductor components are ideal. In other words, when they are on, their resistance is zero; when they are off, their conductivity is zero.
2. The semiconductor components are switched on and off instantaneously. In other words, the switching process takes no time.
3. The resistance of the circuits connecting the components is zero.
4. The resistance of the transformer windings (active and inductive), the energy losses in its magnetic system, and the magnetizing current are all zero.

The electromagnetic processes associated with rectification are considered for two types of static load: active and active–inductive. Such loads are typical of most moderate- and high-power rectifiers.

In this section, we consider thyristors operating with active and active–inductive loads when the firing angle $\alpha > 0$. Obviously, when $\alpha = 0$, the processes in the circuit will be the same as for uncontrollable diode-based rectifiers. As the three-phase bridge circuit is most common, we will consider processes with $\alpha = 0$ in that case.

4.2.2.1 Single-phase circuit with center-tapped transformer

A single-phase full-wave circuit with a center tap is shown in Figure 4.2a. The full-wave circuit is sometimes known as a two-cycle or two-phase circuit, as it rectifies both voltage half-waves. In this circuit, the secondary half-windings of the transformer relative to the tap create a system of voltages with a mutual phase shift $\vartheta = \pi$.

We will consider the operation of the circuit with an active load (when switch S in Figure 4.2a is closed). Suppose that, beginning at $\vartheta = 0$, both thyristors are off, and no current flows. We assume that the potential of point a of the secondary winding is positive relative to the tap (point 0), whereas point b is negative. (In Figure 4.2a, this polarity is noted outside the parentheses.) Obviously, with this polarity of the secondary-winding voltage, the forward voltage $u_{\text{VS1}} = u_{a0}$ will be applied to thyristor VS1, whereas inverse voltage u_{b0} will be applied to thyristor VS2. Suppose that at time $\vartheta = \alpha$ (i.e., with delay α relative to the moment when voltage u_{a0} passes through zero), a control pulse is supplied to the control electrode of thyristor VS1. Then VS1 is switched on, and current $i_d = i_{\text{VS1}}$ begins to flow in load R_d under the action of voltage u_{a0}. Beginning at that moment, inverse voltage u_{ab} will be applied to thyristor VS2. Here u_{ab} is the difference between the voltages in the secondary half-windings u_{a0} and u_{b0}.

Thyristor VS1 will be on until the current flow falls to zero. As the load is active and the current passing through the load—and hence through thyristor VS1—is of the same form as voltage u_{a0}, thyristor VS1 is switched off at time $\vartheta = \pi$. As the voltage at the secondary winding

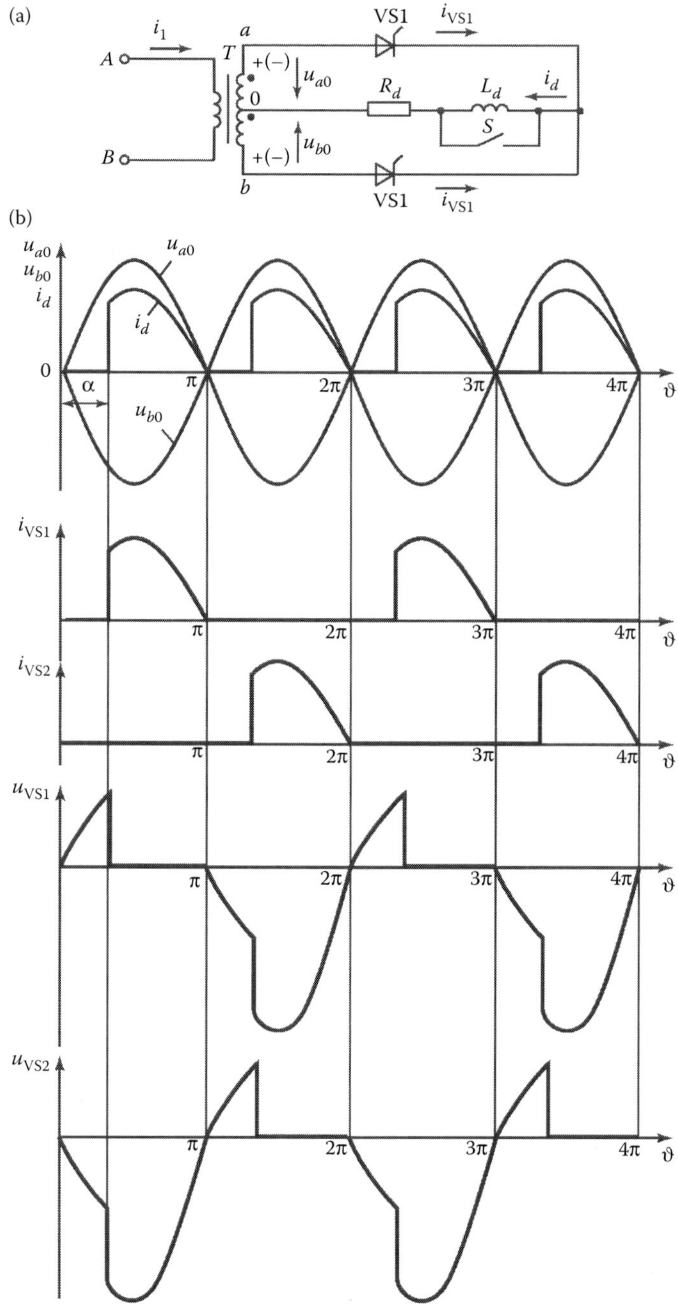

Figure 4.2 (a) A single-phase full-wave rectifier with center-tapped transformer and (b) corresponding voltage and current waveforms.

reverses polarity after half of the period, the supply of a control pulse at time $\vartheta = \pi + \alpha$ switches thyristor VS2 on. These processes are repeated in each period.

The possibility of a specific phase delay α in the moments at which the thyristors are switched on permits change in the output voltage. We measure α from the moments of natural thyristor switching ($\vartheta = 0, \pi, 2\pi, \ldots$), corresponding to the moments when diodes are turned on in an uncontrollable circuit. It is evident from Figure 4.2b that, with an increase in α, the average output voltage U_d will decline. Analytically, this dependence takes the form

$$U_d = \frac{1}{\pi}\int_{\alpha}^{\pi} \sqrt{2}U_2 \sin\vartheta d\vartheta = \frac{\sqrt{2}}{\pi}U_2(1+\cos\alpha), \tag{4.5}$$

where U_2 is the real voltage at the transformer's secondary winding.

If we denote the average rectified voltage for an uncontrolled rectifier ($\alpha = 0$) by U_{d0}, determined from Equation 4.5, we may write

$$U_d = U_{d0}\frac{1+\cos\alpha}{2}. \tag{4.6}$$

According to Equation 4.6, an increase in α from 0 to π reduces the average output voltage from U_d to zero. The dependence of the average output voltage on the firing angle is known as the control characteristic.

The presence of inductance L_d in the dc circuit at $\alpha > 0$ means that current will flow through the thyristor after the voltage at the secondary half-winding passes through zero, on account of the energy stored in the inductance. For example, thyristor VS1 continues to conduct current after the voltage u_{a0} becomes negative (Figure 4.3a). As a result, the period λ of current flow through the thyristor increases, and the rectified voltage will include a section of negative polarity from 0 to ϑ_1, when the current in the thyristors falls to zero. With an increase in L_d, the thyristors will conduct for a time $\lambda = \pi$, and $\vartheta_1 = \alpha$. This corresponds to a boundary-continuous rectified current i_d. In these conditions, each thyristor conducts current for half-period π, with zero current at the beginning and end of each half-period. The inductance L_d at which such behavior is observed is said to be the boundary or critical inductance. With a further increase in L_d or with an increase in the rectifier load, the rectified current remains continuous and its pulsations are smoothed. At large values $\omega L_d/R_d > 5$–10, the current i_d is practically completely smoothed, and rectangular current flows through the thyristors (Figure 4.3b). Obviously, with an increase in α, the area of the negative sections in the rectified voltage increases, and hence the average rectified voltage declines. The average rectified voltage corresponds to its

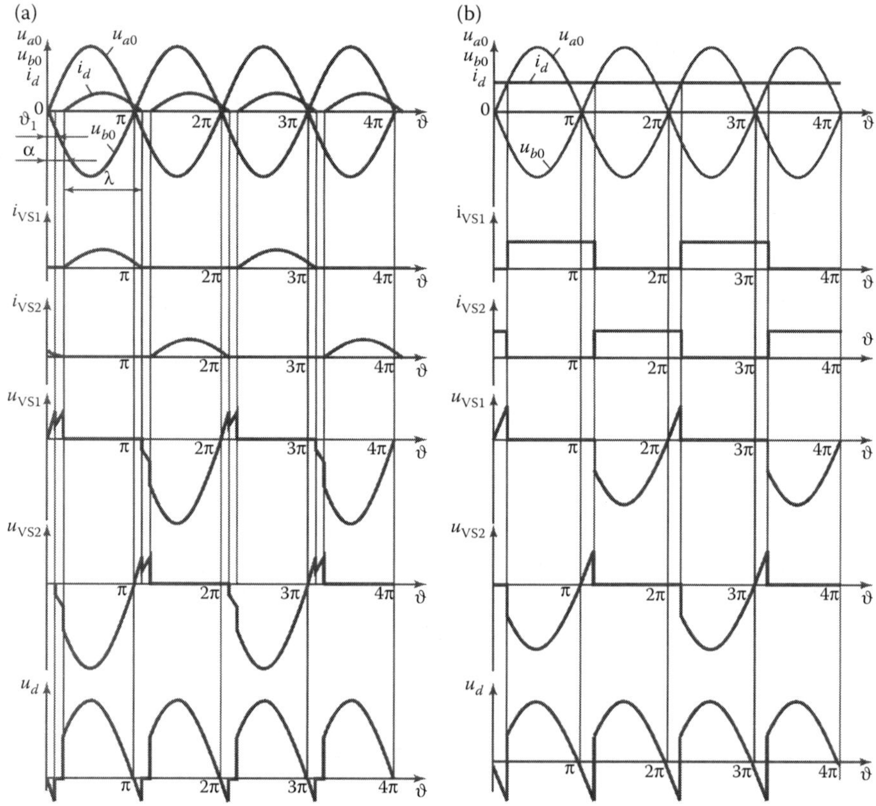

Figure 4.3 Voltage and current waveforms for a single-phase center-tapped full-wave rectifier in the case of a resistive–inductive load ($\alpha = \pi/6$): (a) discontinuous load current and (b) smoothed continuous load current ($\omega L_d = \infty$).

constant component. When $\omega L_d = \infty$, the constant component is applied at resistance R_d and the variable component at inductance L_d.

As the form of the rectified voltage is repeated in the interval from α to $\pi + \alpha$, its average value may be found from the formula

$$U_d = \frac{1}{\pi}\int_{\alpha}^{\pi+\alpha} \sqrt{2}\cdot U_2 \sin\vartheta \, d\vartheta = \frac{\sqrt{2}}{\pi} U_2 \cos\alpha = U_{d0}\cos\alpha. \tag{4.7}$$

According to Equation 4.7, the average rectified voltage is zero when $\alpha = \pi/2$. In that case, the areas of the positive and negative sections in the rectified voltage are equal, and there is no constant component (Figure 4.4). The control characteristic for a resistive–inductive load corresponds to curve 2 in Figure 4.5.

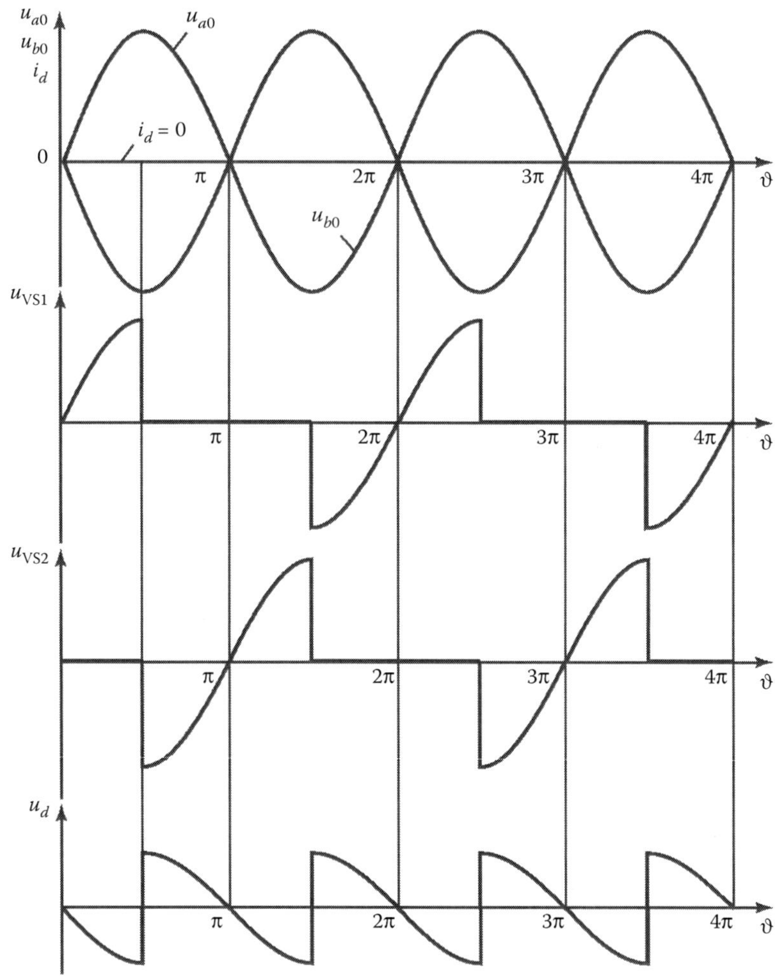

Figure 4.4 Voltage waveforms for a single-phase center-tapped full-wave circuit in the case of a resistive–inductive load and $\alpha = \pi/2$.

If the value of $\omega L_d/R_d$ is such that the energy stored in inductance L_d in the interval when $u_d > 0$ is insufficient for current flow over half of the period, the thyristor that is conducting this current will be switched off before the control pulse is supplied to the other thyristor. In other words, current i_d will be discontinuous. If we compare Figure 4.3a and b, we see that, at the same α, the average rectified voltage will be greater for discontinuous current than for continuous current, due to the decrease in the area of the negative section on the rectified-voltage curve, but less than for rectifier operation with an active load (with

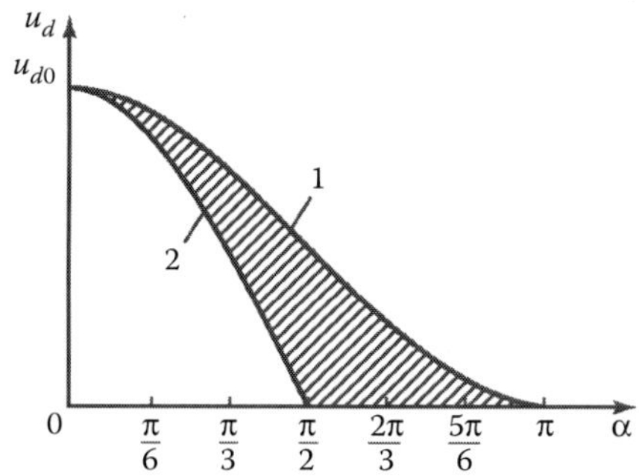

Figure 4.5 Control characteristics of a single-phase full-wave rectifier in the case of an active load (1) and *RL* load (2).

no negative sections). Therefore, with discontinuous current, the control characteristics will be between curves 1 and 2, within the shaded region in Figure 4.5.

With discontinuous current, the transformer and thyristor circuits operate in more challenging conditions, as the effective value of the currents in the circuit elements is greater at the same average rectified current. Therefore, in powerful rectifiers operating with a wide variation of α, the inductance L_d is usually selected so as to ensure the continuous rectified current at near-rated loads.

The parameters of the circuit components are calculated by classical electrical-engineering methods. For example, the average thyristor current is

$$I_{avVS1} = I_{avVS2} = \frac{1}{\pi}\int_0^{\pi} i_{VS}(\vartheta)d\vartheta. \tag{4.8}$$

When $\omega L_d = \infty$, the ideally smoothed constant load current I_d is conducted alternately by the thyristors. Hence

$$I_{avVS1} = I_{avVS2} = \frac{1}{2}I_d. \tag{4.9}$$

If the current i_d is ideally smooth, it is simple to determine the effective and maximum currents and voltages at all the circuit components. Their determination is more complex in the case of poorly smoothed or discontinuous current i_d. In that case, equivalent circuits must be formulated for the thyristor conduction periods.

4.2.2.2 Single-phase bridge circuit

For a single-phase bridge circuit (Figure 4.6) operating with $\alpha > 0$, the voltages and currents at its components are of the same form as in a single-phase center-tap full-wave circuit (Figures 4.2 through 4.5). The basic difference is that a single-phase voltage U_{ab} is supplied, rather than the half-winding voltages U_{a0} and U_{b0}. As a result, two thyristors VS1 and VS3 or VS2 and VS4 participate in the rectification of each voltage half-wave. Therefore, when the firing angle $\alpha = 0$ (or in an uncontrollable diode-based rectifier), the average rectified voltage at the load is

$$U_{d0} = \frac{2\sqrt{2}}{\pi} U_2, \tag{4.10}$$

where U_2 is the effective voltage in the transformer's secondary winding.

Depending on whether the load is active or active–inductive, the average rectified voltage U_d may be calculated as follows.

a. With an active load

$$U_d = U_{d0} \frac{1 + \cos\alpha}{2}, \tag{4.11}$$

where U_{d0} is the average rectified voltage at the output when $\alpha = 0$.

b. With an active–inductive load (when ωL_d is such that the rectified current is continuous)

$$U_d = U_{d0} \cos\alpha. \tag{4.12}$$

The control characteristics of the circuit depend on the ratio $\omega L_d / R_d$ and take the form as in Figure 4.5.

In this case, as in the center-tapped circuit, the power of the components increases with an increase in α in the case of an active load and an active–inductive load with discontinuous currents. This must be taken into account in the calculation and design of the corresponding power components.

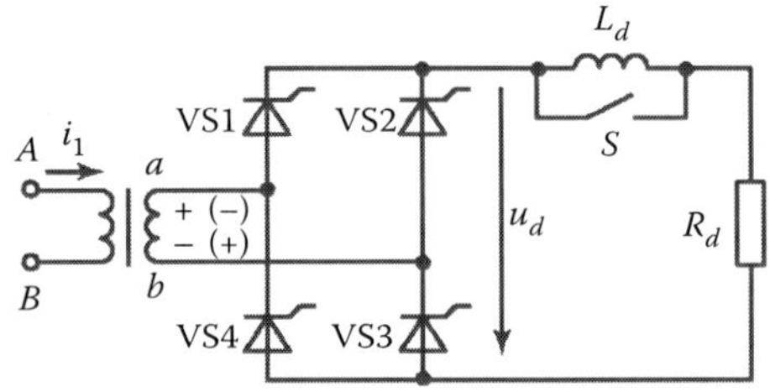

Figure 4.6 A single-phase bridge rectifier.

4.2.2.3 Three-phase circuit with center-tapped transformer

1. *Operation with* $\alpha = 0$. The three-phase circuit with a tap (Figure 4.7) is a three-phase single-cycle circuit, as only one half-wave of the alternating voltage in each phase is rectified. We will consider the operating principle of this circuit for the case in which the primary transformer windings are in a delta configuration, whereas the secondary windings are in a star configuration. First, we assume that switch *S* is closed. In other words, the circuit has an active load. The relations are then refined for the case in which switch *S* is open, on the assumption that $\omega L_d = \infty$.

In the interval $\vartheta_0 < \vartheta < \vartheta_1$ (Figure 4.8), thyristor VS1, connected to phase a, is conducting. Beginning at time ϑ_1, the potential of phase b exceeds that of phase a, and the anode of thyristor VS2 is at a positive voltage relative to the cathode. If a control pulse is supplied to thyristor VS2 at time ϑ_1, it is switched on, whereas thyristor VS1 is switched off under the action of the shutoff voltage u_{ab}. The load current i_d begins to flow through thyristor VS2, which is connected to phase b.

The conducting state of thyristor VS2 continues for a period of 120°, until time ϑ_2, when the potential of phase c exceeds that of phase b and a control pulse is supplied to thyristor VS3. At time ϑ_2, thyristor VS3 begins to conduct, and thyristor VS2 is switched off. Then, at time ϑ_3, current flow resumes at thyristor VS1, and the preceding sequence is cyclically repeated.

Obviously, each thyristor will conduct for a third of the grid voltage period ($2\pi/3$). For the remainder of the period ($4\pi/3$), the thyristor is off and is subject to reverse voltage. Thus, when thyristor VS1 is off, the line voltage u_{ba} is applied to VS1 during the period when thyristor VS2 is conducting, whereas voltage u_{ca} is applied when thyristor VS3 is conducting. As a result, reverse voltage u_{VS1} is applied to thyristor VS1 (Figure 4.8).

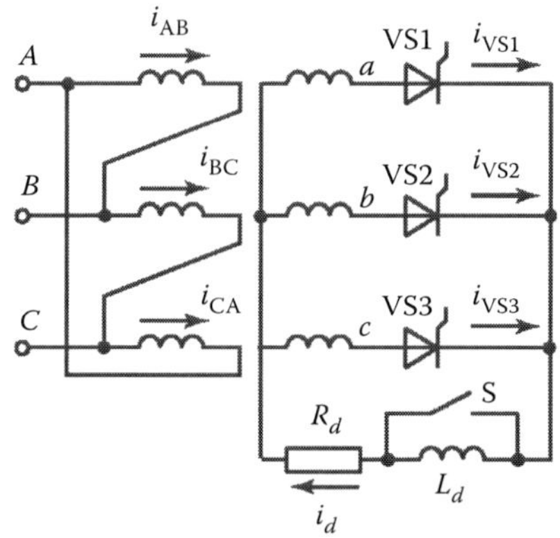

Figure 4.7 A three-phase center-tapped rectifier.

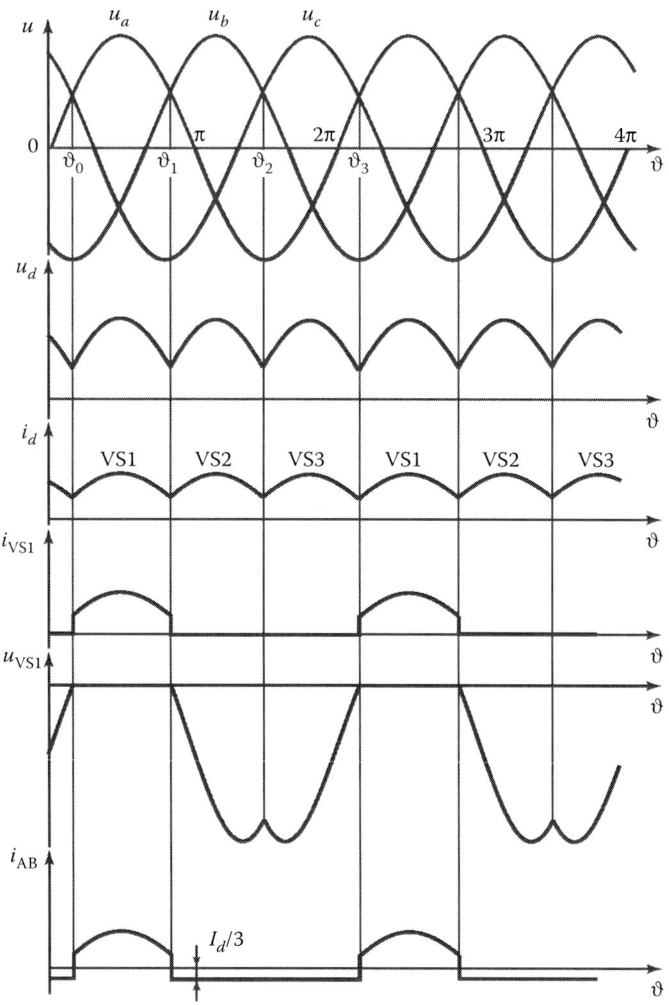

Figure 4.8 Voltage and current waveforms for a center-tapped three-phase rectifier when $\alpha = 0$.

The average rectified voltage is found by integrating the voltage at the transformer's secondary winding over the interval corresponding to repetition of the rectified voltage form

$$U_d = \frac{3}{2\pi} \int_{\pi/6}^{5/6\pi} \sqrt{2} U_2 \sin\vartheta \, d\vartheta = \frac{3\sqrt{6}}{2\pi} U_2 = 1.17 U_2, \tag{4.13}$$

where U_2 is the effective phase voltage at the transformer's secondary winding.

The basic parameters characterizing thyristor operation in the circuit are as follows.

- The circuit coefficient is

$$k = 3\frac{\sqrt{6}}{2\pi}. \tag{4.14}$$

- The maximum reverse voltage at the thyristor (equal to the line voltage at the secondary windings) is

$$U_{\text{Rmax}} = \sqrt{3}\,U_{2m} = \sqrt{6}\,U_2 = \frac{\pi}{3}U_d, \tag{4.15}$$

  where U_{2m} is the amplitude of the phase voltage.
- The maximum thyristor current is

$$I_{\max} = \frac{U_{2m}}{R_d} = \frac{\pi}{3\sqrt{3}}U_d. \tag{4.16}$$

- The average current through the thyristor, given that each thyristor conducts for a third of the period, is

$$I_{\text{avVS}} = \frac{I_d}{3}. \tag{4.17}$$

As the secondary-winding currents in this circuit are pulsating and include a constant component, an induced magnetization flux appears in the transformer magnetic system and can result in magnetic saturation. That entails increase in the calculated transformer power. The currents in the primary windings only contain variable components, as the constant components are not transformed. Therefore, the currents in the primary windings take the form

$$\left.\begin{aligned} i_{\text{AB}} &= \left(i_{\text{VS1}} - \frac{1}{3}I_d\right), \\ i_{\text{BC}} &= \left(i_{\text{VS2}} - \frac{1}{3}I_d\right), \\ i_{\text{CA}} &= \left(i_{\text{VS3}} - \frac{1}{3}I_d\right). \end{aligned}\right\} \tag{4.18}$$

We consider the currents in the thyristors and the transformer windings and the calculated transformer power for a circuit with an active–inductive load, which is most typical of three-phase and multiphase rectification systems. In the case of an active–inductive load, the circuit operation is the same as with an active load, but the current i_d is ideally smoothed, whereas the currents through the thyristors are rectangular. Correspondingly, the currents in the transformer windings are also rectangular. In that case, the curves of the rectified voltage u_d and the reverse voltages at the thyristors are the same as in the case of an active load, and the currents are as follows

$$\left.\begin{aligned} I_{\max} &= \frac{I_d}{3}, \\ I_2 = I_{VS} &= \frac{I_d}{\sqrt{3}}, \\ I_1 &= \frac{1}{k_T}\frac{\sqrt{2}}{3}I_d . \end{aligned}\right\} \tag{4.19}$$

The calculated power of the transformer's primary and secondary windings may be written in the form

$$\left.\begin{aligned} S_1 = 3U_1I_1 &= \frac{2\pi}{3\sqrt{3}}P_d, \\ S_2 = 3U_2I_2 &= \frac{2\pi}{3\sqrt{2}}P_d, \end{aligned}\right\} \tag{4.20}$$

where U_1 and U_2 are the effective phase voltages of the primary and secondary windings, I_1 and I_2 are the effective currents in the primary and secondary windings, respectively, and P_d is the average load power.

2. *Operation with* $\alpha > 0$. In this case, in contrast to an uncontrollable rectifier or a controllable rectifier with $\alpha = 0$, the control pulses reach the thyristors alternately, with a delay α relative to the time at which the sinusoid of the line voltage in the secondary transformer windings passes through zero. The times at which the sinusoidal line voltage passes through zero correspond to intersection of the sinusoidal phase voltages u_a, u_b, and u_c. When the firing angle $\alpha > 0$, different operating conditions may be observed, depending on the type of load and the range of α.

If α varies in the range from 0 to $\pi/6$ (Figure 4.9), the rectified current is continuous for both active and active–inductive loads. The average rectified voltage in this range of α is described as follows:

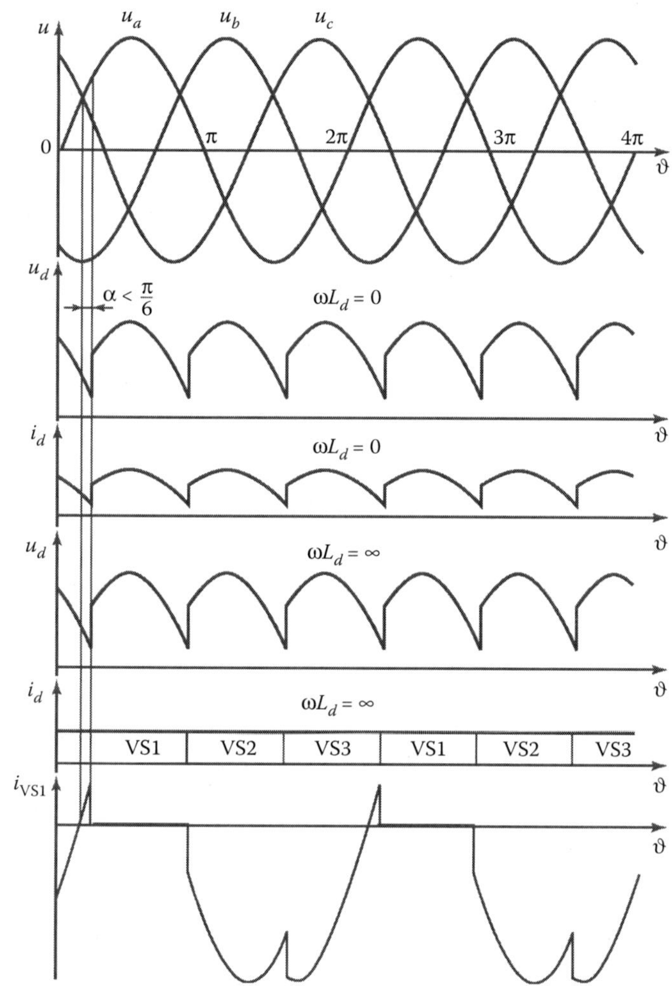

Figure 4.9 Voltage and current waveforms for a center-tapped three-phase rectifier when $\alpha < \pi/6$.

$$U_d = \frac{3}{2\pi}\int_{(\pi/6)+\alpha}^{(5\pi/6)+\alpha} \sqrt{2}U_2 \sin\vartheta \, d\vartheta = \frac{3\sqrt{6}}{2\pi}U_2 \cos\alpha = U_{d0}\cos\alpha. \tag{4.21}$$

When $\alpha = \pi/6$, the instantaneous rectified voltage is zero at the thyristor switching times (Figure 4.10, left). This is said to be a boundary-continuous operation. When $\alpha > \pi/6$, the rectified current i_d becomes discontinuous in the case of active load, and there are sections where the

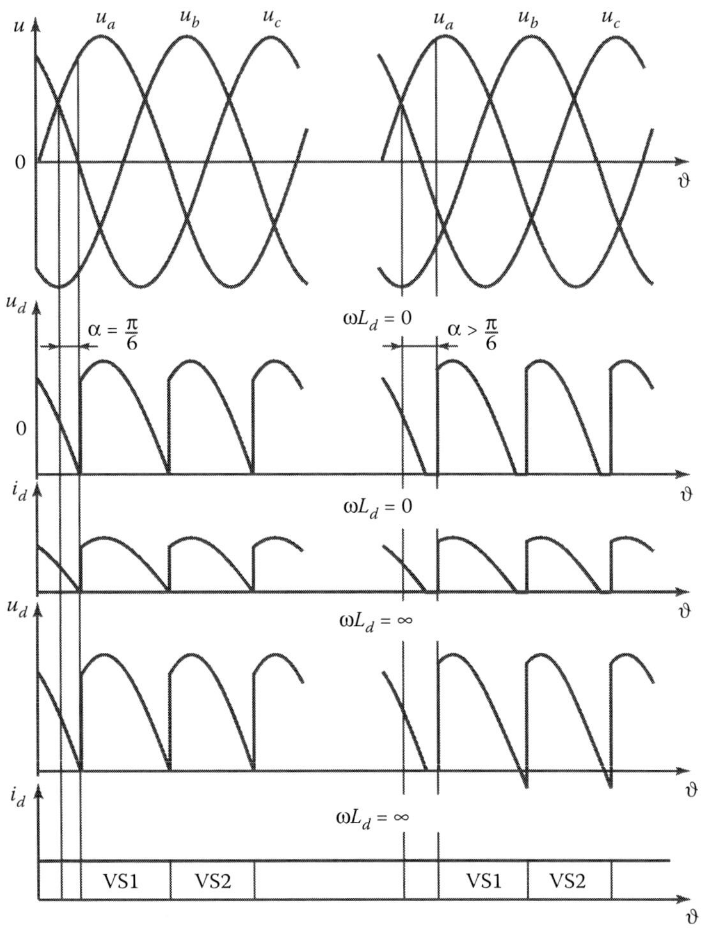

Figure 4.10 Voltage and current waveforms for a center-tapped three-phase rectifier when $\alpha = \pi/6$ and $\alpha > \pi/6$.

rectified voltage u_d is zero (Figure 4.10, right). The interval in which the thyristors are conducting becomes less than $2\pi/3$. In this case, the average voltage is

$$U_d = \frac{3}{\pi}\int_{(\pi/6)+\alpha}^{\pi} \sqrt{2}U_2 \sin\vartheta \, d\vartheta = \frac{3\sqrt{2}}{2\pi}U_2\left[1+\cos\left(\frac{\pi}{6}+\alpha\right)\right]$$

$$= U_{d0}\left[\frac{1+\cos\left(\pi/6+\alpha\right)}{\sqrt{3}}\right]. \quad (4.22)$$

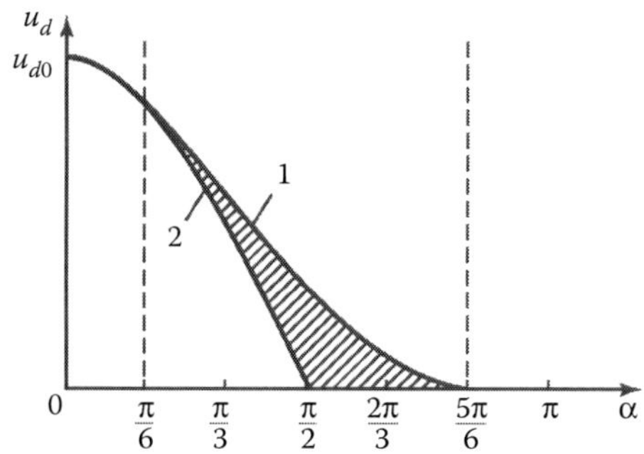

Figure 4.11 Control characteristics for a center-tapped three-phase rectifier with an active load (1) and an *RL* load (2).

With an active–inductive load, the energy stored in the inductance L_d sustains the rectified current i_d in the load even after the rectified voltage becomes negative. If the energy stored in the inductance L_d lasts until the thyristors are switched again, operation with continuous current i_d is possible. When $\omega L_d = \infty$, continuous current will be observed for any α in the range from 0 to $\pi/2$. In that case, the average output voltage U_d may be determined from Equation 4.21. When $\alpha = \pi/2$, the positive and negative sections within the rectified voltage curve are equal in area. That indicates the lack of a constant component in the rectified voltage; in other words, the average value of U_d is zero.

In accordance with the foregoing, we can distinguish two characteristic intervals of α in the control characteristics (Figure 4.11). In the first $(0 < \alpha < \pi/6)$, with either active load or active–inductive load, the control characteristics correspond to Equation 4.21. In the second $(\pi/6 < \alpha < 5\pi/6)$, with an active load, the control characteristic is analytically described by Equation 4.22, and the average value of U_d is zero when $\alpha = 5\pi/6$. In the case of an active–inductive load, with continuous current i_d, the control characteristic in the range $\pi/6 < \alpha < 5\pi/6$ corresponds to Equation 4.21. The shaded region indicates the family of control characteristics in the case of discontinuous current i_d with different values of $\omega L_d/R_d$.

4.2.2.4 Three-phase bridge circuit

1. *Operation with* $\alpha = 0$. The three-phase bridge circuit is shown in Figure 4.12. The corresponding current and voltage waveforms are shown in Figure 4.13 for the case in which $\alpha = 0$. We now consider circuit operation for an active load (with switch *S* closed). Beginning at time ϑ_1, current flows through thyristors VS1 and VS6, whereas the other thyristors are off. In that case, the line voltage u_{ab} is applied to load R_d, and current i_d flows through the circuit consisting of phase winding a, thyristor VS1, load R_d,

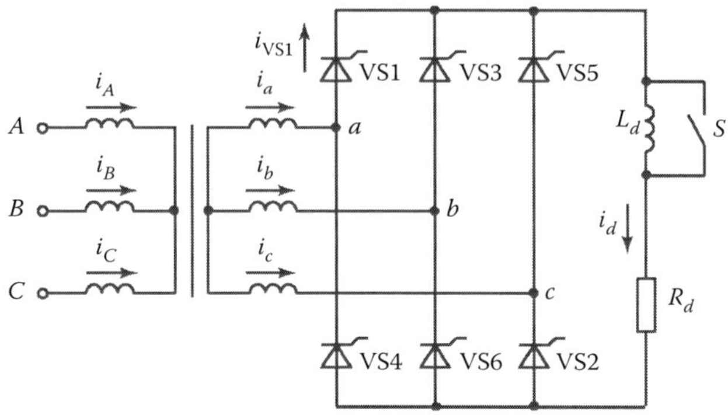

Figure 4.12 A three-phase bridge rectifier.

thyristor VS6, and phase winding b. This process continues until time ϑ_2 (for a period of $\pi/3$), when the potential of phase b becomes more positive than that of phase c. At time ϑ_2, voltage u_{bc} becomes positive; in other words, it is a forward voltage for thyristor VS2. If a control pulse is supplied at that moment to thyristor VS2, it begins to conduct, while thyristor VS6 is switched off. (There is commutation of thyristors VS6 and VS2.) For thyristor VS6, u_{bc} is a reverse voltage. As a result, thyristors VS1 and VS2 are on, while the others are off.

At time ϑ_3, a pulse is supplied to thyristor VS3, which is switched on. Thyristor VS1 is off, as the potential of phase b is greater than that of phase a. Then, at intervals of $\pi/3$, commutation of the following thyristor pairs is observed: VS2–VS4, VS3–VS5, VS4–VS6, and VS5–VS1. Thus, within the period of the supply voltage, there will be six commutations at intervals of $\pi/3$. Three occur in the cathode group of thyristors VS1, VS3, and VS5 (with connected cathodes), and three in the anode group of thyristors VS4, VS6, and VS2 (with connected anodes). Note that the number of thyristors in this circuit is not random but corresponds to the order in which they operate for the specified transformer phase sequence in Figure 4.12.

The sequential operation of different thyristor pairs in the circuit leads to the appearance of a rectified voltage at resistance R_d; it consists of parts of the line voltages of the secondary transformer windings (Figure 4.13). It is evident that, at commutation, the line voltages pass through zero (when two phase voltages—u_a and u_b, say—are equal). Current flows through each thyristor for a time $2\pi/3$; for the remainder of the period, reverse voltage consisting of segments of the corresponding line voltages is applied to the thyristors.

The constant component of the rectified voltage (the average value) is calculated over the interval of repetition of the rectified voltage ($\pi/3$)

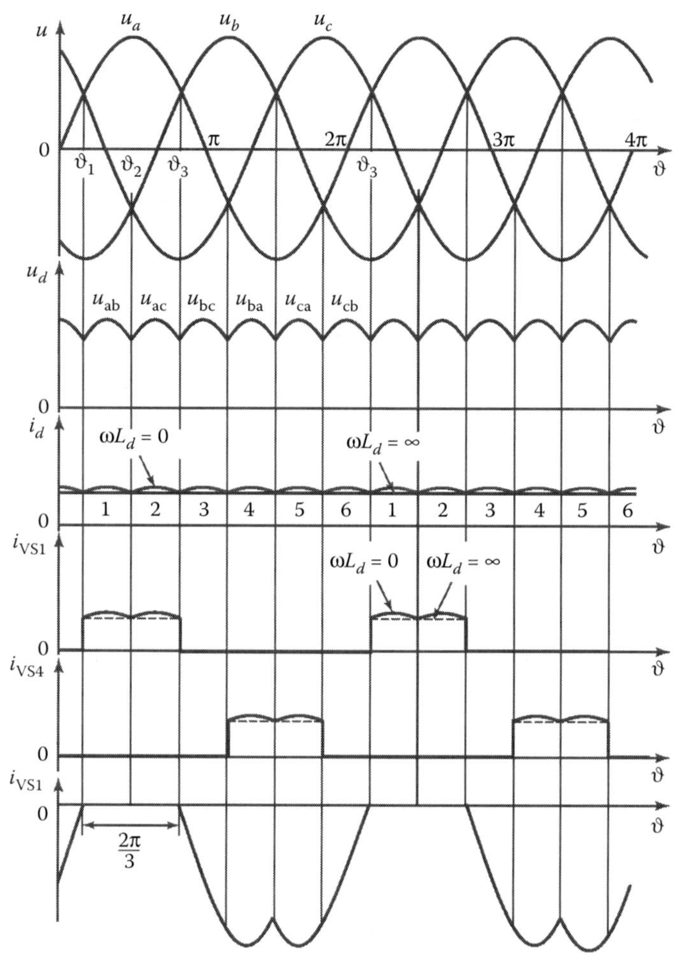

Figure 4.13 Voltage and current waveforms for a three-phase bridge rectifier ($\alpha = 0$).

$$U_d = \frac{3}{\pi}\int_{\pi/3}^{2/3\pi} \sqrt{6}\cdot U_2 \sin\vartheta d\vartheta = \frac{3\sqrt{6}}{\pi}U_2, \tag{4.23}$$

where U_2 is the effective value of the phase voltage at the secondary transformer windings.

Equation 4.23 applies to both active and active–inductive loads. When $\omega L_d = \infty$, the operation of the thyristors in the circuit is characterized by the following parameters.

- The maximum reverse voltage at the thyristor (equal to the amplitude of the line voltage at the secondary winding) is

$$U_{\text{Rmax}} = \sqrt{2} \cdot U_{2\text{line}}. \tag{4.24}$$

- The maximum thyristor current is

$$I_{\max} = I_d. \tag{4.25}$$

- The average thyristor current is

$$I_{\text{avVS}} = \frac{I_d}{3}. \tag{4.26}$$

2. *Operation with* $\alpha > 0$. In a three-phase bridge circuit based on thyristors, the control pulses are sent with delay α relative to the zeros of the line voltages (the moments when sinusoidal phase voltages intersect; Figure 4.14).

Due to the delay α, the average rectified voltage formed from the corresponding segments of the line voltages is reduced. As long as the instantaneous rectified voltage u_d remains above zero (in the range $0 < \alpha < \pi/3$), the rectified current I_d will be continuous, regardless of the load. Therefore, in that range of α, the average rectified voltage for active and active–inductive loads is

$$U_d = \frac{3}{\pi} \int_{-(\pi/3)+\alpha}^{(2\pi/3)+\alpha} \sqrt{3} \cdot U_2 \sin \vartheta d\vartheta = \frac{3\sqrt{6}}{\pi} U_2 \cos \alpha = U_{d0} \cos \alpha. \tag{4.27}$$

With an active load, the firing angle $\alpha = \pi/3$ corresponds to boundary-continuous operation (Figure 4.15, left). When $\alpha > \pi/3$, with an active load, intervals of zero voltage u_d and zero current i_d appear. In other words, operation with discontinuous rectified current begins. In that case, the average rectified voltage is reduced.

When $\alpha = \pi/2$, the positive and negative sections within the rectified voltage curve are equal in area. That indicates the lack of a constant component in the rectified voltage; in other words, the average value of U_d is zero (Figure 4.15, right).

Note that, with discontinuous current i_d, double control pulses separated by some interval or single pulses of length greater than $\pi/3$ must be supplied to the thyristors not only to ensure operation of the circuit, but

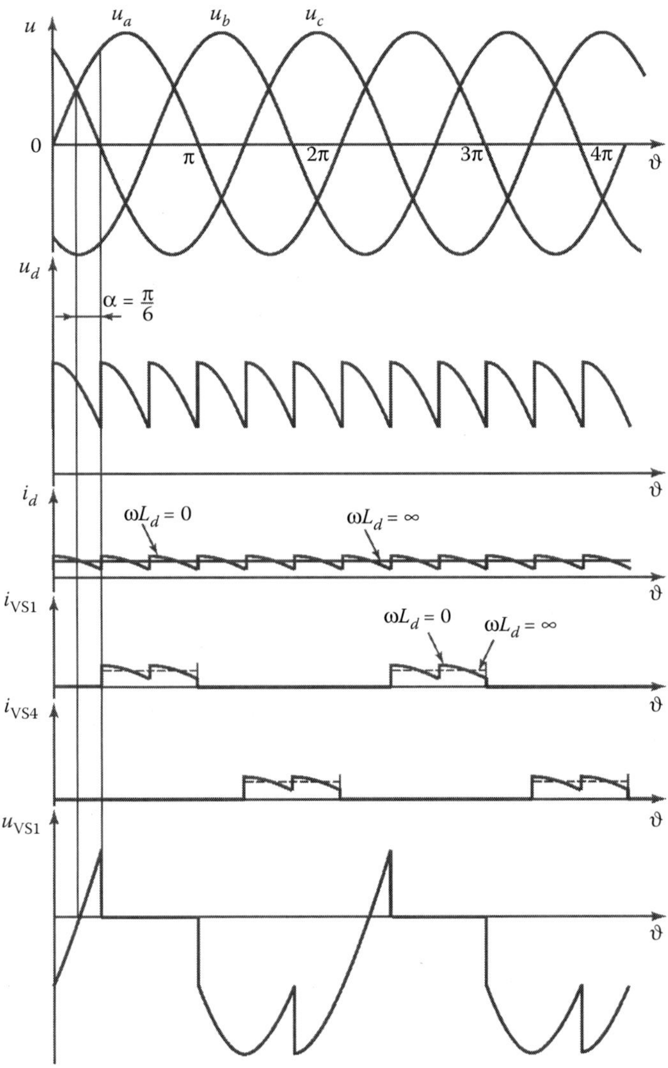

Figure 4.14 Voltage and current waveforms for a three-phase bridge rectifier ($\alpha = \pi/6$).

also for initial startup, because the thyristors of the anode and cathode groups must be switched on simultaneously in order to create a circuit suitable for current flow.

Figure 4.16 shows the control characteristics for a three-phase bridge circuit. With variation in α from 0 to $\pi/3$, the control characteristic for an active or active–inductive load is described by Equation 4.27. When

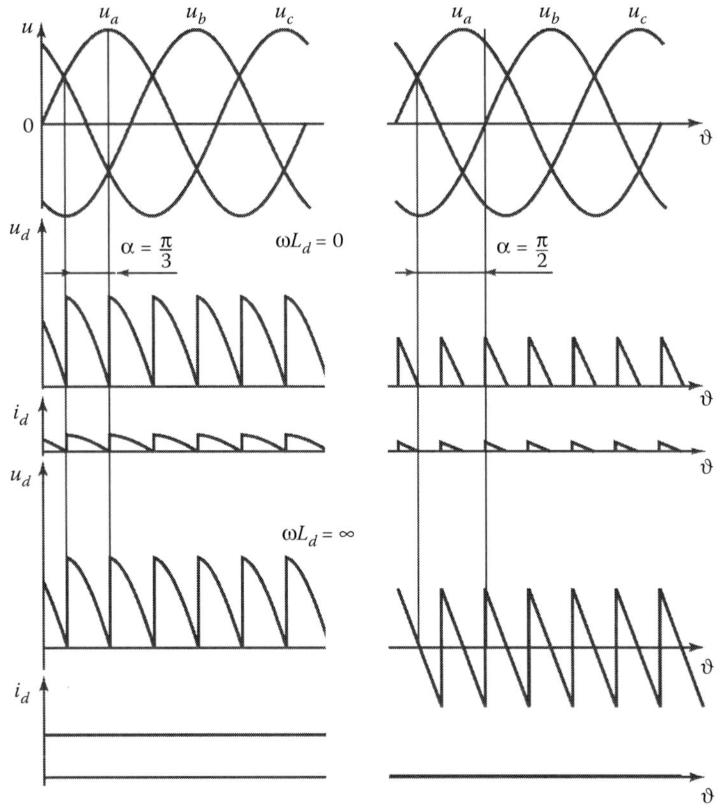

Figure 4.15 Voltage and current waveforms for a three-phase bridge rectifier when $\alpha = \pi/3$ and $\alpha = \pi/2$.

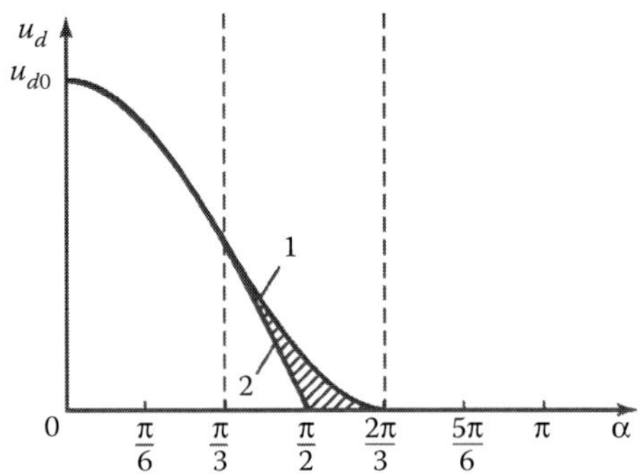

Figure 4.16 Control characteristics for a three-phase bridge rectifier with an active load (1) and an *RL* load (2).

$\alpha > \pi/3$, with an active–inductive load, current i_d is continuous, and the control characteristic is again described by Equation 4.27. The shaded region in Figure 4.16 corresponds to the family of control characteristics with discontinuous current i_d at different firing angles α.

For high-power rectifiers (above 1000 kW), with high voltage and current, we use multiphase circuits consisting of several bridges in series or parallel.

4.2.2.5 Multiple-bridge circuits

We may distinguish between multiple-bridge circuits with a single transformer and those with two or more transformers coupled in different configurations. The main purpose of multiple-bridge circuits is to reduce the ripple of rectified voltage and to improve the form of the current consumed from the grid, so that it is more sinusoidal.

Figure 4.17 shows two types of two-bridge circuits. The first consists of a three-winding transformer in a star/star–delta configuration and two three-phase bridges. The second includes two two-winding transformers and two three-phase bridges. One transformer is in star/star configuration, and the other in delta–star configuration.

In both cases, the phase shift of the transformer secondary voltages is $\pi/6$.

The two circuits operate analogously. Therefore, we will consider only one in more detail: the circuit with two transformers. As the primary windings of transformers T1 and T2 are in different configurations, there will be a phase shift of $\pi/6$ between the pulsations of the rectified voltage U_{d1} of one circuit and U_{d2} of the other circuit. To balance the instantaneous values of the rectified voltages, the bridges are connected in parallel through a compensating inductor. As a result, the total voltage at the load will have a ripple frequency twice that of each circuit. In the present case, each bridge circuit has six pulsations per period, and the total voltage will have 12 pulsations per period. Therefore, this is sometimes known as a 12-phase circuit. (Likewise, in view of the number of pulsations per period, a three-phase bridge circuit is sometimes known as a six-phase circuit.) The difference in instantaneous voltages is experienced by the equalizing reactor, whose two coils are mounted on a single core. The instantaneous values of the rectified voltage may be written in the form

$$U_d = u_{d1} - \frac{u_p}{2} = u_{d2} + \frac{u_p}{2}, \tag{4.28}$$

where u_p is the instantaneous voltage at the compensating inductor.

Figure 4.18 shows the current waveforms for 12-phase circuits (when $\omega L_d = \infty$). It is evident that the current consumed from the grid is more sinusoidal than that for a single-transformer circuit.

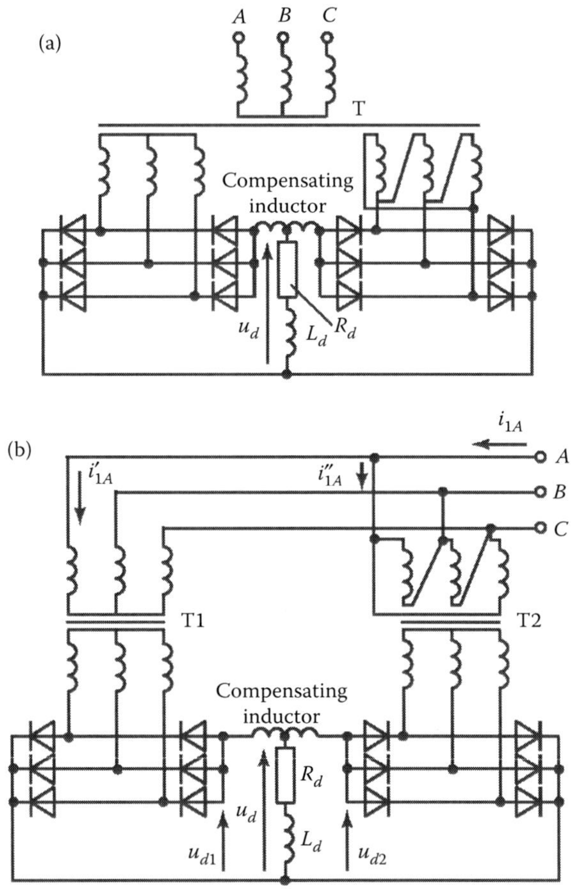

Figure 4.17 A three-phase two-bridge rectifier with parallel bridges.

Note that normal circuit operation requires the selection of the transformation ratios of transformers T1 and T2, so that the average voltages U_{d1} and U_{d2} are equal.

Figure 4.19 shows a two-bridge circuit with the bridges in series. In that case, the average rectified voltage at the load is

$$U_d = U_{d1} + U_{d2}, \tag{4.29}$$

where U_{d1} and U_{d2} ($U_{d1} = U_{d2}$) are the average output voltages of each bridge.

The 12-phase rectification circuit is based on the transformers with different winding configurations.

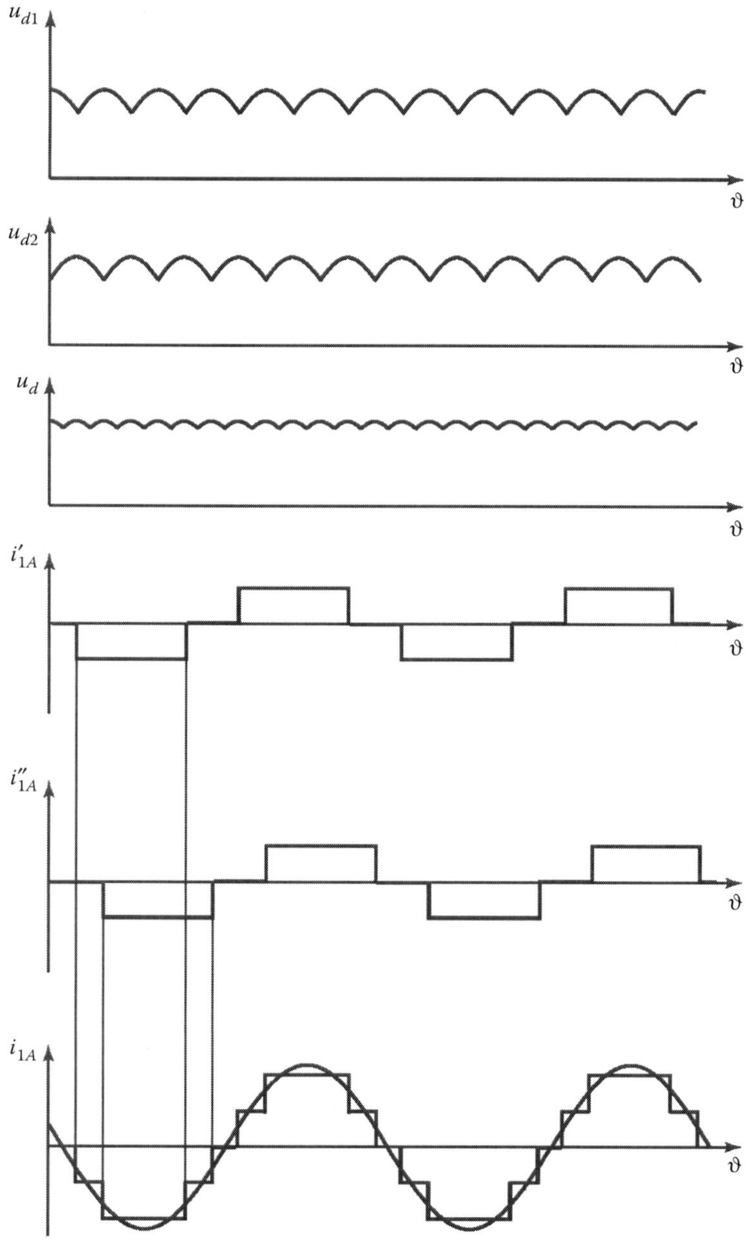

Figure 4.18 Voltage and current waveforms for a two-bridge rectifier.

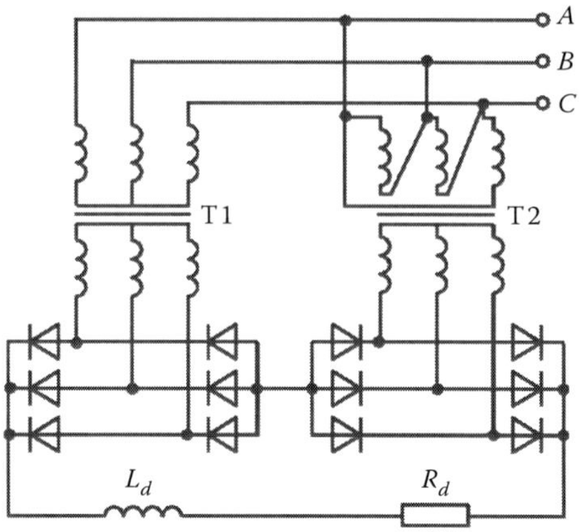

Figure 4.19 A three-phase two-bridge rectifier with series bridges.

In practice, 18- and 24-phase circuits are generally produced, respectively, by connecting three and four bridges in parallel (Rozanov, 1992).

4.2.3 Characteristics of rectifiers

4.2.3.1 Output voltage ripple

The rectified voltage may be represented as the sum of two components: one constant and the other variable. The variable component, in turn, is the sum of harmonic (sinusoidal) voltages

$$u_{\sim}(t) = \sum_{n=1}^{\infty} U_{nm} \sin(nm \cdot \omega t + \varphi_n), \tag{4.30}$$

where n is the number of harmonics, m the number of pulsations in the rectified voltage within the period of the variable component, ω the angular frequency of the grid voltage, U_{nm} the amplitude of harmonic n, and φ_n the initial phase of harmonic n.

The frequency of the rectified-voltage components is

$$f_n = nf_1 = mnf, \tag{4.31}$$

where f is the frequency of the grid voltage and $f_1 = mf$ is the frequency of the first harmonic of the pulsations.

For example, at a grid frequency $f = 50$ Hz, the frequency of the first harmonic ($n = 1$) takes the following values.

a. 100 Hz for a single-phase full-wave circuit ($m = 2$)
b. 150 Hz for a center-tapped three-phase circuit ($m = 3$)
c. 300 Hz for a three-phase bridge circuit ($m = 6$)

The amplitude of the nth voltage harmonic for a circuit with firing angle $\alpha = 0$ is (Rozanov, 1992)

$$U_{nm} = \frac{2U_{\mathrm{d}}}{m^2 n^2 - 1}. \tag{4.32}$$

According to Equation 4.32, the amplitude of the first harmonic ($n = 1$) is the greatest. The others decline in inverse proportion to n^2.

The effective value of the variable component in the rectified voltage may be expressed in the form

$$U_{\mathrm{eff}} = \sqrt{\sum_{n=1}^{\infty} U_n^2}, \tag{4.33}$$

where U_n is the effective value of the nth harmonic.

In practice, the ripple (the content of the variable component) in the rectified voltage is estimated on the basis of the ripple factor K_r. The delay α in supplying the control pulses to the thyristors (with respect to the natural commutation times) changes the harmonics in the rectified voltage. It is evident from plots of the rectified voltage that the variable component (the ripple) increases with α rising. However, the period of the ripple pulsations does not depend on α.

4.2.3.2 *Distortion of the input current*

It follows from the operating principles of the rectifier circuits that they mostly consume nonsinusoidal current from the grid. Only a single-phase full-wave rectifier with an active load, when $\alpha = 0$, consumes sinusoidal current, with zero amplitude of the higher harmonics. With a resistive–inductive load, when $\omega L_d = \infty$, the current is rectangular and may be expressed as the sum of harmonics

$$i_1(\vartheta) = \frac{4I_d}{\pi \cdot k_T}\left[\sin\vartheta + \frac{1}{3}\sin 3\vartheta + \cdots + \frac{1}{n}\sin n\vartheta\right], \tag{4.34}$$

where k_T is the transformation ratio.

It is evident from Equation 4.34 that the primary current of the full-wave circuit ($m = 2$) contains only odd current harmonics. The influence of higher harmonics on the grid is particularly pronounced when the power of the ac source is comparable with the rectifier power.

The harmonic composition of the current consumed from the grid by a controllable rectifier depends significantly on the load. If the load is active or active–inductive but does not ensure continuous current i_d, the amplitude of the higher current harmonics will increase with an increase in α (with constant amplitude of the first harmonic).

With an active–inductive load and ideally smoothed rectified current, the firing angle α has no influence on the harmonic composition of the consumed current. Note that this conclusion assumes zero inductive impedance of the transformer windings.

Usually, passive or active filters are employed to reduce the voltage ripple and the distortion of the rectifier input and output current.

4.2.3.3 The commutation of the thyristors

In theoretical analysis, switching of the current from one thyristor to another (commutation) is assumed to be instantaneous. In practice, commutation will have certain duration on account of inductive impedance in the ac circuit—in particular, the impedance of the transformer windings, which is mainly due to scattering fluxes in the transformer's magnetic system.

This impedance is determined in experimental short-circuiting of the secondary transformer windings and is taken into account in circuit analysis on the basis of general (for each phase) equivalent inductances L_s, which correspond to the total inductance of the secondary winding and the reduced (in terms of the number of turns) inductance of the primary winding. In addition to the inductive impedance, the commutation processes also depend on the active impedance of the windings, but to a much smaller degree, in normal conditions. Therefore, in considering commutation, we only take account of the windings' inductive impedance; the rectified current is assumed to be ideally smoothed ($\omega L_d = \infty$). Given that the commutation processes are qualitatively the same in different circuits, we will consider a simple rectifier: a single-phase full-wave circuit.

Figure 4.20a presents the equivalent circuit of a thyristor-based single-phase full-wave rectifier, together with voltage and current diagrams. The inductive impedance of the windings is taken into account by introducing inductance L_s. Suppose that thyristor VS1 is on. At time ϑ_1, a control pulse is supplied to thyristor VS2. As its anode potential is positive with respect to the cathode at that moment, thyristor VS2 is switched on.

Beginning at time ϑ_1, both thyristors will be on, and the transformer secondary half-windings short-circuit one another. Under the action of emfs e_a and e_b of the secondary half-windings, the short-circuit current i_{sc} appears in the short-circuited circuit (the commutation loop), which is the

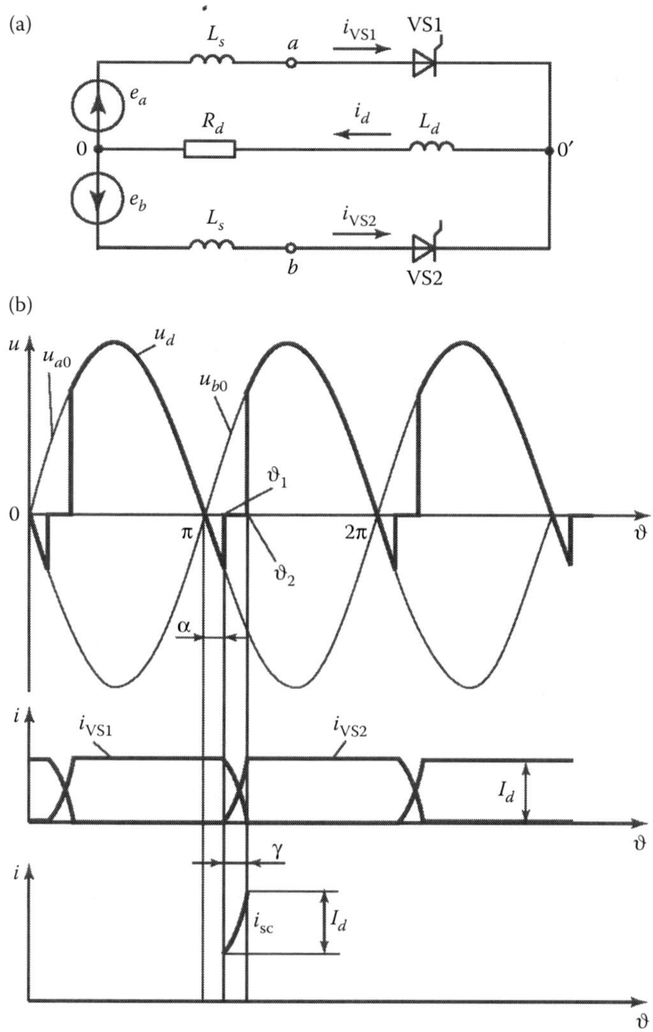

Figure 4.20 (a) Equivalent circuit of a full-wave thyristor rectifier and (b) corresponding voltage and current waveforms.

commutation current. At any time from ϑ_1 on, this current may be determined as the sum of the two components: a steady component i'_{sc} and a free component i''_{sc}

$$i'_{sc} = \frac{2\sqrt{2}}{2x_s} U_2 \cos(\vartheta + \alpha) \tag{4.35}$$

and

$$i''_{sc} = \frac{\sqrt{2} \cdot U_2}{x_s} \cos \alpha, \tag{4.36}$$

where U_2 is the effective voltage of the secondary half-winding and $x_s = \omega L_s$.

Taking account of Equations 4.35 and 4.36, we may write the resultant short-circuit current in the form

$$i_{sc} = i'_{sc} + i''_{sc} = \frac{\sqrt{2}U_2}{x_s}\left[\cos \alpha + \cos(\alpha + \vartheta)\right]. \tag{4.37}$$

When thyristor VS2 is switched on and thyristor VS1 is switched off, the resultant short-circuit current i_{sc} runs from the half-winding b with higher potential to the half-winding a with lower potential. As the rectified current remains constant in the commutation period when $\omega L_d = \infty$, we may write the following current equation for point 0 as

$$i_{VS1} + i_{VS2} = I_d = \text{const}, \tag{4.38}$$

where I_d is the average rectified current or load current.

Equation 4.38 is valid for any time. If the current flows only through thyristor VS1, we obtain $i_{VS1} = I_d$ and $i_{VS2} = 0$. With simultaneous conduction of thyristors VS1 and VS2 (during commutation of the current from thyristor VS1 to thyristor VS2), $i_{VS1} = I_d - i_{sc}$ and $i_{VS2} = i_{sc}$. When the current flows only through thyristor VS2, we obtain $i_{VS2} = I_d$ and $i_{VS1} = 0$.

The length of the commutation interval is characterized by the commutation angle γ, which can be determined from the following equation:

$$I_d = \frac{\sqrt{2}U_2}{x_s}\left[\cos \alpha - \cos(\alpha + \gamma)\right]. \tag{4.39}$$

If the value of γ when $\alpha = 0$ is denoted by γ_0, we can write

$$1 - \cos \gamma_0 = \frac{I_d x_s}{\sqrt{2}U_2}. \tag{4.40}$$

Substituting γ_0 into the initial equation, we obtain

$$\gamma = \arccos\left[\cos \alpha + \cos \gamma_0 - 1\right] - \alpha. \tag{4.41}$$

According to Equation 4.41, γ declines with an increase in α. In physical terms, we may say that an increase in α boosts the voltage under which the current i_{sc} in the commutation circuit develops, and hence i_{sc} reaches I_d more rapidly.

Note that the duration of current flow in the thyristors is greater by γ than in the idealized circuit. It is $\pi + \gamma$.

Commutation has a significant influence on the rectified voltage U_d, as the instantaneous rectified voltage in the given circuit falls to zero in the commutation intervals (Figure 4.20b). As a result, the average rectified voltage is reduced by

$$\Delta U_x = \frac{1}{\pi}\int_{\alpha}^{\alpha+\gamma} \sqrt{2}U_2 \sin\vartheta \, d\vartheta. \tag{4.42}$$

From Equations 4.39 through 4.42, we conclude that

$$\Delta U_x = \frac{I_d x_s}{\pi}. \tag{4.43}$$

Taking account of Equation 4.43, we write the average rectified voltage in the form

$$U_d = U_{d0}\cos\alpha - \frac{I_d x_s}{\pi}. \tag{4.44}$$

4.2.3.4 *External rectifier characteristic*

The rectifier characteristic is the dependence of the average rectified voltage on the average load current: $U_d = f(I_d)$. It is determined by the internal resistance of the rectifier, which results in a decrease in the rectified voltage with an increase in the load. The decrease in the voltage includes components due to the active circuit impedance ΔU_R, the voltage drop in the thyristors ΔU_{VS}, and the inductive impedance ΔU_x that appears in commutation.

Correspondingly, the rectifier characteristic (when $\omega L_d = \infty$) may be written in the form

$$U_d = U_{d0}\cos\alpha - \Delta U_R - \Delta U_{VS} - \Delta U_x. \tag{4.45}$$

According to Equation 4.45, the rectifier's output voltage declines with an increase in the load current I_d on account of the internal voltage drop. The influence of the active and reactive circuit components will depend on the rectifier power. Usually, the active impedance of the

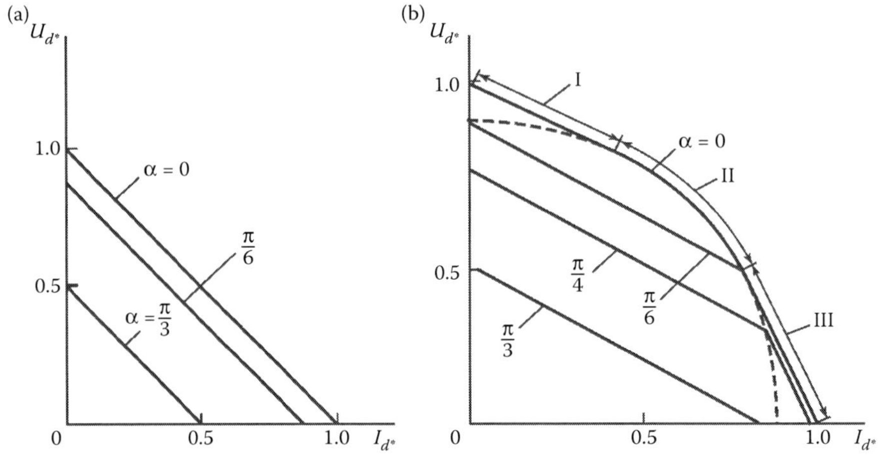

Figure 4.21 (a) External characteristics for a single-phase rectifier and (b) a three-phase bridge rectifier in different operating conditions (I–III).

transformer windings predominates in low-power rectifiers, whereas the transformer's inductive scattering impedance predominates in powerful rectifiers.

Note that, at loads not exceeding the rated value, as a rule, the rectifier's internal voltage drop is not more than 15%–20% of the voltage. However, in the case of overloads and proximity to short-circuits, the internal impedance of the circuit becomes significant. In addition, with overloads in three-phase and multiphase circuits, qualitative changes are observed in the electromagnetic processes that affect the rectifier characteristic. In Figure 4.21, as an example, the characteristics of single-phase and three-phase rectifiers are presented.

4.2.3.5 Energy characteristics of rectifiers

The power factor and efficiency of a rectifier require careful interpretation. We must distinguish between its output power, in which the ripple of the rectified voltage is taken into account, and the power determined by the average output voltage U_d and load current I_d. The latter is usually regarded as the useful power and therefore employed in calculations. When the ripple is slight, the difference between these two quantities may be ignored.

The main losses of active power occur in the following components of power rectifiers: in the transformer (ΔP_T), in the thyristors (ΔP_{VS}), and in auxiliary equipment such as control, safety, cooling, and monitoring systems (ΔP_{aux}). Accordingly, we may calculate the efficiency in the following form for a rectifier with small current pulsations

$$\eta = \frac{U_d I_d}{U_d I_d + \Delta P_T + \Delta P_{VS} + \Delta P_{aux}}. \quad (4.46)$$

For the moderate- and high-power rectifiers manufactured today, the efficiency is 0.7–0.9.

The power factor is the ratio of the active power to the total power. It permits the determination of the total power consumed by a power converter if its active load power and efficiency are known. In determining the rectifier power factor, we must take account of the nonsinusoidal component of the power that it draws from the grid. Figure 4.22 shows the grid voltage u_c and grid current i_c consumed by a single-phase controllable rectifier, on the assumption that the rectified current is ideally smoothed and there is no commutation angle. An analysis of the nonsinusoidal current yields the first harmonic i_{c1}, which lags the voltage u_c by φ_1. Correspondingly, the active power P consumed by the rectifier may be expressed as

$$P = U_s I_{s1} \cos\varphi_1, \quad (4.47)$$

where U_s is the effective grid voltage at the rectifier, I_{s1} the first effective harmonic of the current drawn from the grid, and φ_1 the phase shift of the first current harmonic with respect to the grid voltage.

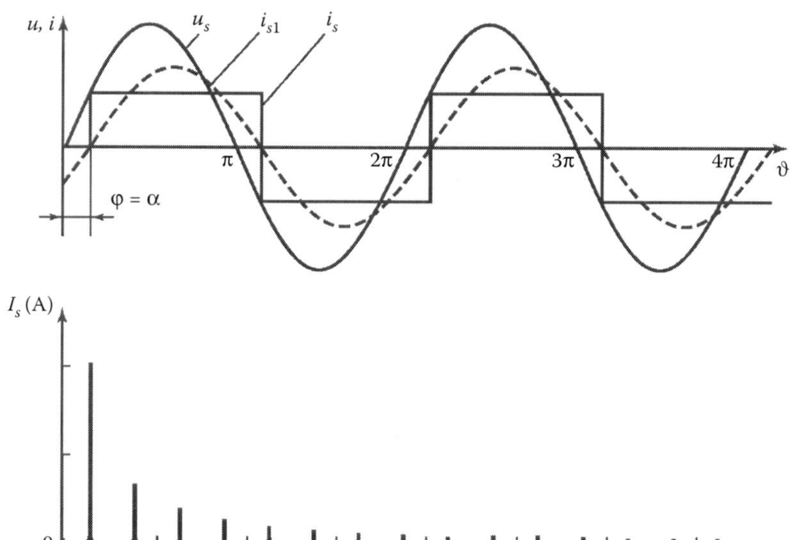

Figure 4.22 Current waveform for a single-phase bridge rectifier ($L_d = \infty$) and its spectrum analysis.

On the basis of the general definition, we can write the apparent power consumed by the rectifier in the form

$$S = U_s I_s = U_s \sqrt{I_{s1}^2 + \sum_{n=3}^{\infty} I_{sn}^2}, \tag{4.48}$$

where I_s is the effective nonsinusoidal current drawn from the grid and I_{sn} is the effective value of its nth harmonic.

According to Equations 4.47 and 4.48, the power factor of the rectifier can be expressed in the form

$$\chi = \frac{P}{S} = \frac{I_{s1}\cos\varphi_1}{\sqrt{I_{s1}^2 + \sum_{n=3}^{\infty} I_{sn}^2}}. \tag{4.49}$$

Controllable rectifiers are characterized by a firing angle α, which is equal to the phase shift of the first current harmonic with respect to the grid voltage, as a rule. Hence, for circuits with ideally smoothed current, according to Equation 4.49, the power factor may be calculated in the form

$$\chi = \nu\cos\alpha. \tag{4.50}$$

For nonsinusoidal current, we must consider not only the active power P and the reactive power Q but also the distortion power T (Chapter 1).

4.3 Grid-tie inverters

4.3.1 Operating principle

Inversion is the conversion of dc electrical energy to ac energy. Linguistically, the term comes from the Latin *inversio*, which signifies overturning. It was introduced in power electronics to denote the process inverse to rectification. In inversion, the flux of electrical energy is reversed and supplied from the dc source to the ac grid. Such a converter is said to be a grid-tie inverter, as it is switched under the action of the alternating voltage in the external grid. As the electrical parameters of the converter are completely determined by the parameters of the external ac grid in this case, it is sometimes referred to as a dependent inverter.

We will consider the operating principle of the grid-tie inverter for the simple example in Figure 4.23a. We assume that the circuit components are ideal, and the internal resistance of the storage battery is zero.

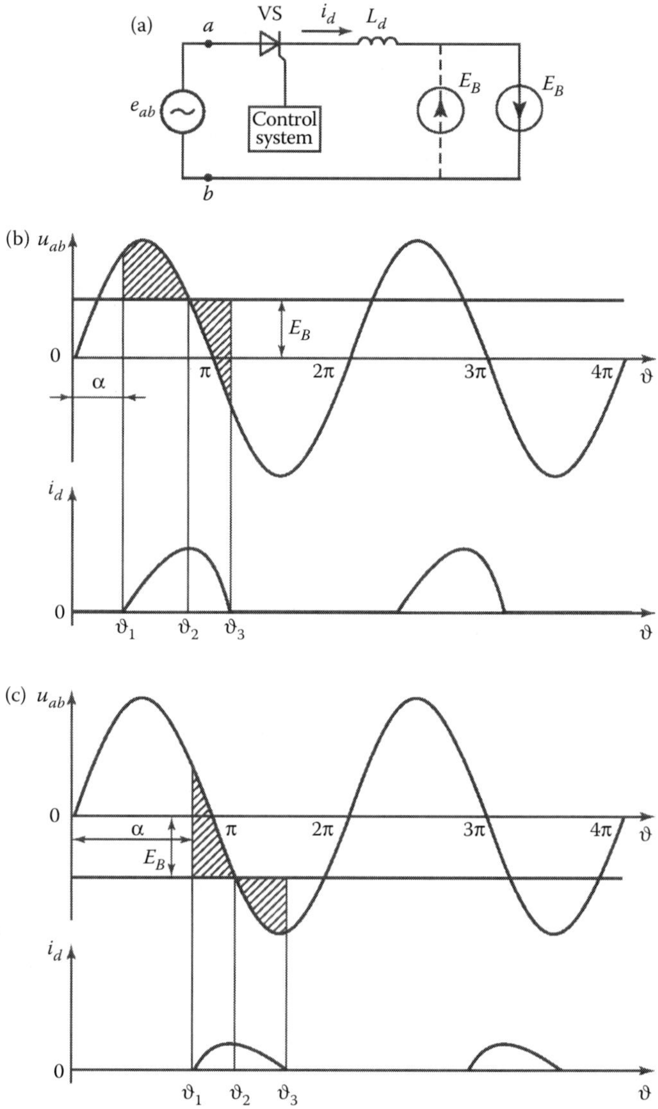

Figure 4.23 (a) A half-wave reversible converter and its voltage and current waveforms in (b) rectifier mode and (c) inverter mode.

Figure 4.23b shows voltage and current waveforms, illustrating the operation of the circuit in the rectifier mode. Under the assumption that the internal resistance of the ac and dc sources is zero, we may conclude that their voltages are equal to the emfs: $e_{\mathrm{ab}} = u_{\mathrm{ab}}$ and $E_B = U_B$. If the positive terminal of the battery is connected in accordance with the dashed line in Figure 4.23, the circuit may operate in the rectifier mode with a counter-emf load, which corresponds to battery charging. With reversal of the battery potential, an operation in the inverter mode is possible. We now consider these processes in more detail.

If a control pulse is supplied to the thyristor at time $\vartheta = \vartheta_1$, determined by firing angle α, the thyristor is switched on. As a result, the supply of current i_d from the grid to the battery begins. Due to the smoothing reactor L_d, the current i_d will vary smoothly over time: it increases when $u_{ab} > U_d$ and decreases when $U_d > u_{ab}$. At time ϑ_3 (when the shaded areas in Figure 4.23b are equal), current i_d is zero and thyristor VS is switched off. The passage of current i_d through the thyristor in the interval from ϑ_2 to ϑ_3, when $U_d > u_{ab}$, is due to the electromagnetic energy stored in the reactor L_d. Subsequently, these processes repeat periodically. As a result, the battery is charged by the rectified current. (Current i_d opposes the emf E_B.)

For conversion of the circuit to the inverter mode, the battery polarity must be reversed.

Energy is transferred from one source to another when the current from the source opposes the emf of the source that is receiving this energy. In the present case, energy will be transferred to the grid from the battery when the grid emf e_{ab} opposes the current i_d. Voltage and current waveforms for the inverter mode are shown in Figure 4.23c. If a control pulse is supplied to thyristor VS at time ϑ_1, the thyristor is switched on, under the action of a positive forward voltage. The forward voltage at the thyristor exists until time ϑ_2. Thereafter, the voltage u_{ab} will exceed the emf E_B in an absolute value. Under the action of the voltage difference $U_B - u_{ab}$, the current i_d will flow in the circuit, in the direction opposite to the grid voltage u_{ab}. The smoothing reactor L_d in the circuit limits the rate of increase and the maximum value of i_d. Due to the energy stored in the reactor, current continues to flow in the thyristor after the voltage u_{ab} exceeds U_d in absolute value. The current i_d is zero at time ϑ_3, when the shaded regions in Figure 4.23c are equal in area.

In terms of circuit structure, dependent inverters are not significantly different from controllable rectifiers. Therefore, they may be regarded as reversible converters, capable of transmitting electrical energy from the grid to a dc source (rectifier mode) or from a dc source to the grid (inverter mode). Therefore, such converters are also known as ac–dc converters (International Electrical-Engineering Dictionary: Power Electronics, 1998).

4.3.1.1 *Operation in the inverting mode*

As already noted, we can switch from the rectifier to the inverter mode and back by reversing the polarity of the ac source relative to the common terminals of the anode and cathode thyristor groups in the bridge circuit.

Figure 4.24a shows the bridge circuit of a single-phase converter. The dashed line corresponds to the connection of the dc source with emf E_{inv}

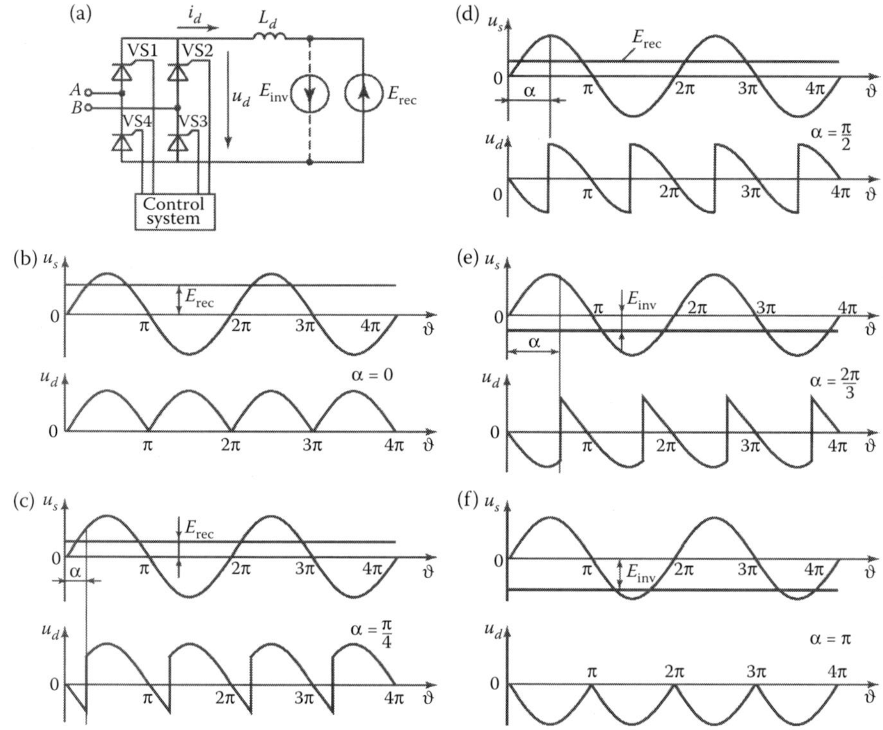

Figure 4.24 (a) A single-phase bridge converter and (b–f) its voltage and current waveforms for rectifier and inverter modes.

in the inverter mode and the continuous line to the connection with emf E_{rec} in the rectifier mode.

We assume that the inductance L_d is relatively large and the dc ripple may be neglected. In other words, we assume that $\omega L_d = \infty$ in the case of steady operation with different values of the firing angle α, which determines when the CS delivers the current pulses to the thyristors.

Figure 4.24 presents instantaneous values of the voltage $u_d(\vartheta)$ on the dc side of the converter (ahead of the reactor L_d). Under the given assumptions, the emfs in the reactor and the inverter mode are equal: $E_{inv} = E_{rec}$. We consider steady operation, with $\alpha = 0$, $\pi/4$, $\pi/2$, $2\pi/3$, and π. The current in the reactor L_d is assumed to be equal to the average value of the current I_d, in all steady operating conditions. With variation in α as in Figure 4.24, this will be the case if the voltage of the dc source (consumer) also varies in accordance with the firing angle α.

In Figure 4.24, the angles $\alpha = 0$ and $\pi/4$ correspond to the rectifier mode. When $\alpha = \pi/2$, the average voltage on the dc side of the converter is $U_d = 0$, and the current I_d stored in the reactor L_d remains constant in view of our assumption that there are no power losses in the circuit components. As a result, when $\alpha = \pi/2$, reactive power is exchanged between the ac sources and the reactor L_d. When $\alpha = 2\pi/3$ and π, the average voltage U_d reverses the sign (so as to oppose current I_d). This corresponds to the inverter mode, that is, transfer of energy flux from the source E_{inv} through the converter's thyristor bridge to the ac grid. Figure 4.25 presents the grid voltage and inverter input current i_s on the grid side, which alternates slightly under the given assumptions. If we only consider the first harmonic of this current, we can plot vector diagrams for different operating conditions (Figure 4.26). We see that a thyristor-based ac/dc converter with natural commutation may operate in two quadrants of the complex plane, corresponding to possible vector values of the first current harmonic.

It is evident from Figure 4.26 that, on switching to the inverter mode, α becomes greater than $\pi/2$. In that case, the thyristor converter produces alternating current from direct current and transfers the energy of a dc source with emf E_{inv} to the grid. On the dc side (ahead of the reactor L_d), an inverter emf with the polarity opposite to the rectifier emf is formed.

4.3.2 Basic circuits operation in the inverting mode

4.3.2.1 Single-phase bridge inverter

An inverter bridge circuit is shown in Figure 4.27a. Suppose that thyristors VS1 and VS3 are conducting. In that case, energy from the dc source E_{inv} is sent through the transformer to the grid, because current i_s in the primary winding of transformer T opposes the voltage there u_{ab}. On the assumption that $L_d = \infty$, the voltage pulsation due to the difference in instantaneous voltages of the secondary transformer windings and the dc source will be applied to the reactor L_d.

To ensure inverter mode, the firing angle α must be greater than $\pi/2$. Therefore, in the circuit analysis, the control angle in the inverter mode is usually measured with respect to the natural commutation times in circuits with uncontrollable diodes (or to the angles $\alpha = 0$, π, 2π, etc. in circuits with thyristors). The angle measured in this way is the lead angle β. The relation between β and α is as follows:

$$\beta = \pi - \alpha. \tag{4.51}$$

Suppose that thyristors VS2 and VS4 conduct in the interval from 0 to ϑ_1. At time ϑ_1, control pulses are sent to thyristors VS1 and VS3. At that moment, the thyristor anode is positive relative to the cathode ($u_{ab} > 0$), and so the thyristors are switched on. The secondary transformer winding is short-circuited. As a result, the short-circuit current i_{sc} appears, opposing the current through the thyristors. In other words, natural commutation begins. When commutation ends at time ϑ_2 (as in the rectifier mode, the

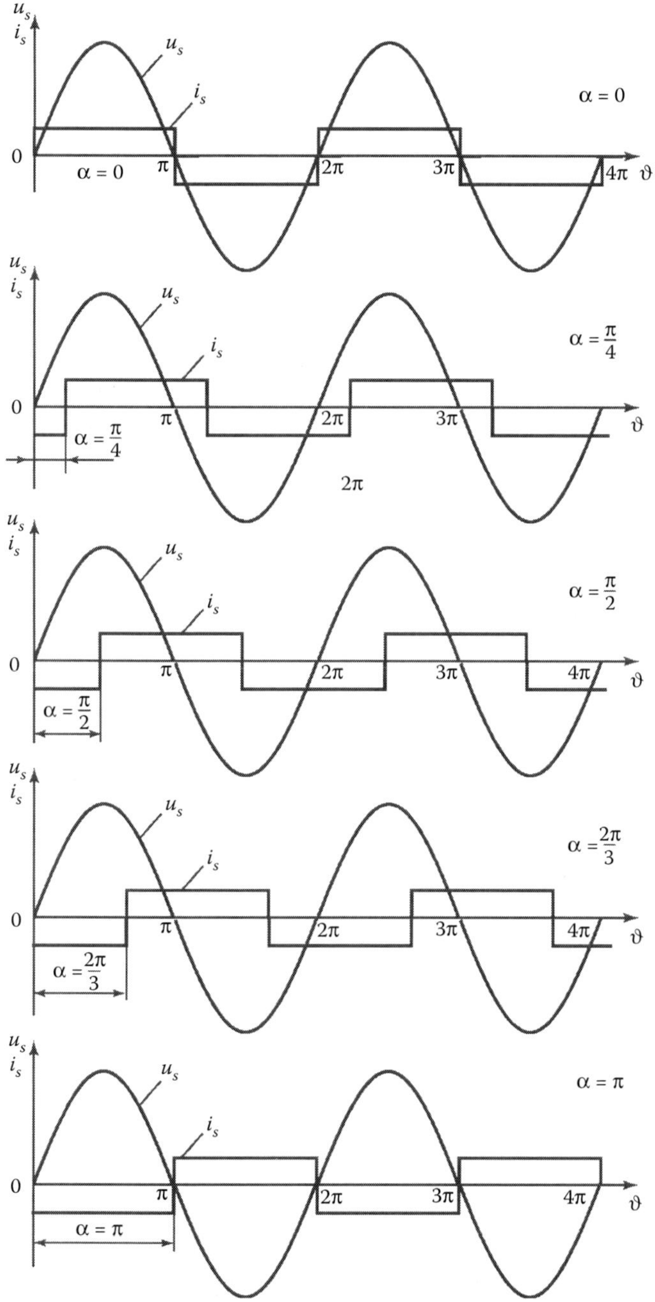

Figure 4.25 Source voltage and consumption current waveforms for a single-phase bridge converter in different values of firing angle α.

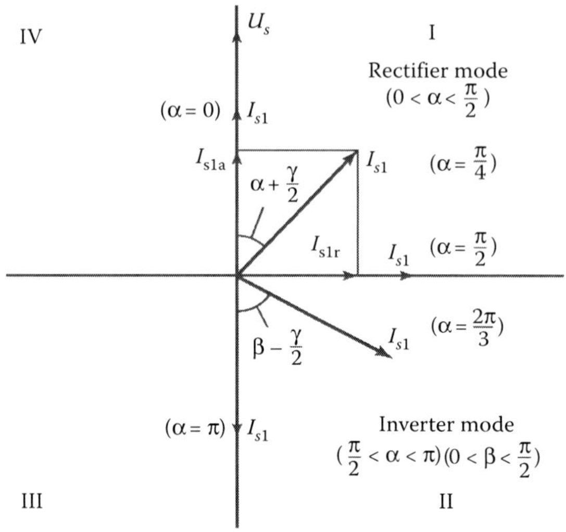

Figure 4.26 Vector diagram for a thyristor-based converter (rectifier and inverter operation).

duration of commutation is expressed by the angle γ), the thyristors are switched off; the inverse voltage u_{ab} is applied to them. As a result, thyristors VS2 and VS4 are able to recover their switching properties until u_{ab} changes sign (when the potential is greater at point b than at point a). The angle corresponding to this interval is the reserve angle δ. The relation between β, γ, and δ is as follows:

$$\beta = \gamma + \delta. \tag{4.52}$$

Thyristors VS1 and VS3 conduct up to time ϑ_4. Before that, at time ϑ_3, control pulses are sent to thyristors VS2 and VS4. As a result, commutation occurs and thyristors VS2 and VS4 are switched on, whereas thyristors VS1 and VS3 are switched off. At that point, the processes periodically repeat.

It is evident from the form of the electromagnetic processes that they are largely similar to rectifier operation with a counter-emf. The main difference is that, in the inverter mode, the dc source is switched on with the opposite polarity to the thyristors and sends energy to the grid. Because the control pulses are sent to the thyristors with lead β relative to the commutation times (with phase shift π), the current i_s sent to the grid passes through zero to positive values before the voltage passes through zero to negative values. Therefore, the first harmonic of the current i_s is ahead of the voltage u_{ab} by an angle of about $\beta - \gamma/2$ (Figure 4.27b).

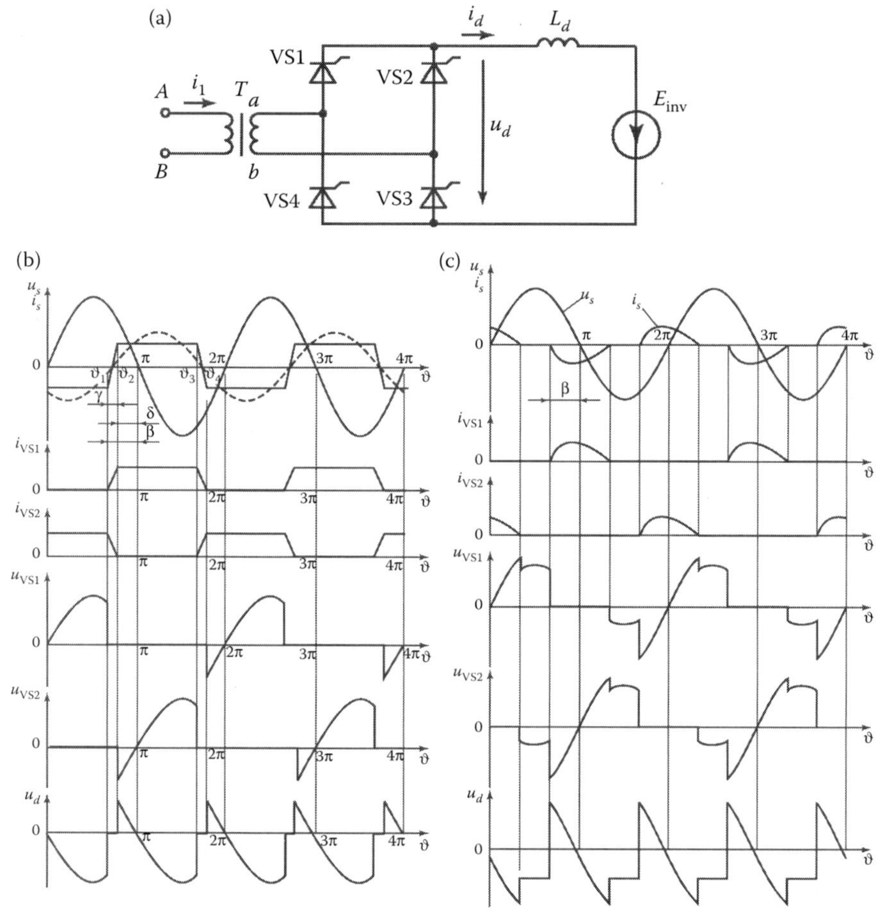

Figure 4.27 (a) A single-phase bridge inverter and voltage and current waveforms in the case of (b) continuous and (c) interrupted load currents.

The vector diagrams of the current i_{s1} and voltage u_{ab} for the rectifier mode and the inverter mode are shown in Figure 4.26. In the rectifier mode, the first current harmonic lags the voltage by about $\alpha + \gamma/2$. It is evident from the vector diagram that, in the inverter mode, the active current component I_{s1a} opposes the grid voltage. This corresponds to the supply of active power to the grid. As in the rectifier mode, the reactive current I_{s1r} lags the grid voltage by $\pi/2$. Hence, in both cases, the converter is a consumer of reactive power. The voltage on the dc side of the converter, known as the counter-emf of the inverter, has pulsations, depending on the angles β and γ. The relations for these pulsations are the same as for the rectifier mode if α is replaced by β. The average voltage U_d is equal to the source voltage E_{inv}.

The relation between the effective voltage U_{ab} at the secondary transformer winding (which depends on the ac grid voltage and the transformation ratio) and the dc source voltage U_d is similar to that for the average rectified voltage of a rectifier. With no inverter load, we obtain

$$U_{d0} = \frac{2\sqrt{2}}{\pi} U_2 \cos \beta, \tag{4.53}$$

where U_2 is the effective voltage at the secondary transformer winding.

The other relations are also similar to those in Section 4.2.2 for a resistive–inductive load of a single-phase rectifier with a continuous reactor current. In the case of discontinuous current i_d (Figure 4.27c), the analytical relations between the circuit parameters are considerably complicated, as in the rectifier mode.

4.3.2.2 Three-phase bridge inverter

Figure 4.28 shows a thyristor-based three-phase bridge inverter, together with voltage and current waveforms in the case of ideally smoothed current I_d. In this circuit, as in a single-phase bridge inverter, control pulses are sent to thyristors with lead β relative to the times corresponding to the onset of thyristor commutation in an uncontrollable rectifier mode ($\alpha = 0$, π, 2π, etc.). At those times, the line voltages of the secondary transformer windings pass through zero; in other words, the sinusoidal phase voltages u_a, u_b, and u_c intersect. In the interval ϑ_0–ϑ_1, under the action of the source voltage U_d, the current I_d flows through the thyristors VS1 and VS2 and the secondary transformer windings (phases a and c). The instantaneous counter-emf of the inverter (Figure 4.28b) is then equal to the difference of u_c and u_a (Zinov'ev, 2003).

At time ϑ_1, determined by the lead angle β, which is specified by the inverter's control signal, a control pulse is supplied to thyristor VS3. This thyristor is switched on. As a result, phases a and b of the secondary transformer windings are short-circuited. The corresponding short-circuit current in those phases opposes the current i_{VS1} in thyristor VS1. In other words, commutation begins, analogous to the processes in a three-phase rectifier bridge circuit (Section 4.2.2). The duration of commutation is γ. The voltage U_d in the commutation interval is calculated as u_c minus the half-sum of voltages u_a and u_b. At the end of commutation, thyristors VS2 and VS3 will transmit current I_d, whereas reverse voltage is applied to the thyristor VS1 for time δ.

Subsequently, the commutation of the thyristors occurs in accordance with their numbering (Figure 4.28b). For each thyristor, the conduction interval is $2\pi/3 + \gamma$.

In both inverter mode and rectifier mode, commutation processes are responsible not only for periodic voltage dips on the dc side, but also

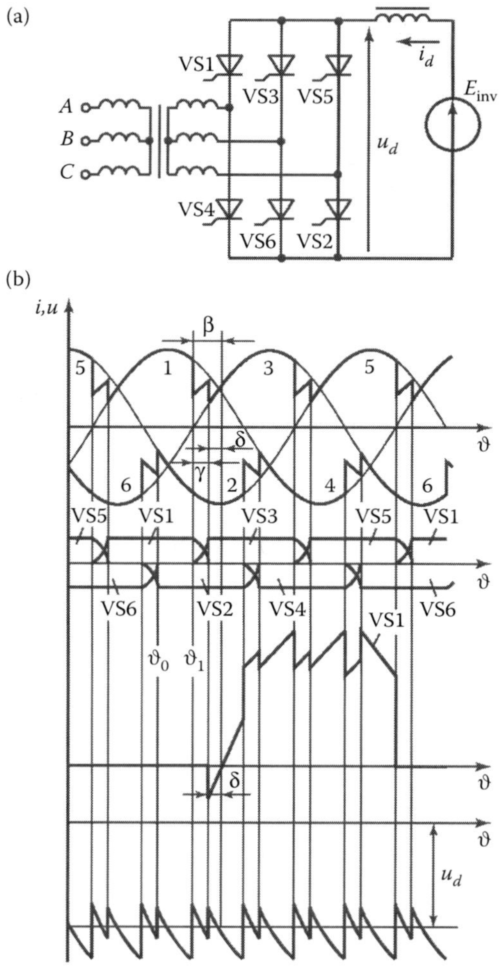

Figure 4.28 (a) A three-phase bridge inverter and (b) its voltage and current waveforms.

for dips and surges in the ac grid voltage. For example, if we assume the equivalent phase inductance (including mainly the transformer's scattering induction), connected directly to the outputs of the converter circuit in Figure 4.29a, the voltage at the outputs will take the waveform as shown in Figure 4.29b. The areas of the voltage dips and surges may be determined as follows:

$$\Delta S_1 = \frac{X_S}{2} I_d, \quad \Delta S_2 = 2X_S I_d \, X_S = \omega L_S. \tag{4.54}$$

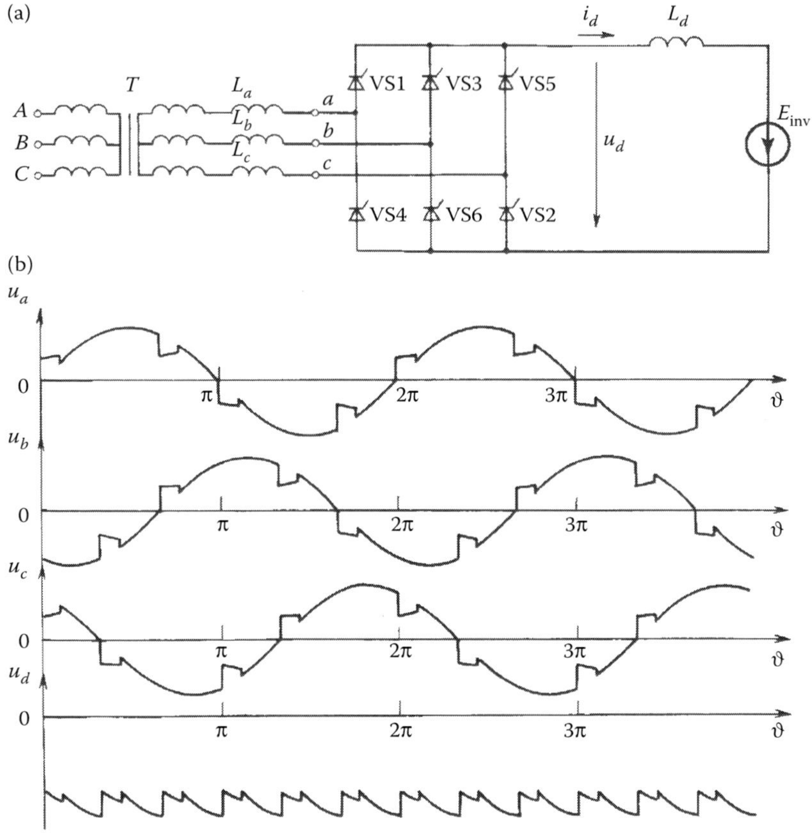

Figure 4.29 (a) A three-phase bridge inverter with equivalent input inductance and (b) voltage waveforms.

Analogous surges and dips are observed in the rectifier mode (Figure 4.2).

The average source voltage U_{d0} with no load is related to the effective phase voltage U_{ph} at the transformer output as follows:

$$U_{d0} = \frac{3\sqrt{6}}{\pi} U_{\text{ph}} \cos\beta. \tag{4.55}$$

The other relations for the inverter mode are similar to those for a three-phase system operating in the rectifier mode in the case of an active–inductive load with continuous current I_d.

4.3.3 Active, reactive, and apparent powers of inverters

In considering the operating principle of the grid-tie inverter, we noted that the first harmonic of the nonsinusoidal grid current is shifted relative to the grid voltage by $\beta - \gamma/2$. As a result, the grid-tie inverter, which transmits the active power from the dc source to the grid, also draws reactive power from the grid. We now consider the power balance in the system consisting of the dc source, a single-phase inverter, and the grid. We assume unit efficiency of the inverter.

The active power consumed by the inverter from the dc source is

$$P = U_d I_d, \tag{4.56}$$

where U_d and I_d are the source voltage and the average current at the inverter input, respectively.

The same power on the ac side (e.g., for a single-phase circuit) may be expressed by the familiar formula if we take into account that the phase shift between the first harmonic of the grid current and the grid voltage is about $\beta - \gamma/2$:

$$P = U_s I_{s1} \cos\left(\beta - \frac{\gamma}{2}\right), \tag{4.57}$$

where U_s and I_{s1} are the effective values of the voltage and the first current harmonic in the grid, respectively.

From Equations 4.56 and 4.57,

$$I_{s1} = I_d \frac{U_d}{U_s \cos\varphi_1}, \tag{4.58}$$

where

$$\cos\varphi_1 \approx \cos\left(\beta - \frac{\gamma}{2}\right).$$

The reactive power of the first current harmonic generated by the grid in the inverter may be expressed in the familiar form as

$$Q = U_s I_{s1} \sin\left(\beta - \frac{\gamma}{2}\right) = P \cdot \tan\left(\beta - \frac{\gamma}{2}\right). \tag{4.59}$$

The inverter also creates higher current harmonics in the grid. For example, for a single-phase center-tap circuit, when $\omega L_d = \infty$, if we neglect the commutation angle γ, the grid current is rectangular and may be described by the harmonic series as

$$i_s = \frac{4I_d}{\pi}\left(\sin\vartheta + \frac{1}{3}\sin 3\vartheta + \frac{1}{5}\sin 5\vartheta + \cdots\right). \tag{4.60}$$

The harmonic composition of the primary current is analogous for a circuit operating in the rectifier mode (Section 4.2.3).

The nonsinusoidal form of the current may be assessed in terms of the distortion factor ν, which depends on the type of the circuit, the angle γ, the inductance L_s, the average current I_d, and other factors (Rozanov, 2009).

The total (apparent) inverter power S on the ac side is

$$S = U_s I_s = U_s\sqrt{I_{s1}^2 + \sum_{n=3}^{\infty} I_{sn}^2}. \tag{4.61}$$

Taking account of higher harmonics, we can write the inverter power factor in the form

$$\chi = \frac{P}{S} \cong \nu\cos\left(\beta - \frac{\gamma}{2}\right). \tag{4.62}$$

The potential for increasing the power factor by reducing β is constrained by the conditions for natural commutation of the thyristors: the angle $\delta = \beta - \gamma$ must always be greater than a specific value δ_{min}. This will be discussed in greater detail later.

Note that if the inverter begins to operate with a lag β, rather than a lead β, it becomes a generator of reactive power, rather than a consumer. If we look again at Figure 4.26, we can isolate two regions on the plane of possible variation in the first harmonic vector of the grid current for a converter with natural commutation of the thyristors.

I. Rectifier mode when the firing angle $\alpha = 0$–$\pi/2$, with the consumption of reactive power from the grid.
II. Inverter mode when $\alpha = \pi/2$–π ($\beta = 0$–$\pi/2$), with the consumption of reactive power from the grid.

4.3.4 Characteristics of inverters

In the analysis of normal inverter operation, it is important to know the inverter's input and boundary characteristics.

The input characteristic is the dependence of the inverter's average input voltage U_d on the average input current I_d.

The inverter's input voltage may be represented by the sum of two components, if we assume zero voltage drop at the thyristors and active impedances in the circuit. The first component is the no-load voltage U_{d0}, which is equal to the input voltage in instantaneous commutation (with $\gamma = 0$). The second component is the average voltage drop ΔU at the commutation intervals. In contrast to rectifiers, for which the voltage drop is subtracted from the no-load voltage, these components are summed in grid-tie inverters

$$U_d = U_{d0} + \Delta U. \tag{4.63}$$

For inverter circuits, U_{d0} and ΔU may be calculated from the relations analogous to those for controllable rectifiers. The voltage drop ΔU depends on the converter's input current: $\Delta U = f(I_d)$. Therefore, the input characteristic of the grid-tie inverter takes the form

$$U_d = \frac{2\sqrt{2}}{\pi} E_2 \cos\beta + \frac{I_d x_S}{\pi}. \tag{4.64}$$

Figure 4.30 shows the input characteristics of a single-phase inverter obtained from Equation 4.64 for different β. We see that, in contrast to the characteristics of a rectifier (shown in the left half-plane of Figure 4.30), these characteristics are ascending: the voltage rises with the current. The rectifier characteristics may be regarded as the continuation of the inverter's input characteristics with the same α and β.

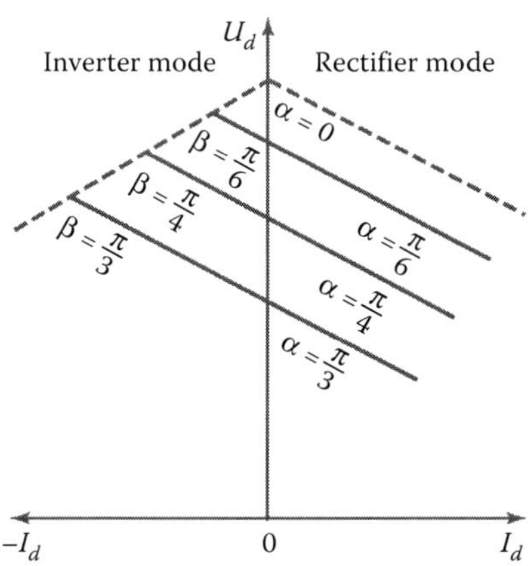

Figure 4.30 Input characteristics of a single-phase converter.

An increase in the input voltage U_d is associated with an increase in the current I_d and hence in the commutation angle γ. In other words, with the constant lead β, the thyristor cutoff angle δ declines. The minimum permissible value δ_{min} is determined by the frequency of the grid voltage and the type of thyristors. It follows from Equation 4.64 that an increase in β is associated with an increase in the permissible commutation angle γ and, hence, in the current I_d. The limiting permissible value of I_d can be determined as follows.

Suppose that the circuit operates in the rectifier mode and that α is numerically equal to δ_{min}. At this value of α, the rectifier characteristic is shown by the dashed line in the section of Figure 4.30, corresponding to the rectifier mode. We continue this characteristic in the section corresponding to the inverter mode. (It is again shown by a dashed line.) The intersection of this characteristic with the inverter's input characteristic will determine the inverter's limiting possible operating conditions in terms of I_d, for different β. For a single-phase inverter, these conditions correspond to the equation

$$U_d = \frac{2\sqrt{2}}{\pi} E_2 \cos \delta_{min} + \frac{I_d x_S}{\pi}. \tag{4.65}$$

As this characteristic indicates the inverter's limiting possible operating conditions, it is known as the boundary characteristic.

The voltage at the converter's dc buses when $I_d = 0$ (with no load) is the same for the rectifier and inverter modes and depends on β (or α). This dependence is usually known as the control characteristic. The converters here considered are reversible. In other words, by adjusting the firing angles and reversing the polarity of the dc source, we may switch between the rectifier mode and the inverter mode. In the rectifier mode, energy is sent from the ac grid to the dc source (which is, thus, a consumer in this case). The reversibility of such converters is utilized in engineering, especially in dc drives.

4.4 *Direct frequency converters (cycloconverters)*

4.4.1 *Thyristor-based ac–ac converters*

Frequency conversion involves the production of alternating current of one frequency from alternating current of another frequency. Numerous power electronic frequency converters exist (Zinov'ev, 2003). In the present section, we confine our attention to frequency converters based on thyristors with a natural commutation. Direct frequency converters—converters with a single transformation of electrical energy—are sometimes also known as direct-coupled converters or cycloconverters.

The number of phases of the input and output voltages in direct-coupled converters is of great importance, as it largely determines the converter structure. Note that multiphase direct-coupled converters are characterized by satisfactory performance and are widely used.

We will consider the operating principle of a direct-coupled converter with natural commutation for the example of a three-phase/single-phase circuit (Figure 4.31a). We can identify two groups of thyristors in the

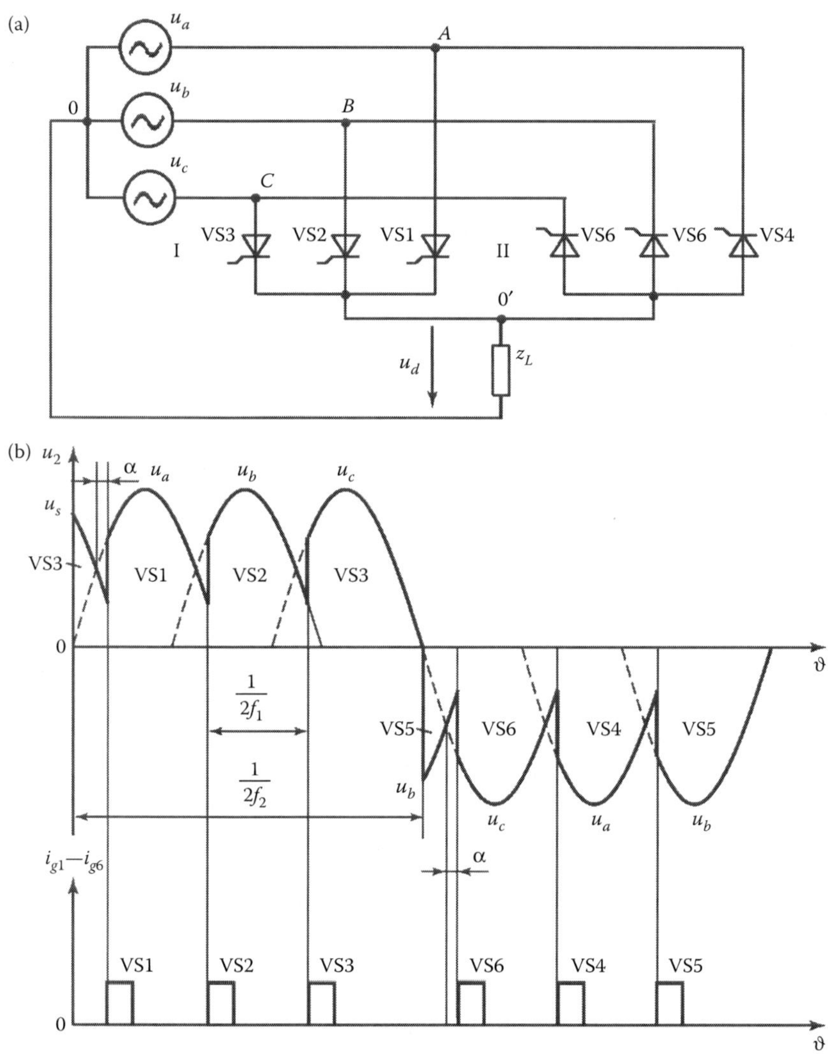

Figure 4.31 (a) A direct-coupled frequency converter and (b) the corresponding voltage waveforms and control pulses in the case of an active load.

converter: cathode group I (VS1, VS2, and VS3) and anode group II (VS4, VS5, and VS6). We assume that the load Z_L is active. In converter operation, control pulses are sent alternately to the anode and cathode groups. When control pulses i_{g1}–i_{g3}, synchronized in frequency with the grid voltage, are sent successively to thyristors VS1, VS2, and VS3 in the cathode group, the system operates in the rectifier mode (as a center-tapped three-phase circuit), and a positive voltage half-wave relative to the transformer tap is formed at the load (Figure 4.31b). The control pulses are sent to the thyristors with a phase shift α relative to the zeros of the grid's line voltage. With the operation of thyristors VS4, VS5, and VS6 in the anode group, a negative voltage half-wave relative to the transformer tap is formed at the load. As a result of the cyclic operation of groups I and II, an alternative voltage with a fundamental frequency f_2 lower than the grid frequency f_1 is created at the load.

The frequency f_2 is determined by the time for which the thyristors of each group are conducting. By adjusting α, the output voltage may be regulated. To eliminate the constant component in the voltage at the load, the operating times of the anode and cathode groups must be equal. The output voltage with an active load is shown in Figure 4.31b. We see that the thyristors of the cathode group operate only after the voltage half-wave formed by the anode group falls to zero and vice versa. That may be explained in that the thyristor is on until its current (in phase with the voltage, in the present case) falls to zero.

In the three-phase/single-phase circuit, commutation within each group (intragroup commutation) lasts for an interval of $\pi/3$. Therefore, if we disregard the commutation interval, we may write the following formula for the length of the output-voltage half-wave:

$$\frac{1}{2f} = \frac{2\pi}{3}n + \left(\pi - \frac{2}{3}(2+1)\right), \tag{4.66}$$

where n is the number of sinusoidal sections in the half-wave and $\pi - 2\pi/3$ is the angle corresponding to zero tail of the output-voltage half-wave.

In general, when the number of grid phases is m, the frequencies of the output voltage f_2 (fundamental frequency) and input voltage f_1 are related as follows:

$$f_2 = \frac{mf_1}{2+m}. \tag{4.67}$$

It is evident from Equation 4.67 that the frequency f_2 of the output voltage may take only discrete values with variation in n ($n = 1, 2, 3, \ldots$). For example, with $m = 3$ and $f_1 = 50$ Hz, f_2 may take values of 30, 23.5, 16.7 Hz, and so on. To ensure smooth frequency variation, we need a pause φ between the end of operation of one group and the onset of operation of the next. In that case, the relation between f_1 and f_2 takes the form

$$f_2 = \frac{f_1 m\pi}{\pi(n+m)+\varphi m}. \tag{4.68}$$

With a resistive–inductive load, the moments at which the output-voltage half-wave passes through zero do not correspond to zero load currents, as the load inductance delays the current relative to the voltage. To ensure that current is sent from the load to the grid in this case (which corresponds to recuperation, the return of the energy stored in the inductance to the grid), the appropriate thyristor group is switched to the inverter mode. For example, if group I operates in the rectifier mode at firing angle α, then, after a specific time, the control pulses are sent to the thyristors of group I with lead β relative to the grid voltage. That pulse sequence corresponds to the inverter mode of the thyristors. In this case, the dc source driving the inverted current is the load or, more precisely, the inductive component of the load. When the group-I thyristors are in the inverter mode, the energy stored in the inductance is returned to the grid and the load current falls to zero. Then the converter CS ensures a pause φ before the group-II thyristors begin to operate in the rectifier mode. Some of those thyristors switch to the inverter mode at a time specified by the control program. Subsequently, these processes repeat periodically.

A direct-coupled three-phase/single-phase converter can be based on two thyristor groups, each with a three-phase bridge configuration. Many direct-coupled circuits produce a three-phase voltage system at the output. Figure 4.32 shows some frequency converters with a three-phase

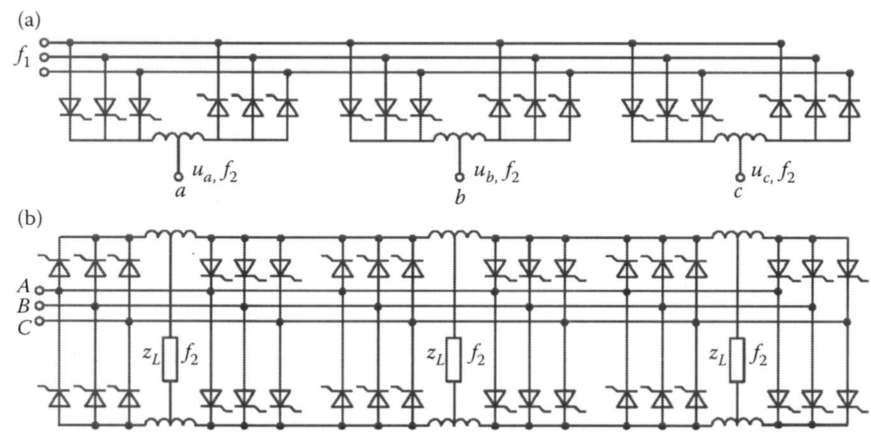

Figure 4.32 A three-phase frequency converter based on (a) center-tapped thyristor groups and (b) bridge groups.

output. In Figure 4.32a, each phase consists of two groups in a three-phase center-tapped configuration; in Figure 4.32b, each phase consists of two groups in a three-phase bridge configuration.

Direct-coupled converters with natural thyristor commutation are relatively simple (at least in terms of the power circuit) and tend to be light and compact. A deficiency is low quality of the output voltage. For example, if each thyristor group operates with constant control angles $\alpha = \beta$ for a half-period, as in Figure 4.32, the output voltage will be seriously distorted, with many higher harmonics. To reduce those harmonics and to ensure a sinusoidal output voltage, regulation of the control angles is required. In addition, the number of pulsations due to thyristor commutation is increased, as in multiphase rectifier circuits.

4.4.2 *Reduction of the output-voltage distortion*

Usually, converters have an active–inductive load. For such converters, in the case of a continuous output voltage, operation of each thyristor group in both inverter and rectifier modes is required.

As the current conduction in the thyristor groups is unidirectional, the positive wave of the current is formed by group I and the negative wave by group II. Therefore, with an active–inductive load, the current will pass through both groups during each half-period of the output voltage. Figure 4.33, as an example, shows the first harmonics of the current i_L and voltage u_L at the converter output, with a power factor cos φ in the case of an active–inductive load. The interval $0–\vartheta_1$ corresponds to an inverter mode II_b, and the current passes through the thyristors in group II. Subsequently, the current begins to pass through the thyristors in group I, which operate in rectifier mode I_a during the interval $\vartheta_1–\pi$. At time $\vartheta = \pi$, the thyristors in group I switch to inverter mode I_b and so on.

To prevent discontinuity on switching from the rectifier mode to the inverter mode, we employ consistent control of the thyristors.

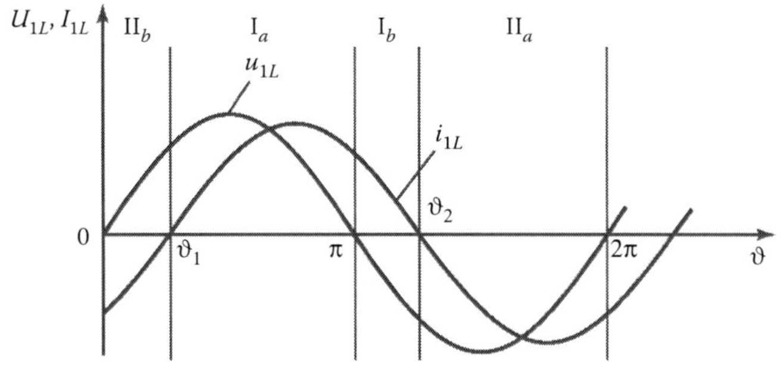

Figure 4.33 Ensuring sinusoidal output voltage in a direct-coupled frequency converter by adjusting the firing angle.

In that case, the delivery of control pulses is such that the thyristors in group I may operate for a single half-period in the rectifier mode with $\alpha \leq \pi/2$, but in the next half-period in the inverter mode with $\alpha = \pi - \beta$. Thyristor group II is prepared for the inverter mode in the first half-period and for the rectifier mode in the second half-period. In such control, considerable balance currents may appear between the two groups. They may be reduced if α and β are selected so that the average voltages in the rectifier and inverter modes are the same, that is, so that $\alpha = \beta$. The balance current due to difference in the instantaneous group voltages is limited by a reactor in the circuit between the two groups.

The output voltage of such a converter is nonsinusoidal, in the general case. Its harmonic composition depends on factors such as the variation in α and β, the number of grid phases, and the frequency ratio of the input and output voltages.

The content of higher harmonics in the output voltage can be reduced if

$$\alpha = \pi - \beta = \arccos(k \sin \omega_2 t), \tag{4.69}$$

where k is the amplitude ratio of the converter's input and output voltages and ω_2 is the frequency of the output voltage.

It follows from Equation 4.69 that, if $k = 1$, α and β must be linear functions of the time. (In that case, the arc cosine becomes a linear function of $\vartheta_2 = \omega_2 t$, which varies from 0 to π.) Figure 4.34 illustrates the control principle more clearly.

In the first half-period (from $\vartheta_2 = \pi/2$ to $\vartheta_2 = \pi$, where $\vartheta_2 = \omega_2 t$), group I is prepared for the rectifier mode, and the control pulses are sent to the thyristors of this group at firing angle α, which varies from 0 to $\alpha - \pi/2$. (Note that the output-voltage half-wave passes through a maximum at $\vartheta_2 = \pi/2$ and through zero at $\vartheta_2 = 0$.) In Figure 4.33, we show the notation for the intervals of thyristor-group operation in the rectifier and inverter modes: I_a and I_b correspond, respectively, to operation of group I in the rectifier and inverter modes during the interval $\pi/2$–π. At this time, the thyristors in group II are prepared for the rectifier mode, with $\beta = 0$–$\pi/2$. Likewise, II_a and II_b correspond, respectively, to operation of group II in the rectifier and inverter modes.

In such control, the content of higher harmonics in the output voltage is considerably reduced, as it becomes close to sinusoidal, with some ripple. The ripple diminishes with an increase in the frequency and in the number of grid phases. With an increase in the number of phases (pulsations) in the output voltage, we note a reduction not only in the higher harmonics of the output voltage, but also in the input current of the frequency converter. Figure 4.35 shows output voltage waveforms for 6 and 12-pulse converters with arc cosine control.

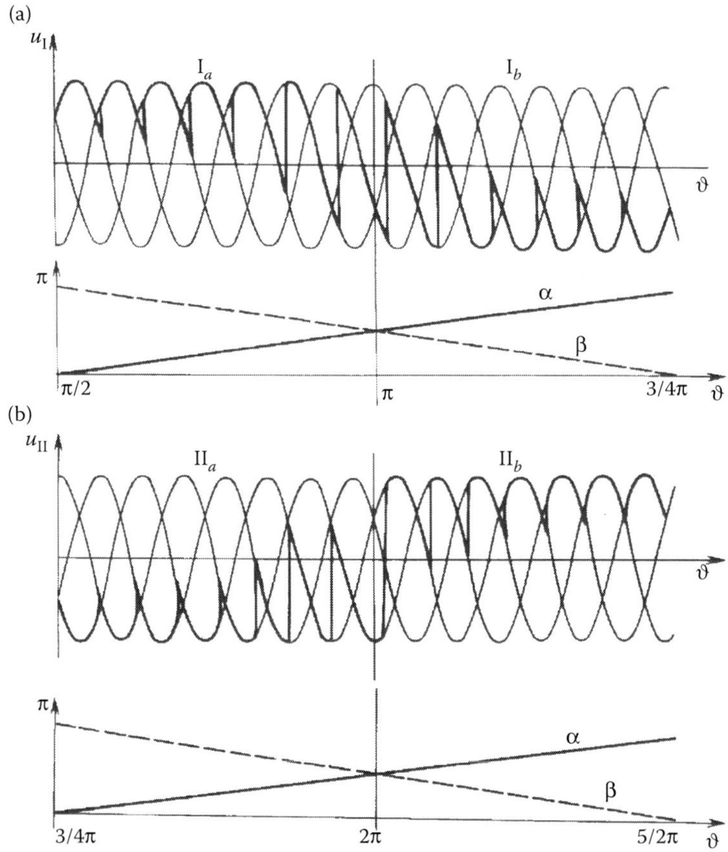

Figure 4.34 Output voltage waveform and firing angle variation for a frequency converter with the operation of thyristor groups (a) I and (b) II.

Note that the power factor of a direct frequency converter is determined not only by the power factor of the load, but also by the ratio of the input and output voltages. With a decrease in the output voltage, the control angles α and β increase, with a consequent decrease in the converter's input power factor. Low input power factor is one of the deficiencies of direct frequency converters with natural thyristor commutation.

4.5 ac voltage regulators based on thyristors

4.5.1 Single-phase ac voltage regulators

According to the IEC definition, an ac thyristor regulator can operate both as a direct ac voltage controller and as an electronic circuit breaker. The

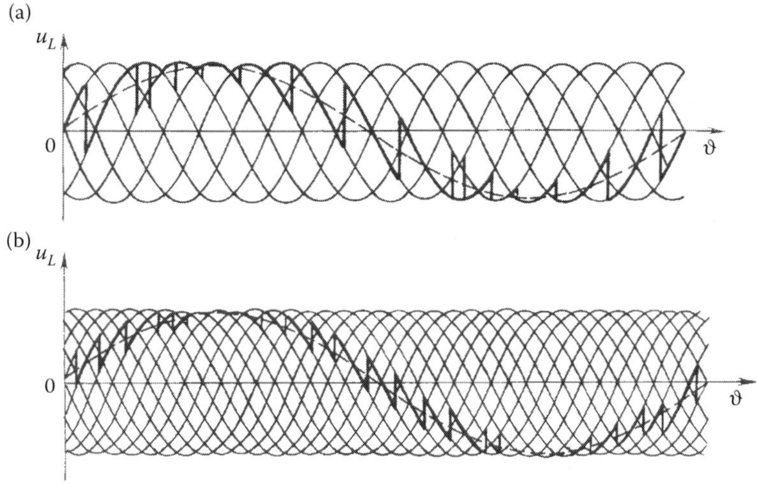

Figure 4.35 Output voltage for a (a) 6-pulse and (b) 12-pulse frequency converter.

function of an electronic circuit breaker is to switch an ac circuit off and on. We now consider regulators based on thyristors with natural commutation in the case of an ac grid.

A single-phase circuit with antiparallel thyristors is shown in Figure 4.36a. This is the basic circuit for thyristor regulators with natural commutation. Obviously, symmetric thyristors may take the place of two opposed thyristors.

We assume that the regulator has an active load. That corresponds to the connection of a resistor R_L to the regulator input. The other circuit components, including thyristors, are assumed to be ideal. In other words, they correspond to the assumptions stated earlier. The thyristors are switched on by the supply of control pulses i_{g1} and i_{g2} to the thyristors' control electrodes. The CS forms the pulses synchronously with the grid voltage $u_s = u_{ab}(\vartheta)$, in a phase corresponding to the firing angle α (Figure 4.36b).

When thyristor VS1 is turned on at time $\vartheta_1 = \alpha$, the input voltage is applied to load resistance R_L. The current i_L in the active-load circuit reproduces the form of voltage $u_{s.}$ When it falls to zero, thyristor VS1 is switched off. At time $\vartheta_1 = \pi + \alpha$, thyristor VS2 is switched on, and the processes repeat periodically if $\alpha = \text{const}$. We can write the following dependence of the effective output voltage $U_{L.\text{eff}}$ on the firing angle α in the case of an active load

$$U_{\text{L.eff}} = \sqrt{\frac{1}{\pi}\int_{\alpha}^{\pi}\left(\sqrt{2}U_{\text{ab}}\sin\vartheta\right)^2 d\vartheta} = U_{\text{ab}}\sqrt{1 - \frac{\alpha}{\pi} - \frac{\sin 2\alpha}{2\pi}}, \tag{4.70}$$

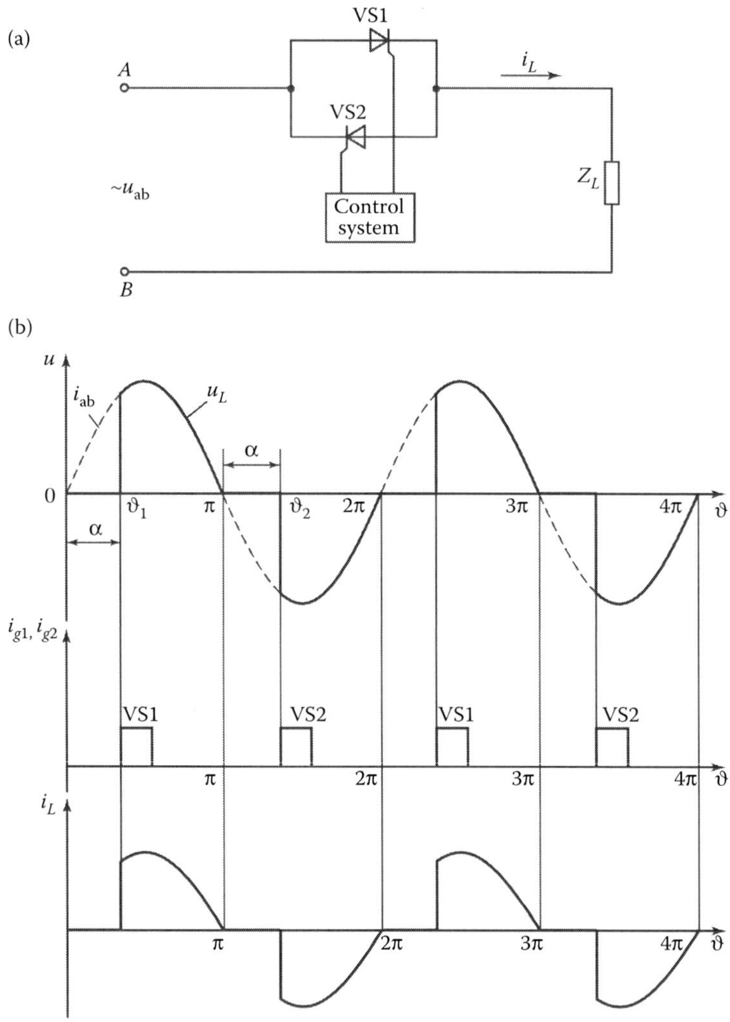

Figure 4.36 (a) An ac regulator based on antiparallel thyristors and (b) voltage and current waveforms.

where U_{ab} is the effective input voltage of the regulator, $\vartheta = \omega t$, and ω is the angular frequency of the grid voltage.

It is evident from Figure 4.36 that adjustment of α from 0 to π permits regulation of the average voltage (and hence also the effective voltage) from the maximum value (equal to the corresponding input voltage) to zero.

Various single-phase regulator circuits are shown in Figure 4.37. Note, in particular, the single-thyristor circuit in Figure 4.37b. However, this

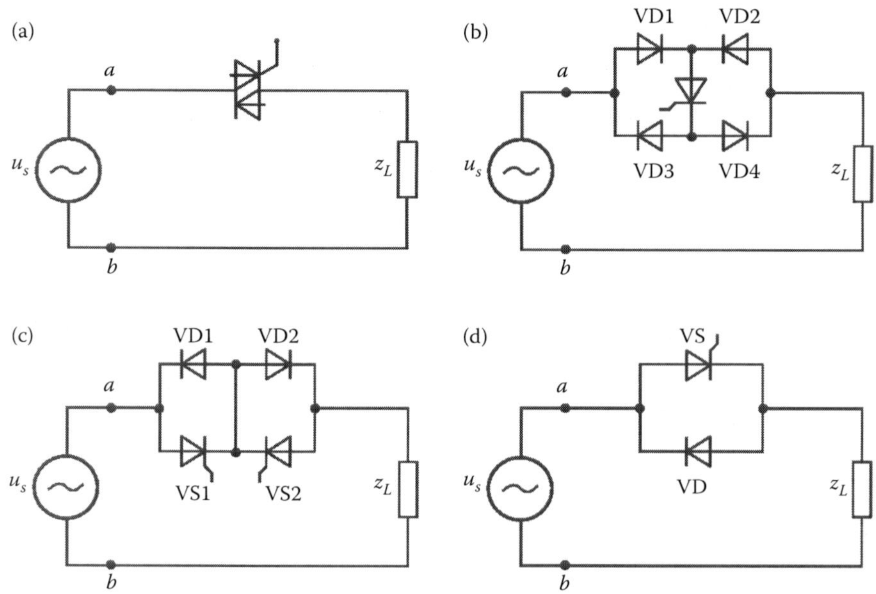

Figure 4.37 Single-phase thyristor regulators: (a) symistor based circuit; (b) single-thyristor circuit with four diodes; (c) circuit based on two thyristors with antiparallel diodes; (d) circuit with thyristor and antiparallel diode.

circuit has a significant drawback. Because the current flows through three semiconductor components (two diodes and a thyristor) in each half-period, the voltage drop will increase, according to the actual volt–ampere characteristics of these components, and hence the power losses will increase. That limits the use of this design at low voltages and low load currents.

The performance of a regulator is largely determined by its control characteristic, which relates the effective output voltage $U_{\mathrm{L.eff}}$ and the thyristor firing angle α. This characteristic depends significantly on the type of load. In the present case, active, active–inductive, and inductive loads are significant. We will consider each in turn.

4.5.1.1 *Operation with active load*

The dependence of the effective output voltage on the firing angle is described by Equation 4.70. It follows from the regulator operating principle that the output voltage $u_L(\vartheta)$ is nonsinusoidal, and the severity of the higher harmonics depends greatly on α. Figure 4.38 shows the harmonic content as a function of the firing angle α for an active load.

Obviously, an increase in α not only distorts the load current and voltage, but also impairs the input power factor χ, which can be determined

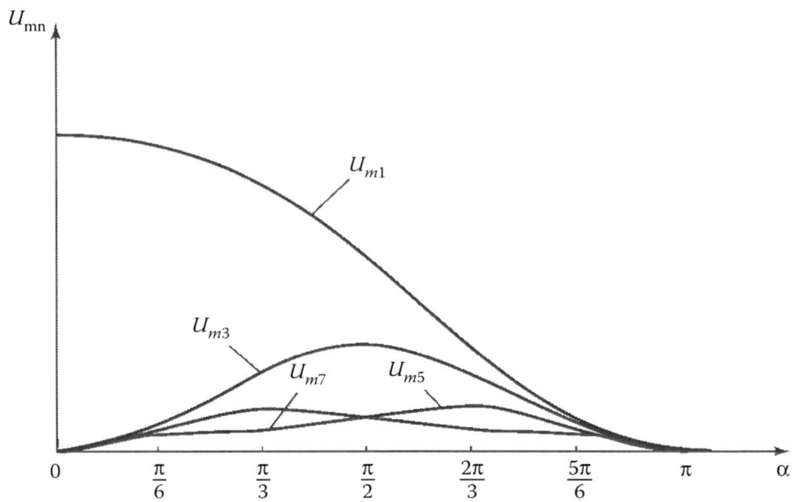

Figure 4.38 Amplitude of the higher harmonics as a function of the firing angle in the case of an active load.

from the series expansion of the current and formulas for the active and total power. The dependence of χ on α for an active load is shown in Figure 4.39.

4.5.1.2 Operation with resistive–inductive load

We assume that the load is a resistor R_L and a reactor (inductance L_L) in series. Turning on any of the thyristors in Figure 4.36a will initiate a transient process in the system consisting of the input-voltage source and the

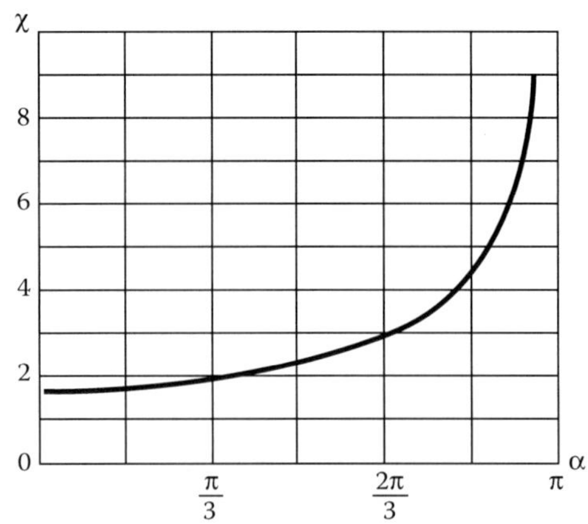

Figure 4.39 Power factor χ of the thyristor regulator as a function of the firing angle for a single-phase regulator with an active load.

load. Under the assumption that the components of the regulator are ideal and also that the input-voltage source is ideal, we may describe the transient process in the form

$$u_{ab}(t) = L_L \frac{di_L}{dt} + i_L R_L, \tag{4.71}$$

where ω is the angular frequency of the voltage source.

The solution is expressed as the sum of a transient component $i_{L.tr}$ and a steady-state component $i_{L.st}$:

$$i_L = i_{L.tr} + i_{L.st} \tag{4.72}$$

in the form

$$i_L(\vartheta) = \frac{\sqrt{2}U_{ab}}{\sqrt{R_L^2 + (\omega L_L)^2}} \left[\sin(\vartheta - \varphi_L) - \sin(\alpha - \varphi_L) e^{-(\alpha-\vartheta)/\omega\tau} \right]. \tag{4.73}$$

We can distinguish between three types of behaviors of $i_L(\upsilon)$, depending on α and φ_L:

$$\begin{cases} \alpha > \varphi_L, & \lambda < \pi, \quad (1) \\ \alpha < \varphi_L, & \lambda > \pi, \quad (2) \\ \alpha = \varphi_L, & \lambda = \pi. \quad (3) \end{cases} \tag{4.74}$$

In case (3), there is no transient component $i_{L.tr}$. Steady-state conditions begin as soon as the thyristor is turned on. In case (2), the transient component extends the first half-wave of the transient process to a duration greater than π. The corresponding conditions in the regulator are such that the second thyristor VS2 will not be turned on at the angle α. Instead, it will be shunted by the first thyristor, which continues to conduct during the second half-period. As a result, regulator operation is asymmetric. That leads to loss of voltage quality and unbalanced thyristor load. These problems may be eliminated by ensuring regulator operation with firing angles $\alpha \geq \varphi_L$.

4.5.1.3 *Operation with inductive load*

If we assume zero active-power losses in the circuit and in the load, regulator operation with an inductive load differs from operation with an active–inductive load, in that there is no damping of the free component. In symbolic form, the time constant $\tau = L_L/R_L \to \infty$.

Figure 4.40c shows the equivalent inductance X_{eq} as a function of α. We see that a regulator with input inductance L_0 may be regarded as the controllable inductance of the electronic CS in the range from $X_0 = \omega L_0$ to ∞.

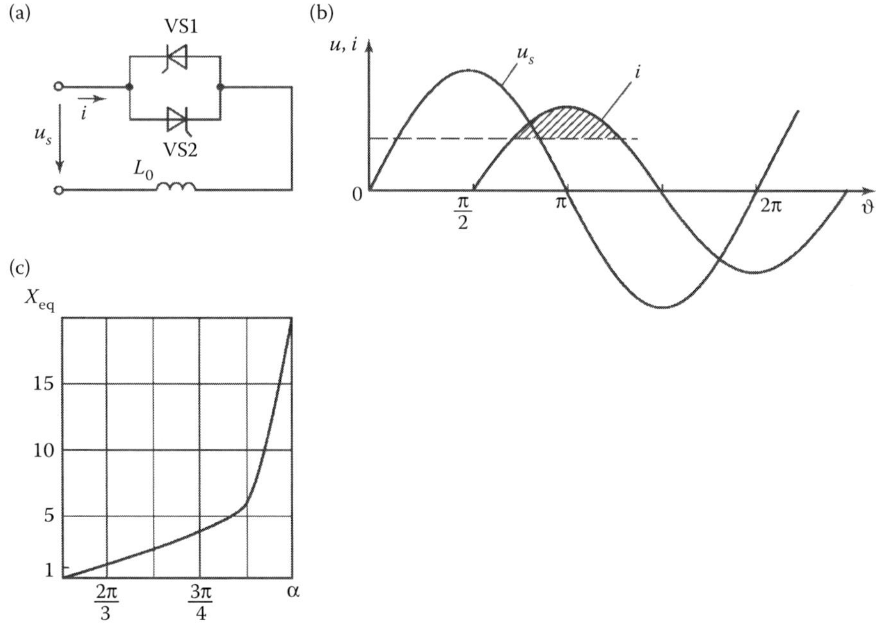

Figure 4.40 A thyristor-based ac regulator with an inductive load: (a) circuit; (b) voltage and current waveforms; and (c) dependence of the equivalent inductance on the firing angle.

This control method is widely used in the power industry for reactive-power compensation in devices consisting of parallel capacitor groups and a resistor antiparallel thyristor.

4.5.2 *Three-phase ac voltage regulators*

Figure 4.41 shows two characteristic designs for three-phase regulators: with a star configuration of the load and an isolated neutral and with a delta configuration. Switching from a single-phase system to a three-phase system complicates the topology and hence the analysis of the processes in the regulator. Some general laws for the operation of three-phase systems may be obtained for individual control-angle ranges (Williams, 1987). Thus, in the star configuration (Figure 4.41a) with an isolated neutral, three different operational modes of the regulator's thyristors may occur in the case of an active load. We will assume that control pulses are supplied at intervals of $\pi/3$ to the thyristors, in accordance with the numbering in Figure 4.41a. The origin $\vartheta = 0$ is the moment at which the phase voltage passes through zero to positive values. The three thyristor modes are as follows.

I. When $0 \leq \alpha \leq \pi/3$, only two thyristors are on in some intervals, whereas three thyristors are on in others: for example, VS5 and VS6; then VS5, VS6, and VS1; and so on.
II. When $\pi/3 \leq \alpha \leq \pi/2$, two thyristors are always on: for example, VS5 and VS6 and then VS1 and VS6.
III. When $\pi/2 \leq \alpha \leq 5\pi/6$, there are alternating periods in which two thyristors are on and all the thyristors are off. For example, there may be current flow through thyristors VS1 and VS6, but then they are turned off and a period follows in which all the thyristors are off. Then VS2 and VS1 are turned on and so on.

It is evident from these examples that, even for the simplest case of an active load, the processes in the three-phase regulator are considerably

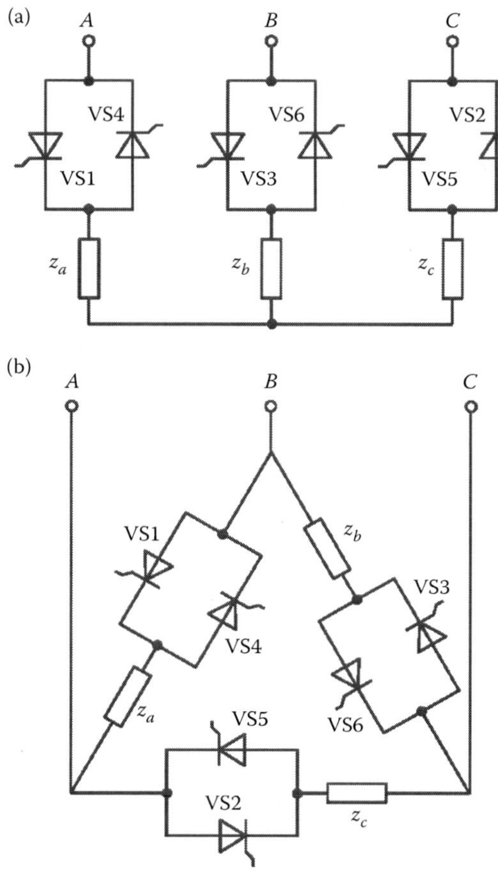

Figure 4.41 Three-phase thyristor regulator circuits: (a) star configuration with an isolated neutral; (b) delta configuration.

more complicated than those in a single-phase regulator. That is also the case for the delta configuration (Figure 4.41b). Note that the regulators with an active–inductive load are most common in practice. In that case, the processes are even more complex, and it is difficult to obtain formulas suitable for practical use. We may note that single-phase regulators are characterized by better spectral composition of the output voltage. In particular, in the delta configuration, current harmonics that are multiples of three do not reach the grid.

To reduce the distortion of the output voltage and output current and also to increase the input power factor χ, voltage stabilization may expediently be based on a thyristor regulator in combination with an autotransformer. Figure 4.42a shows a simplified voltage stabilizer in which the connection to the transformer winding is switched by means of thyristors VS1–VS4. In this system, the output voltage is stabilized by adjusting the switching times. In the positive half-period of the input voltage, thyristor VS1 or VS2 may be on; in the negative half-period, thyristor VS3 or VS4

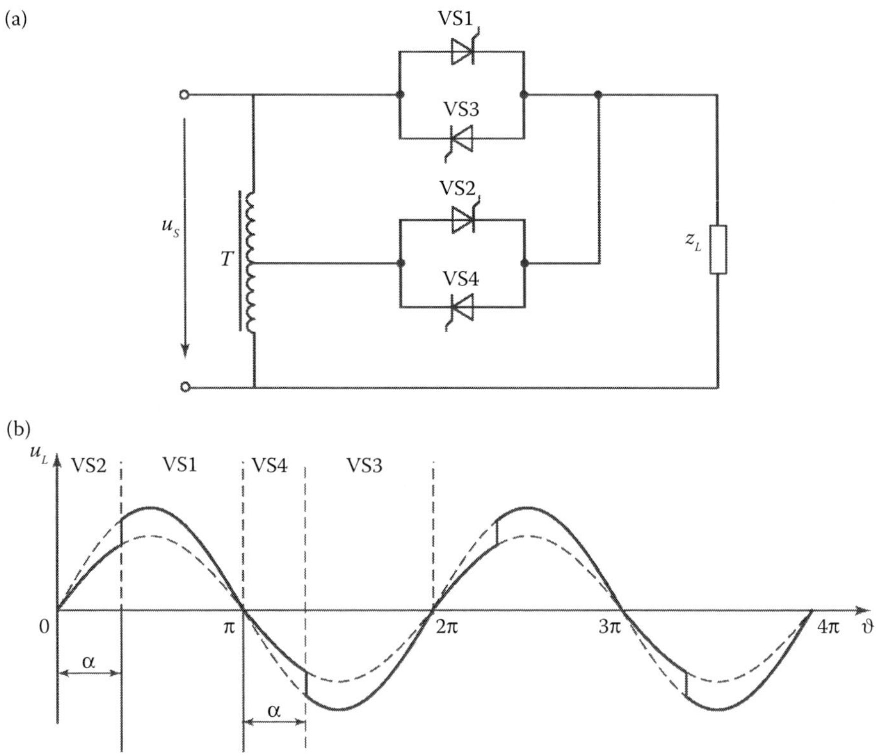

Figure 4.42 (a) A transformer-based voltage stabilizer and (b) its output voltage waveform.

may be on. The commutation of the thyristors in this system depends on the transformer voltage. To ensure natural commutation, switching to the terminals at high potential is required. For example, in the positive half-wave of the output voltage, first thyristor VS2 and then thyristor VS1 will be turned on. In this case, when thyristor VS1 is turned on, a short-circuit is formed. The short-circuit current opposes the load current at thyristor VS2. As a result, thyristor VS2 is turned off, and current begins to flow through thyristor VS1. In this system, the effective value of the output voltage may be regulated by adjusting the switching times of the thyristors. The output voltage of the stabilizer with an active load is shown in Figure 4.42b.

With an active–inductive load, the control of the thyristors is more complicated, because the load current lags the voltage at the transformer winding, and the thyristors are switched off when the load current passes through zero.

References

International Electrical-Engineering Dictionary, IEC BO050-551. 1998. Mezhdunarodnyi elektrotekhnicheskii slovar'. *Power Electronics, Silovaya elektronika, MEK BO050-551*. Moscow: Izd. Standartov (in Russian).

Rozanov, Yu.K. 1992. *Osnovy silovoi elektroniki (Fundamentals of Power Electronics)*. Moscow: Energoatomizdat (in Russian).

Rozanov, Yu.K. 2009. Power in ac and dc circuits. *Elektrichestvo*, 4, 32–36 (in Russian).

Rozanov, Yu.K., Ryabchitskii, M.V., and Kvasnyuk, A.A., Eds. 2007. *Silovaya elektronika: Uchebnik dlya vuzov* (*Power Electronics: A University Textbook*). Moscow: Izd. MEI (in Russian).

Williams, B.W. 1987. *Power Electronics: Devices, Drivers, and Applications*. New York: John Willey.

Zinov'ev, G.S. 2003. *Osnovy silovoi elektroniki (Fundamentals of Power Electronics)*. Novosibirsk: Izd. NGTU (in Russian).

chapter five

Conversion from direct current to direct current

5.1 Introduction: Continuous stabilizers

The conversion of direct current into direct current is intended to improve the power from a dc source and to match the voltage of the source and the consumers. Often, such conversion results in galvanic uncoupling of the source and the load.

The purpose of dc voltage converters is not only the conversion of direct current, but also the regulation or stabilization of the voltage (or current) in the load. Converters used only for stabilization are known as *stabilizers*. We may distinguish between two types of dc converters: continuous converters and pulsed converters.

Continuous converters stabilize the voltage in a dc circuit with variation in the source or load voltage. Continuous converters employ transistors operating in the active region of the output volt–ampere characteristic.

In Figure 5.1a, we show the circuit diagram of a continuous regulator with the transistor and load in series. The voltage at the load is

$$U_{\text{out}\cdot\text{max}} = E_{\min} - U_{\text{sw}\cdot\text{sat}}, \tag{5.1}$$

where $U_{\text{sw}\cdot\text{sat}}$ is the voltage at the transistor in the case of saturation.

In converter operation, the voltage at the load is stabilized: $u_{\text{out}} = U_{\text{st}}$. The current $i_0 = i_{\text{out}} = U_{\text{st}}/R_{\text{lo}}$. The control system (CS), which is based on negative feedback with respect to the output voltage, maintains the collector current at $i_0 = U_{\text{st}}/R_{\text{lo}}$. With an increase in voltage E, the voltage at the transistor $u_{\text{sw}} = E - U_{\text{st}}$ increases.

The excess energy from source E is scattered to the transistor; the power loss in the transistor is $P_{\text{tr}} = u_{\text{sw}} \cdot i_0$.

Neglecting the energy consumption in the operation of the CS, we may write the converter efficiency in the form

$$\eta = \frac{P_{\text{lo}}}{P_{\text{lo}} + P_{\text{tr}}} = \frac{U_{\text{st}} \cdot i_0}{U_{\text{st}} \cdot i_0 + (E - U_{\text{st}}) \cdot i_0} = \frac{U_{\text{st}}}{E}. \tag{5.2}$$

The precision of stabilization is mainly determined by the CS.

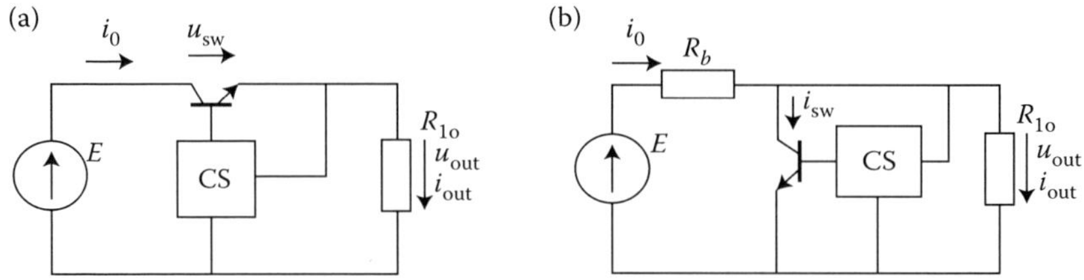

Figure 5.1 Continuous stabilizers: (a) with the load and transistor in series and (b) with a parallel transistor.

In Figure 5.1b, we show a stabilizer with a parallel transistor. The maximum attainable stabilized load voltage is

$$U_{out\cdot max} = E_{min}\frac{R_{lo\cdot min}}{R_{lo\cdot min} + R_b}. \tag{5.3}$$

When $E = E_{min}$, the transistor operates in the cutoff mode; the collector current $i_C = 0$. Due to the CS, characterized by negative feedback with respect to the output voltage, the load voltage is stabilized: $u_{out} = U_{st}$. With an increase in E, the voltage $u_R = E - U_{st}$ is applied to the ballast resistor R_b; the current is $i_0 = u_R/R_b$. The current through the transistor is also increased: $i_{sw} = i_0 - i_{out} = i_0 - U_{st}/R_{lo}$; the voltage at the transistor is $u_{sw} = U_{st}$.

The excess energy from source E is scattered in the ballast resistor R_b and the transistor. The power loss is

$$P_{loss} = P_R + P_{sw} = u_R i_0 + U_{st} i_{sw}. \tag{5.4}$$

The converter efficiency is

$$\eta = \frac{U_{st}^2}{E^2 - U_{st}E}\cdot\frac{R_b}{R_{lo}}. \tag{5.5}$$

The efficiency of the stabilizer with a parallel transistor (Figure 5.1b) is less than that of a stabilizer with the transistor in series (Figure 5.1a). However, the losses in the transistor may be less.

The low efficiency of the continuous stabilizers limits their use to devices whose power is no more than a few watts. Continuous stabilizers and the associated CS are produced in the form of integrated microcircuits.

At present, only pulsed converters are used at high power. In such devices, the regulation and stabilization of the output voltage (or current) may be ensured by adjusting (modulating) the width of the rectangular pulses. The use of switched semiconductor devices considerably increases the converter efficiency and reduces the mass and size of the unit.

The literature contains an enormous variety of circuit designs for pulsed dc voltage converters. In what follows, we consider the most common types.

Pulsed dc voltage converters without transformers are sometimes known as *dc voltage regulators.*

5.2 Basic dc voltage regulators

5.2.1 Step-down dc/dc converter

In the step-down (or buck) dc/dc converter, also known as a voltage regulator of the first kind, the transistor is in series with the source E and the load (Polikarpov and Sergienko, 1989; Ericson and Maksimovich, 2001; Mohan et al., 2003; Rashid, 2004; Meleshin, 2006).

In Figure 5.2, we show the circuit diagram, together with corresponding time diagrams. The circuit includes a smoothing LC filter. (In some

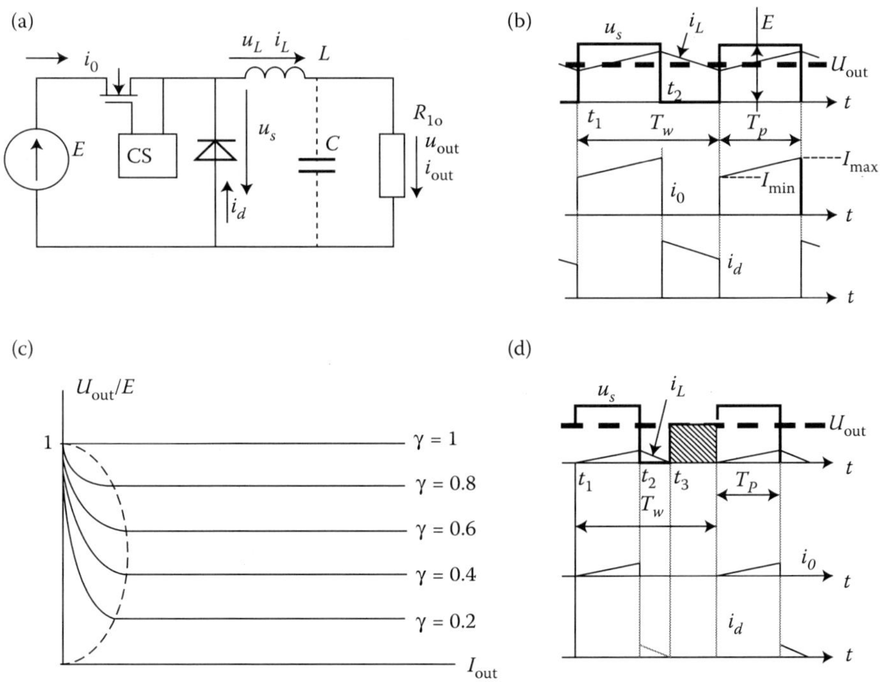

Figure 5.2 Step-down dc/dc converter: (a) circuit diagram; (b) current and voltage diagrams with continuous choke current; (c) characteristics; and (d) current and voltage diagrams with discontinuous current.

cases, the capacitor C is omitted.) When the transistor is switched on for interval t_1–t_2 (pulse length T_p), voltage $u_s = E$ is formed at the output, whereas voltage $u_L = E - u_{вых} > 0$ is applied to choke L. This is accompanied by an increase in the current $i_0 = i_L$ through the source E, the transistor, and choke L; the energy stored in the choke increases.

At time t_2, the transistor is switched off. The current in the choke L passes through the diode. The load voltage $u_L = -u_{out} < 0$ is applied to the choke. The choke current $i_L = i_d$ declines, and the energy stored in the choke declines. The voltage $u_s = 0$; no current is drawn from the source: $i_0 = 0$. At time $t_1 + T_{sw}$, the transistor is switched on again, and the process is repeated. (The repetition period $T_{sw} = 1/f_{sw}$, where f_{sw} is the switching frequency.)

If we neglect the losses in the converter, we may determine the mean output voltage of the converter in the form

$$U_{out} = E\frac{T_p}{T_{sw}} = \gamma E, \tag{5.6}$$

where $\gamma = T_p/T_{sw}$ is the fill factor. In Figure 5.2b, we plot U_{out}. With satisfactory smoothing of the pulsations in the load, $u_{out} = U_{out}$.

The mean current i_L is equal to the mean load current

$$I_{L\cdot me} = \frac{\gamma \cdot E}{R_{lo}}. \tag{5.7}$$

The difference between the minimum and maximum values of i_L is

$$\Delta I = I_{max} - I_{min} = \frac{E(1-\gamma)\gamma}{L \cdot f_{sw}}. \tag{5.8}$$

The converter characteristics are hard. Their slope is determined by the losses in the converter. We present the characteristics in Figure 5.2c, in which the dashed curve corresponds to the boundary between operation with continuous choke current i_L (the basic mode of converter operation, on the rightmost section of the characteristics) and operation with discontinuous current. At this boundary, there is a sharp change in the characteristic.

The boundary in Figure 5.2c corresponds to $I_{min} = 0$ and $\Delta I/2 = I_{L,me}$. Hence

$$I_{out\cdot b} = \frac{E\gamma(1-\gamma)}{2L \cdot f_{sw}}. \tag{5.9}$$

In Figure 5.2d, we show the time diagrams with discontinuous current. At time t_1, the transistor is switched on, and $u_s = E$. The current $i_0 = i_L$ rises, and the energy stored in the choke increases. When the transistor is switched off, at time t_2, the choke current passes though the diode and declines to zero at time t_3; the diode is closed. When the choke current is zero, $u_s = u_{out}$. The part of u_s due to interruption of the choke current is shaded in Figure 5.2d. The appearance of this component increases the mean output voltage; the converter characteristic in the case of discontinuous current falls sharply.

In the case of discontinuous current, the power component of the converter remains operational. However, when developing feedback loops for the CS, the role of discontinuous current must be taken into account.

In basic operation with continuous current, the pulsation of the output voltage may be written in the form

$$\frac{\Delta U_{out}}{U_{out}} = \frac{1-\gamma}{8LC \cdot \gamma \cdot f_{sw}^2}. \tag{5.10}$$

Hence, the switching frequency must be increased in order to reduce the power consumption in the filter's reactive elements. The inductance of choke L is usually selected so that the converter operates solely with continuous current.

With a decrease in γ, the voltage U_{out} increases. At the same time, however, its harmonic composition deteriorates. The ratio of the maximum and mean transistor currents is $1/\gamma$. At small γ, this fact must be taken into account in transistor selection. Consequently, the use of step-down dc/dc converters is inexpedient for considerable reduction in the dc voltage.

At E up to 300–350 V, regulators are based on MOS transistors. If the switching frequency is increased to 100 kHz, the power consumption in the reactive elements may be reduced. At higher voltages, the regulators may be based on IGBTs, with switching frequencies of 20 kHz or more.

The benefit of the step-down dc/dc converter is the continuity of the current supplied to the input of the output filter. Consequently, the parameters of the filter's reactive elements are minimal.

5.2.2 *Step-up dc/dc converter*

In the step-up (or boost) dc/dc converter, also known as a voltage regulator of the second kind, the choke L is in series with the source L. The corresponding circuit diagram is shown in Figure 5.3a, whereas time diagrams with continuous choke current are shown in Figure 5.3b.

When the converter is connected to source E with no control pulses at the transistor, capacitor C is charged through choke L and the diode to

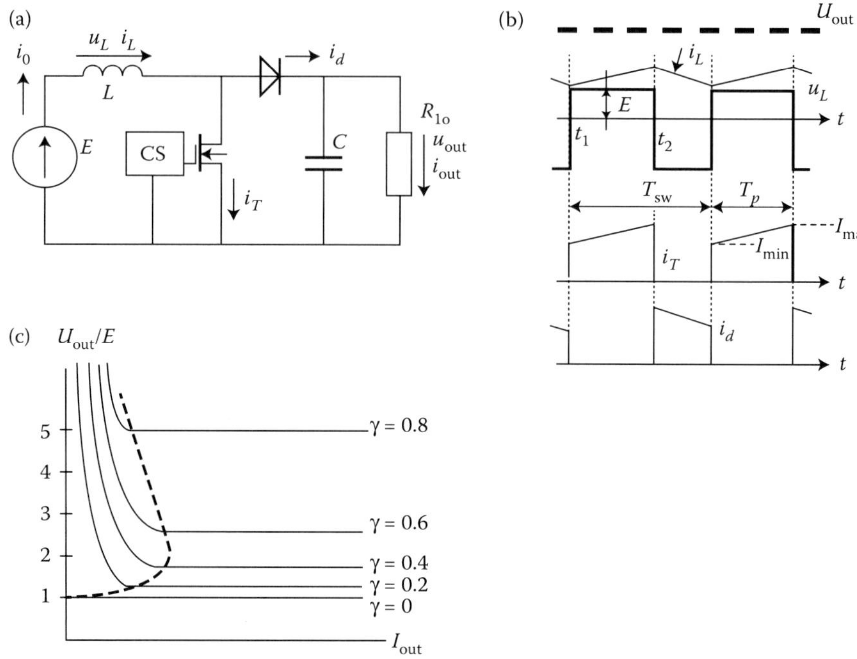

Figure 5.3 Step-up dc/dc converter: (a) circuit diagram; (b) current and voltage diagrams with continuous choke current; and (c) characteristics.

a voltage greater than E. The regulator operates with a mean load voltage $U_{out} > E$.

In steady conditions, when the transistor is switched on at time t_1, the supply voltage is applied to the choke: $u_L = E$. Energy is supplied to the choke, and the current $i_L = i_0 = i_T$ increases. The diode is closed, and the load is disconnected from the source. Then the capacitor C supplies energy to the load.

When the transistor is switched off, at time t_2, the diode is opened, and voltage $u_L = E - u_{out}$ is applied to the choke; the current $i_L = i_0 = i_d$ declines.

If we neglect the losses in the converter, the converter's mean output voltage is

$$U_{out} = \frac{E}{1 - \gamma} > E, \tag{5.11}$$

where $\gamma = T_p/T_{sw}$ is the fill factor. We plot U_{out} in Figure 5.3b.

The converter characteristic is shown in Figure 5.3c. The right side corresponds to the basic mode of converter operation, with continuous choke current. In that case, the characteristics are hard; the losses in the converter are due to inclination of the characteristics. The dashed curve in Figure 5.3c is the boundary of operation with continuous current.

The mean current i_L is

$$I_{L\cdot me} = \frac{E}{(1-\gamma)^2 R_{lo}}. \tag{5.12}$$

The difference between the minimum and maximum values of i_L is

$$\Delta I = I_{max} - I_{min} = \frac{E\cdot\gamma}{L\cdot f_{sw}}. \tag{5.13}$$

The boundary of operation with discontinuous current in Figure 5.3c corresponds to $I_{min} = 0$ and $\Delta I/2 = I_{L,me.}$ Hence

$$I_{out\cdot b} = \frac{E\gamma(1-\gamma)}{2L\cdot f_{sw}}. \tag{5.14}$$

With discontinuous current, when the transistor is off, the choke current flows through the diode. However, the diode closes when $i_L = 0$, corresponding to zero energy stored in the choke. The load is connected to the capacitor, as in the operating interval t_1–t_2, when the transistor current flows. As a result, the load voltage rises, while the characteristic declines sharply (Figure 5.3c). With discontinuous current, the power component of the converter remains operative. However, the role of discontinuous current must be taken into account in developing the feedback loops for the CS. In near-idling conditions, the converter is inoperative on account of the unlimited growth of the output voltage with any value of γ. That may result in failure of the semiconductor devices and the load.

In basic operation, with continuous choke current, the pulsations of the output voltage are described as follows:

$$\frac{\Delta U_{out}}{U_{out}} = \frac{\gamma}{R_{lo}Cf_{sw}}. \tag{5.15}$$

With an increase in U_{out}/E, γ increases, and the quality of the output voltage is impaired. The ratio of the maximum and mean diode currents is $1/(1-\gamma)$; at large γ, that must be taken into account in diode selection. Accordingly, the use of a step-up dc/dc converter to considerably increase a dc voltage is inexpedient.

A benefit of the step-up dc/dc converter is the absence of discontinuities in the current consumed from the source E, with continuous choke current. That permits a decrease in the filter capacitance at the converter input or even its elimination. A deficiency is the discontinuity of the current supplied to the input of the output filter.

5.2.3 Inverting regulator

The inverting regulator—also known as a regulator of the third kind or a buck–boost converter—is shown in Figure 5.4a. It is named as inverting regulator because the source and load voltages are of opposite polarity. A transistor is in series with the source E, whereas the transistor current and source current are the same: $i_T = i_0$. In Figure 5.4b, we show time diagrams in basic operation, with continuous choke current.

At time t_1, when the transistor is turned on, the supply current flows through the transistor and the choke. The supply voltage is applied to the choke: $u_L = E$. Energy is stored in the choke, and the current $i_L = i_0 = i_T$

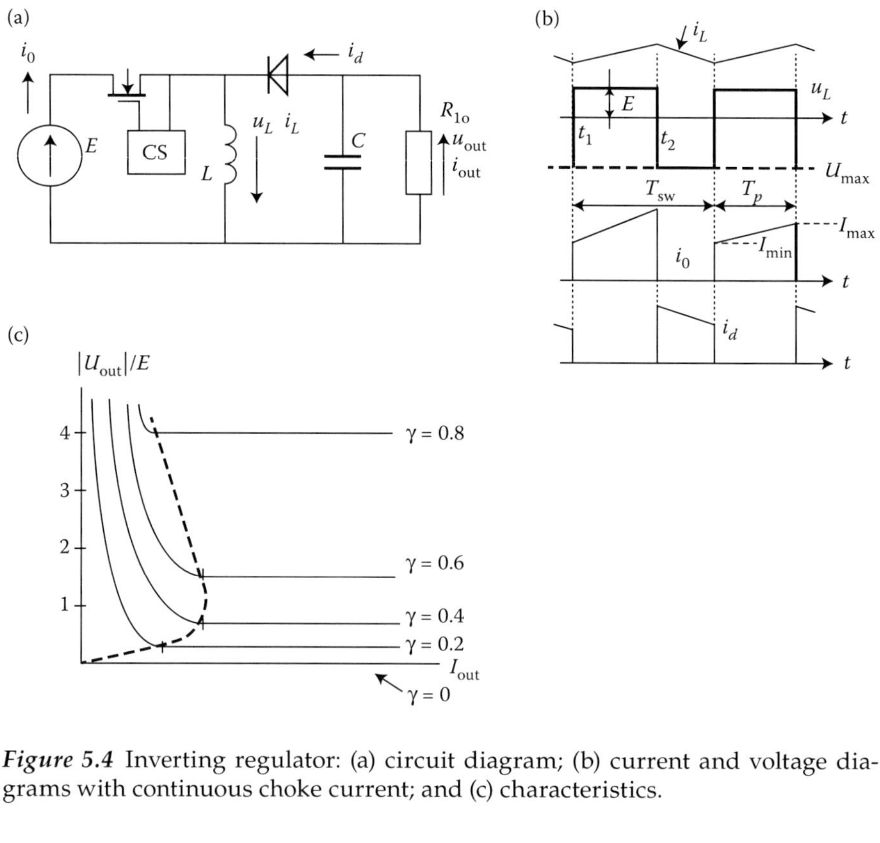

Figure 5.4 Inverting regulator: (a) circuit diagram; (b) current and voltage diagrams with continuous choke current; and (c) characteristics.

rises. The diode is switched off, the load is disconnected from the source, and capacitor C sends energy to the load.

At time t_2, when the transistor is turned off, the diode is turned on as a result of the choke's self-induction voltage. The voltage $u_L = -u_{out}$ is applied to the choke. The choke loses energy, and the current $i_L = i_d = i_{out}$ declines.

If we neglect the losses in the converter, we may determine the mean output voltage of the converter in the form

$$U_{out} = \frac{-E\gamma}{1-\gamma}, \tag{5.16}$$

where $\gamma = T_p/T_{sw}$ is the fill factor. In Figure 5.4b, we plot U_{out}. Depending on γ, the absolute value of U_{out} may be larger or smaller than E.

The maximum voltage at the transistor is $E + |U_{out}|$.

The family of converter characteristics is shown in Figure 5.4c. The right side corresponds to the basic mode of converter operation, with continuous choke current. In that case, the characteristics are hard; the losses in the converter are due to inclination of the characteristics. The dashed curve in Figure 5.4c is the boundary of operation with continuous current.

The mean current i_L is

$$I_{L\cdot me} = \frac{E\gamma^3}{(1-\gamma)^2 R_{lo}} + \frac{E\gamma}{R_{lo}}. \tag{5.17}$$

The difference between the minimum and maximum values of i_L is

$$\Delta I = I_{max} - I_{min} = \frac{E\cdot\gamma}{L\cdot f_{sw}}. \tag{5.18}$$

In Figure 5.4c, the boundary of the discontinuous current corresponds to $I_{min} = 0$ and $\Delta I/2 = I_{L,me}$. Hence

$$I_{out\cdot b} = \frac{E\gamma(1-\gamma)}{2L\cdot f_{sw}}. \tag{5.19}$$

With discontinuous current at time t_2, the choke current passes through the diode. However, when $i_L = 0$, corresponding to zero energy stored in the choke, the diode closes. The load is connected to the capacitor. The load voltage increases, and the converter characteristic is sharply descending (Figure 5.4c). With discontinuous current, the power component of the converter remains operational. However, the role of discontinuous current must be taken into account in developing feedback loops for the CS. In near-idling conditions, the converter is inoperative on account of the unlimited growth of the transistor voltage.

In basic operation, with continuous choke current, the pulsation of the output voltage is

$$\frac{\Delta U_{out}}{U_{out}} = \frac{\gamma}{R_H C f_{sw}}. \tag{5.20}$$

The deterioration of the output voltage as U_{out}/E increases, meaning that the converter cannot effectively be used to produce a considerable increase in the dc voltage.

A benefit of the inverting regulator is that the output voltages higher or lower than the supply voltage may be obtained.

Deficiencies include the discontinuity of the current drawn from the source at low loads and the discontinuity of the current supplied to the input of the output filter. These deficiencies may be eliminated by topological adjustments on the basis of cascade connection of such circuits. To this end, we use a new circuit design, known as the Ćuk converter.

5.2.4 The Ćuk converter

In Figure 5.5, we show the circuit diagram of the Ćuk converter (Mohan et al., 2003; Rozanov, 2007), along with time diagrams. At time t_1, when the transistor is switched off, voltage $u_{L1} = E$ is applied to choke $L1$ and the supply voltage i_0 increases. The negative voltage accumulated at capacitor

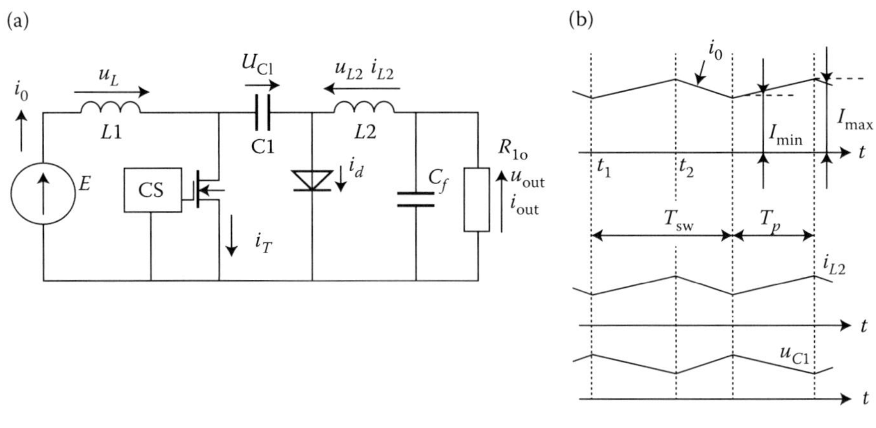

Figure 5.5 Ćuk converter: (a) circuit diagram and (b) current and voltage diagrams with continuous current.

C1 is applied to the anode of the diode, which closes. Capacitor C1 is connected to the input of the LC filter, is discharged by current i_{L2}, and sends energy to the filter and the load.

At time t_2, the transistor is switched on. Choke current $L1$ passes through the capacitor C1 and the diode; the energy and voltage at the capacitor increase. The zero voltage of the open diode is applied to the filter input, and the current of choke $L2$ also passes through the diode. At the end of the interval T_f, the capacitor voltage is restored: $u_{C1}(t_1 + T_f) = u_{C1}(t_1)$.

If we neglect the losses in the converter, we may determine the mean output voltage of the converter in the form

$$U_{\text{out}} = \frac{-E\gamma}{1-\gamma}. \tag{5.21}$$

Depending on γ, the absolute value of U_{out} may be larger or smaller than E.

The corresponding family of converter characteristics is as in Figure 5.4c. The boundary of operation with continuous current is shown by the dashed curve in Figure 5.4c. Discontinuous current is observed with interruptions in the current both at choke $L1$ and at choke $L2$. The corresponding formulas may be found in Polikarpov and Sergienko (1989). In near-idling conditions, the converter is inoperative.

The pulsations of the output voltage take the form

$$\frac{\Delta U_{\text{out}}}{U_{\text{out}}} = \frac{1-\gamma}{8L2 \cdot C_f \cdot f_{\text{sw}}^2}. \tag{5.22}$$

The Ćuk converter has significant benefits.

- It consumes continuous current from the source E, like a step-up dc/dc converter.
- Continuous current is supplied to the filter input, as in the step-down dc/dc converter.

In one version of the Ćuk converter, the two chokes share a single coil and are magnetically coupled. That permits considerable reduction in the pulsations of the current i_0 or the output voltage.

5.2.5 *Regulators with voltage multiplication*

Voltage multiplication is possible in direct-current regulators (Polikarpov and Sergienko, 1989). That considerably increases the ratio of the output voltage and the supply voltage, without the need for transformers. Of the various options, we consider that in Figure 5.6.

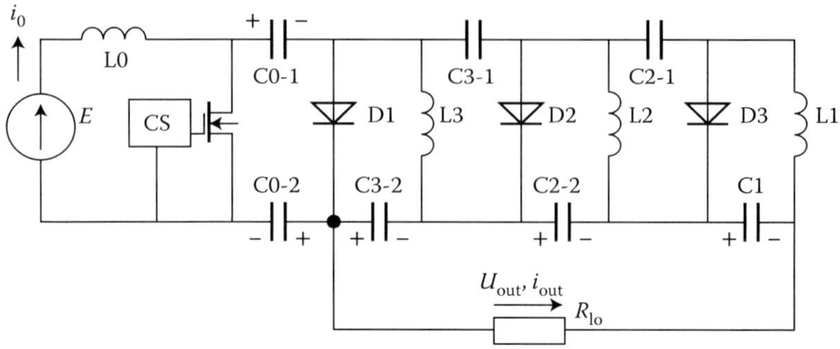

Figure 5.6 Regulator with voltage multiplication.

When the transistor is switched on, energy accumulates in choke L0; diodes D1, D2, and D3 are closed; and the energy from capacitors C0-1 and C0-2 is transmitted to the cells consisting of L-shaped segments C1–L1, C2–L2, and C3–L3.

When the transistor is switched off, current i_0 from choke L0 passes through diode D3; the choke energy is transmitted to capacitors C0-1 and C0-2. Choke L3 sends energy to capacitors C3-1 and C3-2 through diodes D2–D3; choke L2 sends energy to capacitors C2-1 and C2-2 through diodes D1–D2; and choke L1 sends energy to capacitor C1 through diode D1. In this interval, capacitors C3-1–C3-2 and C2-1–C2-2 are connected in parallel through the conducting diodes, whereas chokes L1, L2, and L3 are in series. The mean current in the chokes is equal to the load current I_{out}.

The number of cells of the same type within the converter may be increased or decreased. If we neglect the losses in the converter, we may determine the mean output voltage of the converter in the form

$$U_{out} = \frac{EN\gamma}{1-\gamma}, \tag{5.23}$$

where N is the number of cells in the converter (in Figure 5.6, $N = 3$).

The voltage at the capacitors of each cell is

$$U_C = \frac{E\gamma}{1-\gamma}, \tag{5.24}$$

whereas the inverse voltage at the diodes is

$$U_D = \frac{E}{1-\gamma}. \tag{5.25}$$

As already noted, the losses at reactive elements may be reduced by increasing the switching frequency. However, with an increase in the frequency, the switching losses in the semiconductor devices will increase. Those losses may be reduced by using resonant converters and various types of zero-switching circuits.

5.3 dc voltage regulators with transformer uncoupling of the input and output circuits

When there is a considerable difference between the source voltage and the output voltage at the load, we use dc voltage regulators with built-in transformers (Polikarpov and Sergienko, 1989; Ericson and Maksimovich, 2001; Rashid, 2004; Rossetti, 2005; Rozanov, 2007). In such converters, the transformers operate at high frequency, so as to reduce the power consumption at the reactive components and the overall size and mass of the system. Transformers permit galvanic uncoupling of the input and output circuits and multiplication of the output circuits so as to produce higher voltages. In what follows, we consider only a few of many such devices that exist.

At power up to 1 kW, half-cycle dc voltage converters (flyback and forward converters) are often used. At higher power, full-cycle converters (push–pull converters) are mainly employed.

5.3.1 Flyback converter

In Figure 5.7a, we show the circuit diagram of a flyback converter, characterized by energy transfer during the intervals when the transistor is off; time diagrams of the currents through the switches are shown in Figure 5.7b.

The flyback converter is based on a dc inverting regulator (Section 5.2.3). In split choke L, the turn ratio in the windings is $k = w2/w1$. When the windings are turned on in accordance with Figure 5.7a, the load voltage is positive.

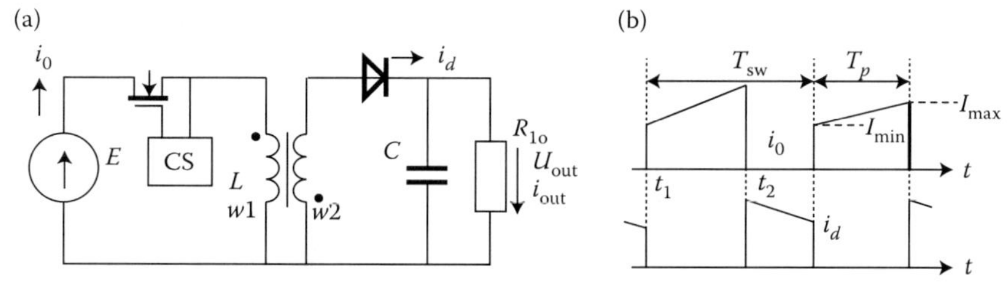

Figure 5.7 Flyback voltage converter: (a) circuit diagram; (b) current and voltage diagrams with continuous choke current.

When the transistor is turned on at time t_1, energy accumulates in the core of the split choke; there is an accompanying increase in the transistor current, which is equal to the source current i_0. The voltage induced in the second winding $w2$ closes the diode, and capacitor C sends its accumulated energy to the load.

When the transistor is turned off at time t_2, the current flow in winding $w1$ stops. The energy stored in the core is sent to the load and replenishes the energy of capacitor C; the diode current i_d falls. Energy is transmitted to the load in the dead time when the transistor is off.

If the standard values of the components on the secondary side (C and R_{lo}) are recalculated for the primary side and the windings are combined, the converter will be identical to an inverting regulator. All the relations and characteristics given in Section 5.3 apply in those conditions. The family of characteristics takes the form as shown in Figure 5.4c; the values of U_{out}/E must be multiplied by $k = w2/w1$.

If we neglect the losses in the converter, we may determine the mean output voltage of the converter in the form

$$U_{out} = \frac{E \cdot k \cdot \gamma}{1 - \gamma}. \tag{5.26}$$

In the interval t_1–t_2, the current flows only in winding $w1$; during the remainder of the period, the current flows only in winding $w2$. The core is magnetized by the whole load current.

The boundary of operation with discontinuous current corresponds to

$$I_{out \cdot b} = \frac{E\gamma(1 - \gamma)}{2k \cdot L \cdot f_{sw}}, \tag{5.27}$$

where L is the inductance of winding $w1$.

In idling (zero load), the converter is inoperative.

The flyback converter has the same deficiency as the inverting regulator: the source current i_0 and the current i_d sent to the load circuit ($R_{lo} - C$) are discontinuous. The following problems may also be noted.

1. For the best converter operation, the magnetic coupling between the windings should be as large as possible. To that end, appropriate technological and design measures are adopted.
2. The core of the split choke operates over an asymmetric portion of the hysteresis loop and acts as an energy store, which increases the consumption at the magnetic elements.
3. In Figure 5.7a, there are no circuits for removing the energy from the choke's scattering inductance. That may lead to impermissible voltage surges at the transistor.

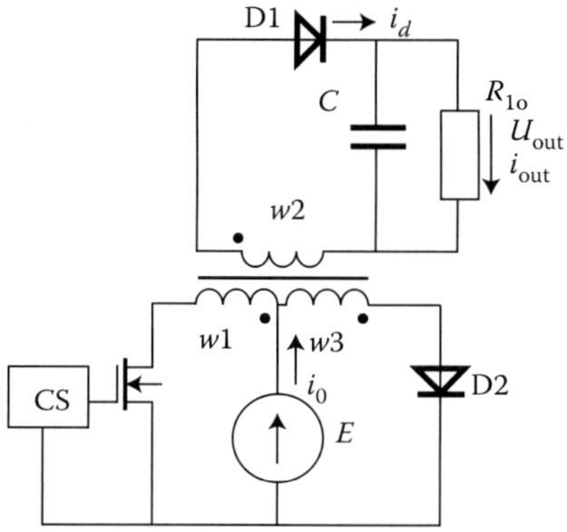

Figure 5.8 Flyback converter with additional recuperative circuit.

To eliminate the latter problem, various circuit designs are employed; one is shown in Figure 5.8. In this case, a recuperative circuit consisting of winding *w*3 and diode D2 is included. Windings *w*1 and *w*3 have the same number of turns; their magnetic coupling is strong.

In interval t_1–t_2, the transistor is on; diodes D1 and D2 are closed by the voltages at windings *w*2 and *w*3. The structure of the circuit's conducting section is as in Figure 5.7a. When the transistor is switched off, the choke current is recuperated to source *E* and then falls to zero as the diode current rises to a steady value. The voltage at the transistor's collector is no more than 2*E*. The operating conditions of the recuperative circuit limit the selection of the transformation ratio *k*:

$$k = \frac{6U_{out}}{E_{min} + E_{max}}. \tag{5.28}$$

5.3.2 *Forward converter*

In Figure 5.9, we show a possible circuit diagram for a forward converter, in which the energy is transmitted to the load during the pulse length T_p.

When the transistor is switched on, the voltage *E* is applied to the winding *w*1. The corresponding voltage induced in secondary winding *w*2 turns on the diode D1. Hence, the voltage *kE* is applied to the input of filter L_fC. Energy is stored in the filter components and in the magnetic field of the transformer core and also transmitted to the load.

When the transistor is switched off, diode D2 is opened. The energy stored in the filter components sustains the current flow in the load. Diode D1 is closed. The energy stored in the transformer core is recuperated to source *E* through winding *w*3 and diode D3.

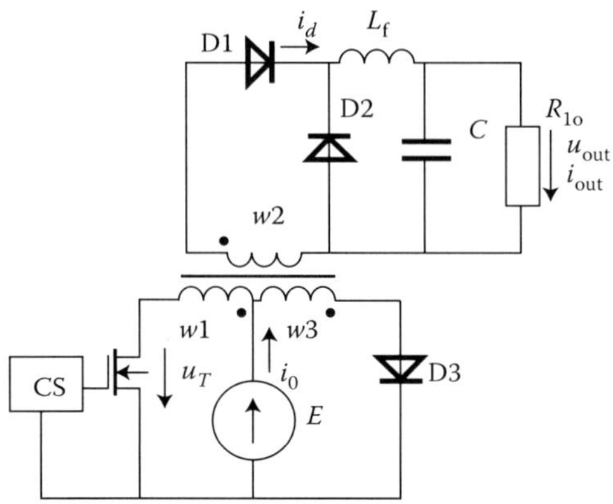

Figure 5.9 Forward voltage converter.

If we neglect the losses in the converter, we may determine the mean output voltage of the converter in the form

$$U_{out} = E \cdot k \cdot \gamma. \tag{5.29}$$

The operation of the forward converter is largely analogous to that of the step-down dc/dc converter (Section 5.2.1); the difference is the presence of the transformer.

In the case of windings with the maximum possible magnetic coupling, the maximum voltage at the collector is $2E$. The transformer operates in the asymmetric mode, without the accumulation of the magnetic flux in the core. When the transistor is switched off, there is a recuperation stage; the magnetic flux falls to its initial value. To that end, we must select the transformation ratio as follows:

$$k = \frac{8U_{out}}{E_{min} + E_{max}}, \tag{5.30}$$

while the fill factor must be $\gamma = 0–0.5$.

Both flyback and forward converters considered consume discontinuous current i_0 from the source.

Benefits of the forward converter include the following:

- Reduced output-voltage pulsations
- A shorter magnetic-flux trajectory in converter operation
- The possibility of operating without a load (idling)

Shortcomings of the forward converter relative to the flyback converter are as follows:

- A more complex circuit design
- A limited range of output voltage

Accordingly, the flyback converter is preferable in compact systems. In that case, problems arise when the load is switched off, and the output-voltage pulsations are more pronounced.

In these half-cycle converter circuits (Figures 5.8 and 5.9), we may introduce additional secondary uncoupled windings with uncontrolled rectifiers, to which various consumers may be connected. The output voltages of the converter may be different in that case.

On the basis of half-cycle dc voltage converters, we may create circuits with voltage multiplication. The principles are the same as in Section 5.2.5.

5.3.3 Push–pull converters

Half-cycle dc voltage converters may be used when power up to 1 kW is required. At larger load power, full-cycle dc voltage converters (push–pull converters) have obvious benefits.

Practically, all push–pull converters include an intermediate ac element and consist of an inverter and an uncontrolled full-wave rectifier. The corresponding circuit diagram is shown in Figure 5.10. Single-phase autonomous voltage inverters are often used (Section 6.1).

The inverter converts dc to ac. A typical inverter output voltage u_p is shown in Figure 5.10. The voltage u_p consists of rectangular pulses with an alternating polarity; the pulses repeat at the cycle duration T_{cy}. The transformer alters the magnitude of the voltage u_p, and the uncontrolled

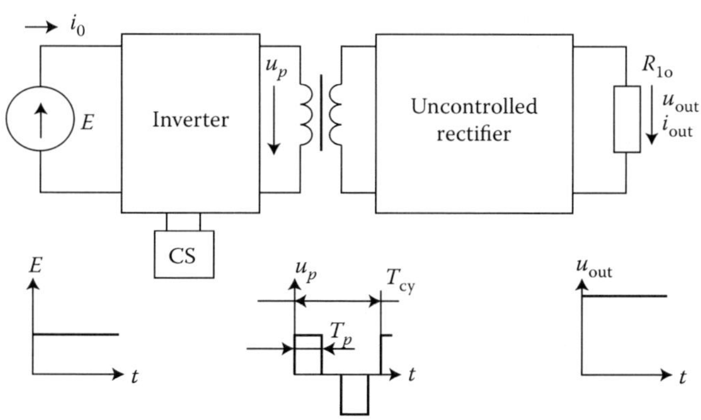

Figure 5.10 Structure of a push–pull dc voltage converter.

rectifier converts the ac voltage to the dc voltage u_{out}. In the converter, U_{out}/E is regulated by means of the CS, which adjusts the relative pulse width T_p/T_{cy} of the inverter's output voltage.

In contrast to half-cycle converters, the transformer in a full-cycle converter operates symmetrically, without bias.

The selection of the inverter and rectifier circuits in the converter largely depends on the primary and secondary voltages; the goal is to maximize the efficiency. At low source voltage E (<50 V), null voltage-inverter circuits are preferable (Section 6.1), as the ratio of the voltage losses at the conducting transistors to the inverter's output voltage is half as large. At large E, half-bridge and bridge circuits are employed.

At low output voltage, the rectifier is based on a center-tapped circuit. In other cases, bridge rectification circuits are also possible.

In full-cycle dc voltage converters, the minimum number of controllable switches is two, as against one for half-cycle converters (Sections 5.3.1 and 5.3.2).

The design principles for single-phase voltage inverters are outlined in Chapter 6. However, inverter operation in full-cycle dc voltage converters presents certain distinctive features. Therefore, we will consider here a typical converter circuit with a low source voltage E (Figure 5.11).

The circuit consists of a single-phase voltage inverter based on transistors VT1 and VT2 in a null circuit, a transformer with transformation ratio k, and an uncontrolled bridge rectifier with an output LC filter. Time diagrams are shown in Figure 5.11b. The losses in the circuit, the magnetizing current, and the scattering induction of the transformer are

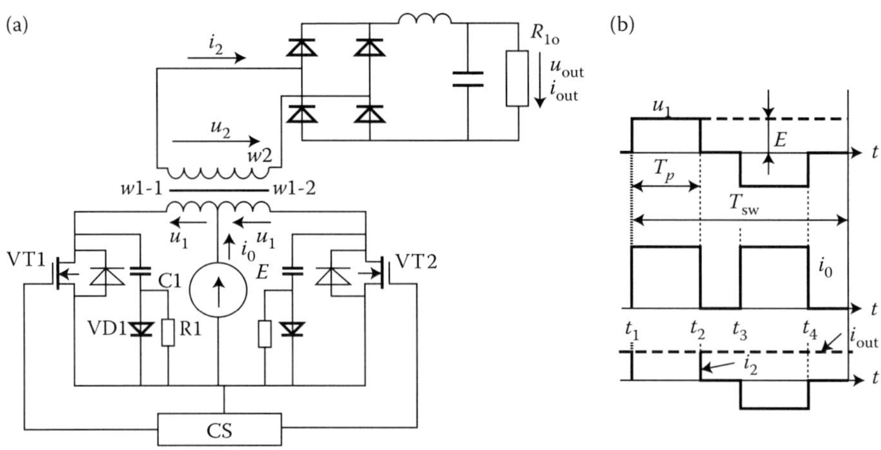

Figure 5.11 (a) Circuit diagram and (b) time diagrams of a push–pull dc voltage converter.

neglected in the analysis. The current and voltage at the load are assumed to be ideally smooth.

In the interval t_1–t_2, transistor VT1 is switched on; and the voltage $u_1 = E$ gives rise to voltage $u_2 = kE$ in the secondary winding. This voltage is sent through the diodes of the uncontrolled rectifier to the input of the LC filter; the energy in the filter's reactive components increases. The current $i_2 = I_{out}$, while $i_0 = kI_{out}$. To closed transistor VT2, the sum of the supply voltage and u_1 is applied; $U_{sw \cdot max} = 2E$.

In the interval t_2–t_3, both transistors are off; the voltage $u_1 = u_2 = 0$. The filter's choke current flows through the rectifier diodes; $i_2 = i_1 = 0$. The filter's reactive components send energy to the load.

In the interval t_3–t_4, transistor VT2 is conducting; $u_1 = -E$ and $u_2 = -kE$. This voltage is rectified and sent to the filter input; the energy in the filter's reactive components increases. In the last interval $t_4 - (t_1 + T_{sw})$, both transistors are turned off; the processes are analogous to those in interval t_2–t_3.

If we neglect the losses in the converter, we may determine the mean output voltage of the converter in the form

$$U_{out} = E \cdot k \cdot \gamma, \tag{5.31}$$

where the fill factor $\gamma = 2T_p/T_{sw}$. The operation of the push–pull dc voltage converter largely resembles that of a step-down dc/dc converter (Section 5.2.1); the differences are the presence of a transformer and the fact that the processes in the filter and the load occur at twice the switching frequency of each of the transistors ($f_{sw} = 1/T_{sw}$).

If we take account of the transformer's scattering inductance, the energy stored in the scattered inductance prevents the transistor from being switched off and results in voltage surges, which may lead to transistor failure. To derive energy from winding *w*1-1, snubber C1–VD1–R1 is connected to transistor VT1. (A similar structure is connected to transistor VT2.) When the transistor VT1 is turned on, the winding current through diode VD1 charges the capacitor C1, and the scattering energy is transmitted to the capacitor. When the winding current falls to zero, the diode is switched off, and the capacitor retains the accumulated energy. When transistor VT1 is again turned on, the capacitor discharges through resistor R1 and transistor VT1, which are in series. The energy stored in the capacitor is dispersed.

Asymmetry in the voltages applied to the transformer windings may lead to unbounded growth in the magnetizing current and saturation of the transformer. To prevent this, appropriate feedback circuits are introduced in the control circuit.

In Figure 5.11a, we may introduce additional uncoupled secondary windings with uncontrolled rectifiers, to which different consumers may be connected. The voltages at the terminals of such converters may be different. Note that the feedback circuit for stabilizing the output voltage will only apply to a single terminal. Whether the output voltage is switched off at the other terminals will depend on the hardness of the characteristic for the relevant channels.

5.4 Multiquadrant direct-current converters

The dc voltage converters considered in Sections 5.2 and 5.3 are able to create current and voltage of the same polarity at the output. In some applications, we need to create dc sources with a variable polarity. If the polarity of the voltage U_{out} and current I_{out} is the same, we are dealing with energy transfer from the source E to the converter output. If the voltage U_{out} and current I_{out} are of opposite polarity, energy from the converter output is returned (recuperated) to the source. When a dc machine is connected to the converter in this case, we are dealing with recuperative braking.

When dc voltage converters are able to change the polarity of the voltage and/or current, they are said to be multiquadrant converters, as their family of characteristics lies in two or four quadrants of the Cartesian coordinate system (Trzynadlowski, 1998). As a rule, multiquadrant dc converters are based on dc voltage regulators (Section 5.2).

5.4.1 Two-quadrant converter

In Figure 5.12a, we show the circuit diagram of a two-quadrant dc converter. The load considered is a dc motor, which is replaced by an equivalent circuit consisting of the motor's counteremf E_m, its resistance R_m, and its inductance L_m.

When an electrical machine operates in the motor mode, only transistor VT1 operates in the converter, and transistor VT2 is always off. When transistor VT1 is turned on, it connects source E to the motor; motor current $i_{out} = i_0$ passes through VT1. When transistor VT1 is turned off, the motor current passes through diode VD2, and the voltage applied to the motor is zero. It is readily evident that the conducting section of the circuit corresponds to a step-down dc/dc converter (Section 5.2.1). In that case, neglecting the losses, we write the output voltage in the form

$$U_{out} = E\frac{T_p}{T_{sw}} = \gamma E. \tag{5.32}$$

In recuperative braking, the motor continues to turn, and the polarity E_m is unchanged. However, on switching to the generator mode, the polarity of current i_{out} will be the opposite of that in Figure 5.12a. In this

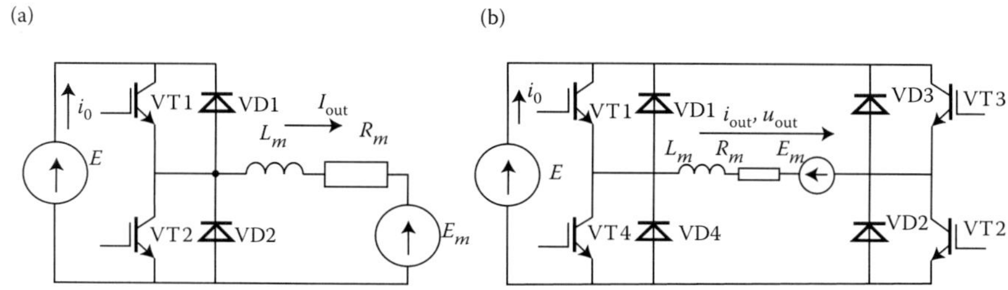

Figure 5.12 (a) Two-quadrant and (b) four-quadrant dc voltage converters.

mode, no control pulses are sent to transistor VT1. When transistor VT2 is off, current flows through diode VD1; the polarity of current i_0 is reversed, and the motor energy is recuperated to source E. When transistor VT2 is turned on, it transmits the current i_{out}. Operation in the recuperation mode corresponds to a step-up dc/dc converter (Section 5.2.2), if we assume that the energy source is the motor's counteremf E_m.

5.4.2 *Four-quadrant converter*

The rotation of a dc electrical machine may be reversed by means of a four-quadrant dc converter (Figure 5.12b).

The converter may operate in four modes.

Quadrant I. The electrical machine operates in the motor mode, with forward rotation. An output-voltage pulse is formed at the motor input when transistors VT1 and VT2 are switched on simultaneously, and $u_{out} = E$. To create an inactive interval, it is sufficient to switch off one of the transistors—say, VT2. Then the motor current flows through VT1 and diode VD3; $U_{out} = 0$ and $i_0 = 0$. In this quadrant, the converter resembles a step-down dc/dc converter: $U_{out} = \gamma E$.

Quadrant II. The motor turns in the same direction, but with recuperative braking. Consequently, the machine operates in the generative mode, and the current i_{out} is reversed. Two modes alternate in the converter.

- An interval of length γT_{sw}, in which all the transistors are on; current passes through diodes VD1 and VD2; and the motor current flows through source E, to which energy is returned.
- An interval of length $(1 - \gamma)T_{sw}$, in which transistor VT3 is on; the load current passes through the circuit VT3–VD1, bypassing the source; and $i_0 = 0$. The same results may be obtained by switching on VT4, which forms a circuit with diode VD2.

In the second quadrant, the converter resembles a step-up dc/dc converter, in which the energy source is the emf E_m.

Quadrant III. The direction of rotation is reversed; the directions of the voltages and currents in the electrical machine are the opposite to those shown in Figure 5.12b. When transistors VT3 and VT4 are switched on simultaneously, $u_{out} = -E$; energy is sent from the source to the motor. When one of those transistors is switched off, current flows through the circuit consisting of a transistor and a diode, bypassing the source: $u_{out} = 0$; $i_0 = 0$; and $u_{out} = -\gamma E$.

Quadrant IV. Recuperative braking occurs. When all the transistors are switched off, the current in the electrical machine, whose direction is as in Figure 5.12b, passes through the circuit VD3–VD4, returning energy to source *E*. When transistor VT1 is turned on, current i_{out} flows through diode VD3, bypassing the source. The same result may be obtained by switching on transistor VT2, which forms a circuit with diode VD4.

We may note the similarities between multiquadrant voltage converters and voltage source inverters (Section 6.1). The circuit in Figure 5.12a corresponds to a half-bridge voltage inverter with asymmetric connection of the load and the circuit in Figure 5.12b to a single-phase bridge inverter.

5.5 Thyristor–capacitor regulators with dosed energy supply to the load

In the case of continuous current, all the dc voltage converters considered here have hard natural characteristics (Bulatov and Carenko, 1982). In other words, their properties correspond to those of emf sources. However, many consumers, especially in electric-drive design and in electrotechnology, require different characteristics. In some cases, the load resistance may fall to zero in operating conditions. The form of the natural characteristics may be corrected by means of CSs with feedback, so as to obtain artificial characteristics of the required configuration (e.g., with stabilization of the load current or power). In highly dynamic conditions, the CS may process abrupt changes with some delay; oscillatory dynamic behavior is possible. In that case, converters with sharply falling natural characteristics are preferable. This category includes converters with dosed energy supply to the load.

Single-throw thyristors are used as switches in converters with dosed energy supply to the load. That permits operation at load voltages where the creation of converters on the basis of completely controllable switches is difficult and expensive.

The circuit diagram of a dc voltage regulator with dosed energy supply to the load is shown in Figure 5.13a.

The switching system in this converter is based on thyristors VT1–VT4, with commutation capacitance C_{co}. If we compare Figures 5.4a and 5.13a, we find that the only difference is the type of switch. In other words, the converter in Figure 5.13a is a dc-inverting regulator (a buck–boost

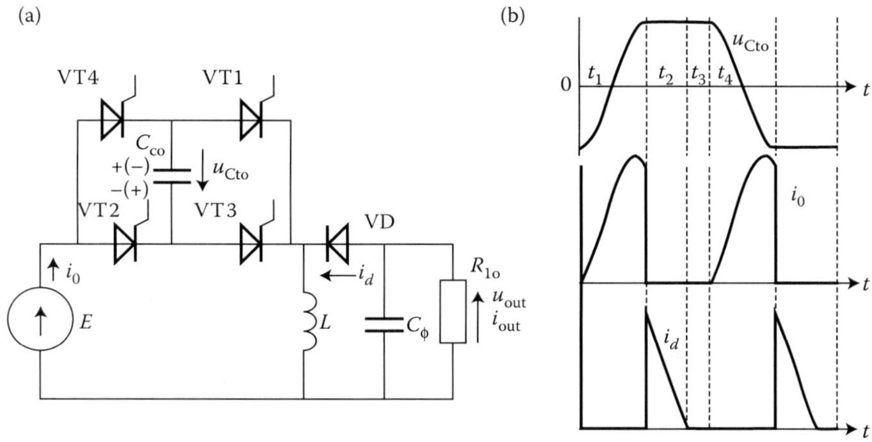

Figure 5.13 (a) Circuit diagram and (b) operational diagrams of a dc voltage regulator with dosed energy supply to the load.

convertor). However, the properties and characteristics of the converter in Figure 5.13a differ from those considered in Section 5.2.3, for the following reasons: the capacitance C_{co} not only ensures thyristor commutation, but also stores energy, and the converter operates with discontinuous current.

We now consider the processes in the converter under steady conditions; time diagrams are shown in Figure 5.13b. Each half-period may be divided into three stages.

Stage I. At time t_1, capacitor C_{co} is charged to voltage $u_{Cco}(t_1) = E + U_{out}$, with the polarity shown outside the parentheses in Figure 5.13a. In interval t_1–t_2, the control pulses are sent to thyristors VT1 and VT2, and capacitor C_{co} is recharged in oscillatory fashion in the circuit E–VT2–C_{co}–VT1–L, which transfers the stored energy to choke L. The length of this stage is determined by the resonant frequency of the C_{co}–L circuit.

Stage II. At time t_2, the voltage at the capacitor reaches $u_{Cco}(t_2) = E + U_{out}$, with the polarity shown in parentheses in Figure 5.13a. The voltage at diode VD becomes positive, and the diode is turned on. Consequently, further charging of capacitor C_{co} is impossible, and the source current i_0 and thyristor currents are discontinuous. The current of choke L passes through diode VD; the choke's energy is transmitted to the load circuit and stored in filter capacitor C_f. The choke current and the diode current i_d decline. The length of this stage depends on the load current.

Stage III. When $i_d = 0$, diode VD is turned off (time t_3). The load receives the energy stored in capacitor C_f; there is no current in the other circuits. The length of this stage is determined by the CS.

At time t_4, the second half-period of operation begins. Control pulses are sent to thyristors VT3 and VT4. Oscillatory recharging of capacitor C_{co}

begins in the circuit E–VT4–C_{co}–VT3–L; from the voltage whose polarity is shown in parentheses in Figure 5.13a, the capacitor is raised to the voltage $E + U_{out}$ whose polarity is shown outside the parentheses in Figure 5.13a. On reaching this voltage, diode VD switches off, and the processes develop as in stages II and III.

Disregarding losses, we may write the converter's mean output voltage in the form

$$U_{out} = E \cdot R_{lo}C_{CO}f(2) + \sqrt{1 + \frac{1}{R_{lo}C_{CO}f}}, \tag{5.33}$$

where f is the repetition frequency, specified by the CS.

As the energy stored in capacitor C_{co} is sent to the load in dosed fashion in each half-period, the converter resembles a power source in terms of its properties, while its natural characteristic is hyperbolic. The system functions with shorting in the load circuit. To ensure regulation or stabilization of the load current, the CS adjusts the frequency f.

References

Bulatov, O.G. and Carenko, A.I. 1982. Thyristor–capacitor converters (Tiristorno-kondensatornye preobrazovateli). Moskva. Jenergoatomizdat (in Russian).

Ericson, R.W. and Maksimovich, D. 2001. *Fundamentals of Power Electronics*. New York: Kluwer Academic Publisher.

Meleshin, V.I. 2006. Transistor converting technics (Tranzistornaja preobrazovatel' naja tehnika). Moskva. Tehnosfera (in Russian).

Mohan, N., Underland, T.M., and Robbins, W.P. 2003. *Power Electronics—Converters, Applications and Design*, 3rd edn. Danvers: John Wiley & Sons.

Polikarpov, A.G. and Sergienko, E.F. 1989. Single-cycle voltage converters in power supply for electronic equipment (Odnotaktnye preobrazovateli naprjazhenija v ustrojstvah jelektropitanija RJeA). Moskva. Radio i svjaz' (in Russian).

Rashid, M. 2004. *Power Electronics: Circuits, Devices and Applications*, 3rd edn. Englewood Cliffs, NJ: Prentice-Hall.

Rossetti, N. 2005. *Managing Power Electronics*, Weinheim: Wiley-IEEE press.

Rozanov, Ju.K., Rjabchickij, M.V., Kvasnjuk, A.A. 2007. *Power electronics: Textbook for universities*. 632 p. Moscow: Publishing hous MJeI (in Russian).

Trzynadlowski, A.M. 1998. *Introduction in Modern Power Electronics*. New York: John Wiley & Sons.

chapter six

Inverters and ac converters based on completely controllable switches

6.1 Voltage inverters

6.1.1 Single-phase voltage inverters

A *voltage source inverter* is an inverter connected to a current source with the predominant properties of a voltage source (Mohan et al., 2003; Rozanov, 2007; Zinov'ev, 2012).

The general structure of a voltage source inverter is shown in Figure 6.1.

Figure 6.2a shows the circuit diagram of a half-bridge single-phase voltage inverter. This inverter may be used independently or in a system with other inverters.

We now introduce the function F_i: $F_i = 1$ if the control electrode of transistor i is at a voltage such that the transistor is switched on, and $F_i = 0$ otherwise. In Figure 6.2a, four combinations of switch states are possible:

1. $F_1 = 1$ and $F_2 = 0$. In that case, $u_{out} = E/2$; the source currents $i_{01} = i_{out}$ and $i_{02} = 0$. When $i_{out} > 0$ (with the current direction corresponding to the arrow in Figure 6.2a), the current flows through transistor VT1; when $i_{out} < 0$, the current flows through diode VD1.
2. $F_1 = 0$ and $F_2 = 1$. In that case, $u_{out} = -E/2$; $i_{01} = 0$ and $i_{02} = -i_{out}$. When $i_{out} > 0$ (with the current direction corresponding to the arrow in Figure 6.2a), the current flows through diode VD2; when $i_{out} < 0$, the current flows through transistor VT2.
3. $F_1 = 1$ and $F_2 = 1$. This case is impermissible, as it leads to short-circuiting of the source.
4. $F_1 = 0$ and $F_2 = 0$. Two scenarios are possible here.
   a. When $i_{out} > 0$, the output current flows through diode VD2 and $u_{out} = -E/2$.
   b. When $i_{out} < 0$, the output current flows through diode VD1 and $u_{out} = E/2$.

In the latter case, the voltage at the load depends on the polarity of the output current but not on the inverter's control signals. However, such conditions are necessary to ensure normal operation of the switches

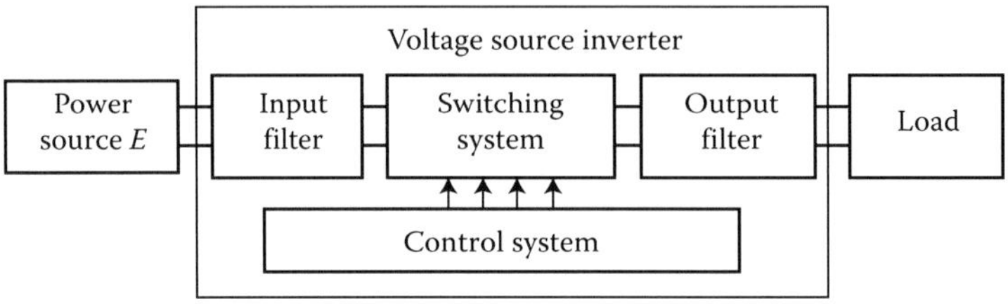

Figure 6.1 General structure of a voltage source inverter.

in the system. When a control pulse turns off a power switch, there is a finite period during which its current falls, and turning on a second switch during that interval leads to short-circuiting of the source through the two switches in series. Therefore, the operation of the second switch must be delayed for a time, ensuring complete decline of the current in the first switch to zero (the dead time). If the dead time is not more than 0.01–0.02 of the interval between commutations of the power switches, the processes in the dead time have little influence on the inverter's output voltage.

Neglecting the dead time, we may conclude that the switches operate alternately. The function $F_i = 1$ if switch i (a transistor or diode) is conducting, while $F_i = 0$ otherwise. In other words, F_i is the switching function of switch i (Section 3.1.3), which conducts current in both directions and consists of a transistor and a diode. In Figure 6.2a, the switches may be represented in a generalized form, as shown in Figure 6.2b.

Neglecting the dead time, we may assume that

$$F_2 = 1 - F_1. \tag{6.1}$$

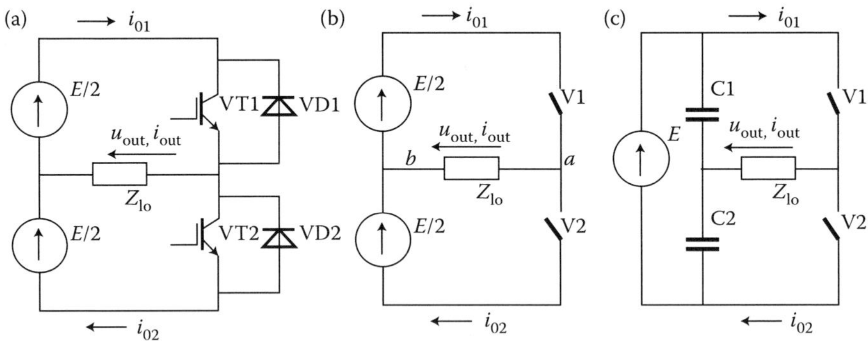

Figure 6.2 (a) Circuit diagram and (b, c) generalized circuits of a half-bridge single-phase voltage inverter.

The potentials of points a and b (Figure 6.2b) relative to the negative pole of the source are

$$\varphi_a = F_1 \cdot E, \quad \varphi_b = \frac{E}{2}, \tag{6.2}$$

whereas the inverter's output voltage is

$$u_{\text{out}} = \varphi_a - \varphi_b = E\left(F_1 - \frac{1}{2}\right). \tag{6.3}$$

The inverter's output voltage takes only two values: $u_{\text{out}} = E/2$ when $F_1 = 1$ and $u_{\text{out}} = -E/2$ when $F_1 = 0$.

The currents drawn from the source are also determined by the switching functions

$$i_{01} = F_1\, i_{\text{out}} \quad \text{and} \quad i_{02} = -F_2\, i_{\text{out}} = (F_1 - 1)\, i_{\text{out}}. \tag{6.4}$$

In Figure 6.3a through d, we show the switching functions and voltage and current curves for a half-bridge inverter with an RL load, when using the simplest control algorithm.

When a control pulse is sent to transistor VT1 during the first half-period (t_1–t_3), the current in the interval t_1–t_2 flows through diode VD1; the load sends energy to the source, and $i_{01} < 0$. In the interval t_2–t_3, the current flows through transistor VT1; energy is transferred from the source to the load, and $i_{01} > 0$. In the second half-period (t_3–t_5), a control pulse is supplied to transistor VT2. In the interval t_3–t_4, the current flows through diode VD2, the load sends energy to the source, and $i_{02} < 0$. In the interval t_4–t_5, the current flows through transistor VT2, energy is transferred from the source to the load, and $i_{02} > 0$.

The output-voltage spectrum is shown in Figure 6.3e. Fourier-series expansion of the output voltage yields

$$u_{\text{out}} = \frac{2E}{\pi} \sum_{k=1,3,5,\ldots}^{\infty} \frac{\sin(k\omega_{\text{out}}t)}{k}. \tag{6.5}$$

The total harmonic distortion is

$$k_{\text{thd}} = \frac{\sqrt{\sum_{k\neq 1} C_k^2}}{C_1} = 0.467, \tag{6.6}$$

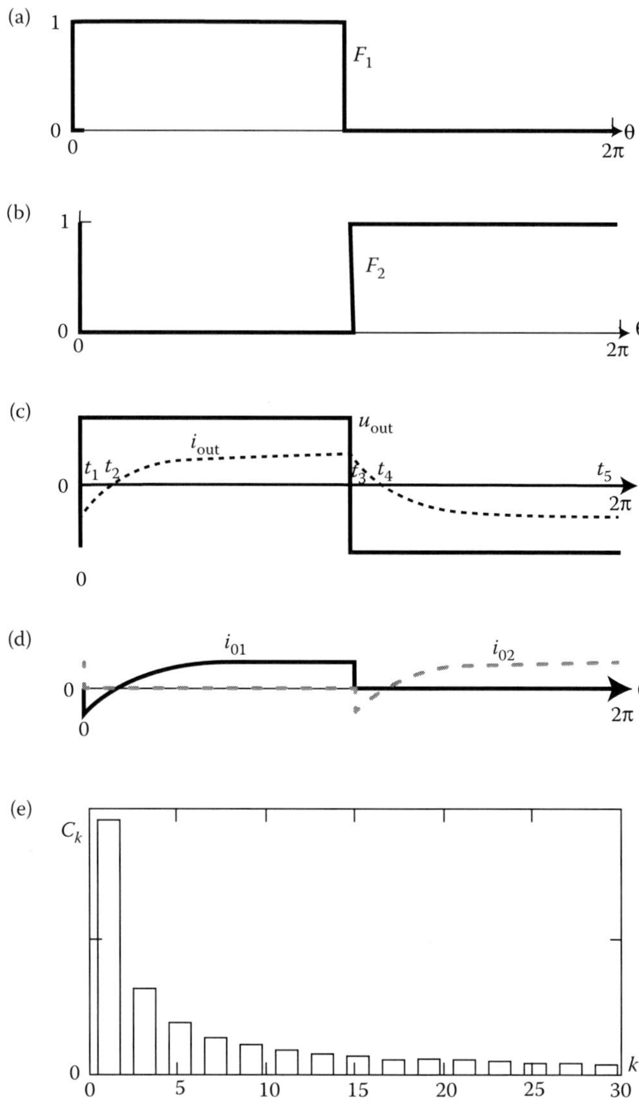

Figure 6.3 Operational diagrams of a half-bridge voltage inverter with an RL load. (a) Function F_1; (b) function F_2; (c) i_{out} and u_{out}; (d) i_{01} and i_{02}; (e) output-voltage spectrum.

and the effective value of the first harmonic (fundamental) is

$$u_{out \cdot 1} = \frac{\sqrt{2}E}{\pi}. \tag{6.7}$$

In Figure 6.2a, the load may be connected through a transformer. This permits an adjustment of the ratio of the fundamental of the output voltage to the voltage E.

A common approach is to connect a half-bridge inverter to the source, as in Figure 6.2c. Capacitors C1 and C2 form the inverter's input filter, eliminating the variable component of the currents i_{01} and i_{02} and ensuring unipolar current at the source E. At the same time, the capacitors split the source voltage. However, two problems arise here.

1. The first harmonics of the currents i_{01} and i_{02} create voltage drops at the capacitors, with deflection of the potential at the midpoint of the capacitors, that distorts the output voltage. To prevent this, the capacitance of the input filter must be considerably increased.
2. Asymmetry of the circuit and control components may lead to the appearance of a constant component and even harmonics in the inverter's output voltage. This problem may be eliminated by using control systems with feedback in terms of the output voltage.

In Figure 6.4, we show the circuit diagram of a *single-phase bridge voltage inverter*. It consists of two half-bridge circuits: (1) transistors VT1 and VT4 and (2) transistors VT2 and VT3. Neglecting the dead time, we may write the following expressions, by analogy with Equation 6.1:

$$F_4 = 1 - F_1 \quad \text{and} \quad F_2 = 1 - F_3. \tag{6.8}$$

The potentials of points a and b relative to the negative pole of the source are

$$\varphi_a = F_1 \cdot E \quad \text{and} \quad \varphi_b = F_3 \cdot E. \tag{6.9}$$

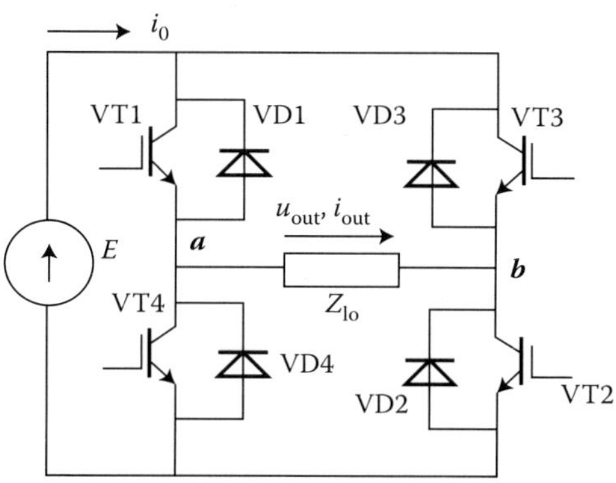

Figure 6.4 Circuit diagram of a single-phase bridge voltage inverter.

Hence, we may find the inverter's output voltage

$$u_{out} = \varphi_A - \varphi_B = E(F_1 - F_3). \tag{6.10}$$

In Equation 6.10, we may note two important differences between the bridge inverter and the half-bridge inverter:

1. The amplitude of the output voltage in the bridge circuit is E, which is twice the amplitude of u_{out} in the null circuit.
2. In the bridge inverter, zero output voltage is possible: when $F_1 = F_3 = 1$, the load is short-circuited by switches VT1 and VT3, and when $F_1 = F_3 = 0$, the load is short-circuited by switches VT2 and VT4.

We now find the relation between the output current and the current i_0 from the source. The current i_0 consists of the currents at transistors VT1 and VT3. Current $i_{out}F_1$ flows through VT1 and current $i_{out}F_3$ through VT3. Thus,

$$i_0 = i_{out}(F_1 - F_3). \tag{6.11}$$

We consider the simplest algorithm for forming the output voltage of a voltage source inverter with an RL load, when $F_1 = 1$ in the first half-period of the output frequency ω_{out} and $F_3 = 1$ in the second half-period. The time diagrams are shown in Figure 6.5.

When control pulses are sent to transistors VT1 and VT2 in the first half-period (t_1–t_3), the current flows through diodes VD1 and VD2 in the interval t_1–t_2, the load sends energy to the source, and $i_0 < 0$. In the interval t_2–t_3, the current flows through transistors VT1 and VT2, energy is transferred from the source to the load, and $i_0 > 0$.

In the second half-period (t_3–t_5), control pulses are sent to transistors VT3 and VT4. In the interval t_3–t_4, the current flows through diodes VD3 and VD4, the load sends energy to the source, and $i_0 < 0$. In the interval t_4–t_5, the current flows through transistors VT3 and VT4, energy is transferred from the source to the load, and $i_0 > 0$.

The output-voltage spectrum is of the same form as in Figure 6.3e, but all the harmonics are doubled

$$u_{out} = \frac{4E}{\pi} \sum_{k=1,3,5,\ldots}^{\infty} \frac{\sin(k\omega_{out}t)}{k}. \tag{6.12}$$

The total harmonic distortion $k_{thd} = 0.467$, and the effective value of the fundamental is

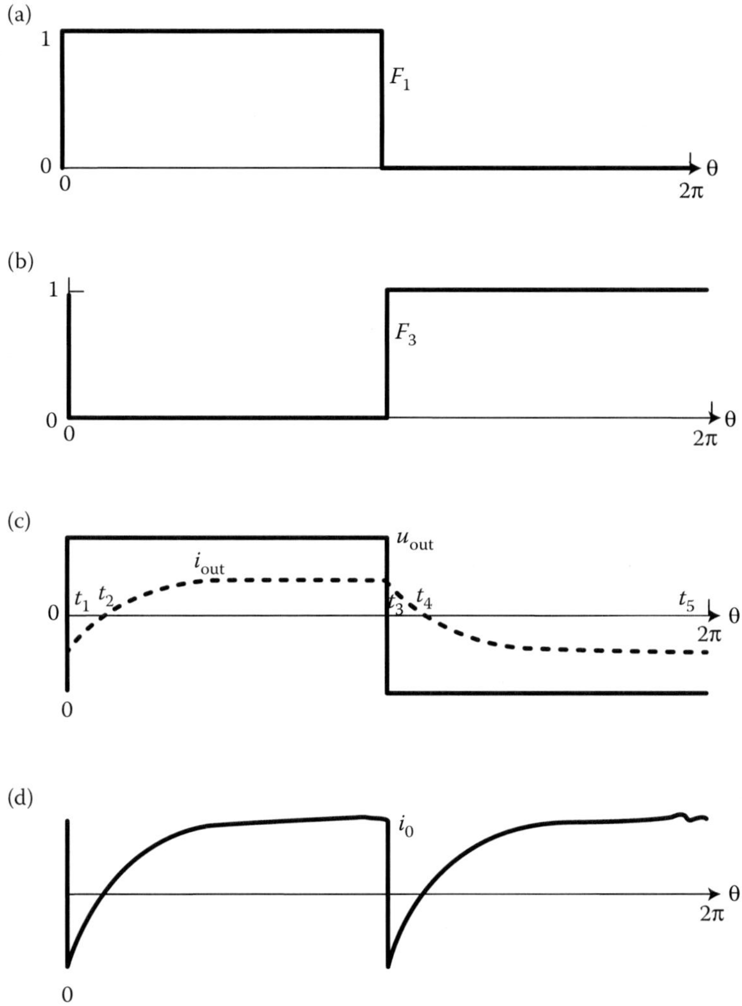

Figure 6.5 Operational diagrams of a bridge inverter with an RL load: (a) step switching function F_1; (b) step switching function F_3; (c) output current i_{out} and output voltage u_{out}; (d) source current i_0.

$$u_{\text{out.1}} = \frac{2\sqrt{2}E}{\pi}. \tag{6.13}$$

The spectrum of the source current i_0 in Figure 6.6 includes a constant component and even harmonics. The second harmonic is the most significant.

The load in Figure 6.4 may be connected through a transformer.

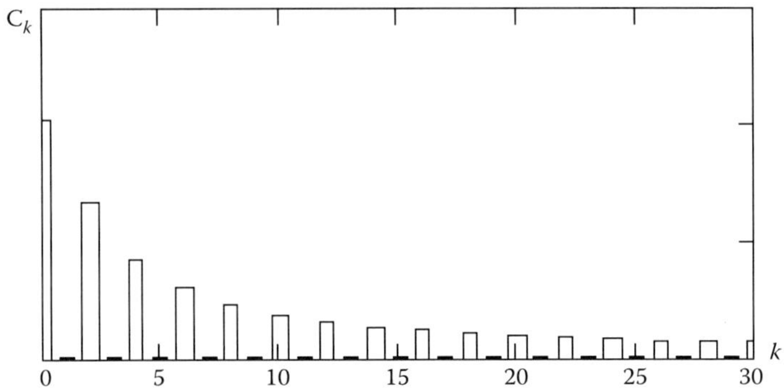

Figure 6.6 Spectrum of the current drawn from the source.

When inverters are connected to low-voltage sources ($E < 30$–50 V), static energy losses in the switches may be reduced, if we use an inverter based on a null circuit with a center tap in the transformer's primary windings (Figure 6.7). As a rule, the inverter employs the simplest control algorithm, in which the control pulse is supplied to transistor VT1 in the first half-period and to transistor VT2 in the second half-period. The time diagrams of the output voltage and current and the source current i_0 are as in Figure 6.5c and d.

When a control signal is supplied to transistor VT1 in the first half-period (t_1–t_3), the current passes through diode VD1 during interval t_1–t_2, the load sends energy to the source, and $i_0 < 0$. In the interval t_2–t_3, the current flows through transistor VT1, energy is transferred from the source to the load, and $i_0 > 0$. In the first half-period, $u_{out} = k_{tr}E$, where k_{tr} is

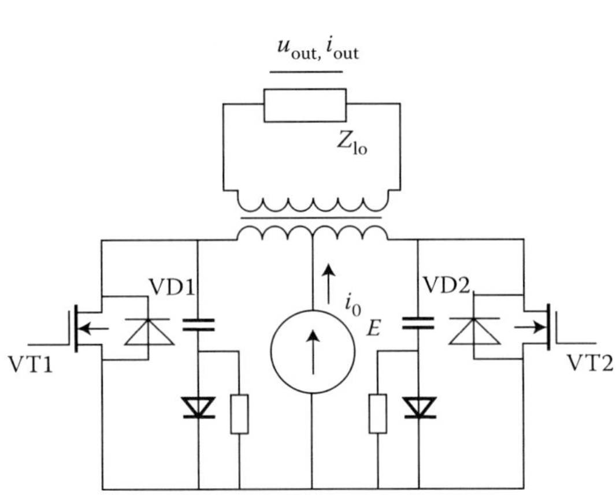

Figure 6.7 Circuit diagram of an inverter with a center tap in the primary transformer windings.

the transformation ratio. In the second half-period (t_3–t_5), a control pulse is sent to transistor VT2. In the interval t_3–t_4, the current flows through diode VD2, the load sends energy to the source, and $i_0 < 0$. In the interval t_4–t_5, the current flows through transistor VT2, energy is transferred from the source to the load, and $i_0 > 0$. In the second half-period, $u_{out} = -k_{tr}E$.

The output-voltage spectrum is shown in Figure 6.3e. Fourier-series expansion yields

$$u_{out} = \frac{4k_{tr}E}{\pi} \sum_{k=1,3,5,...}^{\infty} \frac{\sin(k\omega_{out}t)}{k}. \tag{6.14}$$

The energy stored in the scattering inductances prevents the closure of the transistors and results in voltage surges, which may incapacitate the transistors. To draw energy from the primary windings, snubbers are connected to the transistors. The operating principle of the snubbers was outlined in Section 5.3.3.

6.1.2 *Pulse-width control in single-phase voltage inverters*

The control algorithms already considered for voltage inverters ensure a fixed ratio of the output voltage and the source voltage. Regulation and stabilization of the voltage source inverter's output voltage may be based on external methods (changing the source voltage) or internal methods (changing the switching algorithm).

The simplest method of internal regulation is pulse-width control, in which the inverter's output voltage is synthesized from pulses of equal but controllable length.

As follows from Equation 6.10, the output voltage of a single-phase bridge voltage inverter may take three values: E, $-E$, and 0. In other words, the output-voltage curve may consist of positive and negative pulses separated by various intervals.

We now consider a switching algorithm in which a single pulse is formed in the half-period of the output frequency (single-pulse pulse-width control). The pulse length is $t_p = \gamma/(A_f \cdot f_{out})$, where A_f is the number of pulses per period and γ is the fill factor. In Figure 6.8a through d, we show the switching functions, currents, and voltages in the operation of a voltage source inverter with an RL load. (The dead time is neglected.)

In the period t_1–t_2, we note an interval without pulses. Current flows through switches VD2 and VT4, which short-circuit the load, $u_{out} = 0$, and $i_0 = 0$. The current in the load falls exponentially with time constant $\tau = L_{lo}/R_{lo}$. At time t_2, transistor VT4 is turned off by the control circuit. A control pulse is sent to transistor VT1, and the formation of a positive pulse begins. In the period t_2–t_3, $u_{out} = E$, but $i_{out} = i_0 < 0$. The load sends

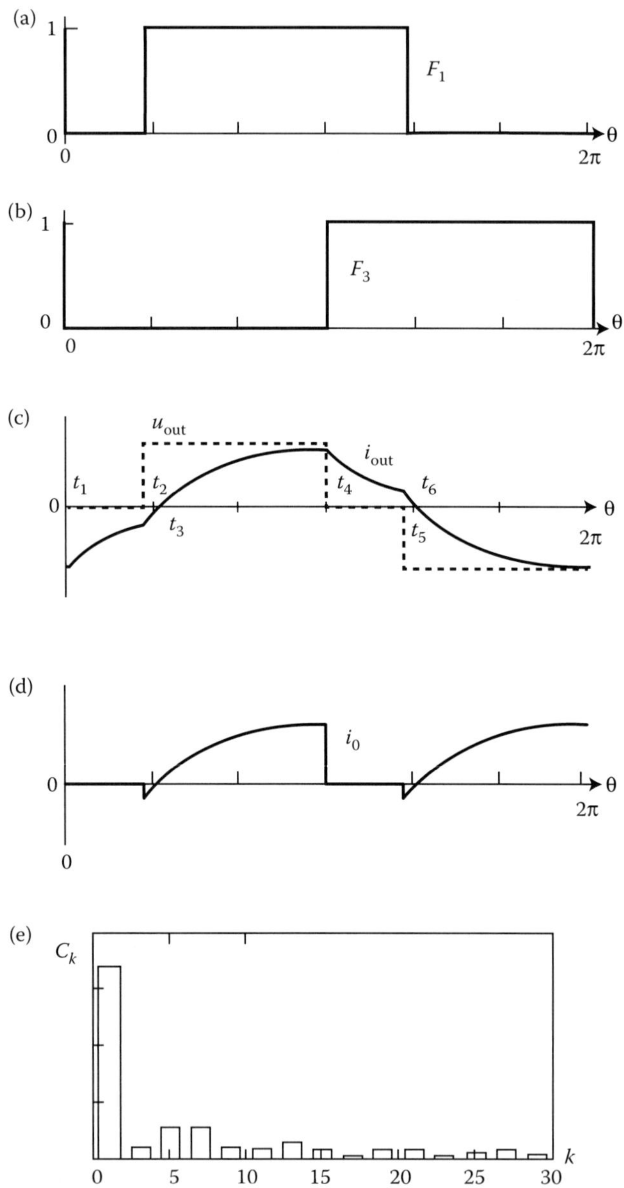

Figure 6.8 (a–d) Operational diagrams of a bridge-type voltage source inverter with an RL load in the case of single-pulse pulse-width control and (e) output-voltage spectrum.

energy to the source, current flows through diodes VD1 and VD2. At time t_3, the current $i_{out} = i_0$ becomes positive and flows through transistors VT1 and VT2; energy is transferred from the source to the load.

In the second half-period, the period t_4–t_5 corresponds to an interval between pulses, current flows through switches VT1 and VD3, $u_{out} = 0$, and $i_0 = 0$. At time t_5, transistor VT1 is switched off. A control pulse is supplied to transistor VT4, and the formation of a negative pulse begins. In the period t_5–t_6, $u_{out} = -E$, but $i_{out} = -i_0 > 0$. The load sends energy to the source, and current flows through the diodes VD3 and VD4. At time t_6, the current $i_{out} = -i_0$ becomes negative and flows through transistors VT3 and VT4; energy is transferred from the source to the load.

The effective value of the output voltage is

$$u_{out} = \sqrt{\frac{1}{T_p}\int_0^{T_p} u_{out}^2 \, dt} = E\sqrt{\gamma}. \tag{6.15}$$

In pulse-width control of multipulse type, $A_f/2$ pulses of equal duration are formed in each half-period, along with corresponding intervals between the pulses. In Figure 6.9a through d, we show the switching functions, currents, and voltages in the operation of a voltage source inverter with an RL load when $A_f = 20$. The load and fill factor γ in Figures 6.8 and 6.9 are the same. It is evident that the VT1–VT4 pair has a *switching frequency* $f_{sw} = A_f \cdot f_{out}$. The effective value of the output voltage is given by Equation 6.15.

In Figures 6.8e and 6.9e, we show the output-voltage spectra. The harmonic composition of u_{out} in single-pulse pulse-width control depends on the fill factor γ, as shown in Figure 6.10.

The amplitude of the fundamental ($k = 1$) is

$$C_1 = \frac{4E}{\pi}\sin\left(\frac{\pi\gamma}{2}\right). \tag{6.16}$$

Table 6.1 presents the dependence of the total harmonic distortion

$$k_{thd} = \frac{\sqrt{\sum_{k\neq 1} C_k^2}}{C_1} = f(\gamma).$$

We see that the harmonic composition of the voltage source inverter's output voltage deteriorates considerably with a decrease in γ.

This problem may be reduced if we use multipulse pulse-width control with an increase in the switching frequency. As a rule, low-frequency

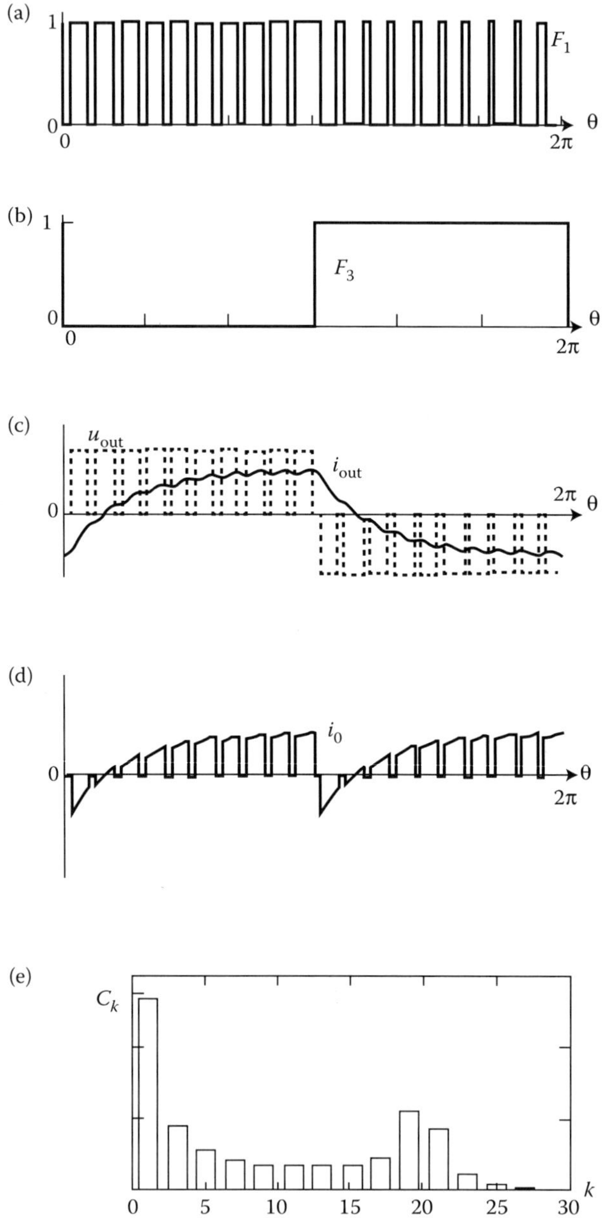

Figure 6.9 (a–d) Operational diagrams of a bridge-type voltage source inverter with an RL load in the case of multipulse pulse-width control and (e) output-voltage spectrum.

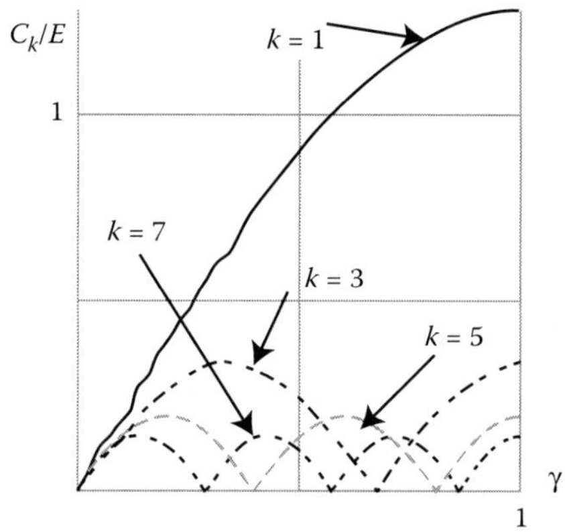

Figure 6.10 Dependence of the amplitude of the output-voltage harmonics on the fill factor in single-pulse pulse-width control.

Table 6.1 Dependence of the Total Harmonic Distortion k_{thd} on the Fill Factor γ

γ	1	0.9	0.8	0.7	0.6	0.5	0.4	0.3	0.2	0.1
k_{thd}	0.47	0.51	0.61	0.75	0.92	1.10	1.27	1.42	1.54	1.61

harmonics have the worst influence on consumers. The spectrum of u_{out} with $A_f > 20–30$ may be divided into a low-frequency region ($k < A_f - 9$) and a high-frequency region, with little effect on the consumer. In the low-frequency region, the amplitude of the harmonics ($k = 1, 3, 5, \ldots$) is

$$\tilde{C}_k = \frac{4E\gamma}{k\pi}, \tag{6.17}$$

whereas the total harmonic distortion is $k_{thd} = 0.47$ and does not depend on γ.

As many consumers are dissatisfied with the quality of the output voltage, numerous methods have been proposed for improving the harmonic composition of u_{out} for voltage source inverters, both by modifying the inverter circuit and by modifying the switching algorithms. With the appearance of powerful high-frequency power transformers and microprocessor control systems, control methods based on pulse-width modulation are dominant (Chapter 7).

6.1.3 *Three-phase voltage inverters*

Three-phase voltage inverters are generally based on a three-phase bridge circuit (Figure 6.11). Each switch V consists of a transistor and an

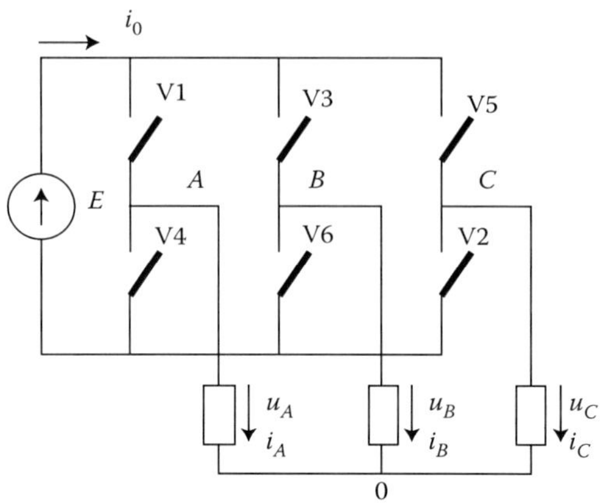

Figure 6.11 Three-phase bridge circuit for a voltage inverter.

antiparallel diode. When a control pulse is supplied to switch i through the transistor or the diode, the switch transmits current and $F_i = 1$.

The inverter consists of three half-bridge circuits (Section 6.1.1). According to Equation 6.1

$$F_2 = 1 - F_5, \quad F_4 = 1 - F_1, \quad F_6 = 1 - F_3. \tag{6.18}$$

The potentials of points A, B, and C relative to the negative pole of source E are as follows:

$$\varphi_A = E \cdot F_1, \quad \varphi_B = E \cdot F_3, \quad \varphi_C = E \cdot F_5. \tag{6.19}$$

With a symmetric load, the potential of the load's midpoint (neutral line 0) is

$$\varphi_0 = \frac{1}{3}(\varphi_A + \varphi_B + \varphi_C) = \frac{E}{3}(F_1 + F_3 + F_5). \tag{6.20}$$

Then, the output phase voltages of the voltage source inverter are as follows:

$$\begin{aligned} u_A &= \varphi_A - \varphi_0 = \frac{E}{3}(2F_1 - F_3 - F_5), \\ u_B &= \varphi_B - \varphi_0 = \frac{E}{3}(2F_3 - F_1 - F_5), \\ u_C &= \varphi_C - \varphi_0 = \frac{E}{3}(2F_5 - F_1 - F_3). \end{aligned} \tag{6.21}$$

These expressions show that the formation of the phase voltage is determined by the operation of switches in all phases of the inverter. That gives rise to important differences between three-phase and single-phase voltage inverters.

The inverter's output line voltages are

$$\begin{aligned} u_{AB} &= \varphi_A - \varphi_B = E(F_1 - F_3), \\ u_{BC} &= \varphi_B - \varphi_C = E(F_3 - F_5), \\ u_{CA} &= \varphi_C - \varphi_A = E(F_5 - F_1). \end{aligned} \quad (6.22)$$

The current i_0 consists of the currents of switches V1, V3, and V5:

$$i_0 = i_A + i_B F_3 + i_C F_5. \quad (6.23)$$

We will consider the simplest switching algorithm, known as 180° conductivity. In Figure 6.12, we show the time diagrams for the switching functions and the voltages in the inverter. The numbering of the switches corresponds to the order in which they operate. The shift between switching functions is 60°, and the duration of the switch's conducting state is 180°.

The phase voltage is a rectangular step function with amplitude $2E/3$. In Figure 6.13, we show the spectrum of the output phase voltage. As in a single-phase voltage source inverter, the spectrum contains odd harmonics, whose amplitude is inversely proportional to the frequency. However, the third harmonic and its multiples are absent.

The effective value of the fundamental is $0.45E$. The effective value of the phase voltage is $0.471E$. The total harmonic distortion $k_{\text{thd}} = 0.29$. The spectrum of the line voltage is of the same form, but all the harmonics are larger by a factor of $\sqrt{3}$.

In Figure 6.14a, we plot the phase voltage u_A and current i_A with an active–inductive load. In Figure 6.14b, we plot the time diagram of the current i_0 drawn from source E, according to Equation 6.23. The spectrum of current i_0 contains a constant component and harmonics that are multiples of $6f_{\text{out}}$.

The algorithm presented here for the formation of the output voltage ensures a fixed ratio of the fundamental of the output voltage to the source voltage. The simplest internal method of regulating the output voltage is *pulse-width control*. In that case, the output voltage consists of pulses of the same duration appearing at constant frequency and separated by specific intervals.

In the formation of output-voltage pulses, the same switches are employed as in 180° conductivity.

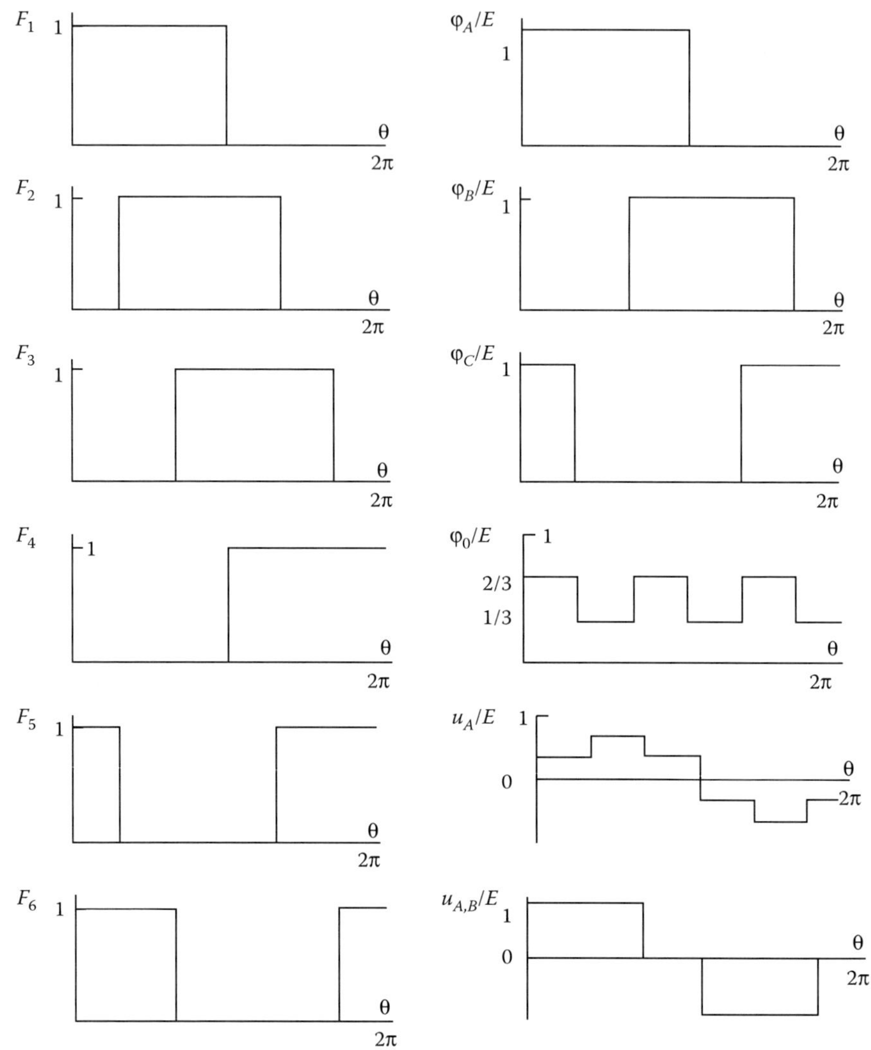

Figure 6.12 Time diagrams of the switching functions and voltages in the case of 180° conductivity.

Two approaches are adopted in creating intervals between the pulses in all phases of the output voltage.

1. Control pulses are supplied to all the transistors of the odd switches: $F_1 = F_3 = F_5 = 1$. In that case, according to Equation 6.21, $u_A = 0$, $u_B = 0$, and $u_C = 0$; the current from the source is $i_0 = 0$.

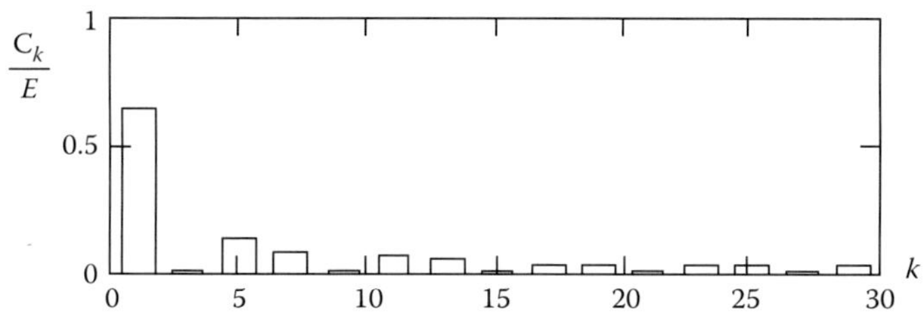

Figure 6.13 Spectrum of the output phase voltage for a three-phase voltage source inverter.

2. Control pulses are supplied to the even switches: $F_2 = F_4 = F_6 = 1$. Once again, $u_A = 0$, $u_B = 0$, and $u_C = 0$; $i_0 = 0$.

To ensure uniform current load on the switches, these approaches are alternated. To ensure symmetry of the phase and line voltages, the number of pulses within the output-frequency period A_f is a multiple of 6.

In Figure 6.15a, we plot the output phase voltage u_A and current i_A of the voltage source inverter. In Figure 6.15b, we plot the current i_0. In Figure 6.15c and d, we present the spectra of the output phase voltage and the current i_0. The inverter operates with a symmetric RL load, $A_f = 24$, and $\gamma = 0.5$.

At small A_f, deterioration in harmonic composition of the voltage source inverter's output voltage is observed with a decrease in γ. We may eliminate this problem by increasing the number of pulses per period and the switching frequency $f_{sw} = A_f \cdot f_{out}$. When $A_f > 20–30$, the spectrum of u_{out} may be divided into a low-frequency region ($k < A_f - 9$) and

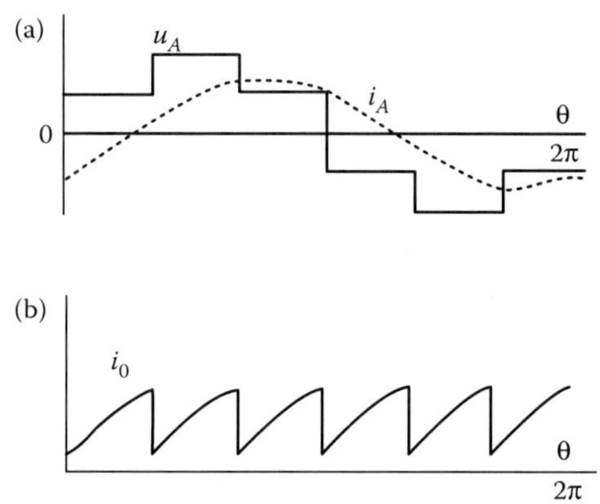

Figure 6.14 (a) Phase voltage and current and (b) current drawn from the source when operating with an RL load in the case of 180° conductivity.

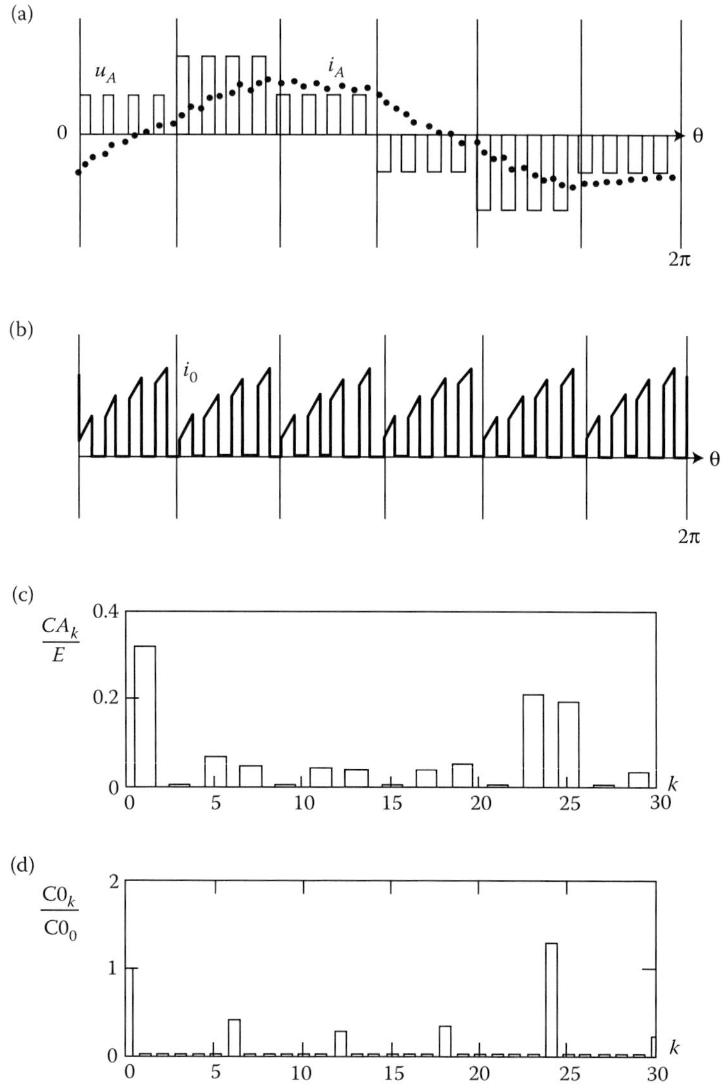

Figure 6.15 (a and b) Current and voltage diagrams and (c and d) spectra of the phase voltage and source current when operating with an RL load in the case of pulse-width control.

a high-frequency region, with little effect on consumers. In the low-frequency region, the amplitude of the harmonics ($k = 1, 5, 7, 11, 13, \ldots$) is

$$C_k = \frac{\sqrt{2}E\gamma}{k\pi}, \tag{6.24}$$

whereas the total harmonic distortion for the low-frequency region is $k_{thd} = 0.29$ and does not depend on γ.

Although this harmonic composition of the output voltage is acceptable for many frequency-based electric drives, we may note a recent trend in which voltage source inverters with pulse-width control are replaced by converters with pulse-width modulation (Chapter 7). That is because, in microprocessor control systems, pulse-width modulation results in better quality of the output voltage with practically no additional costs.

6.1.4 *Three-phase voltage inverters for asymmetric loads*

When the load is connected in a delta configuration, a three-phase bridge voltage source inverter may operate with a symmetric load or with an *asymmetric load* (Chaplygin, 2009). With any load, the line voltages u_{AB}, u_{BC}, and u_{CA} applied to its diagonals remain symmetric. However, they produce asymmetric currents in the diagonals of the load. In Figure 6.16, we show the current i_0 drawn from the source by the voltage inverter in the case of an asymmetric load.

Comparison of Figures 6.14b and 6.16 shows that, with an asymmetric load, the repetition period of the current i_0 is tripled and is equal to half the period of the output frequency. As a result, the spectrum of current i_0 includes additional harmonics—in particular, the component with frequency $2f_{out}$.

In the case of an asymmetric load in a star configuration without a neutral line, the three-phase bridge inverter (Figure 6.11) forms an asymmetric system of phase voltages. This asymmetry cannot be eliminated by means of the control system. In that case, symmetry of the output voltage may be restored by connecting four lines to the load (a star configuration with a neutral line).

In Figure 6.17, we show the circuit diagrams of three-phase inverters for power supply to an asymmetric load.

In Figure 6.17a, we show a circuit consisting of three single-phase half-bridges, with splitting of the supply voltage at capacitors C1 and C2.

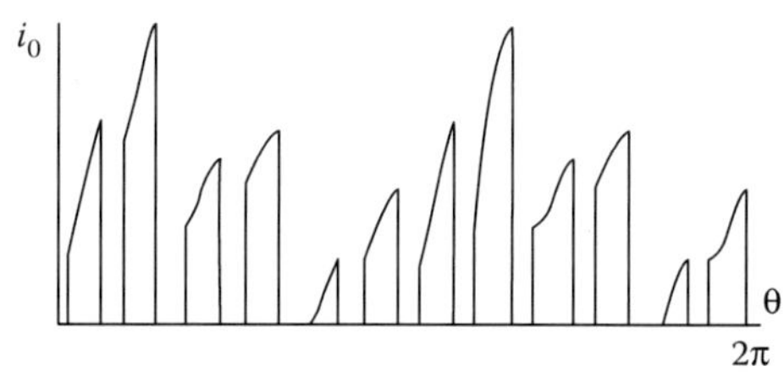

Figure 6.16 Current consumed by a three-phase voltage source inverter with an asymmetric load.

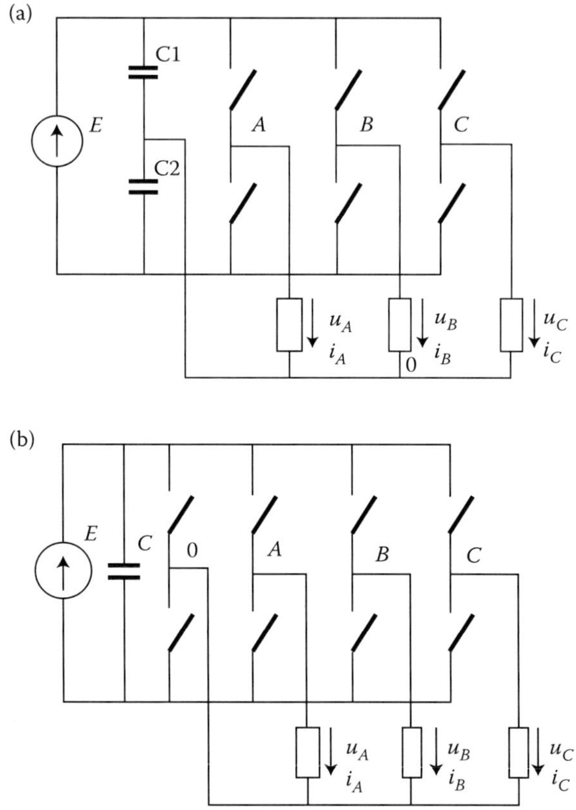

Figure 6.17 Circuit diagrams for three-phase inverters with an asymmetric load.

Each half-bridge operates independently, as in a single-phase half-bridge circuit (Figure 6.2c). The potential of point 0 relative to the negative pole of the source is $E/2$. The 120° mutual shift of the phase voltages is specified by the control system. The phase voltage may take two values: $E/2$ or $-E/2$.

In the circuit with an additional half-bridge (Figure 6.17b) (Chaplygin, 2009; Zinov'ev, 2012), the potential of point 0 relative to the negative pole of the source may be E or $-E$; the mean value over the period of the output frequency is maintained at $\varphi_{0,me} = E/2$. The phase voltages $u_A = \varphi_A - \varphi_0$, $u_B = \varphi_B - \varphi_0$, and $u_C = \varphi_C - \varphi_0$ may take three values: $E/2$, $-E/2$, and 0.

The circuit consisting of three single-phase half-bridges exhibits all the deficiencies of half-bridge voltage source inverters. First, the fundamental of the null sequence of asymmetric phase currents deflects the capacitor's center potential, with distortion of the output voltage. To eliminate this problem, the input-filter capacitance must be greatly increased.

Secondly, the asymmetry of the circuit components and the control system may lead to the appearance of a constant component and even harmonics in the inverter's output voltage. Thirdly, the phase voltage does not have null intervals.

The downside of the circuit with an additional half-bridge is the extra cost of the semiconductor devices and drivers. As shown by calculations, however, the standard power of the input-filter capacitance is an order of magnitude less than the total power of the capacitances in Figure 6.17a. The possibility of generating null intervals in the phase–voltage curve greatly improves the harmonic composition of the output voltage.

6.2 Current inverters

6.2.1 Transistor current inverters

The *current source inverter* is an inverter connected to a dc source with the predominant properties of a current source (Rozanov, 2007).

In Figure 6.18, we show the circuit diagram of a single-phase bridge-type current source inverter. On the dc side, a choke L_d stabilizes the current i_d and evens out the pulsations, in the absence of which $i_d = I_d$. The switches transmit current only in one direction. Diodes protect the transistors from voltages of inverse polarity.

With any switching algorithm, continuous flow of current I_d must be ensured. Therefore, one odd-numbered switch and one even-numbered switch must be turned on.

Accordingly, the switching functions are as follows: $F_3 = 1 - F_1$ and $F_4 = 1 - F_2$.

Rectangular current pulses are sent to the ac circuit. As the load is characterized by inductance, as a rule, we need to connect a capacitive

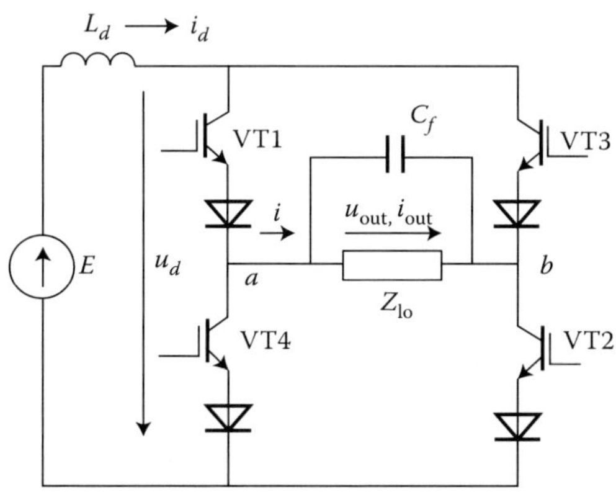

Figure 6.18 A single-phase bridge current inverter.

filter C_f in parallel with the load Z_{lo}. The load and the filter form an ac circuit with impedance Z.

We will consider the simplest operational algorithm of the current source inverter: in that case, transistors VT1 and VT2 are conducting in the first half-period, whereas transistors VT3 and VT4 are conducting in the second half-period. In Figure 6.19, we show current and voltage diagrams in the inverter when the modulus of the load impedance at the output frequency is Z_{lo1} (Figure 6.19a and b) and $Z_{lo2} = 10Z_{lo1}$ (Figure 6.19c and d) and for two values of cos φ at the fundamental (at the output frequency): cos φ = 1 (Figure 6.19a and c) and cos φ = 0.8 (Figure 6.19c and d). On the diagrams, we give voltage values when $E = 300$ V.

In the case of an active load (cos φ = 1), the capacitor voltage is negative at time t_1. When transistors VT1 and VT2 are switched on, the current in the ac circuit is $i > 0$ and that current charges the capacitor. The load voltage varies exponentially. In the period t_1–t_2, the current and voltage

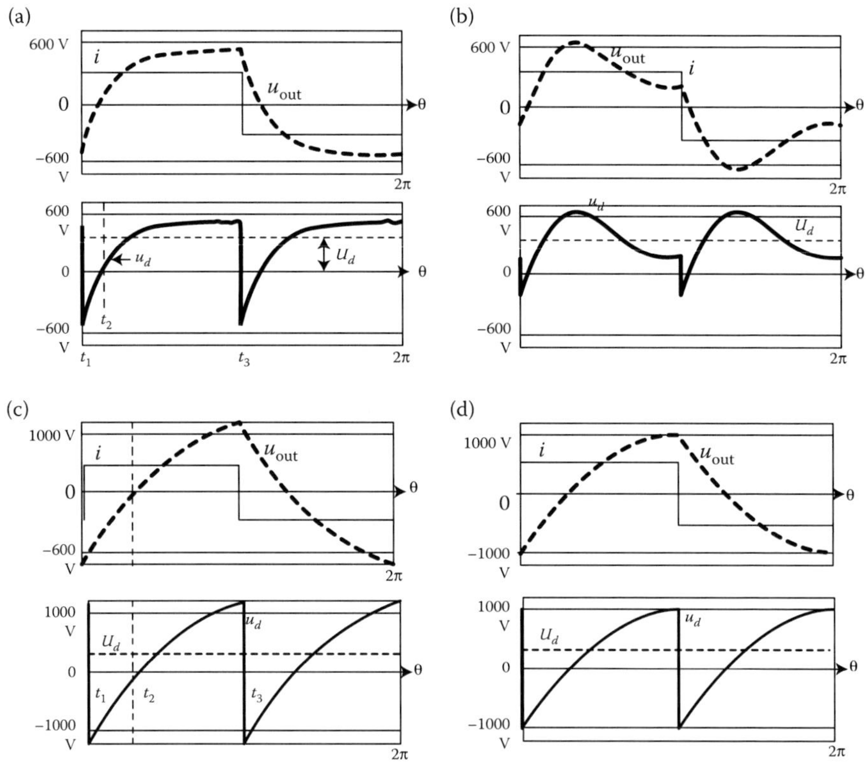

Figure 6.19 Current and voltage diagrams for a single-phase current source inverter with different load parameters: (a) Z_{lo1}, cos φ = 1; (b) Z_{lo1}, cos φ = 0.8; (c) $Z_{lo2} = 10Z_{lo1}$, cos φ = 1; and (d) $Z_{lo2} = 10Z_{lo1}$, cos φ = 0.8.

are in opposite directions; the ac circuit sends energy to choke L_d. In the period t_2–t_3, the voltage and current are of the same polarity; the ac circuit receives energy from the dc circuit. In the second half-period, $i < 0$; the capacitor is discharged. Comparison of Figure 6.19a and c shows that the form of the output voltage changes as a function of the load resistance. As the time constant of capacitor charging changes proportionally, the voltage with a low-resistance load is close to rectangular, with a high-resistance load, u_{out} is close to triangular.

The voltage $u_d = u_{out} \cdot (F_1 - F_3)$ may be of different polarity; $u_d = E$ is its mean value (Figure 6.19). When E = const, the amplitude u_{out} increases with an increase in the load resistance and as u_{out} becomes more triangular, the fundamental of the output voltage $u_{out,1}$ also increases. Therefore, the characteristic $u_{out,1} = f(1/Z_{lo})$ of the current source inverter declines steeply. The family of characteristics may be seen in Figure 6.20.

The current and voltage diagrams with an active–inductive load are shown in Figure 6.19b and d. With a low-resistance load, an additional oscillatory component is seen in the charging and discharging of capacitor C_f. However, the effect of the load on the inverter characteristic is purely quantitative (Figure 6.20). The characteristic takes the form

$$u_{out\cdot 1} = \frac{\pi E}{2\sqrt{2}}\sqrt{1+\left(\frac{Y_C Z_{lo}}{\cos\varphi} - \tan\varphi\right)^2} = \frac{\pi E}{2\sqrt{2}\cos\beta}, \tag{6.25}$$

where Y_C is the modulus of the capacitive filter's conductivity at the output frequency ($Y_{C2} = 2Y_{C1}$), Z_{lo} and φ the modulus and phase of the load impedance, respectively, and β the phase shift between the fundamentals of the voltage u_{out} and current i, which is equal to the phase of the impedance in the ac circuit at the output frequency.

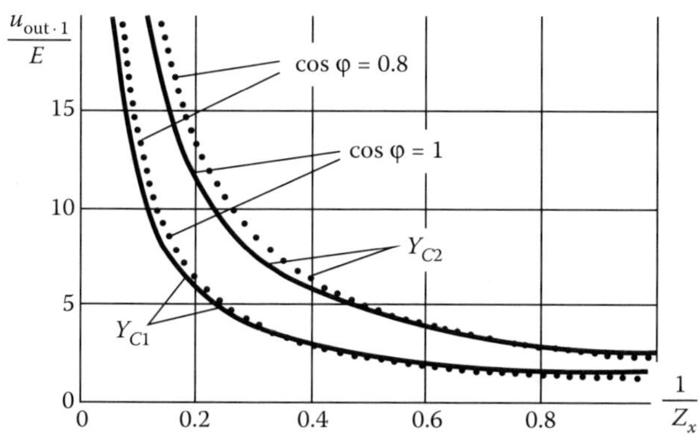

Figure 6.20 Characteristics of a single-phase bridge current source inverter.

The voltage at switch i is $u_{sw} = u_d(1 - F_i)$. The changing polarity of voltage u_{sw} necessitates the use of switches with a blocking diode (Figure 6.18).

In the design of current source inverters, there is an upper limit on the output voltage: $u_{out,1} \leq u_{lim}$. For the required load range, it follows from Equation 6.25 that $Y_C = \omega_{out} C_f$. In operation with low output frequencies, the capacitor costs increase. That limits the use of current source inverters with the switching algorithm described here.

In Figure 6.21, we show a three-phase bridge current inverter. To ensure continuous flow of current I_d, one odd-numbered switch and one even-numbered switch must be on at any time. Hence

$$F_1 + F_3 + F_5 = 1, \quad F_2 + F_4 + F_6 = 1. \tag{6.26}$$

We now consider the simplest switching algorithm: in that case, each switch conducts for one-third of the period, and the sequence in which the switches operate is their numerical order. In Figure 6.22, we show the current and voltage diagrams in the case of a symmetric R load.

The phase currents are

$$\begin{aligned} i_A &= I_d(F_1 - F_4), \\ i_B &= I_d(F_3 - F_6), \\ i_C &= I_d(F_5 - F_2). \end{aligned} \tag{6.27}$$

When these currents flow, the phase capacitors C_f are charged. In intervals without these currents, the capacitors are discharged (Figure 6.22).

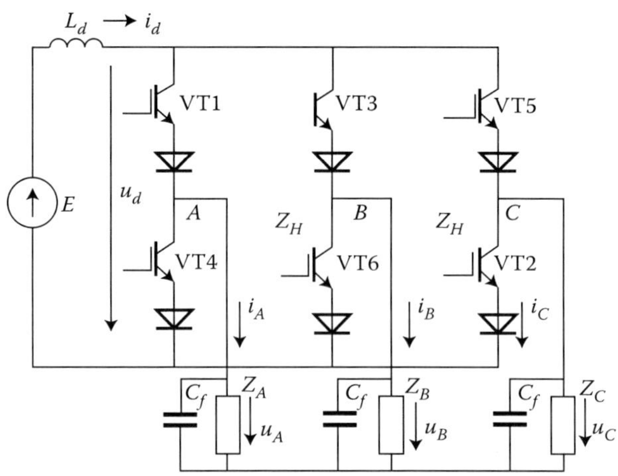

Figure 6.21 A three-phase bridge current inverter.

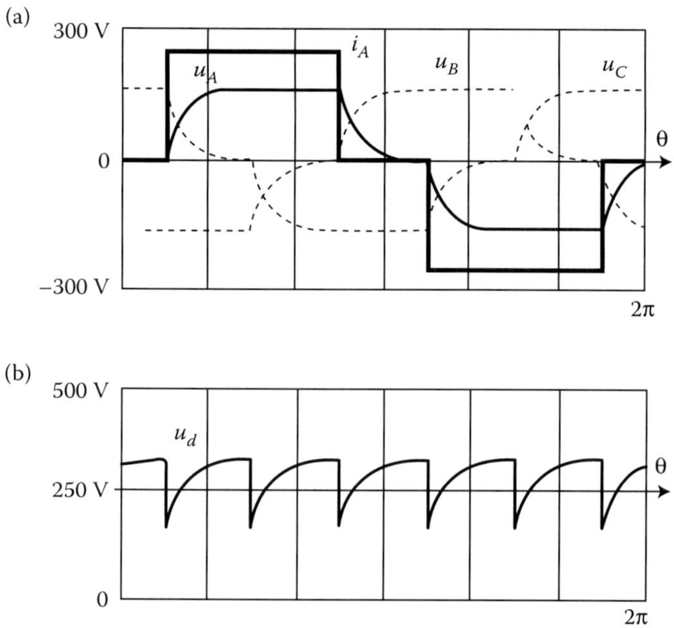

Figure 6.22 Current and voltage diagrams for a three-phase current source inverter with an active load. (a) u_A, u_B, u_C, and i_A; (b) u_d.

The voltage u_d takes the form

$$u_d = u_A(F_1 - F_4) + u_B(F_3 - F_6) + u_C(F_5 - F_2). \tag{6.28}$$

We plot u_d in Figure 6.22. The mean value of u_d is E. As for a single-phase current inverter, the voltage u_d becomes more triangular as the load resistance increases; in addition, the amplitude of u_d increases and it may become negative. The fundamental component of the output phase voltage also increases.

The characteristic of a three-phase current source inverter is sharply descending

$$u_{\text{out}\cdot 1} = \frac{E}{2.34\cos\beta}. \tag{6.29}$$

Here β is the phase shift between the fundamental components of the voltage u_A and current i_A, which is equal to the phase of the impedance in the ac circuit at the output frequency. The effect of the load's phase on the inverter characteristic is purely quantitative, and β changes.

In three-phase current source inverters, the choke inductance L_d required to smooth out the ripple of current i_d may be considerably less than that in single-phase inverters.

The basic features of current source inverters are as follows:

1. Current source inverters contain a choke with high inductance in the dc circuit. As a result, they are larger and heavier than voltage source inverters.
2. Current source inverters do not permit operation in near-idling conditions (with minimal load), on account of the increase in the output voltage and switch voltages.
3. Current source inverters with the simple switching algorithm here described cannot effectively be used to produce low-frequency voltage, on account of the cost of the capacitors.
4. The natural characteristic of current source inverters is sharply descending.
5. Current source inverters are characterized by considerable control inertia, on account of the variation in the source voltage.
6. Current source inverters are stable under the action of short-circuits in the load circuit, on account of stabilization of the source current. In short circuits, the output current may grow very slowly.
7. The transistors in current source inverters may be replaced by single-throw thyristors, without change in circuit design. In that case, C_f also acts as a commutating capacitor. That reduces the inverter cost and permits its use with high-voltage loads (Section 6.2.3).

With the appearance of high-frequency power transistors, current source inverters have become considerably less competitive with voltage inverters. However, they may still prove useful in certain electric drives and in other areas of electrotechnology.

6.2.2 Pulse-width control in current inverters

Regulation and stabilization of the current source inverter's output voltage may be based on external methods (changing the source voltage E) or internal methods. In transistor-based current source inverters, pulse-width control may be employed. We now consider the operation of a single-phase current source inverter (Figure 6.18), with an R load in the case of single-pulse pulse-width control.

To prevent energy supply from the source E during the intervals between pulses, we may supply control pulses to the transistors of a single half-bridge (VT1, VT4 or VT2, VT3). In that case, the load is disconnected from the external circuit; the voltage and the current at Z_{lo} are maintained by the supply of energy stored in capacitor C_f during the pulse. In Figure 6.23, we present the corresponding current and voltage diagrams; the voltages when $E = 300$ V are shown. It is evident that the voltage pulses u_d become shorter with a decrease in the fill factor γ. As the mean value of u_d is E, the amplitude of the pulses u_d increases, along with the output voltage and the fundamental component $u_{out,1}$.

In Figure 6.24, we show the family of characteristics $u_{out,1} = f(1/Z_{lo})$ for a current source inverter with pulse-width control in the case of an R load. The characteristics are sharply descending. In the case of an RL load, the output voltage includes an oscillatory component, which is especially intense at small R. As a result, the characteristics are distorted, and the influence of the load and β on the output voltage becomes more complex.

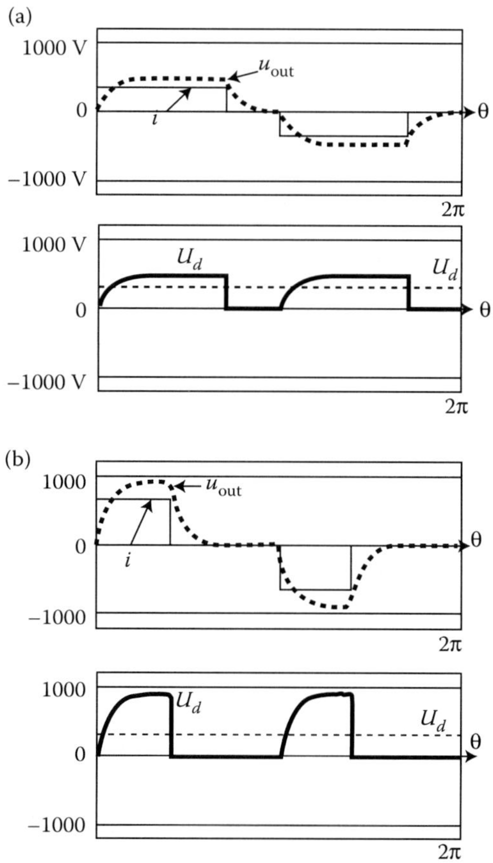

Figure 6.23 Operational diagrams of a single-phase current source inverter with single-pulse pulse-width control in the case of an R load, when (a) $\gamma = 0.7$ and (b) $\gamma = 0.4$.

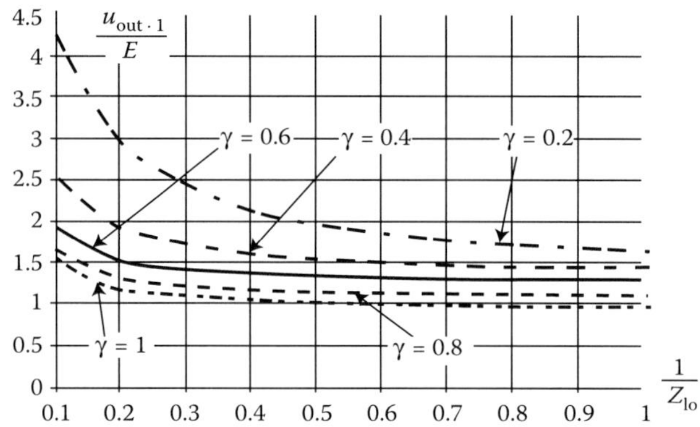

Figure 6.24 Characteristics of a current source inverter with pulse-width control in the case of an active load.

Therefore, pulse-width control cannot expediently be used for regulation and stabilization of the current source inverter's output voltage in the case of an RL load.

The harmonic composition of the output voltage may be improved by pulse-width modulation (Chapter 7).

6.2.3 Current inverters based on single-throw thyristors

We noted in Section 6.2.1 that the transistors in current source inverters may be replaced by single-throw thyristors, without change in the basic circuit design. In Figure 6.25, we show circuit diagrams for single-phase and three-phase current source inverters. These are known as parallel inverters. All the relations and characteristics in Section 6.2.1 apply in this case. The only difference is that the capacitors C_f in the thyristor inverter are commutating capacitors. That entails that the fundamental component of the current i in the ac circuit must lead the fundamental component of the output voltage u_{out} by the phase β. The reactive power created by the capacitor must be greater than the inductive power of the load. In other words, the load's reactive power must be amply compensated.

We now consider the commutation of the thyristors in the first half-period. At time t_1 (Figure 6.19), the voltage at the load and capacitor C_f is negative. Therefore, when pulses are supplied to the control electrodes of thyristors VT1 and VT2, a positive voltage is applied from the capacitor to the cathode of thyristor VT3, whereas a negative pulse is supplied to anode VT4. Through open switches VT1 and VT2, the capacitor voltage is applied to thyristors VT3 and VT4, which consequently are switched off. They remain off as long as the voltage at the capacitor is negative

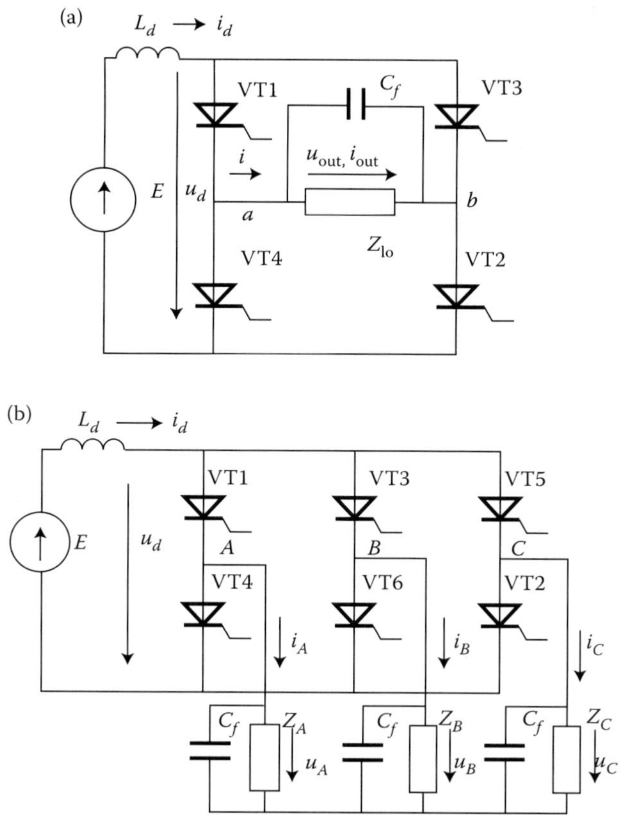

Figure 6.25 (a) Thyristor-based single-phase and (b) three-phase current source inverters.

(interval t_1–t_2). The duration of interval t_1–t_2 is determined by phase β. Therefore, for reliable thyristor commutation, we require that

$$\beta \geq \omega_{out} t_{off}, \tag{6.30}$$

where ω_{out} is the inverter's angular frequency and t_{off} is the time at which the thyristor is switched off.

For the ac circuit at the output frequency

$$\tan\beta = \frac{Y_C Z_H}{\cos\varphi} - \tan\varphi. \tag{6.31}$$

It follows from Equation 6.31 that β declines with a decrease in the load impedance Z_{lo} or with an increase in the load's phase φ. When $Z_{lo} = Z_{lo.min}$,

$\varphi = \varphi_{max}$, and $\beta_{min} = \omega_{out}t_{off}$, we obtain from Equation 6.31 the minimum value of $Y_C = \omega_{out}C_f$ required for reliable thyristor commutation. In any current inverter, there is an upper limit on Z_{lo} on account of the increase in the output voltage and a lower limit on Z_{lo} on account of the commutation conditions in thyristor inverters.

When Equation 6.30 is not satisfied in commutation, both thyristors in the half-bridge are open, and there will be a short in the dc circuit, with a consequent increase in the current through choke L_d. The load is also short-circuited, and the capacitor's unbounded discharge current flows through the thyristors, which may be damaged as a result. Thus, failure to satisfy Equation 6.30—for instance, connection of the inverter to a load $Z_{lo} < Z_{lo \cdot min}$—will result in an accident.

Equations 6.30 and 6.31 are valid for both single-phase and three-phase inverters.

For thyristor-based current source inverters, regulation and stabilization of the output voltage by internal methods require the introduction of additional circuit components. The most common circuit design includes a capacitive-power compensator; a single-phase compensator is illustrated in Figure 6.26. In the ac circuit, a compensator consisting of a choke L_{co} and a bidirectional thyristor switch is introduced in parallel with the load.

When the bidirectional switch is conducting, the fundamental component of the compensator's inductive current is

$$I_{L \cdot 1} = \frac{u_{out \cdot 1}}{\omega_{out} L_{co}}. \tag{6.32}$$

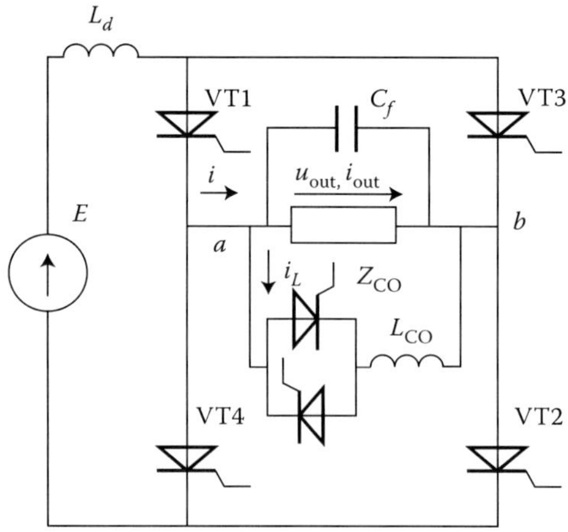

Figure 6.26 A single-phase thyristor current source inverter with additional compensator.

The reactive component of the inverter current is

$$i_r = i_{\text{out.r}} + i_L. \tag{6.33}$$

If delay is introduced in the thyristor switching within the bidirectional switch, discontinuous current flows through choke L_{co}, and the fundamental component of the current i_L declines. In this way, we may regulate the reactive power in the ac circuit and stabilize the phase β, which, according to Equation 6.25, determines the magnitude of the fundamental component in the inverter's output voltage.

Despite the ability to stabilize or regulate the output voltage, the device has serious deficiencies.

- Complication of the power circuit
- Complication of the control system on account of the additional control channel for the thyristors in the compensator
- Increase in the capacitance C_f, as some of its reactive power is consumed in compensation of the compensator's own reactive power.

In the proposed system, the capacitors ensure capacitive load in the ac circuit. The capacitance may be reduced by including cutoff diodes in the circuit (Figure 6.27).

When thyristors VT1 and VT2 and diodes VD1 and VD2 are conducting, the capacitors are charged with the polarity as shown in Figure 6.27. The capacitors are disconnected from the load by diodes VD3 and VD4. When control pulses are supplied to thyristors VT3 and VT4, thyristors

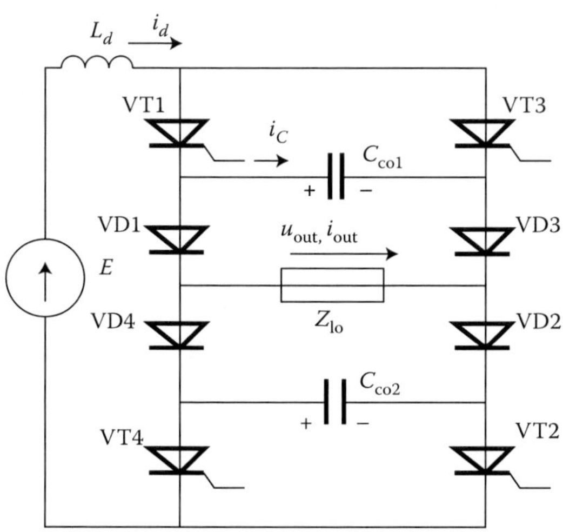

Figure 6.27 A thyristor current inverter with cutoff diodes.

VT1 and VT2 are switched off under the action of the voltage at capacitors C_{co1} and C_{co2}, current i_d flows in the circuit C_{co1}–VD1–Z_{lo}–VD2–C_{co2}–VT4, and the capacitors are recharged. The voltage changes sign and as a result reaches a value equal to the output voltage. Diodes VD1 and VD2 are switched off, whereas diodes VD3 and VD4 begin to conduct. In the second half-period, the processes are analogous.

In this system, the capacitors only participate in commutation processes, and their magnitude is

$$C_{co1} = C_{co2} \geq \frac{I_d t_{off}}{2u_C}. \tag{6.34}$$

On this basis, the inverter may operate with any load and any output frequency.

6.3 ac converters

As shown in Chapter 4, line-commutated ac thyristor converters with profound regulation are characterized by a low power factor. The use of completely controllable switches considerably improves this problem.

6.3.1 ac converters (regulators) without transformers

Such converters are based on dc regulators (Section 5.2). Some circuit diagrams are shown in Figure 6.28a through c. The transistors in the regulator circuits are replaced by bidirectionally conducting switches VT1 and the diodes by bidirectionally conducting switches VT2. In Figure 6.28d through f, we show possible structures for the transistor switches. The regulator input is connected to an ac grid (voltage e). With a positive grid voltage, the direction of the currents through the switches is as shown in Figure 6.28a through c. The switches VT2, replacing the diodes, are on for the first half-period, while the switches VT1 are on for the same periods as the transistors in dc regulators.

Thus, with any polarity of the grid voltage, the converters operate in the same way as dc regulators. The difference is that the converters are connected to an ac voltage source. As the switching algorithms are the same, the formulas for dc regulators may be employed here.

In a step-down ac regulator (Figure 6.28a),

$$u_{out}(t) = \gamma \cdot e(t). \tag{6.35}$$

In a step-up ac regulator (Figure 6.28b),

$$u_{out}(t) = \frac{e(t)}{1-\gamma}. \tag{6.36}$$

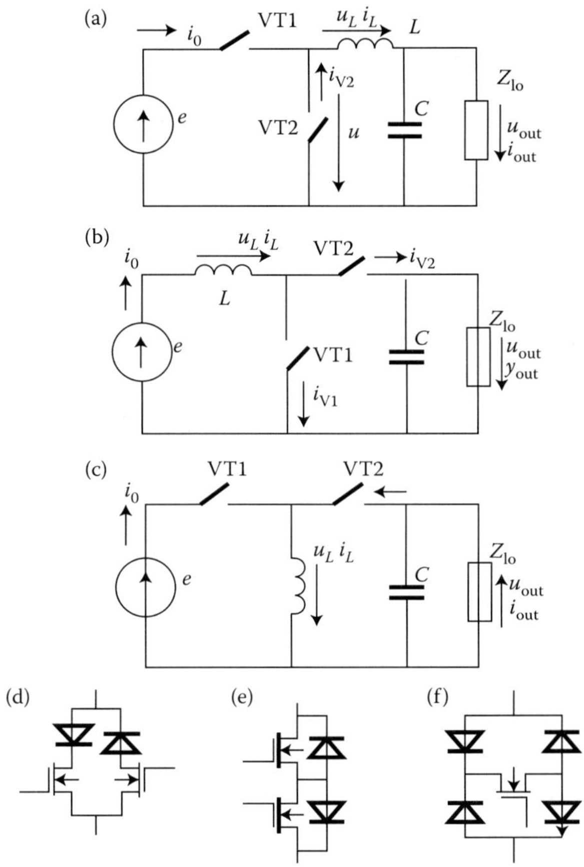

Figure 6.28 (a) Step-down, (b) step-up, (c) buck–boost ac voltage regulators without transformers, and (d–f) circuits for bidirectionally conducting transistor switches.

In a buck–boost regulator (Figure 6.28c),

$$u_{out}(t) = -\frac{\gamma \cdot e(t)}{1-\gamma}, \tag{6.37}$$

where γ is the fill factor.

The fundamental component of the grid current i_0 has a phase shift φ, equal to the phase of the output circuit, which consists of Z_{lo} and C in parallel. In operation with a constant RL load, C may completely compensate the reactive power of the load. In Figure 6.29a and b, we plot the voltages and currents in the step-down regulator (Figure 6.28a). The voltage u consists of a pulse sequence, although the high-frequency components in the

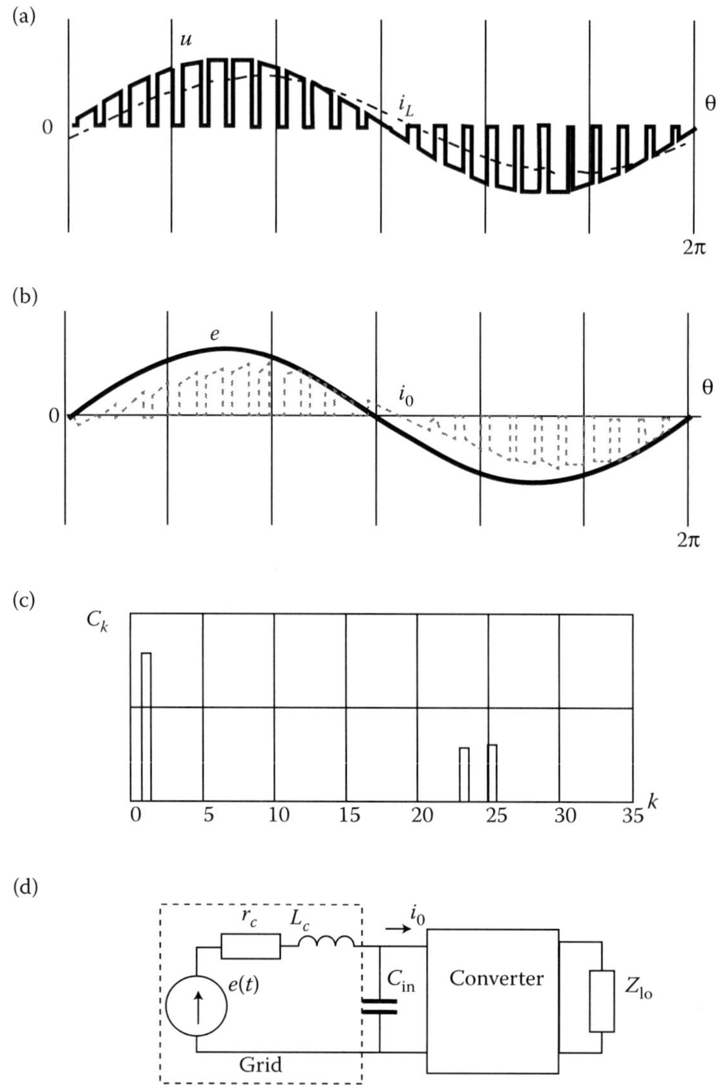

Figure 6.29 (a and b) Current and voltage diagrams for a step-down controller, (c) the grid's current spectrum, and (d) the structure of a converter with a capacitive input filter.

output current i_L are considerably weakened. The grid current i_0 is also pulsed; its spectrum contains high-frequency components (Figure 6.29c). The converter's power factor is $\chi = \nu \cdot \cos \varphi$, where the fundamental factor ν of the grid current, which characterizes the distortion power, depends on the fill factor γ.

To reduce the distortion power, a capacitive filter C_{in} is introduced at the input of ac voltage converters (Figure 6.29d). The filter takes care of the high-frequency components of the converter's input current i_0. To reduce C_{in}, we may increase the switching frequency $f_{sw} = A_f \cdot f_{grid}$, where A_f is the number of pulses on the u curve within the grid period. A step-up converter (Figure 6.28b) may be connected to the grid without a filter, as the converter's input current is continuous.

6.3.2 *ac voltage converter with a voltage booster*

The converter in Figure 6.30a contains a bridge consisting of bidirectionally conducting switches and also a transformer (transformation ratio k_{tr}). In Figure 6.30b, we plot the voltages in this circuit. When switches S1 and S2 are turned on, $u_{out} = e(1 + k_{tr})$. When S2 and S3 (or S1 and S4) are turned on, $u_d = 0$, whereas $u_{out} = e$. By varying the fill factor γ, we may regulate u_{out} ($u_{out} > e$). When switches S3 and S4 are turned on, $u_{out} = -e(1 + k_{tr})$. When S2 and S3 (or S1 and S4) are turned on, $u_d = 0$, whereas $u_{out} = e$. By varying the fill factor γ, we may regulate u_{out} ($u_{out} < e$).

A benefit of this converter is that the harmonic composition of the output voltage and the grid current is better than for the converters in Section 6.3.1. Accordingly, the converter may be used without input and output filters. Deficiencies include greater size and mass on account of the transformer operating at the output frequency and the considerable expenditures on the transistor switches, each of which may be based on the circuits in Figure 6.28d through f.

The converter in Figure 6.30c includes a rectifier R and an inverter I forming the bipolar voltage u_d at the first transformer winding by pulse-width control (Section 6.1.1). The voltage diagrams are shown in Figure 6.30d. Depending on the phase of the inverter bridge's output voltage, the converter may increase or decrease the grid voltage. The rectangular voltage of the voltage booster leads to the appearance of odd output-frequency harmonics in the output voltage; these harmonics are small.

The deficiencies of this converter are increase in mass and size on account of the transformer and the considerable expenditures on the transistor switches.

6.3.3 *Indirect ac voltage converters*

To improve the harmonic composition of the output voltage (without major expenditures on output filters) and to increase the power factor as close as possible to one, we may use indirect ac voltage converters, which consist of a one- or three-phase active rectifier and a one- or three-phase inverter. Both modules form the voltage by pulse-width modulation (Chapter 7).

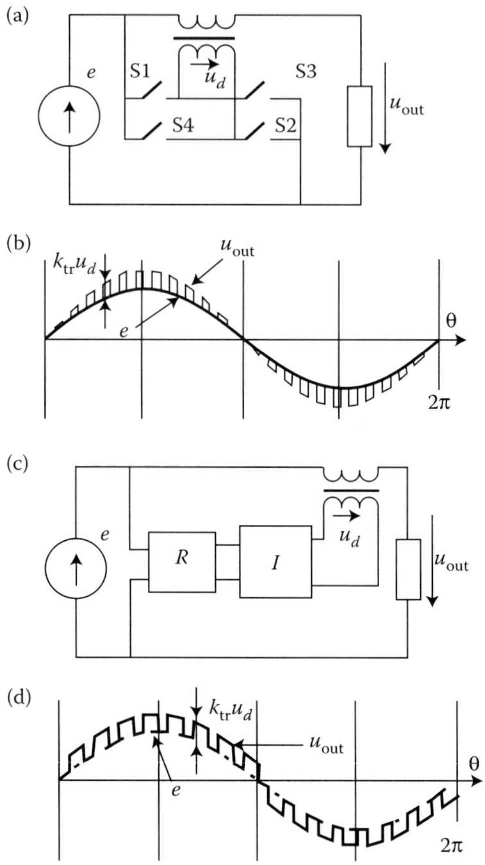

Figure 6.30 (a and c) Circuit diagrams and (b and d) corresponding voltage diagrams for converters with a voltage-boosting transformer.

6.4 Frequency converters

6.4.1 Frequency converters with a dc link

Frequency converters are widely used in electric drives, in the power industry, in optical engineering, and elsewhere. Most frequency converters today are *indirect frequency converters*, consisting of a rectifier and an inverter, and contain a *dc link*. The chosen frequency converter will depend on the number of phases and the frequency variation at the input and output, the requirements on the harmonic composition of the output voltage and current, the requirements on the power factor, and the ratio of the output and input converter voltages. The series of frequency converters must ensure reversal of the energy flux between the grid and the load.

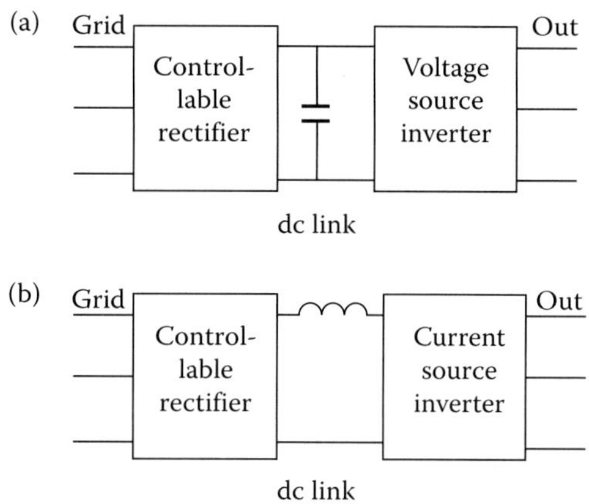

Figure 6.31 Structural diagrams of frequency converters with a dc link based on (a) voltage and (b) current converters.

Frequency converters are based on current and voltage inverters and also resonant inverters (Chapter 8). With strict quality requirements on the output voltage and with broad variation in the output frequency, wide use has recently been made of voltage inverters that form the output voltage by pulse-width modulation (Chapter 7). Current inverters with pulse-width modulation are also employed.

In rectifier selection, we must take account of the requirements on the power factor. Uncontrolled rectifiers with L or LC filters—especially three-phase rectifiers (Chapter 4)—have a high power factor. Higher values of the power factor ($\chi \approx 1$) may be obtained by means of active rectifiers or power-factor correctors (Section 7.3.2). In that case, the frequency converter corresponds to the single-phase/three-phase converters in Figure 6.31. Single-phase/three-phase and three-phase/single-phase converters are of similar structure.

In Figure 6.31a, we use a controllable rectifier, and the dc link contains a capacitive filter (energy store). In Figure 6.31b, we again use a controllable rectifier (Section 7.3), and the dc link contains an inductive filter. The filter in the dc link is the output filter for the rectifier and the input filter for the inverter. The benefits of converters that do not include a transformer are the ability to obtain the output voltages higher or lower than the grid voltage and the ability to reverse the energy flux between the grid and the load.

6.4.2 *Direct frequency converters*

Direct frequency converters are of considerable interest to researchers. In the converter shown in Figure 6.32, each phase (a, b, c) of the load is

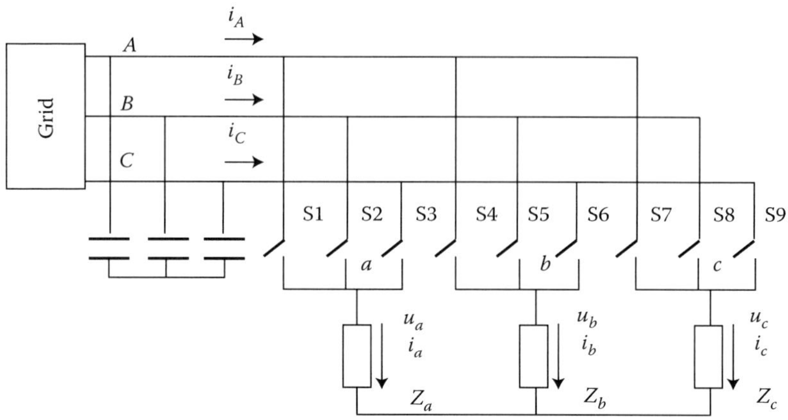

Figure 6.32 A matrix converter of the frequency.

connected to each phase (A, B, C) of the grid by means of bidirectionally conducting switches based on the circuit in Figure 6.28d. This is known as a matrix converter of the frequency (Teichmann et al., 2000; Shrejner et al., 2003; Liu et al., 2005; Chaplygin, 2007; Fedyczak et al., 2009; Manesh et al., 2009).

The input filter of such a matrix converter isolates the input-current components formed by the converter whose frequency is close to or above the commutation frequency of the converter switches. Therefore, the system is light and compact. The use of pulse-width modulation in this structure ensures very high quality of the output voltage when current, whose fundamental component is in phase with the grid voltage, is drawn from the grid. The matrix converter facilitates energy transmission in two directions between the grid and the load. In Figure 6.33, we plot the voltages and currents at the input (u_A, i_A) and output (u_a, i_a) of the matrix converter. The input frequency is 50 Hz, and the output frequency is 100 Hz.

The benefits of the matrix converter are the absence of filters (energy stores) in the dc link and lower voltage losses at the switches.

However, the matrix converter also has a number of serious drawbacks.

1. The matrix converter contains nine bidirectionally conducting switches, that is, 18 completely controllable unidirectionally conducting switches. In contrast, three-phase/three-phase converters with a dc link (Figure 6.31) have only 12 switches.
2. The matrix converters require considerable expenditures on drivers and their power sources.
3. Additional current sensors are required for commutation of the switches in the matrix converter.

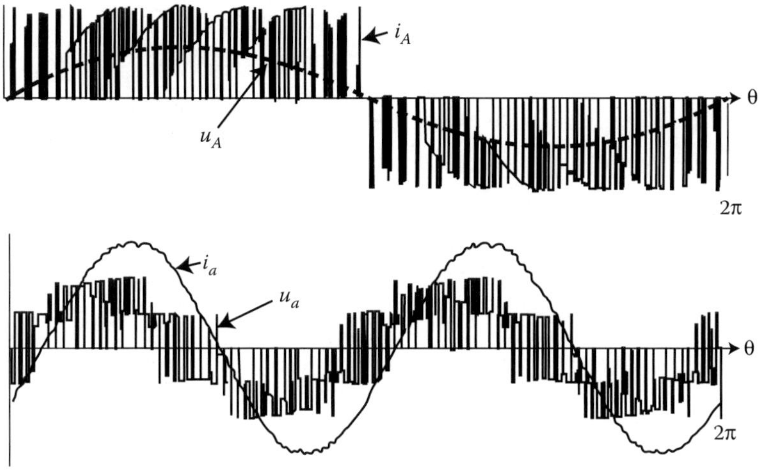

Figure 6.33 Current and voltage diagrams at the input and output of the matrix converter.

4. The transformation ratio of the matrix converter, that is, the ratio of the maximum possible effective value of the output phase voltage to the effective value of the grid phase voltage, is $k_{U,mcf} = 0.86$–0.95, whereas $k_U > 1$ is possible in indirect frequency converters (Figure 6.31).
5. The matrix converter's output voltage contains intense parasitic harmonics in the low-frequency region. (The amplitude of the fifth harmonic may be 10% of the fundamental component, and the amplitude of the second harmonic may be 1.5%–4.6% of the fundamental. Harmonics with frequencies below that of the fundamental component may appear, as well as a constant component.) Thus, the matrix converter falls far short of frequency converters with a dc link in terms of the quality of the output voltage. With asymmetry of the grid, the matrix converter's output voltage deteriorates, on account of the appearance of low-frequency components that lie below the output frequency. This effect is especially pronounced at the maximum output frequencies.
6. The power factor of the matrix converter is high. Nevertheless, the spectrum of the grid current contains low-frequency components that are not present in converters with a dc link.

However, for a very large class of low-power rectifiers (say, single-phase rectifiers), the methods outlined here cannot be used to improve the compatibility with the grid, on account of the expenses involved. To that end, we use correction of the power factor on the dc side of the rectifiers.

References

Chaplygin, E.E. 2007. Analysis of distortion of network current and output voltage of matrix frequency converter (Analiz iskazhenij setevogo toka i vyhodnogo naprjazhenija matrichnogo preobrazovatelja chastoty). *Jelektrichestvo*, 11, 24–27 (in Russian).

Chaplygin, E.E. 2009. Two-phase pulse-width modulation in three-phase voltage inverters (Dvuhfaznaja shirotno-impul'snaja moduljacija v trehfaznyh invertorah naprjazhenija). *Jelektrichestvo*, 6, 56–59 (in Russian).

Fedyczak, Z., Szczesniak, P., and Koroteev, I. 2009. Generation of matrix-reactance frequency converters based on unipolar PWM AC matrix-reactance choppers. *Proc. ECCE 2009*, San Jose, pp. 1821–1827.

Liu, T.-H., Hung, C.-K., and Chen, D.-F. 2005. A matrix converter-fed sensorless PMSM drive system. *Electr. Power Comp. Syst.*, 33(8), 877–893.

Manesh, A., Ankit, R., Dhaval, R., and Ketul, M. 2009. Techniques for power quality improvement. *Int. J. Recent Trends Eng.*, 1(4), 99–102.

Mohan, N., Underland, T.M., and Robbins, W.P. 2003. *Power Electronics—Converters, Applications and Design*, 3rd edn. John Wiley & Sons.

Rozanov, Ju.K. 2007. Power electronics: Tutorial for universities (Silovaja jelektronika: Uchebnik dlja vuzov) Rozanov, Ju.K., Rjabchickij, M.V., Kvasnjuk, A. A. Moscow: Izdatel'skij dom MJeI (in Russian).

Shrejner, R.T., Krivovjaz, V.K., and Kalygin, A.I. 2003. Coordinate strategy of controlling of direct frequency converters with PWM for AC drives. Russian Electrical Engineering, No. 6.

Teichmann, R., Oyana, J., and Yamada, E. 2000. Controller design for auxiliary resonant commutated pole matrix converter. *EPE-PEMS 2000 Proceedings*, Kosice, Slovak Pepublic, Vol. 3, pp. 17–18.

Zinov'ev, G.S. 2012. Bases of power electronics (Osnovy silovoj jelektroniki). Izd-vo Jurajt, Moskva (in Russian).

chapter seven

Pulse-width modulation and power quality control

7.1 Basic principles of pulse-width modulation

According to the definition in IEC 551-16-30, pulse-width modulation (PWM) is a pulse control in which the width or frequency of pulses is modulated within the period of the fundamental frequency so as to create a specific output voltage waveform. In most cases, PWM is used to ensure a sinusoidal voltage or current, that is, to reduce the magnitude of the higher harmonics relative to the fundamental component (the first harmonic). We may distinguish the following basic methods of PWM (Rashid, 1988; Mohan et al., 1995; Rozanov et al., 2007):

- Sinusoidal PWM (and its variants)
- Selective elimination of higher harmonics
- Hysteresis or Δ-modulation
- Space-vector (SV) modulation

The classical technique of sinusoidal PWM entails modifying the width of the pulses forming the output voltage (current), by comparing a given voltage signal (the reference signal) with a triangular voltage signal of higher frequency (the carrier signal). The reference signal is the modulating signal and determines the required waveform of the output voltage (current). In the present case, this signal is sinusoidal and is at the same frequency as the fundamental component of the voltage or current. Many versions of this method exist, with special nonsinusoidal waveforms of the modulating signals. By such means, the level of specific harmonics may be successfully reduced.

Selective harmonic elimination permits the suppression of high-frequency harmonics of the output voltage (current) by adjusting the pulse width and the number of switching operations within each half-period. At present, this method is successfully implemented by means of microprocessor controllers.

Hysteresis modulation is based on the relay tracking of a reference signal, for example, a sinusoidal signal. In the simplest case, this method combines the principles of PWM and pulse-frequency modulation.

Various approaches permit the stabilization of the modulation frequency or the limitation of its variation.

SV modulation is based on the transformation of a three-phase voltage (current) system to a two-phase stationary frame, with the derivation of a generalized SV of voltage (current). Within each cycle determined by the modulation frequency (in each sampling period), the switching occurs between the basic vectors, which correspond to valid states of the converter switches. This results in the formation of the required modulating vector, corresponding to the reference voltage (current) in the three-phase system.

PWM-based control enables us to generate the fundamental component of voltage or current that has the desired frequency, amplitude, and phase. As force-commutated switches are used in the ac–dc converters in this case, it can operate in the four quadrants of the complex plane. In other words, operation in both rectifier and inverter modes is possible, with any value of the power factor cos φ in the range from –1 to +1. In addition, with an increase in the modulation frequency, the ability to reproduce the specified voltages (currents) at the inverter output improves. Accordingly, on the basis of force-commutated converters, active power filters to eliminate harmonics can be implemented (Section 7.3.3).

The main definitions are presented in the following by example of sinusoidal PWM in a single-phase half-bridge voltage source inverter (Figure 7.1a). Switches S1 and S2 are regarded as force-commutated devices connected in series and parallel to diodes. The serial diodes correspond to unidirectional conduction of the switches (e.g., transistors or thyristors), and the parallel reverse diodes ensure the conduction of negative currents.

Figure 7.1b shows the reference modulating signal $u_m(\vartheta)$ and the carrier signal $u_c(\vartheta)$. The control pulses for switches S1 and S2 are formed by the following principle. When $u_m(\vartheta) > u_c(\vartheta)$, switch S1 is on and S2 is off. When $u_m(\vartheta) < u_c(\vartheta)$, the states of the switches are reversed: S2 is on and S1 is off. Thus, the output voltage waveform is the bidirectional pulses (Figure 7.1b). In practice, to eliminate states in which both switches S1 and S2 are on, there must be some interval between the instant at which one switch is turned on and the instant at which the other is turned off. Obviously, the pulse width depends on the amplitude ratio of the signals $u_m(\vartheta)$ and $u_c(\vartheta)$. The parameter characterizing this ratio is the modulation index defined as

$$M_a = \frac{U_m}{U_c}, \tag{7.1}$$

where U_m and U_c are the amplitudes of the modulating signal $u_m(\vartheta)$ and the carrier signal $u_c(\vartheta)$, respectively.

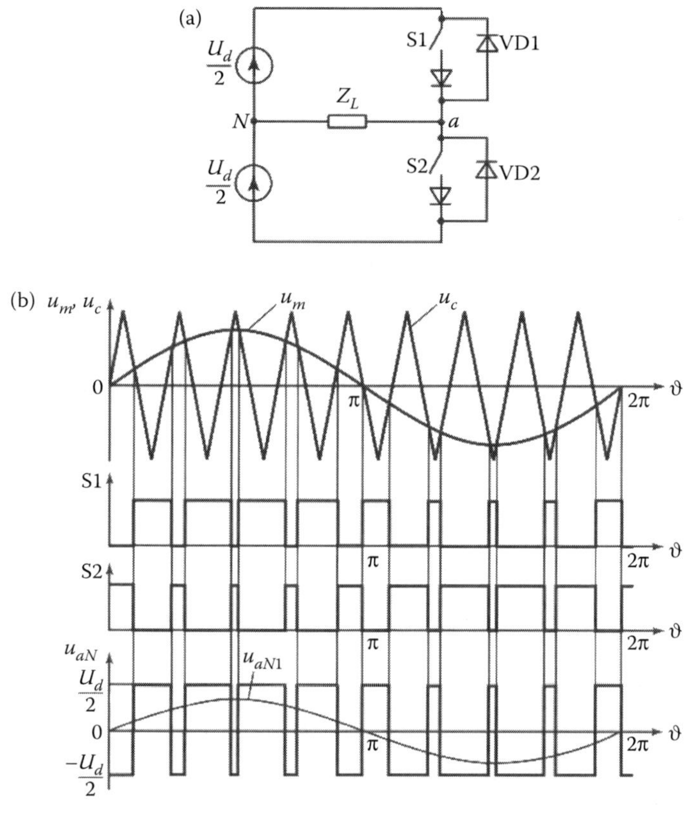

Figure 7.1 (a) Circuit diagram of a single-phase half-bridge voltage source inverter and (b) voltage waveforms for the sinusoidal PWM.

The frequency f_c of the triangular carrier signal is equal to the switching frequency of S1 and S2 and is usually considerably greater than the frequency f_m of the modulating signal. The ratio of f_m and f_c is an important parameter of the modulation efficiency and is known as the frequency-modulation ratio

$$M_f = \frac{f_c}{f_m}. \tag{7.2}$$

For small values of M_f, the signals $u_m(\vartheta)$ and $u_c(\vartheta)$ must be synchronized, so as to eliminate undesirable subharmonics in the output voltage. If these signals are synchronized, M_f is an integer. As recommended by Mohan et al. (1995), maximum value of M_f that requires the synchronization is 21.

Taking into account Equation 7.1, the amplitude of the fundamental harmonic of the output voltage U_{aN1} (Figure 7.1b) can be written as

$$U_{aN1} = M_a \frac{U_d}{2}. \tag{7.3}$$

According to Equation 7.3, the amplitude of the fundamental component of the output voltage is equal to its pulse height $U_d/2$ when $M_a = 1$. A typical dependence of the first harmonic of the output voltage (in the dimensionless form) on modulation index M_a is shown in Figure 7.2. In the range of M_a from 0 to 1, U_{aN1} varies linearly as a function of the modulation index. The limiting value of M_a is determined by the relevant modulation principle, according to which the maximum value of U_{aN1} is the height of a rectangular half-wave ($U_d/2$). Further increase in M_a leads to a nonlinear increase in U_{aN1} to the maximum value.

From the Fourier expansion of the rectangular function, the maximum amplitude of the first harmonic of voltage is

$$U_{aN1} = \frac{4}{\pi}\frac{U_d}{2} = \frac{2}{\pi}U_d. \tag{7.4}$$

The value $M_{a\,\max}$ corresponding to the maximum value of U_{aN1} depends on M_f and is approximately 3.2. In the range $M_a = 1–3.2$, U_{aN1} depends nonlinearly on the modulation index (curve piece A–B in Figure 7.2). This nonlinear region is overmodulation one.

The value of M_f depends on the selected frequency of the carrier signal $u_c(\vartheta)$ and significantly affects converter performance. With an

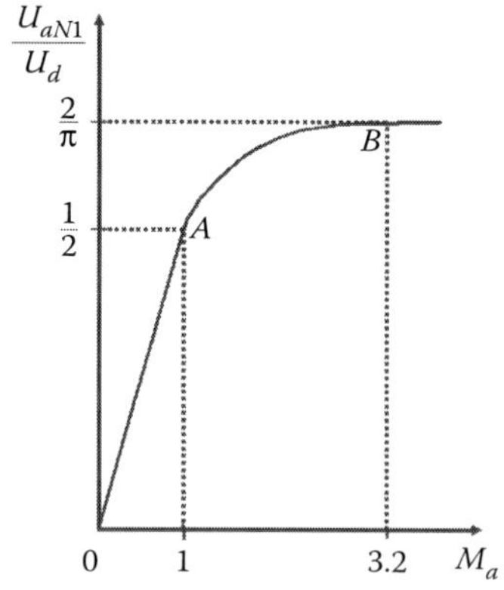

Figure 7.2 Dependence of the relative amplitude of the fundamental component of output voltage on the modulation index in a half-bridge circuit.

increase in the frequency, the switching losses in the converter increase, but the output voltage spectrum is improved and it is easier to filter the high-frequency harmonics due to modulation. In many cases, f_c must be selected so as to reach the upper limit of the acoustic range (20 kHz). In the f_c selection voltage levels, power of the converter and others should be taken into consideration. Therefore, the selection of M_f is a multiple-factor optimization problem. The general trend is the higher M_f for low-power converters and low voltages and vice versa.

7.1.1 *Pulse modulation with a stochastic process*

The use of PWM in converters is associated with the appearance of higher harmonics in the modulated voltages and currents. Their strongest components in the spectrum appear at frequencies that are multiples of M_f. Harmonics also appear at frequencies grouped around multiples of M_f. In that case, the amplitudes of the harmonics decrease on moving away from those values. Higher harmonics may have the following consequences:

- Acoustic noise
- Deterioration of electromagnetic compatibility with other electrical devices

The main sources of acoustic noise are electromagnetic components (inductors and transformers) under the action of voltages and currents containing harmonics with acoustic frequencies. Note that noise may appear at specific frequencies where the higher harmonics are a maximum, although the phenomena that produce noise, such as magnetostriction, complicate the problem solution. Problems with electromagnetic compatibility may arise over a broad frequency range and will be differently expressed for different devices. Traditionally, design and technological measures are adopted to reduce noise, although passive filters and various design methods are used to ensure electromagnetic compatibility.

A promising approach is to modify the spectrum of the modulated voltages and currents—specifically, to equalize the frequency spectrum and to reduce the amplitude of the strongest harmonics by distributing them stochastically over a broad frequency range. It allows reducing the concentration of the noise energy at frequencies at which the harmonics may be a maximum. A benefit of this approach is that it does not require modifying the converter power components and generally relies on software, with a little change in the control system. The main principles of these methods are considered in the following.

PWM is based on the adjustment of the duty ratio $\gamma = t_{on}/T$, where t_{on} is the switch on time and T is the period of pulses. Usually, under steady-state conditions, these parameters, as well as the position of the pulse within period T, are constant, and the results of PWM are determined as the integral average values. In this case, the values of t_{on} and T, as well as the pulse position, result in an undesirable spectrum of the modulated parameters. If we assume that t_{on} and T are random, while keeping the required value of γ, the processes become stochastic, and the spectrum of the modulated parameters is modified. For example, the position of the pulse within period T can be random, or the stochastic variation of T can be ensured. For this purpose, a random-number generator can be used to affect the generator of modulation frequency $f = 1/T$. Also, the position of the pulse within period T can be changed with zero mathematical expectation. The average value of γ must remain at the level specified by the control system. As a result, the spectrum of the modulated voltages and currents is equalized.

7.2 PWM techniques in inverters

7.2.1 Voltage source inverters

7.2.1.1 Single-phase full-bridge voltage source inverter

A simplified topology of a single-phase full-bridge voltage source inverter is shown in Figure 7.3. In this circuit, as in a single-phase half-bridge inverter, sinusoidal PWM is based on comparison of the sinusoidal modulating signal with a triangular carrier signal. In the full-bridge circuit, in contrast to the half-bridge circuit, either unipolar or bipolar modulation techniques can be used. The switches S1–S4 are analogous to those in Figure 7.1. To simplify the consideration of modulation, the node N is marked between two equal dc-side capacitances C_d (Figure 7.3).

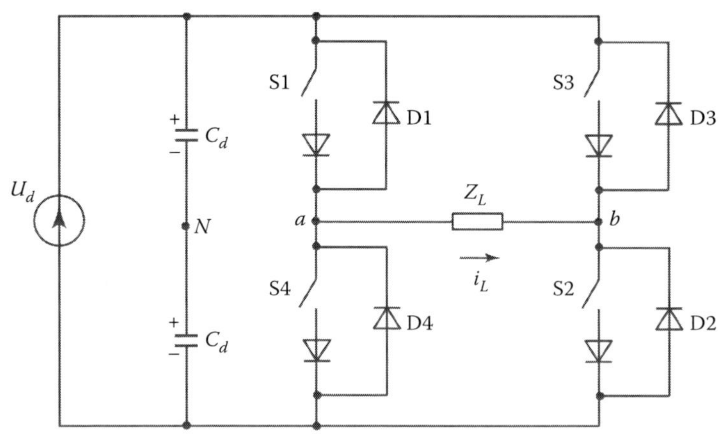

Figure 7.3 A single-phase full-bridge voltage source inverter.

In a full-bridge inverter, there are four combinations of switch as summarized in Table 7.1, where on and off states of switch S are denoted by 1 and 0, respectively. Table 7.2 presents the voltages u_{aN}, u_{bN}, and $u_{ab} = u_{aN} - u_{bN}$ for these states. Depending on the type of modulation, different combinations of switch states may be used.

In *unipolar modulation,* two modulating signals $u_m(\vartheta)$ and $-u_m(\vartheta)$ are used at the same time (Figure 7.4). In that case, there are two sequences of switch control pulses. One sequence controls switches of one leg (S1 and S4), whereas the other controls the other switches (S3 and S2). The pulse sequence generated by comparing the reference signal $u_m(\vartheta)$ and the triangular signal $u_c(\vartheta)$ determines the voltage u_{aN}. When the negative modulating signal $-u_m(\vartheta)$ is compared with the carrier signal $u_c(\vartheta)$, the pulse sequence for u_{bN} is generated. This results in simultaneous modulation of the potential of node *a* relative to *N* (switches S1 and S4 are used) and modulation of the potential of node *b* (switches S3 and S2 are used). In that case, the potential of node *a* relative to node *N* is $U_d/2$ when switch S1 is on (states I and III) and $-U_d/2$ when switch S4 is on (states II and IV). The potential of node *b* relative to node *N* is equal to $U_d/2$ when switch S3 is on (states II and III) and $-U_d/2$ when S2 is on (states I and IV). The conditions of commutation of switches are

Table 7.1 Combinations of Switch States for a Single-Phase Full-Bridge Inverter

	Switch states			
State	S1	S2	S3	S4
I	1	1	0	0
II	0	0	1	1
III	1	0	1	0
IV	0	1	0	1

Table 7.2 Conducting Components and Voltage Values for the Switch States

	Conducting switches and diodes		Voltage		
State	$i_L > 0$	$i_L < 0$	u_{aN}	u_{bN}	u_{ab}
I	S1, S2	D1, D2	$U_d/2$	$-U_d/2$	U_d
II	D3, D4	S3, S4	$-U_d/2$	$U_d/2$	$-U_d$
III	S1, D3	D1, S3	$U_d/2$	$U_d/2$	0
IV	S2, D4	S4, D2	$-U_d/2$	$-U_d/2$	0

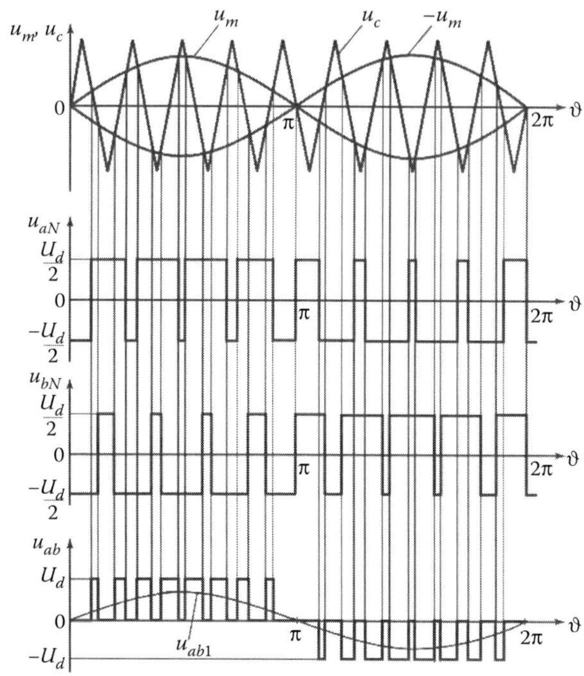

Figure 7.4 Waveforms of output voltages and modulating and carrier signals for the unipolar sinusoidal PWM in a single-phase full-bridge inverter.

$$\begin{aligned} &u_m(\vartheta) > u_c(\vartheta)\text{—S1 is on, S4 is off,} \\ &u_m(\vartheta) < u_c(\vartheta)\text{—S4 is on, S1 is off,} \\ &-u_m(\vartheta) > u_c(\vartheta)\text{—S3 is on, S2 is off,} \\ &-u_m(\vartheta) < u_c(\vartheta)\text{—S2 is on, S3 is off.} \end{aligned} \tag{7.5}$$

As a result, a waveform of the inverter output voltage $u_{ab}(\vartheta)$ is a sequence of unipolar pulses in each half-period of the sine wave specified by the modulation signal $u_m(\vartheta)$, as shown in Figure 7.4.

In the case of *bipolar PWM technique*, the switching algorithm differs. In that case, only two states I and II periodically alternate in accordance with the following conditions:

$$u_m(\vartheta) > u_c(\vartheta)\text{—state I,} \quad u_m(\vartheta) < u_c(\vartheta)\text{—state II.} \tag{7.6}$$

The switching of states I and II corresponds to the modulation in a half-bridge voltage source inverter implemented by switches S1 and S2 (Figure 7.1). The full-bridge and half-bridge inverters differ in the amplitude of the output voltage pulses. This amplitude is U_d in the full-bridge circuit, as against $U_d/2$ in the half-bridge circuit. Correspondingly, a change in the voltage also alters the maximum amplitude of the fundamental component of the output voltage, which is U_d when $M_a = 1$, according to Equation 7.3. In the case of overmodulation, when $M_a > 1$, the modulated voltage transforms to a rectangular form, when $M_a > 3.2$, and an amplitude of the output voltage first harmonic

$$U_{ab1} = \frac{4}{\pi}U_d. \tag{7.7}$$

Considering the influence of an active-inductive load on the electromagnetic processes in a voltage source inverter circuit, we notice that the fundamental component of the load current lags the fundamental component of the output voltage. That entails need of the energy flow changing from the load to the source. As a result, after reversing the fundamental component of voltage, the current continues to flow in the previous direction. Accordingly, the circuit contains diodes D1–D4, in parallel with switches S1–S4. We assume that the positive direction of the load current ($i_L > 0$) corresponds to flow from node *a* to node *b*; or, in a half-bridge circuit, from node *a* to *N*. Then, in intervals in which the energy stored in the load inductances is extracted, the current i_L is negative and passes through the antiparallel diodes, returning to the dc-voltage source U_d (Table 7.2). The time at which the current passes through zero (reverses its direction) depends on the load parameters. Taking only the fundamental components of the load current and output voltage into consideration, this instant is determined by the load angle

$$\varphi_L = \text{arctg}\frac{\omega_1 L}{R}. \tag{7.8}$$

where ω_1 is the angular frequency of the fundamental components and L and R are the load inductance and resistance, respectively.

Obviously, the value of φ_L directly affects the distribution of the load current between the switches and diodes. For example, in the case of active load, no current flows through the diodes. If the load is inductive, in contrast, the mean value of the switch current is equal to the mean value of the diode current.

The spectrum of the output voltage is an important feature of PWM. For sinusoidal PWM, the output voltage contains harmonics of order n, determined by the frequency-modulation ratio M_f (Espinoza, 2001)

$$n = eM_f \pm k, \tag{7.9}$$

where $k = 1, 3, 5, \ldots$ when $e = 2, 4, 6, \ldots$ and $k = 0, 2, 4, 6, \ldots$ when $e = 1, 3, 5, \ldots$ for bipolar PWM and for unipolar PWM $k = 1, 3, 5, \ldots$ when $e = 2, 4, 6, \ldots$.

Thus, the frequency spectrum of the output voltage in single-phase inverters contains not only the first harmonic (frequency f_1), but also multiples of M_f with associated frequencies, in accordance with Equation 7.9. The advantage of unipolar modulation is that the lowest order modulation harmonics will be of higher frequency, as in this case the harmonics are multiples of $2M_f$. With an increase in M_f in the range from 0 to 1, the distortion of the output voltage falls significantly. That permits the use of light passive LC filters to obtain sinusoidal voltage practically.

The first harmonics of the output voltage are determined by the inverter input voltage U_d and, according to Equation 7.1, by the modulation index M_a. For inverter operation with $0 \leq M_a \leq 1$, the amplitude of the fundamental component of the output voltage is $U_{ab1} < U_d$. To increase this value without seriously impairing the spectrum of the output voltage, modified methods of sinusoidal PWM are employed. For example, various methods are based on comparison of a triangular carrier signal with a nonsinusoidal modulating signal u_m. Examples include the following (Rashid, 1988):

- A trapezoidal modulating signal (Figure 7.5a)
- A stepped modulating signal (Figure 7.5b)
- A signal with harmonic injection (Figure 7.5c)

All these methods allow 5%–15% increase in the amplitude of the output voltage, relative to that obtained by the sinusoidal PWM. The spectrum of the output voltage then permits an effective filtration of the higher harmonics.

With significant constraints on the modulation frequency, a selective harmonic elimination technique is successfully used. Usually, in this case, the low-frequency part of the voltage spectrum is eliminated: the third, fifth, and seventh harmonics. That follows from the dependence of the amplitude of higher harmonics on the pulse width of the output voltage, when its waveform features one pulse per half-cycle. If the width of voltage pulses is controlled, the spectrum of the output voltage satisfies the following expression:

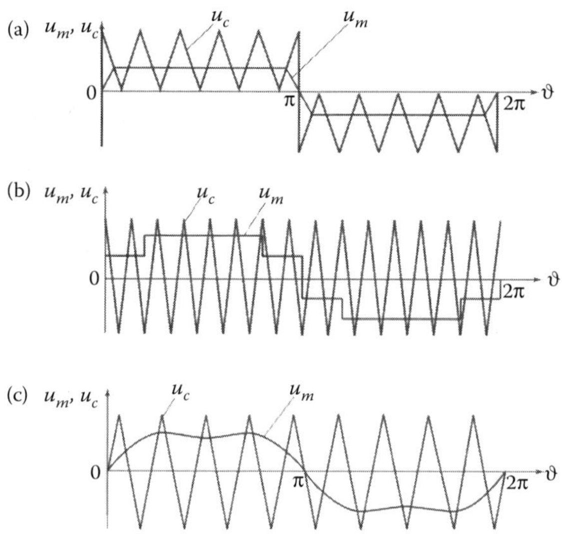

Figure 7.5 Modified methods of sinusoidal PWM with (a) trapezoidal and (b) stepped modulating signals and (c) with harmonic injection.

$$u_{ab}(\vartheta) = \sum_{n=1,3,5}^{\infty} \frac{4U_d}{\pi n} \sin\frac{n\delta}{2} \sin n\vartheta, \tag{7.10}$$

where U_d is the input voltage of the inverter, n the number of harmonics, and δ the angular width of the voltage pulse within a half-period.

From Equation 7.10, we can see that, when $\delta = 2\pi/3$, the third harmonic is eliminated from the spectrum of the inverter output voltage. Then switching occurs twice in a half-cycle. If the number of switching and voltage pulses (N) per half-period increases, more harmonics may be suppressed. For single-phase inverters we can eliminate ($N - 1$) harmonics if output voltage has N pulses during half-period. For example, Figure 7.6

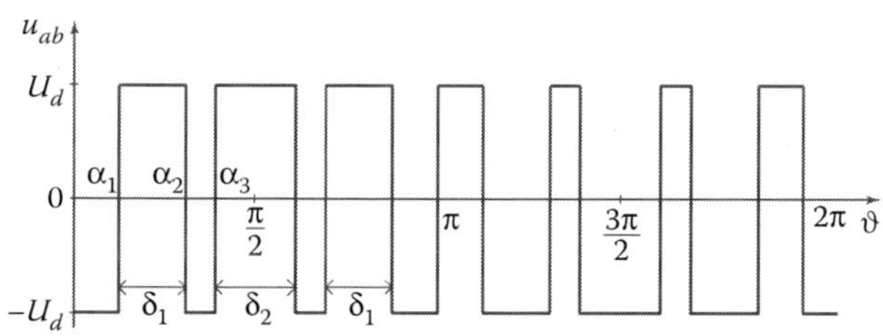

Figure 7.6 Waveform of single-phase inverter output voltage for selective elimination of the third and fifth harmonics.

shows the inverter voltage waveform for bipolar modulation with $N = 3$ and $M_a = 0.5$. In this case, the angles $\alpha_1 = 22°$, $\alpha_2 = 55°$, and $\alpha_3 = 70°$ ($\delta_1 = 33°$ and $\delta_2 = 40°$) and harmonics with $n = 3$ and $n = 5$ are eliminated from the spectrum. Correspondingly, when $N = 4$, harmonics with $n = 3$, 5, and 7 are removed. With unipolar modulation, α and δ differ from the values for bipolar modulation. General techniques of selective harmonic elimination are considered in detail (Espinoza, 2001).

Hysteresis or Δ-modulation is widely used in practice. This provides (Hossein, 2002) relatively simple means of obtaining voltages and currents equal to a given reference. The simplest application of Δ-modulation is a pulse control used in dc–dc converters. Essentially, this involves tracking the controlled signal to follow a given reference signal within a permissible deviation. In the case of load current control in the voltage source inverter circuit, when the current becomes greater than the reference $i_m \pm \Delta i_L$, the state of the converter switches is changed as shown in Figure 7.7. To receive a sinusoidal load current, the sinusoidal reference signal with the same frequency as the fundamental component is used. Usually, the width of the deviation band depends on the width of the hysteresis loop for the relay comparator that generates the switching pulses. A drawback of the technique is the switching frequency variation, which depends on the rate of change of the controlled signal (di_L/dt). Thus, with a sinusoidal modulating signal i_m, the switching frequency is greater in

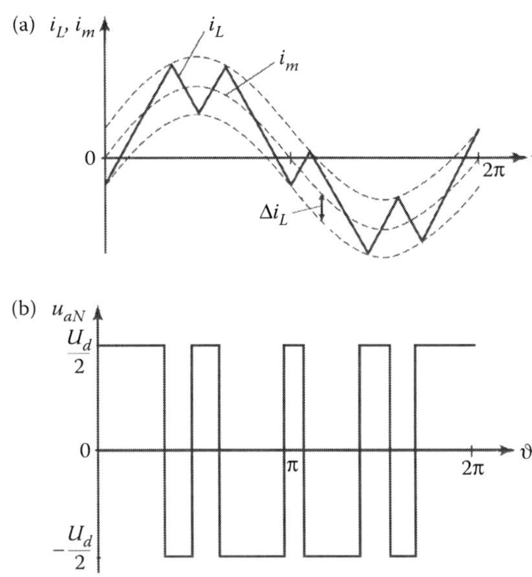

Figure 7.7 (a) Waveforms of the load current and (b) output voltage for hysteresis current modulation in a half-bridge voltage source inverter.

intervals where the sinusoid is close to its maximum value than in intervals where it passes through zero (Figure 7.7). Special methods are used to stabilize the switching frequency. To prevent the appearance of low-frequency harmonics, hysteresis modulation is usually employed at high M_f values.

7.2.1.2 *Three-phase voltage source inverter*

Figure 7.8 shows a topology of a three-phase voltage source inverter. Switches S1–S6 are identical to those in the single-phase inverters considered. The phase voltages u_{aN}, u_{bN}, and u_{cN} may be considered relative to node N, where capacitors C_d are connected. In a balanced three-phase system, the voltages and currents are the same in each phase (except the phase shift) and add up to zero. The sum of their values for any two phases determines the voltage and current in the third phase. Thus, only two voltages and currents are independent. That must be taken into account in specifying the reference signals. In three-phase inverters, as in single-phase inverters, the carrier-based PWM can be used by comparison of three sinusoidal modulating signals and triangular carrier signal.

For a three-phase inverter, there are eight valid switch states given in Table 7.3. States I–VI produce nonzero output voltages, whereas states VII and VIII correspond to zero phase-to-phase voltages. All these states are used to generate a given voltage by PWM. Changing of switch states depends on the ratio of the carrier and reference signals. So as to consider modulation in three-phase and single-phase systems within the same framework, we focus on the phase-to-phase voltage u_{ab}, equal to the difference between the voltages of phases a and b relative to node N (Figure 7.8). As the base node, we can choose any point of circuit. For example, it can

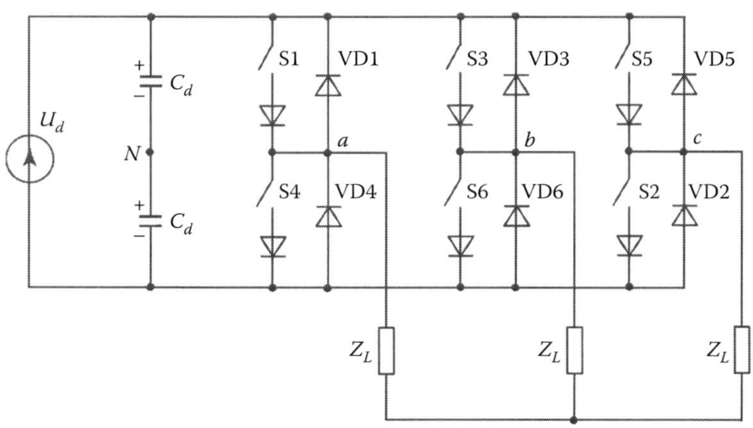

Figure 7.8 A three-phase voltage source inverter (bridge circuit).

Table 7.3 Switch States and Voltage Values for a Three-Phase Voltage Source Inverter (0, off; 1, on)

	Switch state						Voltages		
State	S1	S2	S3	S4	S5	S6	u_{aN}	u_{bN}	u_{ab}
I	1	0	0	0	1	1	$U_d/2$	$-U_d/2$	U_d
II	1	1	0	0	0	1	$U_d/2$	$-U_d/2$	U_d
III	1	1	1	0	0	0	$U_d/2$	$U_d/2$	0
IV	0	1	1	1	0	0	$-U_d/2$	$U_d/2$	$-U_d$
V	0	0	1	1	1	0	$-U_d/2$	$U_d/2$	$-U_d$
VI	0	0	0	1	1	1	$-U_d/2$	$-U_d/2$	0
VII	1	0	1	0	1	0	$U_d/2$	$U_d/2$	0
VIII	0	1	0	1	0	1	$-U_d/2$	$-U_d/2$	0

be a point of the negative dc-source terminal. Considering the node N, the voltage $u_{ab} = u_{aN} - u_{bN}$. According to Table 7.3, values of phase voltages u_{aN} and u_{bN} are as follows:

- When switch S1 is on, phase a is connected to the $+U_d$ terminal, and $u_{aN} = U_d/2$; when S4 is on, phase a is connected to the $-U_d$ terminal, and $u_{aN} = -U_d/2$.
- When switches S3 and S6 are on, $u_{bN} = U_d/2$ and $-U_d/2$, respectively.

From Table 7.3, we can see that the states of switches S1, S3, and S5 are inverse to the states of switches S4, S6, and S2, respectively. That simplifies the realization of the control algorithm in voltage source inverters.

Figure 7.9 shows the waveforms of a three-phase inverter with sinusoidal PWM. For PWM of the voltages u_{aN} and u_{bN}, the conditions of change in states of the switches can be written as

When $u_{mA}(\vartheta) > u_c(\vartheta)$, S1 is on, S4 is off; when $u_{mA}(\vartheta) < u_c(\vartheta)$, S1 is off, S4 is on

When $u_{mB}(\vartheta) > u_c(\vartheta)$, S3 is on, S6 is off; when $u_{mB}(\vartheta) < u_c(\vartheta)$, S3 is off, S6 is on

In accordance with these conditions, during the positive and negative half-cycles of the modulating signals u_{mA} and u_{mB}, the potentials of phases a and b are $U_d/2$, when S1 and S3 are on, and $-U_d/2$, when S1 and S3 are off (S4 and S6 are on).

The amplitude of the fundamental component of the line voltage U_{ab1}, with $0 \le M_a \le 1$, may be written as

$$U_{ab1} = \sqrt{3}\, M_a \frac{U_d}{2}. \tag{7.11}$$

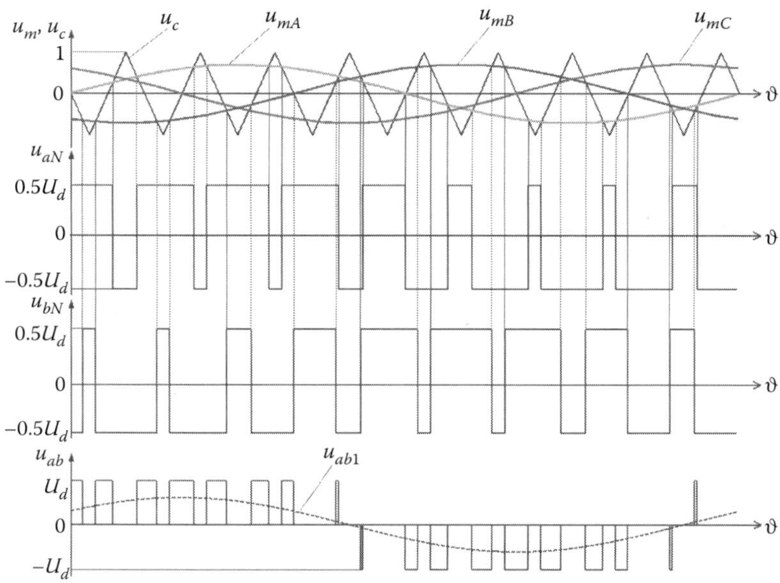

Figure 7.9 Waveforms of modulating and carrier signals and output voltages for sinusoidal PWM in a three-phase inverter.

In the case of overmodulation ($M_a > 1$), the amplitude of the first harmonic increases to the maximum value corresponding to the square-wave operation

$$U_{ab1} = \frac{4}{\pi}\frac{U_d}{2}\sqrt{3} = \frac{2\sqrt{3}}{\pi}U_d. \tag{7.12}$$

The frequency spectrum of the output line voltages has no harmonics which frequencies are multiple of three. The value of the carrier signal frequency should be selected as an odd multiple of three ($M_f = 9, 15, 21, \ldots$), rounding to an integer at small values of M_f.

In three-phase inverters, as in single-phase systems, the current load of the switches may be estimated from the mean value of current within a single period. Obviously, the mean current in switches S1–S6 and diodes D1–D6 correspond to segments of phase currents. Note that for an active-inductive load, the current of the phase is distributed between the corresponding switch and the parallel diode. Thus, the current flows from the dc source to the load through the switch and, on reversal, is returned to the source through the diode. The time of current reversal is determined by the phase-shift factor (cos φ) of the first harmonic of phase current and voltage. Therefore, if we consider only the fundamental component of current, it is simple to calculate the static power losses in the switches and the diodes.

7.2.2 Current source inverters

Before the practical use of self-controlled power electronic devices, pulse modulation for current source inverters (CSIs) practically has no application, because it was very difficult to realize modulation of dc current by use of the forced commutation of thyristors. However, self-controlled switches such as power transistors or gate turn-off thyristors (GTO) allow PWM in CSIs, as well as in voltage source inverters. The duality of CSIs and voltage source inverters enables us to apply practically identical PWM techniques. The duality of inverter circuits means the identity of output current and voltage waveforms using the same modulating techniques.

The significant difference between the PWM algorithms for voltage source inverters and CSIs is producing zero intervals in the output voltage and current waveforms, respectively. For example, zero-voltage intervals in a single-phase voltage source inverter shown in Figure 7.3 are produced when the upper (S1 and S3) or lower (S4 and S2) switches are on. Then the load is shorted and disconnected from the dc source U_d. Such switching is not applicable in a CSI circuit shown in Figure 7.10a, as the dc-side current I_d must have short circuit at all times. Otherwise, the break in the circuit with inductor L_d results in a high-voltage surge.

Figure 7.10a shows a single-phase CSI based on unidirectional switches (e.g., GTOs). In the case of transistor switches, the series-connected diodes are used for reverse voltage blocking. Since we can have an inductive load, a capacitor should be connected to the output terminals of the CSI. The parallel capacitor absorbs energy when the switch states change. The equivalent circuit for steady inverter operation is shown in Figure 7.10b, in which the voltage source U_d and inductor L_d are replaced by the dc current source I_d, although the switches and capacitive load are represented by a voltage source U_c. Zero load currents are produced when either switches S1 and S4 or switches S3 and S2 are on. The current I_d freewheels through these switches, and the voltage U_c is reduced to zero; in other words, the current source is shorted by the conducting switches. When the switches S1 and S2 are on, the load current is positive. The current $i_L = -I_d$ when S3 and S4 are on. Thus, the load current is equal to $+I_d$, 0, and $-I_d$. In order to produce a required current waveform, we should switch these values of current, according to a specific algorithm. Using PWM techniques similar to the one developed for voltage source inverters, we can generate sinusoidal ac current. To reduce the modulating harmonics, the output CL filters are used, and the

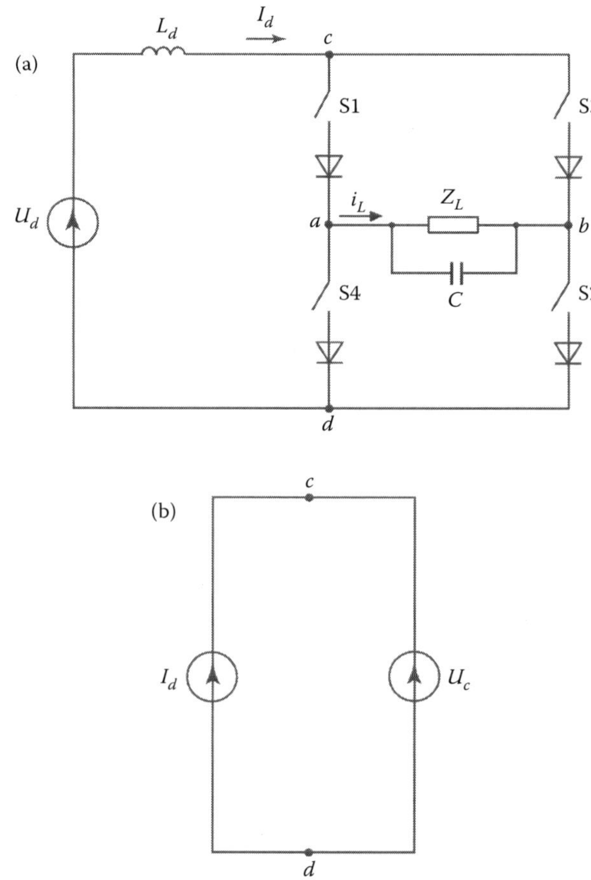

Figure 7.10 A single-phase CSI based on (a) self-commutated switches and (b) its equivalent circuit.

capacitors are connected to an ac side of the inverter. The output current waveform for sinusoidal PWM is shown in Figure 7.11.

Single-phase CSIs are not widely used in practice. However, three-phase CSIs (Figure 7.12) are used in electric drive applications. Therefore, it is important to ensure sinusoidal output voltage. For this purpose, the sinusoidal PWM of output currents is used. Table 7.4 presents the valid switch states and the current values for a three-phase CSI. In contrast to a voltage source inverter, there are three zero states (VII, VIII, and IX) that correspond to zero values of the phase currents. In these states, the switches of one leg are on, that is, S1 and S4, S3 and S6, or S5 and S2.

Another significant difference between a CSI and a voltage source inverter is that the output voltage depends directly on the load. The

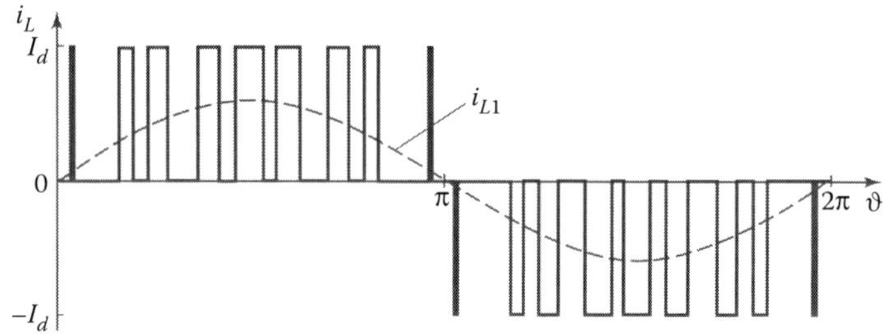

Figure 7.11 Phase current waveform for sinusoidal PWM in a single-phase CSI.

simplest way to control load voltages is by adjusting the modulation index M_a. That entails taking account of the following expressions for the amplitude of fundamental output phase current I_a:

$$
\begin{aligned}
I_{a1} &= \sqrt{3}M_a\frac{I_d}{2} \quad \text{when } 0 < M_a \le 1, \\
\frac{\sqrt{3}}{2}I_d \le I_{a1} &\le \frac{2\sqrt{3}}{\pi}I_d \quad \text{when } M_a > 1.
\end{aligned}
\tag{7.13}
$$

As we can see from Equation 7.13, in accordance with duality of CSIs and voltage source inverters, the first harmonic of phase current in the

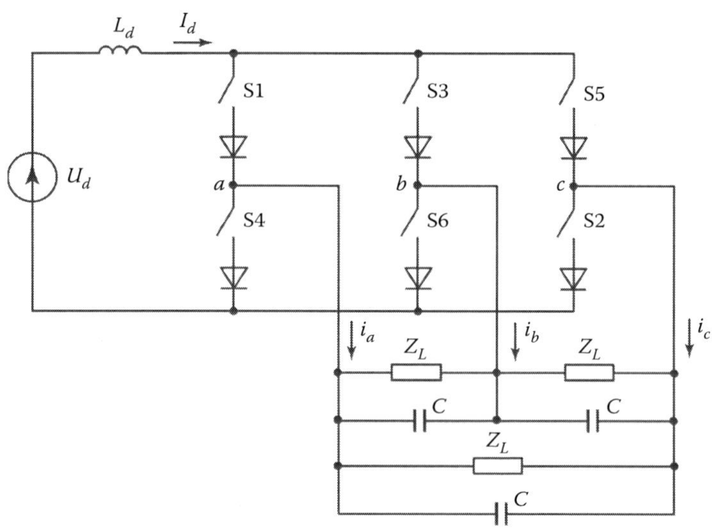

Figure 7.12 A three-phase CSI.

Table 7.4 Switch States and Current Values for a Three-Phase CSI (0, off; 1, on)

State	Switch state						Currents		
	S1	S2	S3	S4	S5	S6	i_a	i_b	i_c
I	1	0	0	0	0	1	I_d	$-I_d$	0
II	1	1	0	0	0	0	I_d	0	$-I_d$
III	0	1	1	0	0	0	0	I_d	$-I_d$
IV	0	0	1	1	0	0	$-I_d$	I_d	0
V	0	0	0	1	1	0	$-I_d$	0	I_d
VI	0	0	0	0	1	1	0	$-I_d$	I_d
VII	1	0	0	1	0	0	0	0	0
VIII	0	0	1	0	0	1	0	0	0
IX	0	1	0	0	1	0	0	0	0

CSI and the first harmonic of line voltage in the voltage source inverter depend on M_a in the same way.

Note that control of zero interval width in the phase current waveform can be used to selectively eliminate low-order harmonics in the current spectrum, similar to the one performed in voltage source inverters.

7.2.3 SV modulation

Transformation of three-phase signals to a two-phase frame simplifies voltage and current control in three-phase converters. In addition, such transformation of coordinates is used in digital control systems of ac motors by frequency converters (ac voltage controllers). Therefore, SV modulation technique in three-phase voltage-source converters is widely used (Kazmierkowski et al., 2001; Akagi, 2002). The SV PWM is based on representation of three-phase voltage signals in the stationary αβ frame as vectors with discrete phase change, in accordance with the switch states of the converter (Table 7.3). There are eight valid states that correspond to defined values of output line voltages. In the case of SV modulation, the inverter model in a complex plane is used, instead of the model in phase coordinates. Thus, it is necessary to generate space voltage vector, which corresponds to the required line voltages. In order to produce three sinusoidal voltages during one cycle (the rotating reference SV), six nonzero vectors (switch states) are used consecutively in each sampling period. The sampling frequency can be regarded as the carrier frequency, as in sinusoidal PWM. The SV-modulation strategy using the αβ stationary frame (Section 3.1.4) is considered in detail in the following.

Table 7.5 presents the switch states for a bridge circuit of a voltage source converter (Figure 7.13a) and the corresponding values of SVs for

Table 7.5 Switch States and SV Values for a Three-Phase Voltage Source Converter

State no.	Switch states S1	S3	S5	SV U_k
I	1	0	1	$\frac{2}{3}U_d\, e^{j5\pi/3}$
II	1	0	0	$\frac{2}{3}U_d\, e^{j0}$
III	1	1	0	$\frac{2}{3}U_d\, e^{j\pi/3}$
IV	0	1	0	$\frac{2}{3}U_d\, e^{j2\pi/3}$
V	0	1	1	$\frac{2}{3}U_d\, e^{j\pi}$
VI	0	0	1	$\frac{2}{3}U_d\, e^{j4\pi/3}$
VII	1	1	1	0
VIII	0	0	0	0

output voltages u_a, u_b, and u_c (Figure 7.13b). The waveforms shown in Figure 7.13b illustrate 180° switching algorithm, when each switch is on during an interval π. The on state of each switch in the upper group (S1, S3, S5) corresponds to an off state of switches in the lower group (S4, S6, S2) and vice versa (Figure 7.13a). The numbering of states is of no fundamental significance, but their sequence must strictly correspond to the alternation of voltages u_a, u_b, and u_c. Each state corresponds to an interval of π/3. In other words, there is a discrete change in the states of switches S1–S6 at the boundary of these intervals. Note that only the six active (nonzero) states are used; the zero states, in which the output voltages are zero, are disregarded. Therefore, in the αβ plane, we can identify six vectors U_1–U_6 corresponding to the position of the voltage SV at the boundaries where the states of switches S1–S6 change. As a result, six sectors bounded by the vectors U_1–U_6 are formed on the αβ plane (Figure 7.14a). The ends of the vectors can be connected by straight lines forming a regular hexagon. The center of the hexagon, which combines the starting points of the vectors, corresponds to zero vectors U_7 and U_8.

During the cycle of fundamental output voltages u_a, u_b, and u_c, consecutive change in sectors occurs, as shown in Figure 7.13c, and the SV U_S (Figure 7.14a) is moved within each sector from one state to another (U_1–U_2–U_3–U_4–U_5–U_6). Modulus of the rotating vector U_S is equal to the amplitude of sinusoidal voltages u_a, u_b, and u_c. The active (nonzero) vector

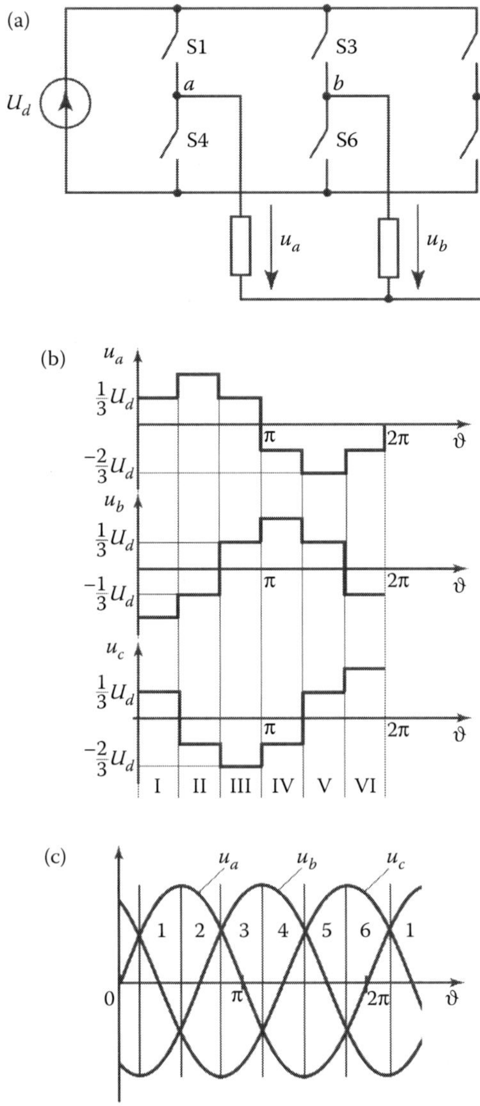

Figure 7.13 Bridge circuit of (a) a voltage source converter, (b) waveforms of output phase voltages, and (c) its fundamental components.

modulus is determined by the maximum value of the square-wave (staircase waveform) phase voltage, i.e., it is equal to $2U_d/3$, where U_d is the input dc-side voltage of the inverter.

In sinusoidal PWM, switching occurs when the modulating signals are equal to the triangular signals at the carrier frequency f_c. The amplitude of sinusoidal modulating signals is directly proportional to

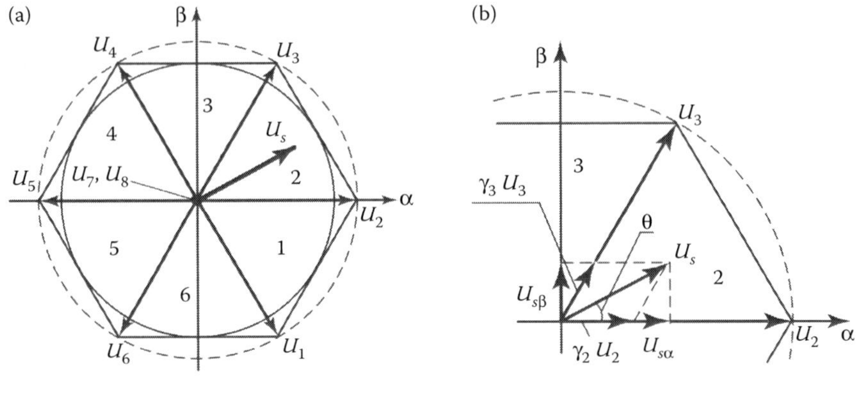

Figure 7.14 SV diagram in the αβ plane.

the modulation index M_a. In general, these modulating signals are generated for each phase in order to produce a three-phase system of converter output voltages. Otherwise, in SV PWM, essentially only the reference SV is a resulting modulating signal to generate all output voltages of the three-phase converter. In sinusoidal PWM, the switching signals can be received using an analog comparator. In SV PWM, in contrast, a digital control system is used to calculate the sampling period and the switching times. The range of f_c is approximately the same in the SV and the sinusoidal PWM techniques.

To produce the required voltages in SV PWM, the reference voltage SV is averaged within each of six sectors. For sector k, this is done by switching of nearest active vectors (U_k and U_{k+1}) and one zero vector (U_7 or U_8). In this case, the voltage vectors are summed, but the time of using for each vector should be taken into account. The relative value of used time γ_k for SV U_k can be regarded as the duty ratio in PWM dc–dc converters. Correspondingly, the time of vector U_k using is defined as $t_k = \gamma_k \cdot T_S$, where $T_S = 1/f_c$ is the sampling period of the carrier signal. As only three states are used in each sampling period, we can write the following expression:

$$t_k + t_{k+1} + t_7 + t_8 = T_S, \tag{7.14}$$

where t_k and t_{k+1} are the times during which vectors U_k and U_{k+1} are used and t_7 and t_8 are the using times for zero vectors U_7 and U_8.

The reference SV U_S rotates at the angular frequency (ω_m) of the output voltage in the αβ stationary frame. For a balanced phase voltage, the modulus and phase of this vector are completely determined by its components $U_{S\alpha}$ and $U_{S\beta}$. In that case, the times t_2 and t_3 in the second sector can be calculated from the trigonometric relations for vectors $U_{S\alpha}$,

$U_{S\beta}$ and $\gamma_2 U_2$, $\gamma_3 U_3$. These vectors of sector 2 are shown in Figure 7.14b, which allows us to calculate t_2 and t_3 taking into account the phase angle θ of the vector U_S. Since the modulation frequency f_c is considerably greater than the voltage frequency f_m, we can regard the parameters of vector U_S as constant over the period T_S. According to Figure 7.14b,

$$U_{S\beta} = \gamma_3 U_3 \sin 60° = U_S \sin \theta. \tag{7.15}$$

Hence, considering that $\gamma_3 = t_3/T_S$ and

$$M_a = \frac{2}{\sqrt{3}}\frac{U_S}{U_3},$$

we find that

$$t_3 = \frac{\sqrt{3}}{2}\frac{M_a}{\sin 60°}T_S \sin\theta = M_a T_S \sin\theta.$$

Analogously, to find t_2, we can write

$$U_{S\alpha} = U_S \cos\theta = \gamma_2 U_2 + \gamma_3 U_3 \cos 60° \text{ or}$$

$$M_a\frac{\sqrt{3}}{2}\cos\theta = \gamma_2 + \gamma_3 \cos 60°. \tag{7.16}$$

Using Equation 7.15, we find that

$$t_2 = \frac{\sqrt{3}}{2}\frac{M_a}{\sin 60°}T_S \sin(60° - \theta) = M_a T_S \sin(60° - \theta),\ (t_7 + t_8) = T_S - t_2 - t_3.$$

Similarly, to calculate t_2 and t_3 values for phase angle θ in sector 2, switching times t_k and t_{k+1} are defined for sector k. Thus, the reference voltage SV U_S is the sum of the averaged values of the nearest active vectors and zero vector U_z during the sampling period in each sector k: $U_S = \gamma_k U_k + \gamma_{k+1} U_{k+1} + \gamma_z U_z$. The amplitude of the modulating vector U_S is controlled by variation in the modulation index M_a. Modulation of the SV allows us to generate the given converter output voltages in accordance with the inverse abc/αβ transformation (Section 3.1.4)

$$\begin{vmatrix} u_a(\vartheta) \\ u_b(\vartheta) \\ u_c(\vartheta) \end{vmatrix} = \begin{vmatrix} 1 & 0 \\ -\frac{1}{2} & \frac{\sqrt{3}}{2} \\ -\frac{1}{2} & -\frac{\sqrt{3}}{2} \end{vmatrix} \begin{vmatrix} U_{s\alpha}(\vartheta) \\ U_{s\beta}(\vartheta) \end{vmatrix}. \tag{7.17}$$

There are many variants of SV PWM algorithms, in which different vector sequences and different zero vectors can be used within a sampling period. Often, the method of symmetrical position of active and zero vectors within the modulation period T_S is shown in Figure 7.15. During the first half-period, the sequence U_z–U_k–U_{k+1}–U_z is used, whereas in the second half-period, we have the reverse sequence U_z–U_{k+1}–U_k–U_z. The zero SVs U_z is alternately chosen among U_7 and U_8. In that case, we can write the following expression for the second sector (with appropriate modification for other sectors):

$$\frac{T_S}{2} = t_2 + t_3 + t_7 + t_8, \quad t_7 = t_8. \tag{7.18}$$

Another switching algorithm features using only one zero vector (U_7 or U_8) in each modulation period. Any method should provide reducing of the switching frequency and undesirable harmonics. Some switching algorithms enable to reduce the switching frequency f_c by 33%, without deterioration of the modulation efficiency. In addition, the switching losses may be reduced by 30%, depending on the load power factor.

In SV modulation, the range of the output voltage is limited by the maximum value of the modulation index. In Figure 7.14a, the region bounded by an inscribed circle of hexagon ($M_a \leq 1$) corresponds to the linear functional dependence between the output voltage and M_a. The maximum value of voltage SV $U_S = U_d/\sqrt{3}$, where U_d is the dc-side voltage.

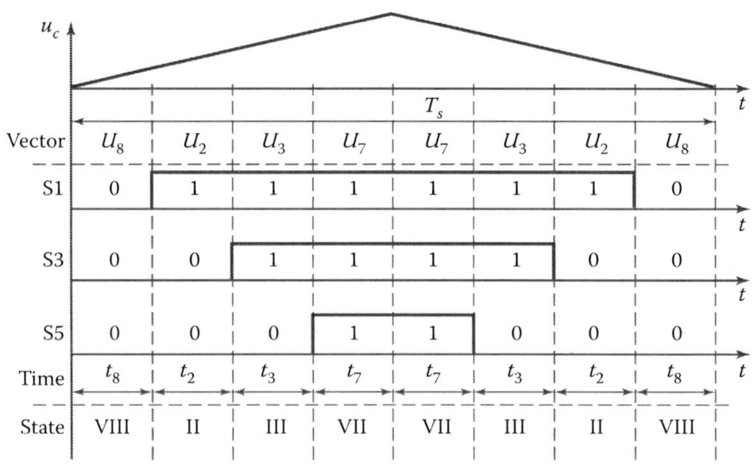

Figure 7.15 Switching algorithm for SV PWM (sector 2).

In the case of overmodulation within the region bounded by a circumscribed circle ($M_a \geq 1$), we can increase M_a to $2/\sqrt{3} \approx 1.15$. In this case, the zero vectors on-time (t_7 or t_8) is equal to zero (only nonzero SVs are used). Further increase in M_a leads to distortion of the output voltage waveforms that results in square-wave nonmodulated voltages.

7.3 Power quality control on the basis of PWM converters

7.3.1 Functional capabilities of PWM converters

The advance in self-commutated fast-speed power electronic switches permits pulse control of currents and voltages and their modulation, according to any given reference signal. This fundamentally changes the functional capabilities of converters. Previously, converters based on silicon-controlled rectifiers (thyristors) were predominantly used for conversion from one type of electric power to another, for example, from ac to dc or from dc to ac. That was accompanied by distortion of the mains current and consumption of reactive power. The PWM control and the ability to operate in four quadrants (in the rectifier and inverter mode, with any phase shift between the mains current and voltage) significantly allow expanding of functional capabilities of the ac–dc converters. Such converters have no drawbacks featured to line-commutated converters based on thyristors. Thus, PWM converters based on self-commutated switches become the basis for development of multifunctional power quality controllers. Thus, power electronics today provides to solve many urgent problems of electrical power engineering, referred to energy saving.

The capabilities of ac–dc pulse-width modulated converters based on self-controlled power electronic devices include the following:

1. Ensuring of sinusoidal mains currents in the rectifier mode and sinusoidal load voltages in the inverter mode
2. Correction of the power factor of consumers, power line, or source
3. Compensation of reactive power (inductive or capacitive)
4. Stabilization of the load voltage
5. Active filtering of current or voltage harmonics in the power system
6. Hybrid filtering (combination of active and passive filters) of the current or voltage harmonics in the power system
7. Compensation of unbalanced currents and voltages in a three-phase power system on the basis of an elimination of reverse-sequence and zero-sequence currents or voltages
8. Transfer of electric energy between the power system and energy storages of different types

9. Effective integration of renewable energy sources and utility power system
10. Effective control of electric drives and electrical processing equipment
11. Increase of efficient application of hybrid and electric vehicles as a result of power quality improvement

7.3.2 Operation modes of ac–dc PWM converters

7.3.2.1 Inversion

In Chapter 4, we considered the operation of line-commutated ac–dc converters implemented with thyristors, without modulation of the output parameters. The ac of such converters is nonsinusoidal, which deteriorates their compatibility with the ac mains and limits their application. The development of high-frequency self-commutated power electronic devices permits the elimination of such problems using PWM. The advantages of ac–dc grid-connected PWM inverters are considered in this chapter.

Traditionally, converters used for inversion have a smoothing inductor on the dc side, which is a characteristic feature of CSIs. This inductor is a required element of the converter circuit because of the difference in the instantaneous values of dc side and ac voltages (Figure 7.16). With high values of inductance L_d and without PWM, we have square-wave phase converter currents. This can result in distortion of the mains voltage. Because of the high amplitudes of low-frequency current harmonics, heavy filters must be used. The use of sinusoidal PWM of phase currents significantly reduces the required power of the output filters in current source converters.

In the case of a voltage source converter circuit, its inverter operation mode requires an inductive output filter, because of the periodic connection of ac and dc sources with different voltages (Figure 7.17). In CSIs, these circuits are separated by the dc-side inductor L_d, which limits the rate of current change. In voltage source inverters, these inductors are components of the output LC filters, smoothing the ripples due to PWM at high frequency. In the given converters, the inductance L_f is used to smooth the current waveform due to the instantaneous difference in voltages U_d and u_S. At high modulation frequency, the low value of inductance is necessary. In inverters without sinusoidal PWM, the high inductance is required, which results in significant deterioration of converter performance. Therefore, ac–dc converters based on a voltage source inverter without PWM are of no practical use. The main characteristics of ac–dc converters based on current-source and voltage-source inverters are considered in more detail in the following. As an example, single-phase converter circuits are compared.

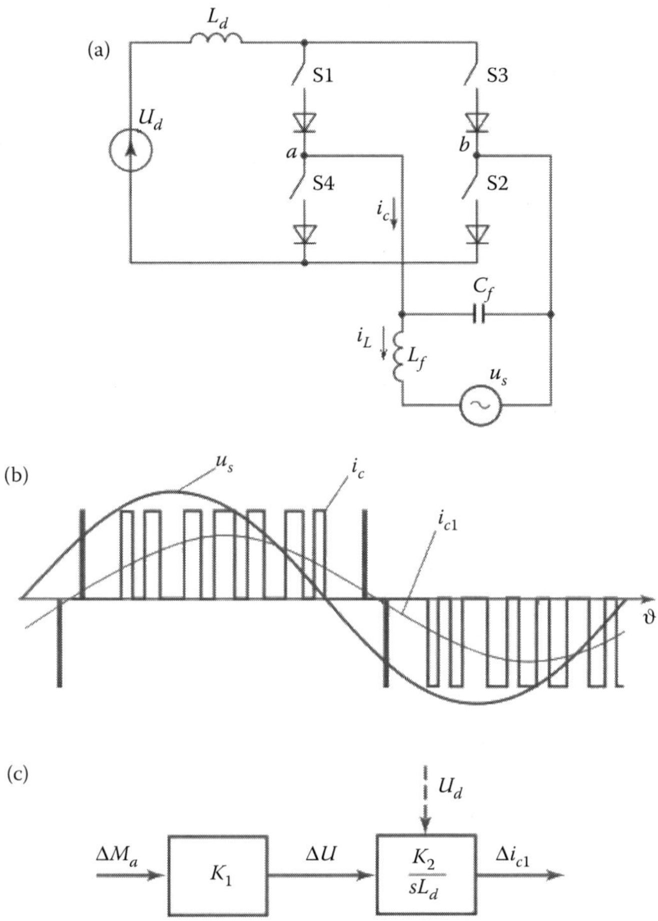

Figure 7.16 (a) Current source ac–dc converter circuit; (b) waveforms of ac voltage source and converter current; and (c) block diagram of fundamental current control.

7.3.2.2 Current source converters

In this case, a CL filter is connected on the ac side of the converter to remove modulating harmonics of the output current i_c, as shown in Figure 7.16a. Figure 7.16b shows the waveforms of sinusoidal mains voltage and modulated converter current i_c. This current can be written as a harmonic series

$$i_c(\omega_1 t) = \sum_{n=1}^{\infty} I_{cn} \sin(n\omega_1 t + \varphi_n), \tag{7.19}$$

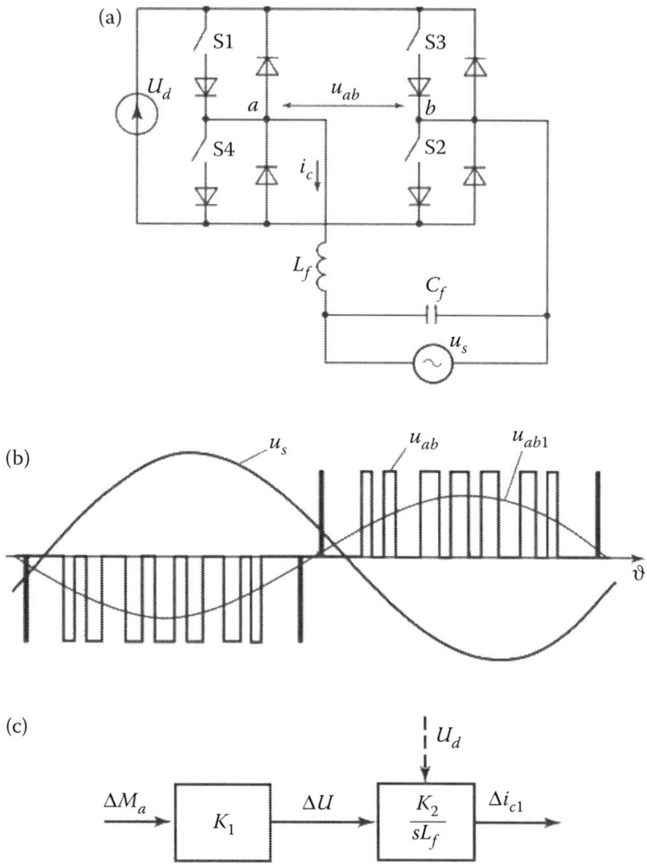

Figure 7.17 (a) Voltage source ac–dc converter circuit; (b) waveforms of ac voltage source and converter output voltage; and (c) block diagram of fundamental current control.

where ω_1 is the angular frequency of the fundamental component, I_{cn} the amplitude of the nth current harmonic, and φ_n the initial phase of the nth harmonic.

The harmonic amplitude of mains current (in inductor L_f) is

$$I_{Ln} = \frac{I_{cn}}{\omega_n^2 L_f C_f - 1}. \tag{7.20}$$

As we can see from Equation 7.20, the harmonics with frequencies $\omega_n \gg \omega_1$ are successfully filtered by capacitor C_f and low value of inductance L_f. Thus, the current between the ac and dc voltage sources is mainly smoothed by the high value of inductance L_d. Consequently, in the current source converter, the fault protection of the switches—for example, in the case of a short-circuit on the ac side—can be ensured relatively simply. Another benefit of the circuit is that it operates effectively with inductive energy storages. However, the drawback of the current source converter is the inertial variation of dc-side current due to the inductance $L_{d.}$ Figure 7.16c shows a simplified block diagram of the converter current control as a function of the modulation index M_a. This figure corresponds to small deviations of parameters (ΔM_a and Δi_{c1}). The dynamic characteristics of the converter in this case are mainly determined by the value of the inductance L_d, as shown in Figure 7.16c as the integral unit in the operator form. The effect of the CL filter on the dynamics is significantly less than that of the inductance L_d and is not shown in Figure 7.16c.

7.3.2.3 Voltage source converters

In this case, the filter inductor L_f is the main element smoothing the current due to the difference in the modulated converter voltage u_{ab} and mains voltage u_S (Figure 7.17a and b). When the ac-source power considerably exceeds the converter power, the converter current harmonics (except the fundamental) can be approximately presented by a harmonic series

$$i_{cn}(\omega_1 t) = \sum_{n \neq 1}^{\infty} \frac{U_{abn}}{n\omega_1 L_f} \sin(n\omega_1 t + \varphi_n), \tag{7.21}$$

where U_{abn} is the amplitude of the nth voltage harmonic.

As we can see from Equation 7.21 in the present case, in contrast to the current source converter, the amplitudes of the current harmonics are determined by the inductance L_f. Accordingly, this filter inductance must be greater than that in a CSI. At the same time, because of high modulation frequency, the inductance L_f is considerably less than the inductance L_d in Figure 7.16a. Therefore, the control of the output current is less inertial. The block diagram of the converter current control is shown in Figure 7.17c. This is often the main benefit of the voltage source converter when compared with the current source converter.

Concluding this brief comparison of different ac–dc converters, both types of PWM converters based on self-controlled switches can ensure the four-quadrant operation. The rectifier mode of ac–dc converters is considered in the following.

7.3.2.4 Rectification

PWM in the rectifier mode is used to increase the power factor, by reducing both the reactive power of the fundamental current and the distortion power (i.e., the decrease in mains current harmonics). For this purpose, self-commutated ac–dc converters can be used. In the rectifier mode, the phase of the fundamental current is shifted in accordance with positioning of the current vector in the second and third quadrants of the complex plane. Note that rectification and inversion in thyristor-based ac–dc converters generally were implemented with current source converter circuits. Grid-connected voltage source converters (Figure 7.17a) have been adopted with the development of self-controlled devices. The relation between the fundamental current and voltage is best seen in a vector diagram. The fundamental current and voltage are represented as complex amplitudes, assuming the sinusoidal waveform of mains voltage and zero current of the filter capacitor C_f:

$$\dot{U}_S + j\omega_1 L_f \dot{I}_{c1} = \dot{U}_{ab1}, \tag{7.22}$$

where $\dot{U}_{ab1}$ is the complex amplitude of the fundamental converter voltage and I_{c1} is the complex amplitude of the fundamental converter current (Figure 7.17a).

Assuming the mains voltage vector U_S as the base, the vector diagram for four-quadrant operation is obtained (Figure 7.18) using Equation 7.22. Depending on the phase of converter current $\dot{I}_{c1}$, we have rectification regions (II and III), whereas regions I and IV correspond to the inverter operation mode. In Figure 7.18, circle 1 is the locus of the ends of the vector $\dot{U}_{ab1}$ with phase variation of the current vector $\dot{I}_{c1}$ in the range $0 < \varphi < 2\pi$. Circle 2 is the locus of the ends of the current vector $\dot{I}_{c1}$. It follows from Equation 7.22 that the radius of circle 1 is equal to the inductor L_f voltage $\Delta U_L = \omega_1 L_f I_{c1}$. From Figure 7.18, we can see that, when the current $\dot{I}_{c1}$ leads the mains voltage (regions III and IV), the converter voltage $\dot{U}_{ab1}$ is less than the mains voltage U_S due to the inductor voltage $\Delta\dot{U}_L$, but when the current $\dot{I}_{c1}$ lags the mains voltage (quadrants I and II), the converter voltage $\dot{U}_{ab1}$ is increased by the value $\Delta\dot{U}_L$.

7.3.2.5 Reactive power control

Converter operation at the boundary of regions I–II and III–IV corresponds to the exchange of reactive power with the mains. The generated power is inductive at the boundary I–II, and at the boundary III–IV, the converter generates the capacitive reactive power. Such operation is used to regulate the reactive power in the mains or to compensate reactive power of a specific type. For example, in power transmission, the inductance of the line produces reactive power, whose compensation requires capacitive power. If there is an excess of capacitive reactive power in the mains, a source of inductive power is required for its compensation. In converters operating as reactive power compensators, the dc source can be a capacitor or inductive energy storage. In that case, of course, active power is required only for compensation of the active

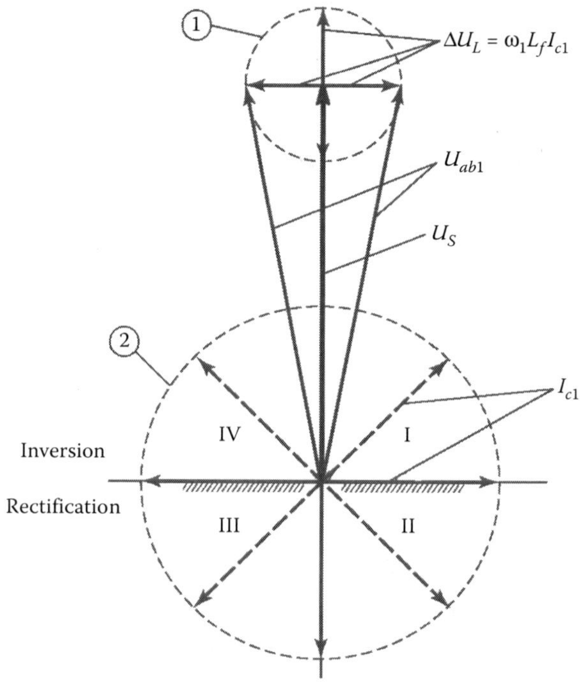

Figure 7.18 Vector diagram for rectifier and inverter modes of a voltage source converter (four-quadrant operation).

power losses in the converter circuit, including the energy storage. The converter consumes this low active power from the mains. Therefore, in the reactive power control mode, the converter operation region in Figure 7.18 is close to the boundaries I–II and III–IV in the rectification area. In Figure 7.18, this is represented by the shading along the boundaries in quadrants II and III.

Thus, the application of force-commutated switches and PWM enables ac–dc converter operation in the rectifier or inverter mode, with leading or lagging power factor. This is possible for the voltage source converter as well as for the current source converter. This capability is important for the development of power electronic devices with low environmental impact that do not produce neither current and voltage harmonics nor reactive power at the fundamental frequency.

The ability to regulate capacitive reactive power is very important in reducing the power losses in distribution power systems that feature inductive load currents. In other words, reactive power compensators allow an increase in the power factor.

7.3.3 *Active power filters*

7.3.3.1 *Principle of active filtering*

Traditionally, in electrical engineering, undesirable current or voltage harmonics were reduced by means of passive filters. Such filters are a combination of inductive (*L*) and capacitive (*C*) elements connected in different circuits with specified topology. As the electrical impedance of the passive filters is frequency-dependent, they change the harmonic composition of nonsinusoidal currents and voltages. In systems connected to a power converter, the filters mainly ensure sinusoidal voltage (current) in the ac circuits and reduce the ripple in the dc circuit. Examples of such filters have been considered in sufficient detail in previous chapters. Although passive filters are simple in design, they have two primary disadvantages:

1. Limited functional capabilities on account of the circuit topology and the parameters of its components
2. Uncontrollability

These problems significantly reduce the effectiveness of passive filters in changing the harmonic composition of the current (voltage), especially with frequency variation and with variation in the grid parameters. In addition, transient processes in the system may give overvoltage and overcurrent. A deficiency of passive filters is their sensitivity to variation in their own parameters, for example, on account of aging.

In contrast to passive filters, active power filters contain controllable components such as transistors, so that the frequency characteristics can be regulated. However, until recently, no components were available for the creation of active power filters in power electronics.

At the beginning of the 1970s, the first low-power active power filters were produced on the basis of analog integrated circuits. They were used in microelectronic components for information and control systems.

The modern generation of self-controlled high-speed switches (based on transistors and GTOs) provides the basis for new devices: active power filters (Akagi, 2002).

7.3.3.2 *Active filters for power conditioning*

According to IEC standards, an active power filter is defined as a converter for filtration. This term is very general and does not characterize the significant features of the filter. However, before we offer a more specific definition, we need to delineate the boundaries within which the corresponding devices fall. First, we are talking about ac filters, as active power filters are not widely used in dc power circuits and are found predominantly in secondary power sources of relatively low power. As a rule, moreover, dc filters are based not on converters but on electrical signal amplifiers. In what follows, we will refer to such filters as active dc filters. In view of the foregoing, we may present a more specific definition of active power filters as follows. An active (power) filter is an ac–dc converter with capacitive or inductive energy storage on the dc side, which produces a current (voltage) equal to the difference between the nonlinear current (voltage) and the fundamental sinusoidal current (voltage), by pulse modulation. Of course, active filtering can also be performed by devices with more number of functions. For example, a converter can compensate the fundamental reactive power and filter out higher harmonics. In that case, the expanded functions should be reflected by a hybrid term (e.g., a filter–compensator). We now consider the operation principle of active power filters in more detail.

On the basis of their circuit and control principles, we divide active power filters into current sources and voltage sources. In Figure 7.19, we show simplified equivalent circuits with active power filters in the form of voltage sources (u_{AF}) and current sources (i_{AF}). In Figure 7.19a, the power source u_s has a nonsinusoidal voltage. To ensure sinusoidal voltage u_L at the load terminals, we connect an active power filter, corresponding to an equivalent voltage source u_{AF}, in series with the power source. For this circuit, we can write the following general relations:

$$\begin{aligned} u_L(\vartheta) &= U_1 \sin(\vartheta - \varphi_1), \\ u_S(\vartheta) &= \sum_{n=1}^{\infty} U_n \sin(n\vartheta - \varphi_n), \\ u_{AF}(\vartheta) &= \sum_{n\neq 1}^{\infty} U_n \sin(n\vartheta - \varphi_n), \end{aligned} \tag{7.23}$$

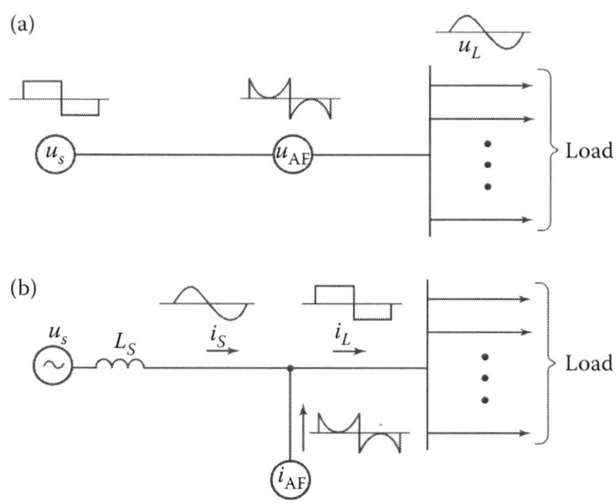

Figure 7.19 (a) Series and (b) shunt active power filter topologies.

or

$$u_L(\vartheta) = u_S(\vartheta) - u_{AF}(\vartheta),$$

where ϑ is the phase of the fundamental voltage and φ_n is the initial phase of the nth harmonic.

Assuming zero losses in the active power filter and a linear load, we obtain the active power of the active power filter over the period of the fundamental harmonic

$$P_{AF} = \frac{1}{2\pi}\int_0^{2\pi}\left[\sum_{n\neq 1}^{\infty} U_n \sin(n\vartheta - \varphi_n)\right] \times I_{L1} \sin(\vartheta - \varphi_{i1})d\vartheta = 0, \quad (7.24)$$

where I_{L1} and φ_{i1} are the amplitude and initial phase of the sinusoidal load current, respectively.

It follows from Equation 7.24 that, under the given assumptions, the active power filter has no influence on the active power balance in the source–load system. At the same time, it is directly involved in the exchange of distortion power with the nonsinusoidal voltage source. The distortion power is inactive and circulates, in the present case, between the distortion source and the section of line connecting the source and the active power filter. The active power filter, which accepts and emits energy due to voltage distortion, is electric energy storage: a capacitor or inductor.

The nonsinusoidal current i_L, induced by a nonlinear load, as a rule, is removed by means of an active power filter that generates a nonsinusoidal current equal to the difference between its input current i_L and the fundamental harmonic i_{L1}. This active power filter is usually connected in parallel to the nonlinear load. The point of filter connection is usually selected as close as possible to the load terminal (Figure 7.19b). Under our assumption that there are no power losses in the active power filter, we can write

$$\begin{aligned}
i_L(\vartheta) &= \sum_{n=1}^{\infty} I_n \sin(n\vartheta - \varphi_{in}),\\
i_{AF}(\vartheta) &= \sum_{n\neq 1}^{\infty} I_n \sin(n\vartheta - \varphi_{in}),\\
i_S(\vartheta) &= i_L(\vartheta) - i_{AF}(\vartheta) = I_1 \sin(\vartheta - \varphi_{i1}),\\
P_{AF} &= \frac{1}{2\pi}\int_0^{2\pi}\left[\sum_{n\neq 1}^{\infty} I_n \sin(n\vartheta - \varphi_{in})\right] \times U_1 \sin(\vartheta - \varphi_1)d\vartheta = 0,
\end{aligned} \quad (7.25)$$

where I_n and φ_{in} are the amplitude and initial phase of the nth current harmonic, and U_1 and φ_1 are the amplitude and initial phase of the sinusoidal load voltage.

According to Equation 7.25, a shunt active power filter generating current to compensate the load distortion power has no influence on the active power balance in the source–load system. However, in contrast to a series active power filter, the distortion power is exchanged between the nonlinear load and the active power filter.

It follows from Equations 7.24 and 7.25 that active power filters can be ac–dc converters, capable of producing a given nonsinusoidal ac current or voltage. Then, for the exchange of the nonactive power with the mains, capacitive or inductive energy storages can be used. Obviously, in the general case, such active power filters may ensure the exchange of nonactive power, including the first-harmonic reactive power. As, under the given assumptions, the average value of power on the ac side of the filter over the period of the fundamental harmonic is zero, there is no need for a source or consumer of active power on the dc side of the active power filter. Obviously, converters generating the required nonsinusoidal current or voltage must be based on self-controlled high-speed switches for which PWM is possible.

7.3.3.3 Active power filter circuits

Depending on the type of storage, the basic circuit of the active power filters will include a current source converter or a voltage source converter. We now consider the operation of an active power filter for the example of single-phase circuits in which energy is stored in an inductor (Figure 7.20a) or a capacitor (Figure 7.21a).

Figure 7.20a shows the active filter based on CSI with inductor L_d on its dc side. Switches VT1–VT4 are controlled by pulse modulation. As a result, the sequence of current pulses i_{AF} is produced at the output of the active power filter, according to the specified control principle. After this current is passed through a CL filter, a waveform of current i_{AF} corresponds to the current-modulating reference signal. In Figure 7.20b, as an example, the mains voltage u_{ab} and the current of the active power filter are shown in the case of generating an output current i_{AF} equal to the sum of the third and fifth harmonics. In Figure 7.20b, we assume a conventionally idealized circuit, neglecting the ripple of current I_d ($L_d = \infty$). By adjusting the modulating function, we can obtain different waveforms of the output current. The basic limit on the precision in reproducing required current is the converter modulation frequency f_m. In the first approximation, it is assumed that f_m is at least 10 times greater than the maximum frequency of harmonic being generated. Another demand is the absence of active power at the output of the active power filter. The active power unbalance between the active filter and the mains leads to

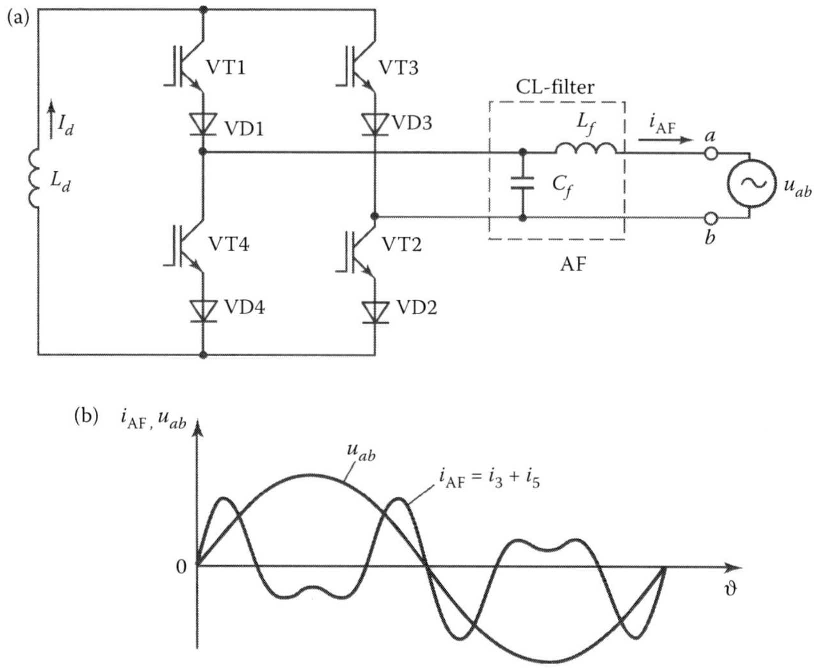

Figure 7.20 (a) Active power filter based on current source converter and (b) waveforms of current and voltage.

an increase or a decrease in the inductor storage current I_d (or the voltage U_d of a capacitive storage).

An active power filter with a capacitive storage (Figure 7.21a) is based on a voltage source converter. Therefore, it is dual with respect to the topology shown in Figure 7.20a. Accordingly, the output voltage u_{ab} produced in Figure 7.21a and the output current i_{AF} produced in Figure 7.20a will be similar. Figure 7.21b shows the output current waveform in the case of generating the third and fifth current harmonics. The control system of the converter must regulate the output voltage, which depends on the reference nonsinusoidal current.

In Figures 7.20 and 7.21, the active power filters are connected in parallel to the mains and can be represented as equivalent sources of nonsinusoidal current. The same circuits can be connected in series with the mains and will then be presented as equivalent sources of nonsinusoidal voltage. When an active power filter with an inductive storage is connected in series with the mains (usually through a transformer), an impedance Z_{AF} must be connected to the filter output, so as to ensure the flow of the fundamental load current i_L in the circuit consisting of the voltage source u_S and the load Z_L (Figure 7.22). That is a consequence of the high internal

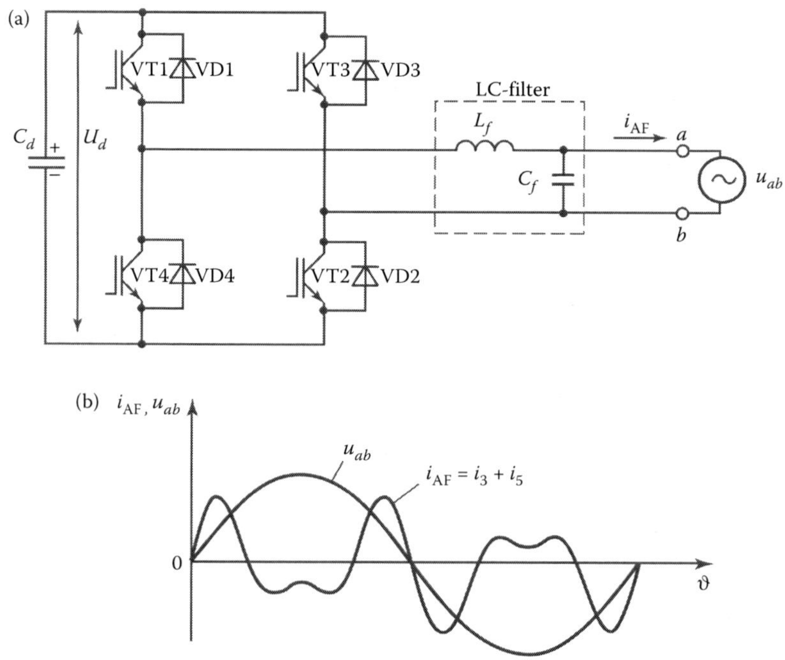

Figure 7.21 (a) Active power filter based on a voltage source converter and (b) waveforms of current and voltage.

impedance of the active power filter with an inductive storage. Of course, the impedance added will slightly reduce the fundamental mains voltage. Thus, we can distinguish between four active filter circuits: shunt and series topologies with an inductive or a capacitive storage. Filters with a capacitive storage are most widely used in practice, on account of their

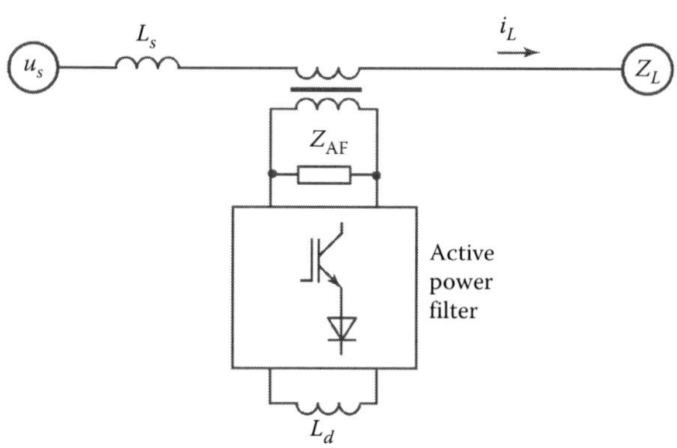

Figure 7.22 Topology of series active power filter with an inductive storage.

high speed and performance. Note, however, that filters with a superconducting inductive storage are unsurpassed in the compensation of reactive power or distortion power and also provide a power reserve when the mains voltage disappears. In such conditions, of course, we consider high-power filter-compensation systems (Rozanov and Lepanov, 2012).

In Figures 7.23 and 7.24, single-phase active power filters with a capacitive storage are shown. The shunt topology (Figure 7.23) is usually employed to eliminate current distortion due to nonlinear load, for example, a rectifier with considerable inductance on the dc side. For more efficient filtering, the active power filter is connected directly to the terminals of the nonlinear load. In the absence of an active power filter, the distorted load current results in a voltage drop at the mains impedance. As a result, the voltage at the terminals of various consumers will be nonsinusoidal. To eliminate or reduce the current distortion, the active power filter connected to the terminals of the load generates a current i_{AF}, which is subtracted from the distorted load current i_L. The resultant mains current is sinusoidal and equal to the fundamental component of the load current i_{L1}, in accordance with Equation 7.25. Figure 7.23b illustrates these current waveforms.

As already noted, the active power filter may not only eliminate current harmonics, but also compensate the reactive power of the fundamental current at a nonlinear consumer. In that case, the maximum power of the active filter is determined by the maximum sum of the compensated and filtered currents. The parameters of basic circuit components are calculated by the methods used for PWM voltage source inverters.

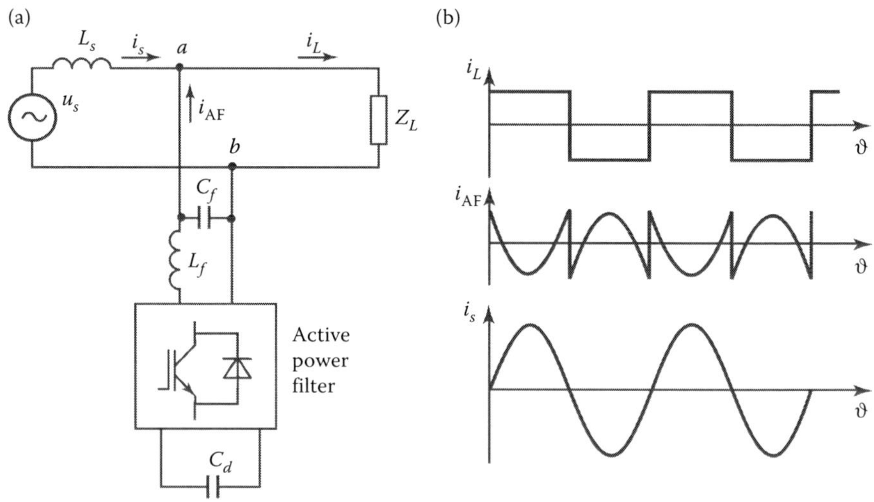

Figure 7.23 (a) Shunt active power filter with a capacitive storage and (b) current waveforms.

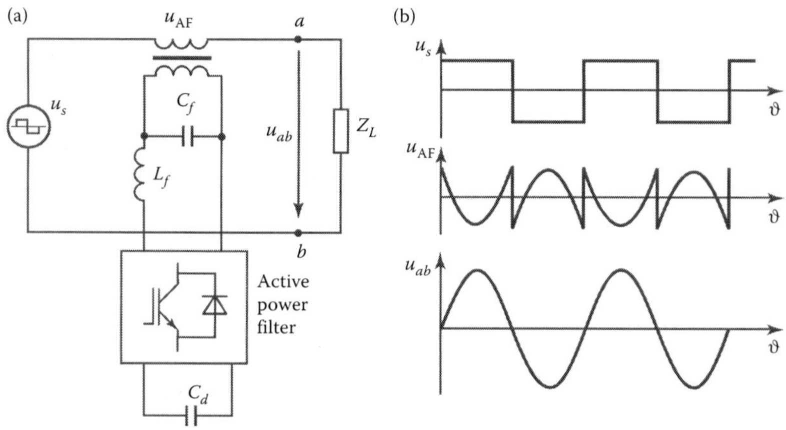

Figure 7.24 (a) Series active power filter with a capacitive storage and (b) voltage waveforms.

Figure 7.24 shows a series active power filter with a capacitive storage. In this case, the active filter ensures a sinusoidal voltage at the load in the case of a distorted (nonsinusoidal) source voltage u_S. Essentially, the series active power filter is a high-frequency voltage booster, which permits the required voltage waveform to be produced. Note, however, that an active filter with energy storage cannot generate or consume active power over a long period, on account of the lack of corresponding sources or sinks. At the same time, the series active filter enables to eliminate low-frequency voltage fluctuations or brief voltage drops. That permits expansion of its functions. For example, a structure consisting of shunt and series active power filters is widely used (Figure 7.25). In this structure, the shunt active filter is used not only for filtering and reactive power compensation, but also as a dc voltage source U_d, ensuring active power exchange between the ac mains and the series active filter. Accordingly, the series filter can be used as a voltage regulator at the ac terminals.

7.3.3.4 dc active filters

As already noted, dc active filters are not widely used, mainly because there are many other options for eliminating dc ripple. At the same time, such active filters can be successfully employed with low-power dc sources. For example, they can be used in circuits with ac voltage generators in counterphase with the ripple voltage. Figure 7.26 shows equivalent circuits including active filters based on ac generators of current- or voltage-source type. In these circuits, if we assume ideal filtering of the ripple, the following conditions must be satisfied:

$$i_{AF} = i_d, \quad u_{AF} = u_d, \tag{7.26}$$

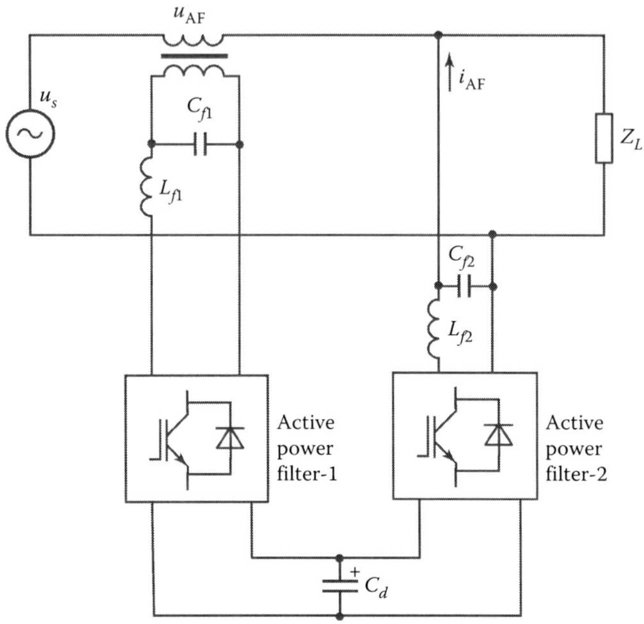

Figure 7.25 Combination of series and shunt active filters.

where i_d and u_d are the current and voltage ripple of a dc source with average voltage U_d, respectively.

7.3.4 Hybrid filters

7.3.4.1 Characteristics of passive filters

Traditionally, passive filters based on inductive and capacitive components are used to ensure sinusoidal current and voltage in power-supply systems. The filtration principle is based on the fact that the impedance of the filter components depends on the current frequency and usually also on

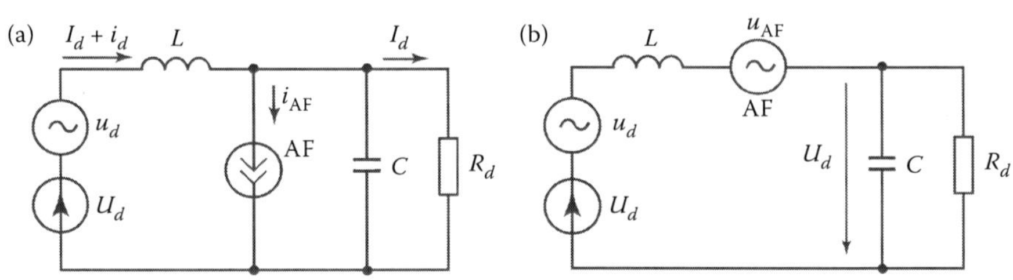

Figure 7.26 Equivalent circuits of (a) shunt and (b) series dc active filters.

resonance phenomena in series and parallel circuits containing capacitors and inductors. Passive filters differ in circuits and correspondingly in frequency characteristics. It is conventional to distinguish between detuned and tuned passive filters. Detuned filters have a resonance below the frequency of the harmonic being filtered out. The degree of detuning is calculated as

$$\delta = \left(\frac{\omega_1}{\omega_n}\right)^2 \cdot 100\%, \tag{7.27}$$

where ω_1 and ω_n are the angular frequencies of the fundamental component (the first harmonic) and the filtered (nth) current or voltage harmonic.

Usually, $\delta = 5\%–15\%$, depending on the distortion of the filtered parameters. Detuned filters are designed not only for the elimination of higher harmonics, but also for reactive power compensation at the fundamental harmonic. As the frequency of the filtered harmonic is above the tuned frequency in detuned filters, resonance in the grid–filter system is practically impossible. Usually, the primary function of a detuned filter is reactive power compensation at the fundamental harmonic; the filter capacitance is selected on that basis. The filter inductor can be considered as a capacitor current limiter and provides its protection in transient processes associated with fluctuation in the grid voltage.

Tuned filters are used to eliminate the harmonics at the filter resonant frequency. Although in this case there is a reactive power compensation at the fundamental frequency, it is not the main factor in selecting the filter parameters. Rather, the selection criterion is minimum filter size or cost. In this case, the capacitance is considerably less than that in a detuned filter, which is chosen for purposes of reactive power compensation. Tuned filters are widely used to ensure sinusoidal voltage and current and to reduce pulsations in power converters.

Passive filters are simple and reliable devices for enhancing power quality. However, they have a significant drawback: their parameters are not controllable. As a result, it is impossible to meet all the contradictory requirements in different operating conditions and to correct filter parameters.

An important parameter of a passive filter is the quality factor Q, which is the ratio of the maximum energy stored in the reactive components (capacitor or inductor) to the energy dissipated in the filters' active elements. It follows from this definition that the quality factor can be described by various analytical expressions. In particular, for a series-resonant circuit (Figure 7.27a)

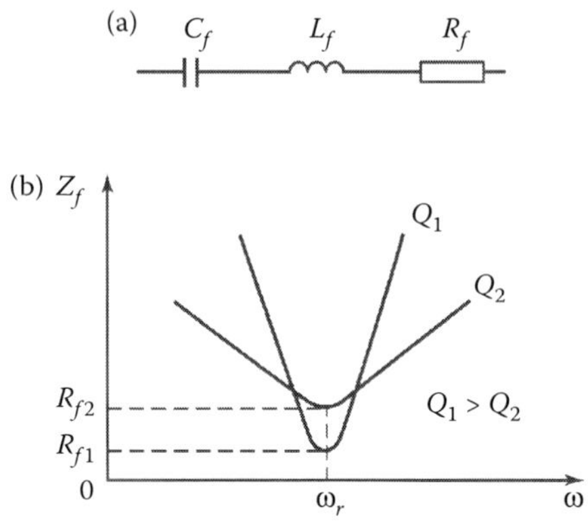

Figure 7.27 (a) Equivalent circuit of a single-frequency passive filter and (b) frequency dependence of the filter impedance at different values of the quality factor Q.

$$Q = \frac{\rho}{R_f}, \tag{7.28}$$

where $\rho = \sqrt{L_f/C_f}$ is the characteristic circuit impedance (filter reactance) and R_f is the equivalent resistance of the filter.

The quality factor determines the increase in voltage at the capacitor C_f and inductor L_f, relative to the voltage applied to the filter circuit. With an increase in Q, the slope of frequency dependence of the total filter impedance increases, whereas the bandwidth relative to the resonant frequency ω_r decreases (Figure 7.27b). This gives the contradictory requirements to filter parameters, especially in steady-state operation. On the one hand, an increase in the quality factor Q results in more efficient filtering at the tuned frequency ω_r; on the other hand, the negative impact of frequency detuning on the filter impedance is increased. Such frequency deviation can result from aging of the filter components or the impact of the ambient temperature. The frequency of the harmonic can also deviate from ω_r. In addition, besides loss of efficiency, antiresonance can emerge, in which the filter impedance increases at the antiresonance frequency, with an increase in the corresponding voltage harmonic at the terminals of the power system. This is associated with a current resonance in a parallel circuit containing grid inductance L_S, to which the filter is connected, as in Figure 7.28a. In that case, neglecting the active impedance R_f, we can write the antiresonance frequency ω_{ar} as

$$\omega_{ar} = \frac{1}{\sqrt{(L_f + L_S)C_f}}. \tag{7.29}$$

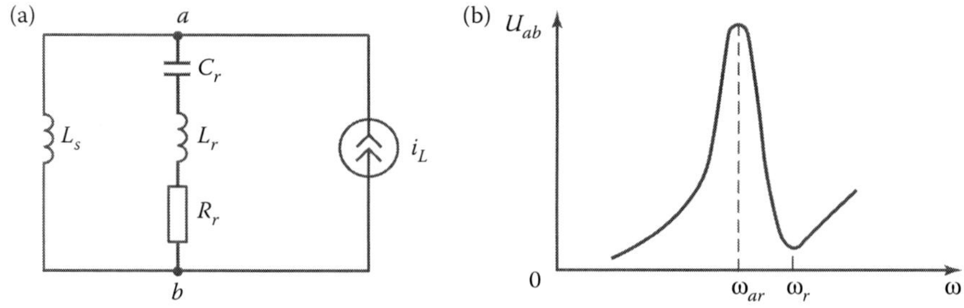

Figure 7.28 (a) Equivalent circuit in the case of antiresonance and (b) frequency dependence of the voltage.

It is evident from Equation 7.29 that when the filter is connected to a grid with a high-power source—in other words, with low L_S—the antiresonance frequency is close to the resonance frequency ω_r. That can result in a sharp increase in the corresponding voltage or current harmonics in the grid and filter (Figure 7.28b) and hence lead to various extremely undesirable consequences such as failure of the filter capacitor or breakdown of the insulation. The negative consequences of antiresonance are worse at high values of Q.

Thus, even under steady-state conditions, we need to find compromises in selecting the quality factor Q of a passive filter. In addition, the quality factor has a considerable influence on the transient processes in the power supply system. Perturbations in this system due to commutation of the load, external voltage surges, or change in the operating conditions will be associated with transient processes, in which the voltages and currents deviate considerably from their steady values. The presence of passive filters with reactive components not only prolongs such processes, but also produces considerable voltage and current surges. With weak damping of the resonance circuits—in other words, high value of Q—these phenomena can result in failure of the whole power supply system. Therefore, passive filters sometimes include additional resistive components, which reduce the quality factor but eliminate overvoltages and current surges in the system.

7.3.4.2 Regulation of passive filters

Theoretical and practical developments in active filtering offer the possibility of control of the parameters of passive filters. The suitability of active filters for these purposes is obvious if the following factors are considered. First, purely active power filters are based on high-power components, as the filter power is determined by the total power of the nonlinear load. Therefore, the high cost of active power filters constrains their use.

Secondly, passive filters, which are traditional devices for enhancing the power quality, do not require new manufacturing technologies and are compatible with practically all existing power systems. In recent years, various methods have been developed for control passive filters on the basis of active power filters, which permit us to adjust their parameters. In such applications, the power of the active filters employed is at least an order of magnitude less than that of an active power filter. In addition, automatic correction of the passive filter parameters in the course of operation is feasible. The combination of passive and active power filters is known as hybrid filters.

The operation of a hybrid filter is based on the generation current or voltage in order to modify frequency characteristics of the passive filter and to improve its compensation characteristics. This device is based on active power filters, that is, on self-controlled ac–dc converters with PWM. Figure 7.29 shows the main hybrid filter topologies. Parallel connection of active and passive filters is most commonly employed. The producing of modulated voltage or current of an active filter can be considered as change in its instantaneous input impedance $z_{AF}(t)$ due to the average

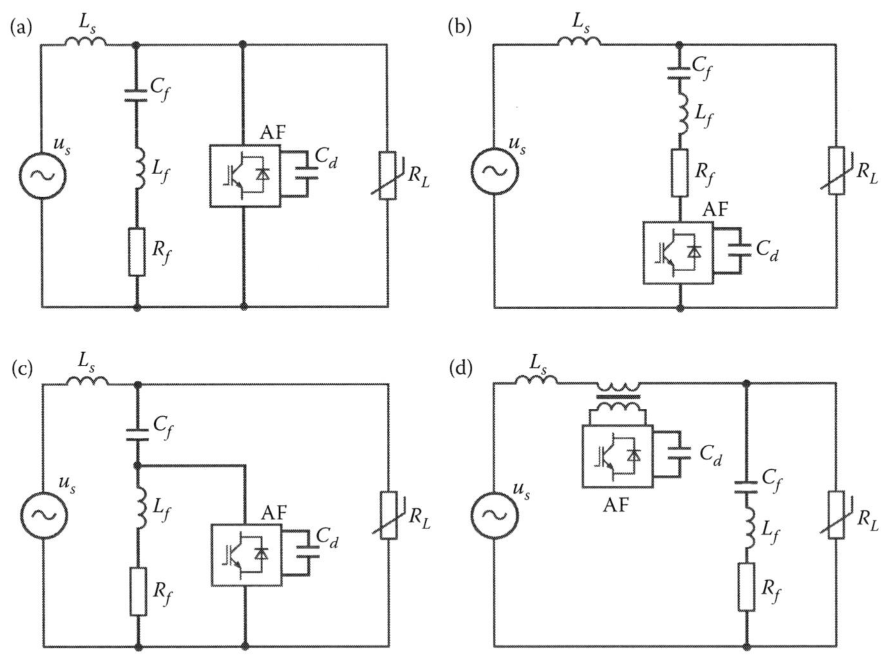

Figure 7.29 Hybrid filter topologies: (a) a parallel configuration; (b) a series configuration; (c) a combined structure; and (d) a combination of series active filter and shunt passive filter.

modulated values $i_{AF}(t)$ and $u_{AF}(t)$. For an active filter with a capacitive storage, we can write

$$z_{AF}(t) = \frac{u_{AF}(t)}{i_{AF}(t)} = \frac{U_d \cdot m(t)}{i_{AF}(t)}, \tag{7.30}$$

where $m(t)$ is the modulating function (smooth component) and U_d is the average voltage on the dc side of the active filter (at the capacitive storage).

It follows from Equation 7.30 that, with an appropriate modulating function $m(t)$, we can ensure that $dz_{AF}(t)/dt = 0$. This is equivalent to an active impedance R_e, which can be positive or negative. Positive R_e corresponds to energy consumption from the grid and negative R_e to power generation. Variation in $z_{AF}(t)$ with zero average value over the period of the fundamental frequency corresponds to the exchange of nonactive power (reactive and distortion power) between the active filter and the system with passive filter. Thus, by adjusting $m(t)$, we can vary $z_{AF}(t)$ so as to obtain the required frequency characteristic of the hybrid filter. Such control can be considered as the introduction of equivalent impedance in the passive filter. The basic constraints in the present case are the frequency characteristics of the switches and the storage capacitance. That capacitance limits the rate of energy variation to realize the equivalent impedance. The character of the equivalent impedance $z_{AF}(t)$ is determined by the topology of the hybrid filter, the regulator input signal, and the modulating function. Note that the connection point of this impedance in the equivalent circuit need not be the same as its actual position in the circuit.

Active filter-based regulator in the structure of a hybrid filter results in the following possibilities:

- More efficient filtering under steady-state operation, by correction of the frequency characteristic
- Reduced impact of fluctuation in the filter parameters and the frequency of the filtered harmonic
- Elimination of antiresonance in the power-supply system at frequencies close to the filtered harmonic
- Damping of undesirable resonant phenomena due to the passive filter elements
- Decrease in the grid current harmonics from various sources

The improvements made possible by the hybrid filter depend on the spectral composition of the regulator input signal. To improve the operation of a passive filter at the tuned frequency, it is sufficient to track the input signal harmonic of that frequency. Accordingly, the power of the active filter components can be considerably less than when monitoring the whole frequency spectrum. In addition, signal modulation is simplified. At the same time, the damping of resonant phenomena in the system is possible only when the input signal of the hybrid filter regulator has a broad spectrum. The control methods and circuits for hybrid filters take very diverse forms but have much in common with active power filters (Rozanov and Grinberg 2006).

7.3.5 *Balancing of currents in a three-phase system*

When the load contains single-phase consumers, unbalanced phase currents appear in three-phase power systems. That leads to voltage unbalance, overloading of the phase lines, and increased power losses in the transmission lines. For load balancing, we can effectively use a compensator based on a four-quadrant converter with self-controlled switches (Kiselev and Rozanov 2012; Kiselev and Tserkovskiy 2012). In reactive power compensation and harmonics filtering, the compensation of unbalanced currents does not require active power, as the unbalance is characterized by nonactive power.

Figure 7.30 shows the connection of an unbalanced current compensator to a four-wire power system. The compensator is based on a three-phase ac–dc converter with a capacitive storage on the dc side. To eliminate unbalance of the phase currents in a three-phase system without a neutral line, the compensator must generate reverse-sequence

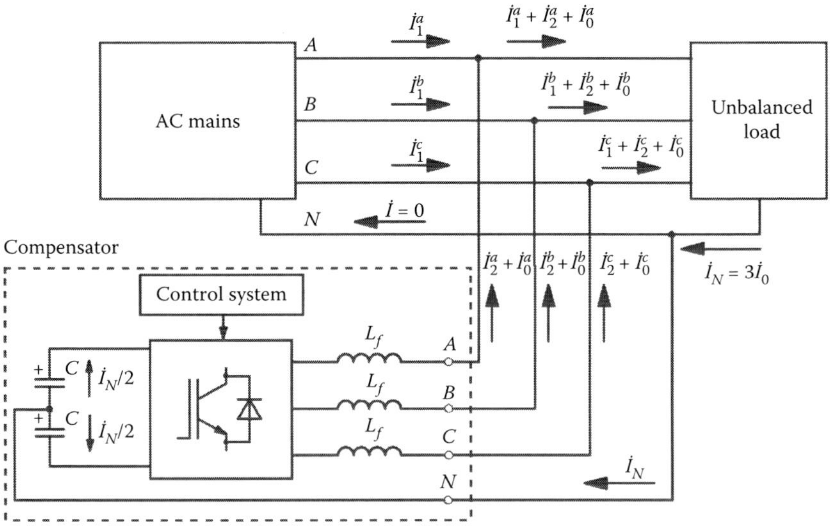

Figure 7.30 Compensation of unbalanced currents in a three-phase four-wire system: I_1, direct-sequence current; I_2, reverse-sequence current; I_0, zero-sequence current; and I_N, neutral current.

currents in opposite phase to the reverse-sequence currents of the load. In that case, the oscillatory component (100 Hz) of the instantaneous power is exchanged between the compensator capacitors (C) and the load. In the case of a four-wire system with a neutral, the compensator additionally generates zero-sequence currents. A loop for zero-sequence current flow between the load and the compensator is made by connecting the neutral between the compensator capacitors. Complete compensation of the zero-sequence currents results in zero current in the neutral between the compensator and the ac mains.

7.4 *Basic control systems of ac–dc converters with PWM*

In general terms, control methods with PWM in ac converters can be divided into two groups:

1. With direct voltage control, in the absence of current feedback
2. With current feedback

The methods differ in the method of PWM and in the components employed. The first group is mainly employed for voltage source inverters, for example, in uninterruptible power systems. Both traditional and new methods of PWM can be utilized (e.g., the SV method).

Figure 7.31 shows the block diagram of a control scheme based on sinusoidal PWM for a three-phase voltage source inverter. The control pulses for the converter switches are produced at times determined by comparing the reference phase voltages u_a^*, u_b^*, and u_c^* with a triangular carrier signal from a carrier-frequency generator. The required modulation index M is taken into account here; it is determined by the control system (not shown in Figure 7.31). The control pulses for the converter switches are generated by output control cascades of Driver 1–Driver 6, at the triggering times of comparison units Comp 1–Comp 3, which depend on the level ε of the control signals. The voltage source converter is controlled by switches S1, S3, and S5, which are on when switches S4, S6, and S2 are off and vice versa (not elements in the control channels). The dashed line in Figure 7.31 shows possible links that would improve the converter characteristics. For example, to expand the linear region of the voltage-modulation range, a signal with the frequency of the third harmonic u_3 is added to the reference signal. In addition, to smooth the frequency spectrum, signals from the "random generator" are supplied to the carrier-frequency generator. Taking account of the probability distribution density function, these random signals determine the period of the carrier signal. A channel with a "voltage sensor" may be used to correct changes or monitor pulsations in the input

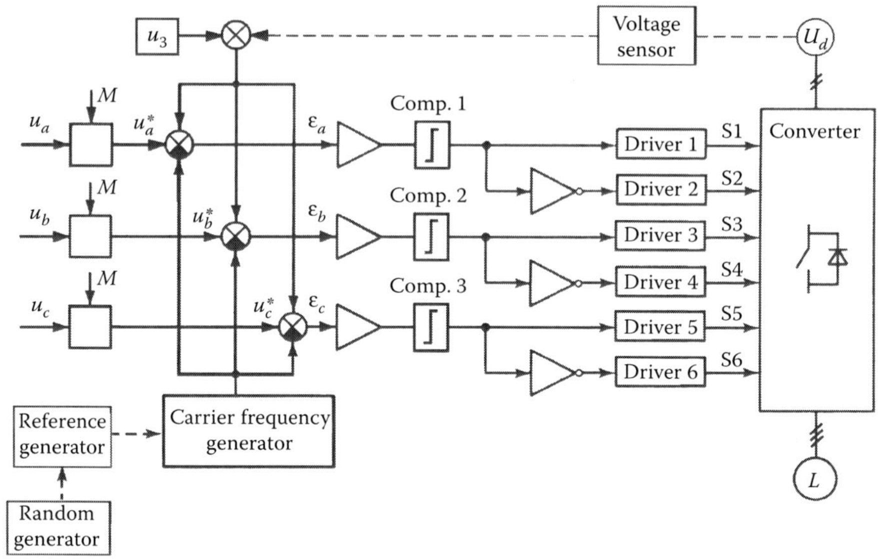

Figure 7.31 The block diagram of a control system based on sinusoidal PWM.

voltage at the dc side. This structure permits modulation directly by comparison of analog signals or by comparison of digital signals formed by a microprocessor controller. Digital methods improve the stability of the control system under standard perturbations due to various errors.

Figure 7.32 shows a simplified control structure based on SV PWM. In this case, the system is regulated with respect to a single SV (reference

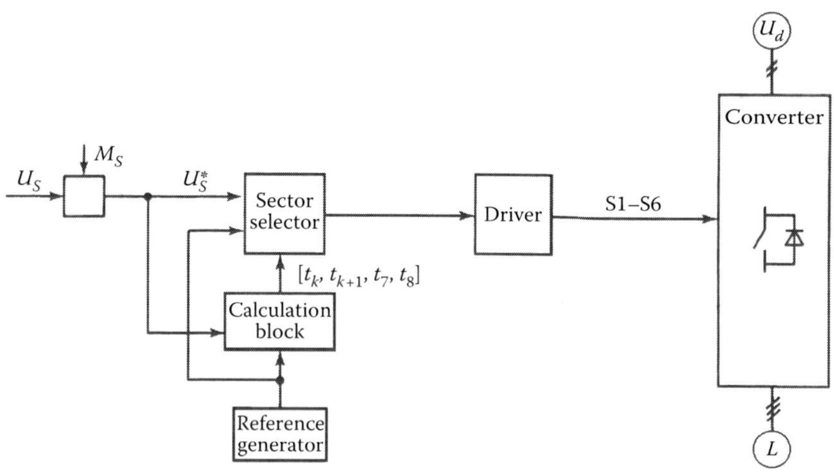

Figure 7.32 A simplified control scheme for SV PWM.

vector) determined by the microprocessor for each modulation cycle. Taking account of the modulation index, we specify the reference vector U_S^* as the reference value of the output voltage. For example, the required voltage can be supplied from the automatic-control channel for the output parameters of a converter operating in an electric drive. At the same time, perturbations from the input voltage U_d on the dc side can be taken into account. The sample frequency corresponding to the modulation frequency of the converter is determined by the "reference generator." This frequency determines the number of switch commutations per period of the fundamental output voltage frequency. Within each cycle (or half-cycle), the "calculation block" computes the coordinates of vector U_s^* for a single sector. Then, the "sector selector" determines the sector number corresponding to the state of vector U_S^* at the given time and its coordinates are recalculated for the αβ frame. As a result, the driver generates the control signals for the converter switches.

Current control methods based on PWM without current feedback permit voltage modulation over a broad range, including overmodulation, with limitations of switching losses and higher harmonics.

In contrast, systems with current feedback are faster with load perturbations, which results in better dynamic characteristics of the converters. The use of current feedback improves the control precision with change in the instantaneous current, thereby offering effective protection against current surges. Control methods based on PWM with current feedback are widely used in converters for electric drives.

Figure 7.33 shows a simplified control structure for an ac–dc voltage source converter with load current feedback. Numerous such methods exist. We can note systems with direct tracking of the actual current i_L with respect to the reference value i_L^* within a bandwidth δ that determines the switching frequency (hysteresis control). Such systems are widely used because of their practical simplicity. A drawback is that the switching frequency changes in tracking a nonlinear sinusoidal signal. The available methods for limiting the frequency range are not very effective and are constrained by the frequency difference between the modulated signal and the carrier frequency, which is more than an order of magnitude. That is especially important in the case of modulation in active power filters to eliminate higher-order harmonics. Note also that, in contrast to analog systems, the application of the hysteresis principle in digital control systems requires a significant increase in the microcontroller frequency and in the speed of the analog–digital converter so as to ensure the specified control precision. In such cases, it is expedient to use predictive modulation, in which the rate of parameter variation is taken into account.

The development of active power filters and reactive power compensators entails the creation of systems with PWM that can regulate individual harmonics or harmonic spectrum of nonsinusoidal currents and voltages.

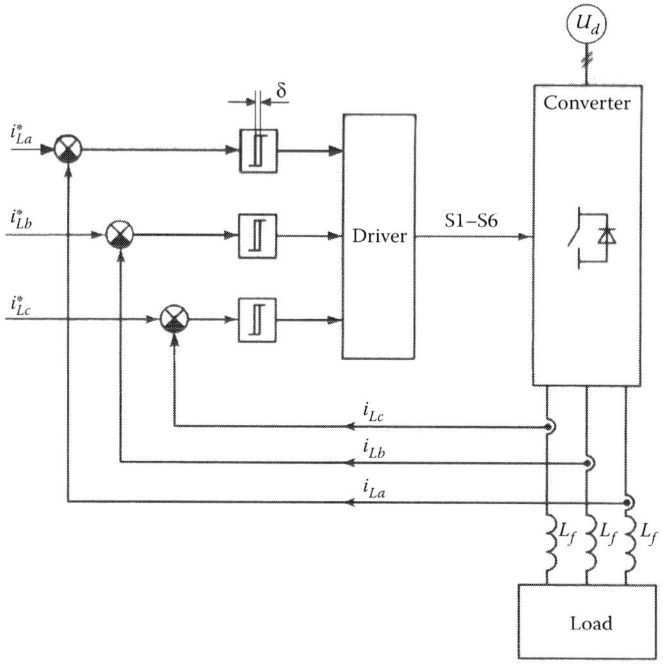

Figure 7.33 A simplified control scheme of ac–dc converter with load current feedback (hysteresis current control).

That follows from the operating principle of various reactive power compensators, including active power filters. We now consider in more detail some standard structures and components of control systems with PWM for that purpose; as an example, we look at ac–dc voltage source converters. We also assume that they can function as nonactive power regulators of the following types:

- Active power filters
- Hybrid filters
- Reactive power compensators at the fundamental harmonic
- Unbalanced currents compensators

In all these cases, the converter contains a capacitor for the exchange of nonactive power between the converter and the ac mains (in regulators based on voltage source converters). A structure of such a regulator is shown in Figure 7.34, in which the power switches of the converter are controlled from the module that generates the control pulses (Driver). A "digital control system" implements the control algorithms; signals i_c and i_L from the load current and converter current sensors, the mains voltage u_s, and the capacitor voltage U_d are the inputs of this module. Filters

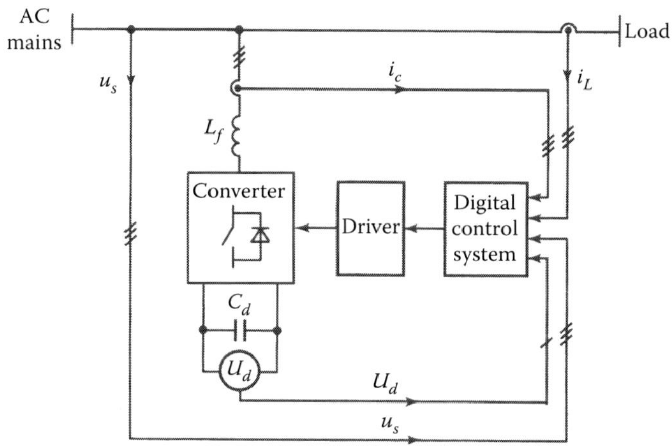

Figure 7.34 The general block diagram of a nonactive power regulator.

are introduced in the digital control system to obtain the information regarding individual harmonics or part of the spectrum of nonsinusoidal signals. Analysis of various filtration methods shows that the most effective is the digital filtration of signals converted from the coordinates of a three-phase system to a two-phase orthogonal rotating frame. The resulting coordinate system must be synchronized with the mains voltage in terms of phase and frequency. The control system includes a phase-locked loop (PLL) module, based on an automatic phase–frequency tuning system. In operational terms, the phase–frequency tuning system is an automatic control system, ensuring high-precision synchronization.

Figures 7.35 and 7.37 through 7.39 show simplified block diagrams of control systems for different regulators. The block diagram of a digital

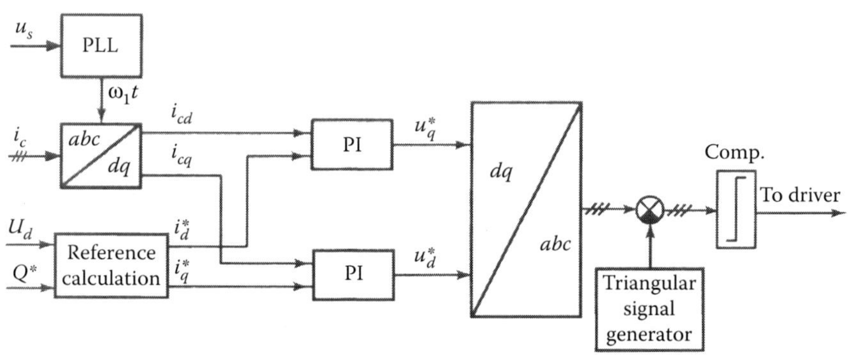

Figure 7.35 The block diagram of a digital control system for reactive power compensator.

control system for the reactive power compensator is shown in Figure 7.35. External signals from the parameter sensors and the signal specifying the reactive power are their inputs. There are three main units of the control system:

1. The PLL, which generates individual sinusoidal signals synchronized in frequency and phase with the mains voltage.
2. The synchronous frame transformation (*abc/dq*), which calculates the *dq* components of converter currents using synchronization signal.
3. The reference currents calculation unit, which receives the signal U_d from the voltage sensor of capacitor C_d and the required reactive power signal from the external control system. The capacitor voltage signal is used for compensation of the active power losses in the compensator.

The signals corresponding to the actual (measured) and required (calculated) currents in the *dq* frame go to proportional–integral controllers, which generate the regulating voltage *dq* signals for control of the current. After inverse coordinate transformation *dq/abc*, the signals corresponding to the required converter voltages are compared with the carrier signal from a "triangular signal generator," producing a signal at the switching frequency. At the output of the comparator ("Comp"), control pulses are generated for the switches in each converter phase and go to Driver. In Figure 7.36, waveforms of current signals are shown, illustrating the control of a reactive power compensator.

Figure 7.37 shows the structure of a digital control system for an active power filter. This system realizes the hysteresis control of the converter currents. It contains two basic calculation blocks determining the reference current i^*.

1. The voltage controller that receives a signal from the voltage sensor $U_{d.}$ This unit calculates the active current i_1^* required for compensation of the power losses in the converter and hence for maintenance of the given voltage at capacitor C_d.
2. The harmonics calculation unit based on the use of digital filters. It receives the signal from the load current sensor i_L and calculates the required part of the current harmonic spectrum $\sum_{n\neq 1}^{k} i_n$.

The sum of currents i_n and i_1^* determines the reference current i^*. The converter output current i_L and the reference current i^* go to the comparison unit Comp, which generates control pulses for converter switches.

The methods of harmonics calculation can be different. Figure 7.38 presents the block diagram of harmonics extraction using two phases of

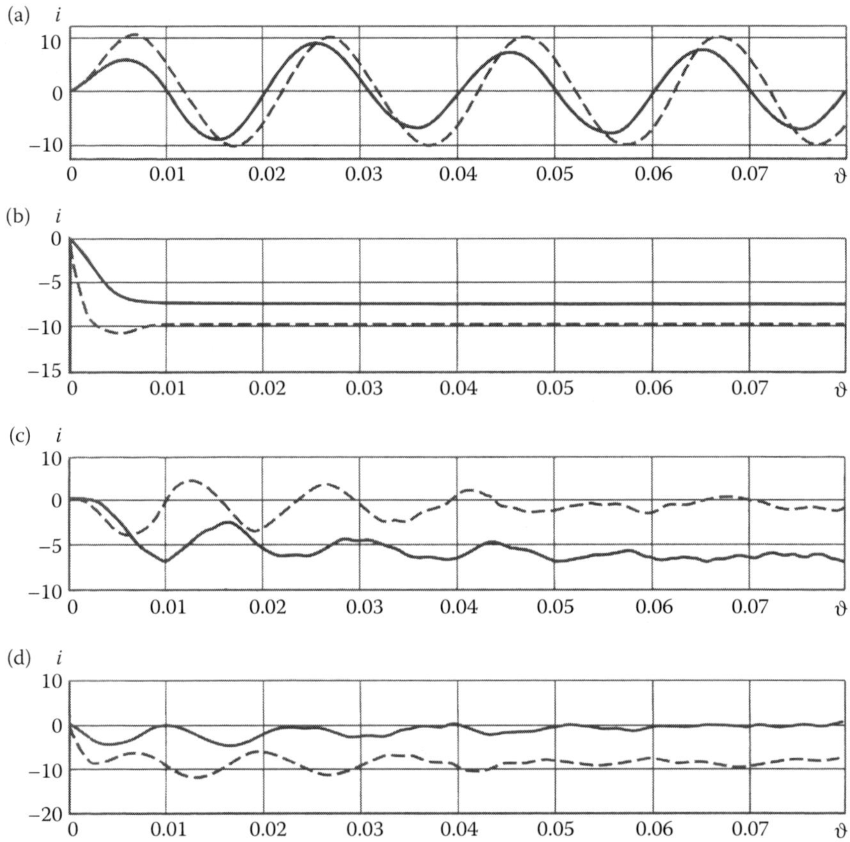

Figure 7.36 (a) Waveforms of load (– –) and mains current (—); current waveforms in *dq* coordinates: (b) *d*-component (– –) and *q*-component (—) of load current; (c) *d*-component (– –) and *q*-component (—) of converter current; (d) *d*-component (– –) and *q*-component (—) of mains current.

load currents. For this purpose, the fundamental harmonics of the load current in phases *A* and *B* are obtained by digital low-pass filters and then they are subtracted from the measured load currents. The current in phase *C* is calculated from the results of the other two phases. In Figure 7.39, the *abc/dq*-transformation-based harmonic calculation method is shown. The fundamental harmonics in the synchronous *dq* frame are constant values. Therefore, they can be calculated as a mean value. The *dq* components of current pass through a low-pass filter. The received constant values are eliminated from *dq* components of the load current. The higher harmonics of the load current are obtained by inverse transformation *dq/abc*. The waveforms of load current and its harmonics are illustrated in Figure 7.40.

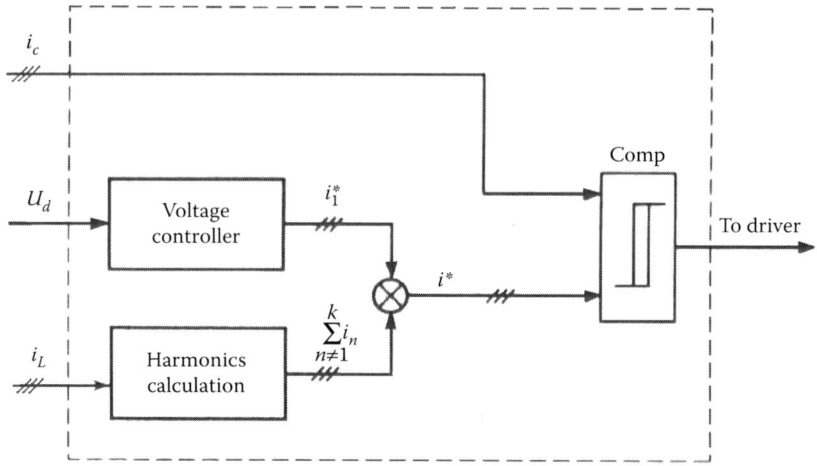

Figure 7.37 The block diagram of the digital control system for an active power filter.

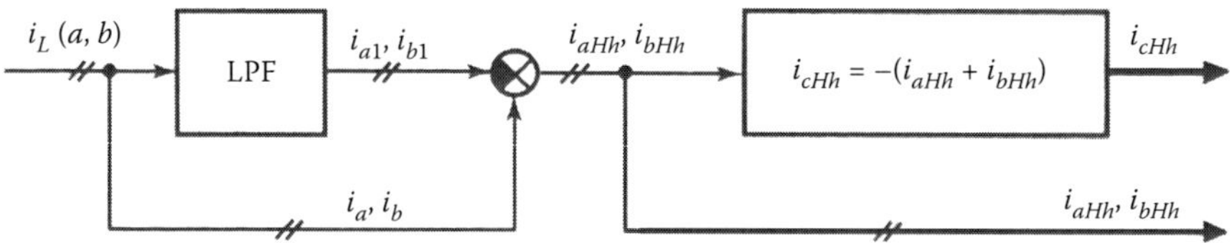

Figure 7.38 The block diagram of higher harmonics extraction.

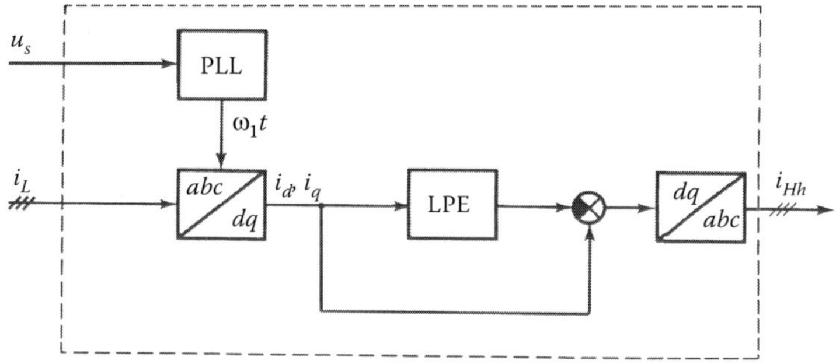

Figure 7.39 The block diagram of higher harmonics calculation based on *abc/dq* transformation of the load current.

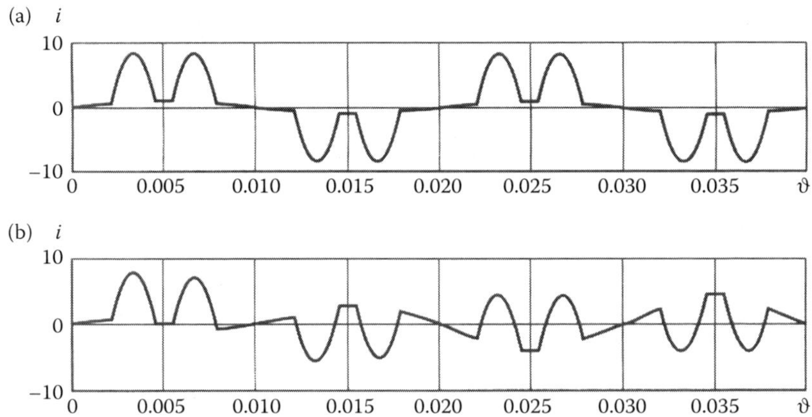

Figure 7.40 (a) Waveforms of the load current and (b) its higher harmonics sum.

References

Akagi, H. 2002. Active filters for power conditioning. *The Power Electronics Handbook*, Skvarenina, T.L., Ed., Part III, Section 17.4. USA: CRC Press.

Espinoza, J.R. 2001. Inverters. *Power Electronics: Handbook*, Rashid, M.H., Ed., pp. 225–267. USA: Academic Press.

Hossein, S. 2002. Hysteresis feedback control. *The Power Electronics Handbook*, Skvarenina, T.L., Ed., Part II, Section 7.7. USA: CRC Press.

Kazmierkowski, M.P., Krishnan, R., and Blaabjerg, F. 2001. *Control in Power Electronics*. USA: Academic Press.

Kiselev, M.G. and Rozanov, Yu.K. 2012. Analysis of the operation of a static reactive-power compensator with balancing of the load. *Elektrichestvo*, 3, 63–69 (in Russian).

Kiselev, M.G. and Tserkovskiy, Y.B. 2012. Analysis of the static reactive power compensator operating in mode of load balancing. Proceedings of 15th International Power Electronics and Motion Control Conference EPE PEMC 2012 ECCE Europe, 3–6 September, 2012, Novi Sad, Serbia.

Mohan, N., Underland, T.M., and Robins, W.P. 1995. *Power Electronics: Converters, Application and Design*. New York: John Wiley & Sons.

Rashid, M.H. 1988. *Power Electronics*. USA: Prentice-Hall.

Rozanov, Y.K. and Lepanov, M.G. 2012. Operation modes of converters with SMES on dc-side used for improving of electrical systems efficiency. Proceedings of 15th International Power Electronics and Motion Control Conference EPE PEMC 2012 ECCE Europe, 3–6 September, 2012, Novi Sad, Serbia.

Rozanov, Yu.K. and Grinberg, R.P. 2006. Hybrid filters for decrease of nonsinusoidal current and voltage in power supply systems. *Elektrotekhnika*, 10, 55–60 (in Russian).

Rozanov, Yu.K., Ryabchitskii, M.V., and Kvasnyuk, A.A. 2007. *Power Electronics: A University Textbook*, Rozanov, Yu.K., Ed. Moscow: Izd. MPEI (in Russian).

chapter eight

Resonant converters

8.1 Introduction

In resonant converters, resonant circuits are used to reduce the switching power losses (IEC, 551-12-26). Resonant circuits were first used in thyristor converters, in which they permit an increase in the working frequency. Different circuits have been proposed for resonant converters, and various classification systems exist (Rozanov, 1987; Rashid, 1988; Kazimierczuk and Charkowski, 1993). Three groups of resonant converters can be distinguished:

1. Converters with resonant circuits that include a load
2. dc converters in which components of resonant circuits are connected to the converter switches so as to ensure soft switching
3. Inverters with a common resonant element on the dc side to ensure soft switching

As resonant converters are associated with the same electromagnetic processes, we employ the following parameters and notation for second-order resonant circuits of series and parallel types (Figure 8.1):

- The resonant angular frequency of an ideal resonant loop (with active impedance $R = 0$) $\omega_0 = 1/\sqrt{LC}$.
- The characteristic impedance $\rho = \sqrt{L/C}$.
- The Q-factor of a series circuit $Q_S = 1/\omega_0 CR = \omega_0 L/R$.
- The Q-factor of a parallel circuit $Q_P = R/\omega_0 L = \omega_0 CR$.
- The damping factor in series and parallel circuits $d_S = 1/2Q_S$ and $d_P = 1/2Q_P$, respectively.
- The eigenfrequency of oscillation ω_R (sometimes known as the free frequency), which is the oscillation frequency if the active load impedance R is disregarded: in a series circuit $\omega_{RS} = \omega_0\sqrt{1 - R^2/4\omega_0^2 L^2}$ and in a parallel circuit $\omega_{RP} = \omega_0\sqrt{1 - 1/4\omega_0^2 C^2 R^2}$

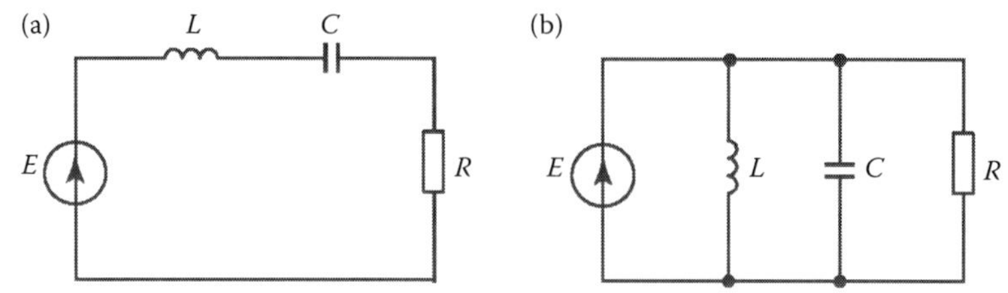

Figure 8.1 (a) Series and (b) parallel resonant circuits.

8.2 Converters with a load in resonant circuit

8.2.1 Converters with serial connection of the load

In most cases, such converters are used for direct dc–ac conversion. In other words, they are inverters. However, they are also used for indirect dc–dc conversion. In that case, they consist of two elements: an inverter and a rectifier. In addition, they may be based on special single-switch circuits, when they are said to operate in class E (Rashid, 1988). In that case, they can serve as inverters or rectifiers. In what follows, we use a common term for such converters: series resonant inverters or converters. Where necessary, their distinguishing features will be noted.

The converters can be divided into two groups:

1. Those based on unidirectional switches
2. Those based on bidirectional switches

Figure 8.2a shows a series resonant inverter based on thyristors. We will have oscillating process in the resonant circuit when thyristors VS1 and VS2 are on, if we ensure that $R < 2\sqrt{L/C}$, where $L = L1 = L2$. For natural commutation, we require that the next thyristor be switched on after the preceding transient process has ended. In other words, current flow in the L–R–C circuit must be discontinuous. In that case, the frequency of the thyristor control pulses ω_s must be less than the eigenfrequency of the circuit ω_R: $\omega_S < \omega_R$. Equivalent circuits for the inverter in different operating periods are shown in Figure 8.2b. Waveforms of current i_C and capacitor voltage u_C are shown in Figure 8.2c. We assume that, when thyristor VS1 is switched on, capacitor C has been charged to voltage U_{C0}. After thyristor VS1 is switched on, the processes in the inverter correspond to the equivalent circuit for interval I and can be described as follows:

$$L\frac{di_C}{dt} + i_C R + \frac{1}{C}\int i_C \, dt = E. \tag{8.1}$$

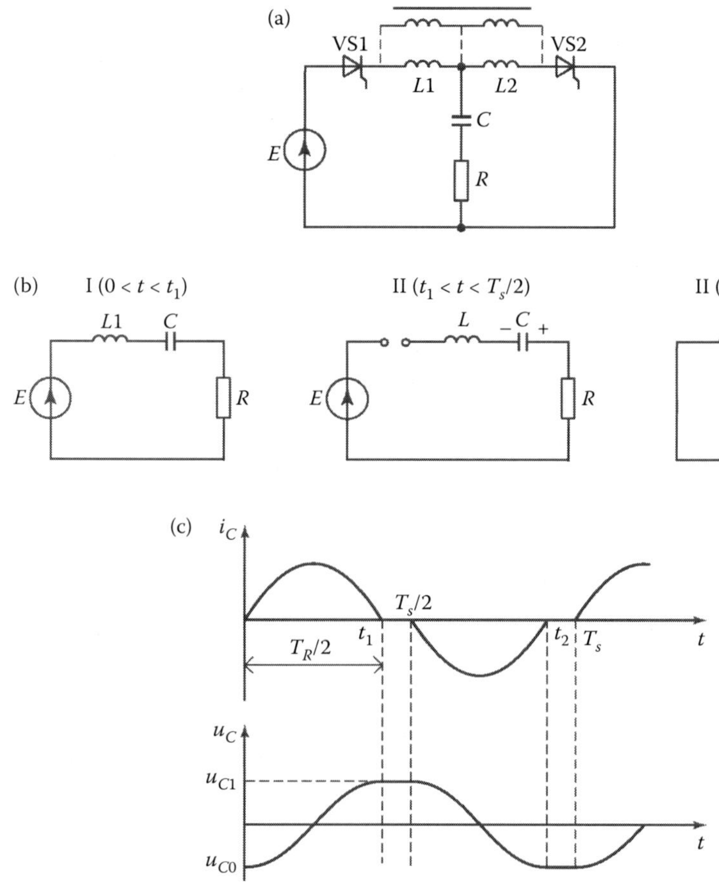

Figure 8.2 A series resonant inverter based on thyristors: (a) circuit; (b) equivalent circuits; and (c) current and voltage waveforms with switching frequency $\omega_S < \omega_R$.

Taking account of the initial condition $U_C(0) = U_{C0}$, we can write the solution of Equation 8.1 in the form

$$i_C(t) = \frac{E + U_{C0}}{\omega_{RS}L} e^{-\delta t} \sin \omega_{RS} t, \tag{8.2}$$

where

$$\delta = \frac{R}{2L} \quad \text{and} \quad \omega_{RS} = \sqrt{\omega_0^2 - \delta^2}.$$

From Equation 8.2, taking into account that $i_C(t_1) = 0$, we find the time t_1 beyond which the circuit corresponds to the equivalent circuit in interval II.

During this interval, the current $i_C(t) = 0$, and the voltage is $u_C(t) = u_C(t_1) = U_{C1}$. This value can be determined on the basis of Equation 8.2.

At time $t = T_s/2$, thyristor VS2 is turned on. The inverter begins to operate in accordance with the equivalent circuit for interval III. The processes in this interval are again determined by Equation 8.1, but without the emf E and with a new initial condition for the capacitor voltage, which is U_{C1}, as in interval II. In interval III, the solution of Equation 8.1 takes the form

$$i_C(t) = \frac{U_{C1}}{\omega_{SR}L} e^{-\delta t} \sin \omega_{SR} t. \tag{8.3}$$

Note that, in steady inverter operation, the capacitor voltage is the same at the beginning and end of the period (at the beginning of interval I, $u_C = U_{C0}$). These processes then periodically repeat. The inverter's working frequency in discontinuous operation is limited by the circuit's resonant frequency ω_0, which theoretically may be attained when $R = 0$ and thyristors VS1 and VS2 are turned on instantaneously. In practice, we need to ensure that the current-free period in interval II is greater than the time required to switch off the thyristor: $t_q < (T_S - T_R)/2$.

In contrast, with discontinuous current i_C, the ratio of the maximum instantaneous current to the mean current drawn from the source is increased, and its harmonic composition is impaired. At the same time, the harmonic composition of the inverter's output voltage u_R at load R deteriorates. These problems may be alleviated by introducing a transformer between the reactors $L1$ and $L2$, as shown by the dashed line in Figure 8.2a. In that case, when one of the thyristors is switched on, an emf will appear at the inductance connected to the other thyristor. As a result, the thyristor that is on in the preceding interval will be forcibly turned off. In that case, we can ensure a higher commutation frequency of the thyristors: $\omega_S > \omega_R$.

A more effective means of reducing the input-current ripples and the distortion of the output voltage is to use bridge and half-bridge circuits (Figure 8.3). In turn, the half-bridge circuit can be formed with a center tap in the power circuit (Figure 8.3a) or with a tap between the capacitors in series (Figure 8.3b). Note that the bridge circuit increases the inverter power by doubling the number of thyristors. Also for adjusting the output voltage, pulse–width control can be used.

The use of bidirectional switches (Figure 8.4a) permits the formation of series resonant converters on the basis of circuits with the properties of voltage–source inverters. In that case, reactive energy can be exchanged between the resonant circuit and the input voltage source. As a result, the converter can operate with continuous circuit current over a relatively

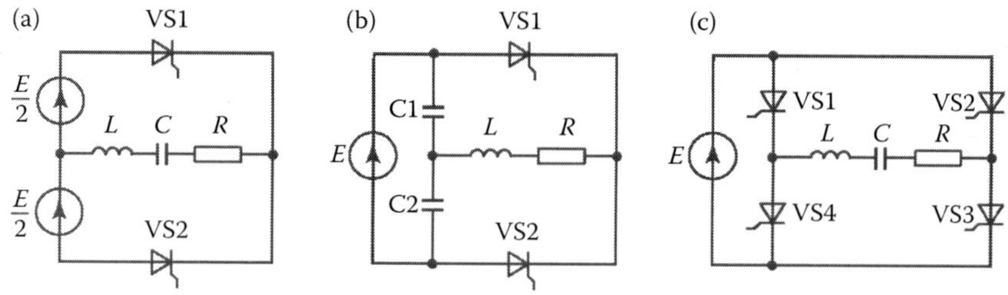

Figure 8.3 Series resonant inverters: (a) half-bridge with a center tap in supply circuit; (b) half-bridge with a tap between the capacitors in series; and (c) bridge circuit.

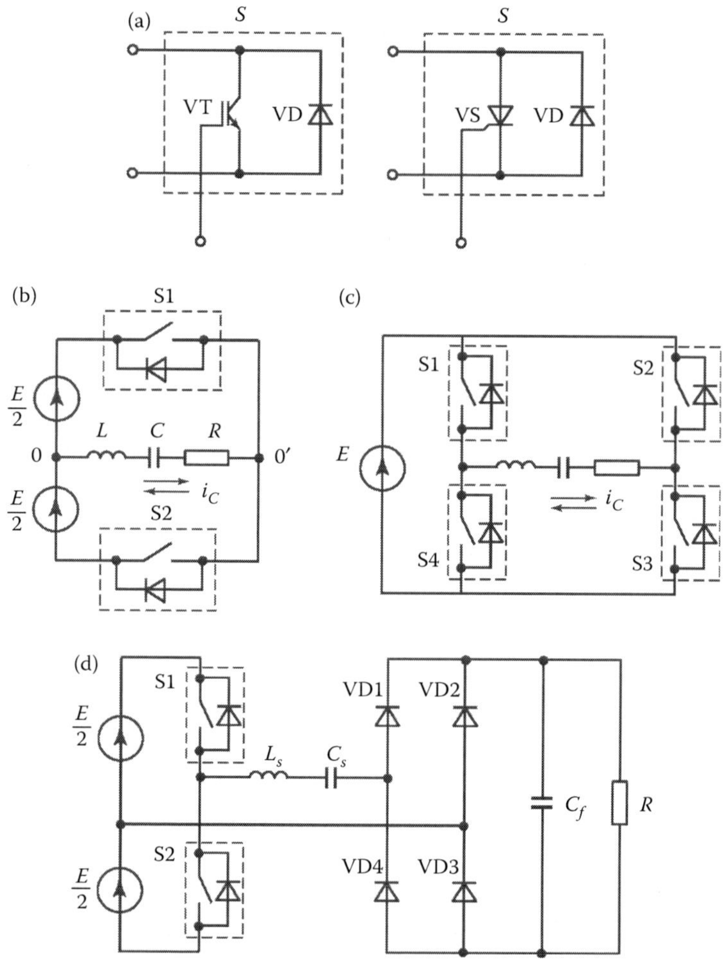

Figure 8.4 Resonant converters based on bidirectional switches: (a) switch circuit; (b) half-bridge inverter; (c) full-bridge inverter; and (d) dc converter.

broad frequency range. The losses when thyristors are turned on or off are reduced.

Figure 8.4b and c shows typical resonant inverters based on bidirectional switches and have the properties of a voltage source on the dc side. The dc converter shown in Figure 8.4d consists of two elements: a resonant half-bridge inverter and a rectifier with output filter capacitance C_f. Assuming that C_f is sufficiently large, we can regard the rectifier output voltage as ideally smoothed and equal to the average value of the load voltage. In the converter, ac voltage is formed at the output of the half-bridge inverter connected to a resonant LC circuit, whose current is rectified by a single-phase diode bridge. The basic means of regulating the output voltage in such circuits is to adjust the frequency. Note here that the processes that determine the instantaneous current and voltage depend on the switching frequency ω_S and the eigenfrequency ω_R of the resonant circuit. Taking account of the similarity of the processes that occur in the circuits in Figure 8.4, we now consider them in more detail for the example of a transistor-based half-bridge circuit presented in Figure 8.5a. To obtain the relations between the basic circuit parameters in simpler form,

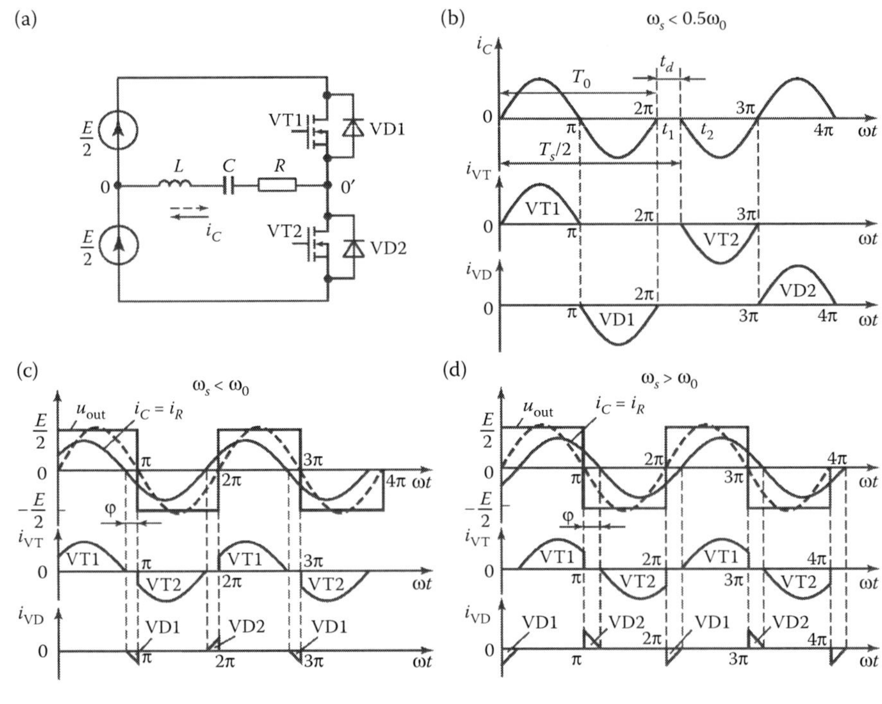

Figure 8.5 A series half-bridge inverter: (a) circuit and (b–d) current and voltage waveforms.

we assume a high Q-factor of the resonant circuit, in which the current damping within a single period is small and may be neglected.

There are three types of steady operating modes in the given circuits (Hui and Chung, 2001):

- Discontinuous current with frequency $\omega_S < 0.5\omega_0$
- Continuous current with frequency $0.5\omega_0 < \omega_S < \omega_0$
- Continuous current with frequency $\omega_S > \omega_0$

8.2.1.1 Discontinuous current mode ($\omega_S < 0.5\omega_0$)

We assume that, at time $t = 0$ (Figure 8.5b), the voltage at capacitor C is $U_C(0) = 0$, and all the switches are off. When a control pulse is supplied to transistor VT1, it is turned on, and current i_C flows for the half-period $T_0/2 = \sqrt{LC}/2$. As the switch is bidirectional, the vibrations in the circuit are not interrupted, and current i_C continues to flow in the opposite direction through diode VD1. At the end of the period T_0, when $t = t_1$, the switch is turned off, and the oscillatory process is terminated. The next oscillation period begins when a voltage pulse is supplied to transistor VT2 (for uniform current loading of the switches) at time $t_2 = T_S/2$. If we neglect the time for the transistor to be switched on and off, the period t_d can be zero when $\omega_S = 0.5\omega_0$ (this is the boundary—continuous mode). With a switching frequency $\omega_S < 0.5\omega_0$, the pause t_d increases, and the effective load current I_R can be regulated, but with a discontinuous current i_C, whose spectral composition deteriorates with an increase in t_d. Despite the deficiencies of operating with discontinuous current, the switching losses are practically eliminated, on account of soft switching with zero current. In other words, the transistor is switched off while the antiparallel diode is on, and hence the current and voltage in the transistor are zero.

In this circuit, the transistor may be replaced by a thyristor. In that case, the thyristor must be switched off within the period t_d, in accordance with the reliable cutoff condition

$$\frac{T_S}{2} - T_0 \geq t_q. \tag{8.4}$$

8.2.1.2 Continuous current mode ($0.5\omega_0 < \omega_S < \omega_0$)

Simplified analysis of operation with continuous current i_C in the resonant circuit may be based on the fundamental component, disregarding all other harmonics of the voltage sent to the LCR circuit; this voltage takes the form of a meander. Below the resonant frequency, the current in the resonant circuit will be capacitive and will lead the first voltage harmonic $u_{\text{in}1}$ at the circuit. In this case, the current and voltage diagrams for the converter are shown in Figure 8.5c. The transistors are switched off when their currents pass through zero. The power losses due to switching are practically zero, and the currents are smoothly switched to the inverse diodes. However, the currents will be nonzero when the transistors are switched on and the diodes are switched off. Accordingly, there will be power losses, which decline with a decrease in the instantaneous current. Thus, the switching losses are somewhat reduced in this circuit. In other words, the losses will only be practically zero within a single switching interval in the period T_S. If the inverter operates at $\omega_S = \omega_0$, the phase shift between the fundamental components of the current i_C and voltage u_{in1} will be zero. That corresponds to the resistive character of the circuit impedance. In this case, the inverse diodes do not transmit the current i_C; rather, each transistor in the circuit will conduct for one half-period of the circuit current and will be turned on and off at zero current. The switching losses will be zero.

8.2.1.3 Continuous current mode ($\omega_S > \omega_0$)

In this case, the current i_C will be inductive and will lag the fundamental component of the circuit voltage (Figure 8.5d). Therefore, soft switching of the current i_C from the diodes to the transistors will be observed, as i_C will pass through zero at switching. Conversely, when the transistors are switched off, i_C will be abruptly switched to the diodes, with corresponding power losses. Thus, when $\omega_S > \omega_0$, when the phase shifts of the fundamental voltage and current components are of the same magnitude but different sign, the power losses will be approximately the same but will occur in different switching operations. If $\omega_S < \omega_0$, the losses will be small when the transistors are turned off; if $\omega_S > \omega_0$, they are small when the transistors are turned on.

In operation with continuous current, one benefit is improvement in the waveform of the inverter output voltage.

A common deficiency of inverters whose series resonant circuit contains a load is the limited scope for regulating the output voltage with broad variation in the load impedance. This problem is obvious if we take into account that the circuit's oscillatory properties disappear with an increase in R, although the inverter becomes inoperative as $R \to \infty$. Therefore, inverters with a series resonant circuit are used with a load that is more or less constant. Another perturbing factor is variation in the inverter source voltage, which destabilizes the inverter load voltage.

Various methods can be used to regulate the output voltage.

- Control of the source voltage
- Control of the switching frequency in the inverter
- Pulse-width regulation of the voltage at the resonant LCR circuit

The first method is obvious and requires no explanation. Control of the switching frequency is widely used and is universal in terms of the topology of circuits with a series resonant circuit. Therefore, we will now consider this case in more detail. As an example, Figure 8.4b shows the bridge circuit of an inverter based on bidirectional switches consisting of transistors and antiparallel diodes. A square-wave voltage with amplitude E is applied to the resonant circuit in the bridge. Its fundamental component can be written as

$$u_{out1} = \frac{4}{\sqrt{2}\pi} E \sin \omega_S t. \tag{8.5}$$

The equivalent load circuit of this inverter is shown in Figure 8.6 for a resistive load R. As the equivalent circuit is linear, the fundamental voltage component at the load is

$$u_{R1} = \frac{u_{out1} R}{\sqrt{R^2 + \left(\omega_S L - \frac{1}{\omega_S C}\right)^2}}. \tag{8.6}$$

If we assume that the Q-factor of the series circuit is Q_S at resonant frequency ω_S, as in Equation 8.1, and introduce the ratio $\nu = \omega_S/\omega_0$, the modulus of the transfer function of the fundamental input-voltage component $U_{out1}(\omega_S)$ and fundamental output-voltage component $U_{R1}(\omega_S)$ can be written as follows, based on Equations 8.5 and 8.6:

$$\left|W(j\omega_S)\right| = \left|\frac{U_{R1}(j\omega_S)}{U_{out1}(j\omega_S)}\right| = \frac{1}{\sqrt{1 + Q_S^2\left(\nu - \frac{1}{\nu}\right)^2}}. \tag{8.7}$$

On the basis of Equation 8.7, we can draw the dependence of the output voltage on ν. To that end, it is expedient to use dimensionless

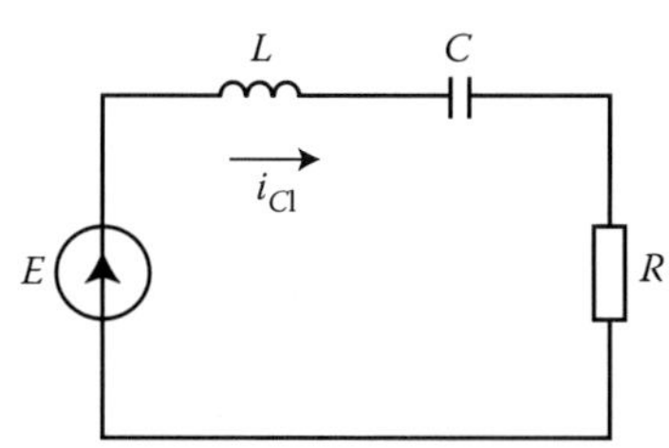

Figure 8.6 Equivalent circuit of a series bridge resonant inverter.

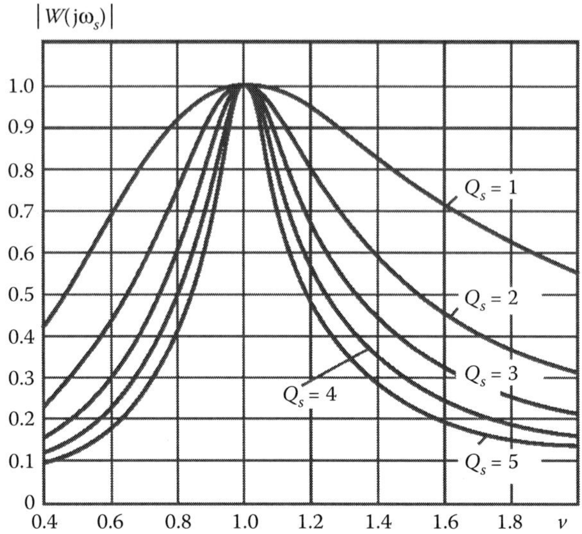

Figure 8.7 Dependence of the modulus of the transfer function $|W(j\omega_S)|$ on the frequency ratio $\nu = \omega_s/\omega_0$ at different Q_S values.

quantities, expressed relative to the effective value of the fundamental input-voltage component U_{out1} and the resonant frequency ω_0. On that basis, we may plot the modulus of the transfer function $|W(j\omega_S|$ against $\nu = \omega_S/\omega_0$. Figure 8.7 shows several curves for different values of Q_S (Hui and Chung, 2001; Nagy, 2002). Regulation of the output voltage by adjusting the frequency is an effective approach, especially at high Q_S. At low Q_S, however, the range of regulation is relatively broad, which impairs inverter performance. To eliminate this problem, we employ pulse-width control of the output voltage, on the basis of a bridge inverter circuit (Figure 8.4c). In pulse-width control, a rectangular output voltage is formed, with half-wavelength $\lambda = \pi - \alpha$ (Figure 8.8). Here α is the control angle, generated by the inverter switching algorithm. The effective value of the fundamental voltage component at the inverter RLC circuit can then be written in the form

$$U_{out1} = \frac{4E}{\sqrt{2}\pi}\cos\alpha. \tag{8.8}$$

A deficiency of such control is deterioration of the current harmonic composition in the circuit. However, its use together with frequency regulation is a good compromise, in which the deficiencies of the two methods do not significantly impair inverter performance.

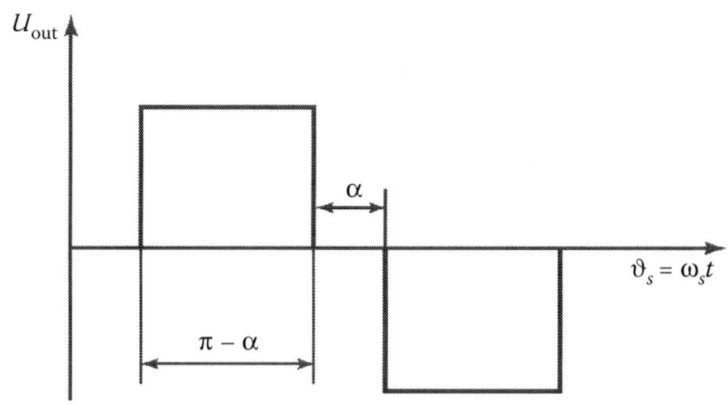

Figure 8.8 Output voltage waveform for the inverter bridge (Figure 8.4c) with pulse-width control.

8.2.2 Converters with parallel connection of the load

The classical version of an inverter with the oscillatory circuit and the load in parallel is a thyristor-based current-source inverter (Figure 8.9a). The first resonant inverters were created for inductive heating of metal by high-frequency current. The input reactor in this system has large inductance L_d, which ensures continuous and well-smoothed current over a wide range of active loads. Therefore, its equivalent circuit can be regarded as dual, according to conventional electrical engineering terminology (Figure 8.9b). Hence, the variation in the output voltage in the current-source inverter is similar to the output current in voltage-source inverters. According to the equivalent circuit, the processes in the system will correspond to the equation

$$C\frac{du_{out}}{dt} + \frac{u_{out}}{R} + \frac{1}{L}\int u_{out}dt = I_d, \tag{8.9}$$

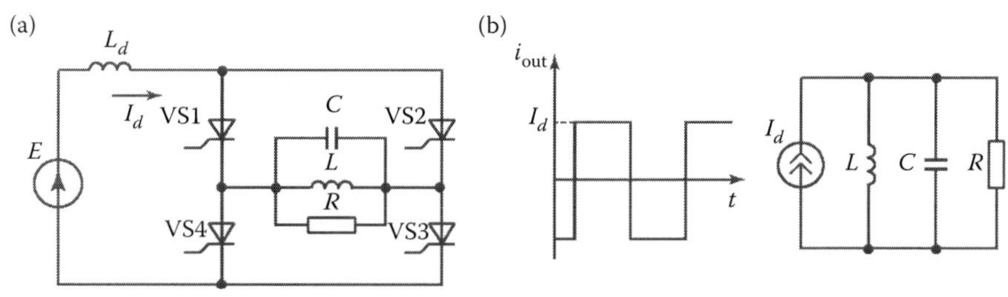

Figure 8.9 A resonant inverter with an oscillatory circuit and the load in parallel: (a) circuit and (b) equivalent circuit in steady-state operation.

where u_{out} is the instantaneous output voltage (at the load R) of the current inverter and I_d is the amplitude of the input current in meander form, in steady operation.

Taking account of the initial conditions, we may write the solution in the form

$$u_{\text{out}} = \frac{I_d}{\omega_{\text{RP}} C} e^{-\delta t} \sin \omega_{\text{RP}} t, \tag{8.10}$$

where $\delta = R/2L$ and $\omega_{\text{RP}} = \sqrt{\omega_0^2 - \delta^2}$.

It follows from Equations 8.2 and 8.10 that the variation in current i_C in the series resonant voltage inverter is analogous to the variation in the voltage u_{out} in a parallel resonant current inverter; accordingly, they are dual. The current inverter must be based on unidirectional switches. If the switches are unable to withstand the inverse voltage, a diode is introduced in series with them. Therefore, the circuit is well suited to thyristor use, and consequently, it was widely used in the early days of resonant inverters. A parallel current inverter based on ordinary thyristors may operate not only when discontinuous voltages arise in the resonant circuit under the action of the input current, but also with continuous voltage when $\omega_S > \omega_0$. In that case, the following condition must be satisfied:

$$t_q \geq \frac{\beta}{\omega_S}, \tag{8.11}$$

where t_q is the thyristor's cutoff time and β is the capacitive phase shift between the first current and voltage harmonics. In practice, the minimum permissible value $\beta_{\min}$ is not large, especially for fast thyristors, which are used in such applications. Therefore, we may assume operation with $\omega_S = \omega_0$.

The benefits of such inverters include the following:

- The input reactor limits the maximum input currents, and hence the maximum currents switched by the thyristors
- The inverters have good filtration properties, due to the parallel capacitor
- They may operate at low loads, including zero load when $R \to \infty$

Analysis of the dependence of the output voltage u_{lo} on the switching frequency ω_S yields functions similar to those in Figure 8.7 for a series resonant inverter with discontinuous current. In the present case, however, the variable plotted vertically must be the modulus of the inverter's input impedance $|Z(j\omega_S)|$:

$$|Z(j\omega_S)| = \left|\frac{U_{out}}{I_{out1}}(j\omega_S)\right|, \tag{8.12}$$

where $I_{out1} = 4I_d/\sqrt{2}\pi$ is the effective value of the fundamental input-current component, which is of rectangular form.

If we disregard the losses in the system components, the amplitude of the alternating current in steady conditions is a function of the active load power

$$I_d = \frac{U_{out}^2}{RE}, \tag{8.13}$$

where U_{out} is the RMS voltage at load R.

Of course, the practical implementation of such dependences at high frequency (above 5 kHz) is only possible in current inverters based on completely controllable switches, for example, transistors. As already noted, however, the use of switches that cannot withstand inverse voltage, such as transistors, requires the introduction of a diode in series. An example of a transistor-based bridge circuit for a current-source inverter, with a parallel resonant loop, is shown in Figure 8.10. Obviously, in this circuit, we can use not only frequency control of the output voltage, but also pulse-width control of the input current.

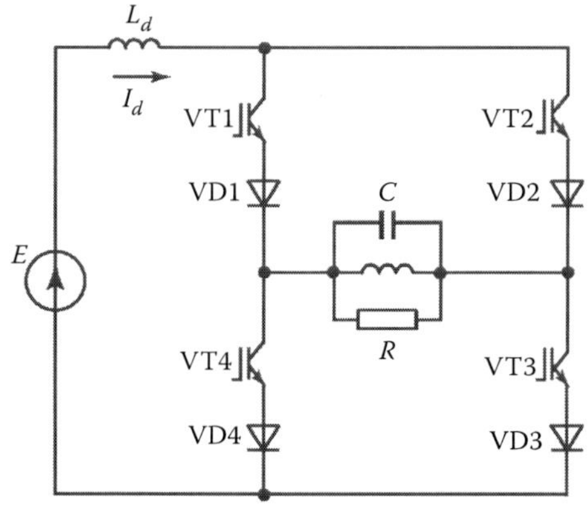

Figure 8.10 A bridge resonant inverter with parallel connection of resonant circuit elements and load.

A deficiency of a current inverter with a parallel resonant loop is that it cannot operate in conditions resembling a short-circuit, in contrast to an inverter with a series resonant circuit, whose operation is constrained at very low loads.

A current-source inverter with a resonant circuit at the output (Figure 8.9a) is considered under the assumption of large input-reactor inductance L_d, when we may assume that $\omega_S L_d \to \infty$. However, without changing the circuit design of such inverters, we may significantly change their characteristics and their current and voltage variation if we significantly reduce the inductance of their input reactor, to values supplementing the inductance of the output resonant loop. In addition, small inductance of the input circuit can create discontinuous input current at the inverter. That will reduce the switching losses in the inverter. However, the permissible ranges of load variation and output-voltage regulation will be limited in that case.

The voltage regulation of the current-source inverter can be significantly improved without resorting to frequency or pulse-width regulation. To that end, we use regulation of a reactor inductance connected in series with antiparallel thyristors. The load can also be connected in parallel to one of the elements in the series resonant circuit. Usually, the load is connected directly to the capacitor of the oscillatory circuit or else through transformer T (Figure 8.11a). In those cases, as a rule, voltage-source inverters with bidirectional switches are used. That ensures a broader range of operation and regulation of the output voltage. The equivalent circuit of an inverter with a parallel load is shown in Figure 8.11b. We see that the inverter can operate at small loads, including zero

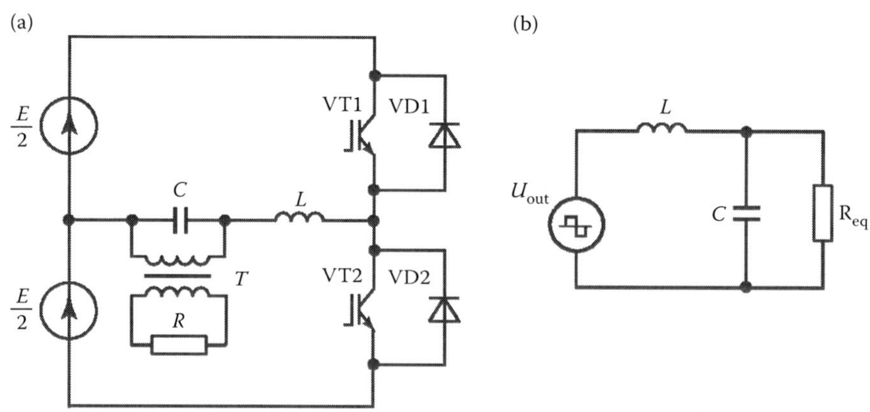

Figure 8.11 (a) A series resonant inverter with parallel connection of the load and the capacitor of the oscillatory circuit and (b) its equivalent circuit.

load. The output voltage u_{out} is a maximum for operation at the resonant frequency ω_0, when the switching frequency $\omega_S = \omega_0$. The maximum value of the output voltage is determined by the inverse Q-factor of the series circuit $1/Q_S$. To regulate the voltage at small loads, the working frequency is raised above the resonant frequency.

8.2.3 Series–parallel resonant inverters

The series–parallel resonant inverter combines the benefits of voltage inverters with series and parallel resonant circuits. The circuit topology is determined by including an additional capacitor or reactor in the resonant element. By adjusting the magnitude and point of introduction of the supplementary elements, we may obtain a wide range of circuits with different topologies and operational characteristics. A common design is a circuit formed by selecting capacitors corresponding to about a third of the total capacitance in the series circuit and connecting them in parallel to the inverter load (Figure 8.12a). The resulting circuit will have some of the properties of both series and parallel resonant inverters. The equivalent circuit is shown in Figure 8.12b. Analysis of the dependence of the output voltage on the switching frequency shows that the circuit may operate normally at low loads (including zero load). The output voltage may be regulated by varying the frequency ω_S. In contrast, overloads and short-circuit currents at the output are limited by the series circuit at a level determined by the working frequency. It follows from the circuit's operating principle that, by adjusting the ratio of the capacitors C_1 and C_2, we may ensure the best inverter operation for specified technical characteristics.

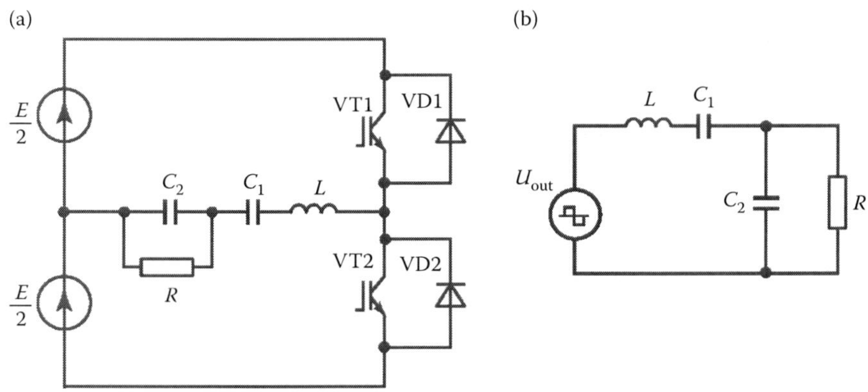

Figure 8.12 (a) A parallel–series resonant inverter and (b) its equivalent circuit.

8.2.4 Converters of class E

8.2.4.1 Inverters of class E

Such converters are connected to a dc source through reactor L_d, whose inductance ensures the properties of a current source at the inverter input. In inverters of class E, current pulses at elevated frequencies are formed by a single transistor and sent to a series resonant circuit L_sC_s with a high Q-factor ($Q_S \geq 7$), which is connected to load R. The switching frequency ω_s of the inverter is much greater than the resonant frequency of the series circuit ω_0. The inverter circuit with a single transistor ensures soft switching and is characterized by high efficiency. Usually, it is used for inverters of relatively low power (<100 W) with variable output voltage and practically constant load R. They are typically used as lamp ballast (Rashid, 1988).

Figure 8.13a shows the inverter circuit. Two basic modes of operation can be distinguished: optimal operation and near-optimal operation. In optimal operation, the transistor is switched when the voltage u_{VT} and current i_{VT} pass through zero. In that case, there is no need for the inverse diode VD; a dashed line shows the connection to the diode in Figure 8.13a. We observe optimal operation at certain parameters of the circuit elements, including the load R, which must remain constant in this case. In optimal operation, the losses in the inverter are a minimum, and its efficiency is a maximum.

Figure 8.13b shows equivalent circuits for different states of transistor VT in optimal operation. The equivalent circuits for intervals I and II apply when the transistor is on and off, respectively. Figure 8.13c shows waveforms of current and voltage in the case of steady inverter operation. In interval I, transistor VT is on. When it is turned on ($t = t_0$), current $i_{VT} = i_d + i_R$ flows through transistor VT. The component i_d corresponds to the current of the input reactor L_d, whose magnitude in steady conditions can be assumed approximately constant and equal to the average value of I_d. This assumption is possible because capacitor C_s of the series oscillatory circuit blocks the constant current component I_d in steady conditions. The other current component i_R corresponds to the current in the circuit containing the load R. As a result, as we see from the i_{VT} diagram in Figure 8.13c, the current of transistor VT begins at zero, as the initial condition for the switching on of transistor VT in optimal operation is that $i_{VT}(t_0) = I_d + i_R(t_0) = 0$. This corresponds to the shunting of transistor VT by two sources: current source I_d created by the input voltage E and inductance L_d and the nonzero current of the series circuit L_sC_sR. In other words, the voltage at capacitor C_d at the end of the preceding interval and, hence, at the beginning of the given interval ($t = t_0$) is zero, that is, the capacitor C_d is completely discharged at $t = t_0$. Note that the mean values $U_{VT} = U_{Cd} = E$, as the constant voltage component at the other circuit

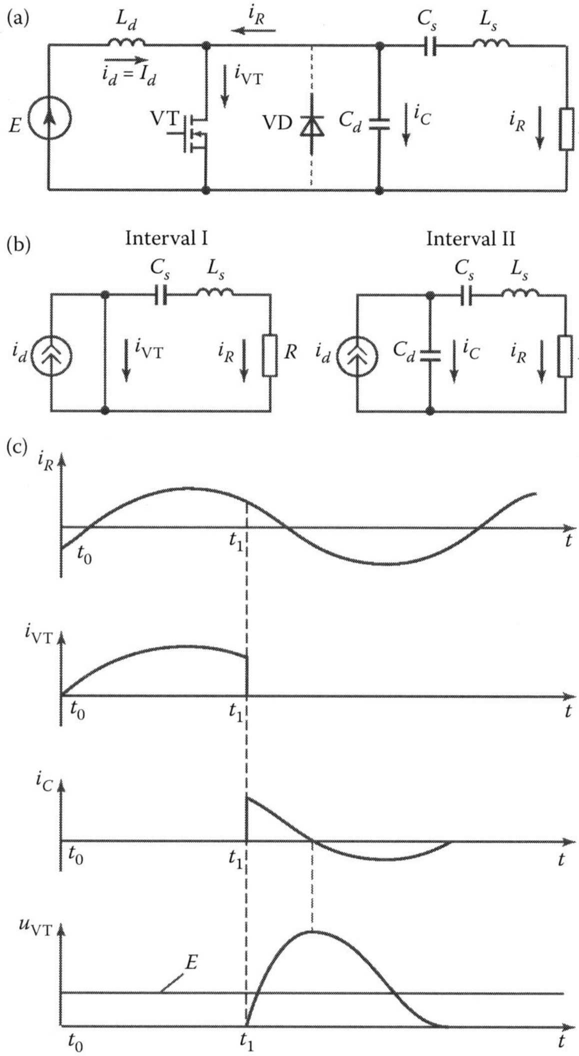

Figure 8.13 (a) An inverter of class E; (b) equivalent circuit; and (c) current and voltage waveforms.

elements is zero in steady operation. At $t = t_1$, transistor VT is turned off, and the processes in the inverter then correspond to the equivalent circuit for interval II (Figure 8.13b). As a result, current i_{Cd} is supplied to capacitor C_d. However, as the voltage at the capacitor is an integral function of the current that is present, it begins to smoothly increase from zero. Hence, there are practically no power losses when transistor VT is switched off.

As long as $i_{Cd} > 0$, the voltage u_{Cd} rises, reaching a maximum when i_{Cd} passes through zero. Then the voltage u_{Cd} begins to fall, reaching zero at $t = t_0 + T$ in optimal operation, that is, at the end of the transistor's switching period. Then transistor VT is switched on again, and the process periodically repeats. Thus, in optimal operation, the transistor is switched off at zero voltage and is switched off at zero current—in other words, with practically no switching losses. Note, however, that optimal operation and near-optimal operation correspond to strictly defined parameter ratios of the circuit elements, including the load (Middlebrook, 1978). At the same time, the elimination of switching losses is associated with an increase in the maximum transistor voltage ($u_{Cd} \approx 3E$) and current ($i_{VT} \approx 3I_d$).

With an increase in the inverter load, that is, when the resistance R is reduced below the optimal value or other circuit parameters are adjusted with disruption of optimal operation, the processes in the circuit are modified, and the switching losses increase. In such nonoptimal operation, the voltage $u_{Cd} = u_{VT}$ changes sign before current i_{Cd} is zero. In that case, to eliminate inverse voltage at transistor VT, we must connect inverse diode VD (as shown by the dashed line in Figure 8.13a). In the presence of diode VD, reversal in the voltage polarity switches current i_{Cd} to the diode. As a result, the switching losses in the circuit increase. Therefore, the benefits of the circuit are most evident in optimal or near-optimal operation. In that case, the output voltage can be regulated within a narrow range by slight variation in the switching frequency.

8.2.4.2 Rectifiers of class E

Structurally, the dc converter often consists of an inverter and a rectifier. Therefore, the converter efficiency can be increased by reducing the losses in each component. For example, an inverter of class E can be combined with a rectifier of class E. A simple rectifier circuit of this type was proposed in Lee (1989). It is based on a single diode and contains a resonant series element, ensuring that the rectifier's diode is switched off at zero current (Figure 8.14a). High-frequency voltage $u_{in}(t) = U_{in\,max} \sin \omega_{in} t$ is applied to the rectifier input. Capacitor C is connected in parallel with diode VD and is chosen so as to create resonance with inductance L_d at the input-voltage frequency ω_{in}. The rectified voltage is smoothed by filter C_f, whose capacitance is determined by the permissible ripple and may be relatively large. The presence of a nonlinear element—diode VD—hinders rigorous analysis of the processes in the system. However, qualitative conclusions may be obtained if we consider equivalent circuits for two states of the diode VD (Figure 8.14b): when it is off (interval I) and on (interval II). The load R and the filter C_f constitute a dc voltage source with mean voltage U_R. In interval I, the current in inductance L_d is equal to the difference of the currents corresponding to the input ac voltage $u_{in}(t) = U_{in\,max} \sin \omega_{in} t$ and the dc voltage U_R of the source equivalent

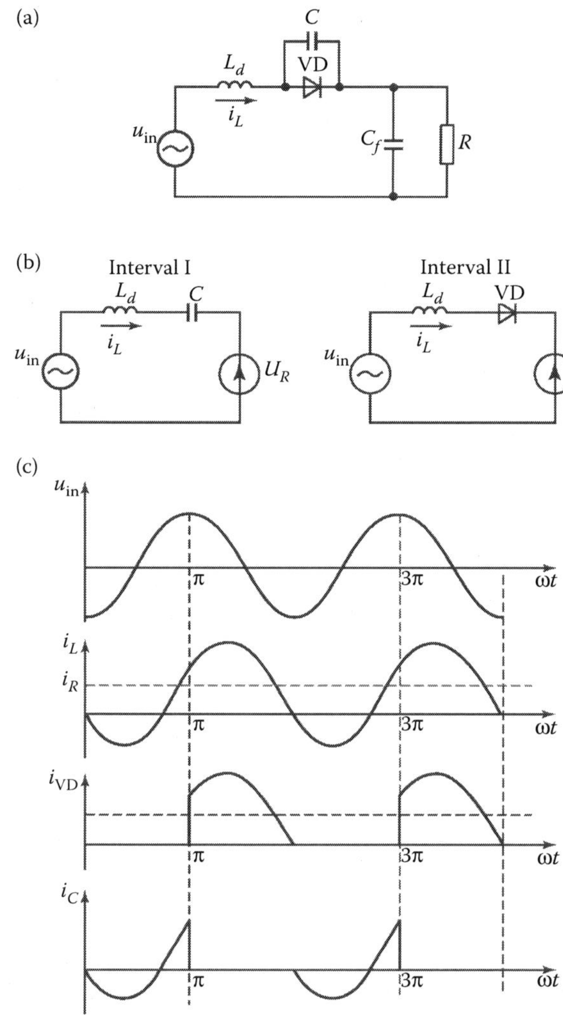

Figure 8.14 (a) A rectifier of class E; (b) equivalent circuit; and (c) current and voltage waveforms.

to the output filter C_f and load R. Assuming that the output voltage is relatively smooth, we may regard the load current as constant and equal to the mean value $I_R = U_R/R$. In interval I, diode VD is off, and the difference between voltages $u_{in}(t)$ and U_R is applied to the LC circuit tuned to resonance with the input-source frequency ω_{in}. In interval I, the circuit current flows through capacitor C, at which the voltage is the inverse of that for diode VD. When the input voltage reaches the maximum value $U_{in\,max}$, equal to U_R, diode VD is switched on, and the inductance current

i_L is switched from capacitor C to diode VD. The operation of the system corresponds to interval II. Diode VD is switched off when current i_L falls to zero; when this occurs will depend on the difference between $u_{in}(t)$ and U_R. Taking account of the assumptions adopted, we may use the approximate formula

$$i_L = \frac{U_{in\,max}}{R}\sin(\omega_{in}t - \varphi) - I_R, \tag{8.14}$$

where $\varphi = f(R)$.

If the diode is switched off at zero current, there will be practically no switching losses.

8.3 Quasi-resonant converters

8.3.1 Basic circuits with quasi-resonant switching

In quasi-resonant switching, smooth switching trajectories are formed by means of resonance in inductive and capacitive components connected to the controllable semiconductor switch. As resonance is associated with oscillatory processes in the resonant LC circuit, conditions may be created for switching at zero current or voltage, depending on the circuit and the type of quasi-resonant switch. Given that a semiconductor switch is a significantly nonlinear element in the switching intervals, we conclude that the current or voltage in the circuit that switches on the semiconductor device will be nonsinusoidal and will not correspond to strict resonant specifications. In addition, the duration of the oscillations is usually no more than one or two half-periods, on account of the transition of the system on switching to a new state with different variables describing the trajectory of parameter variation. Therefore, such circuits are said to be characterized by a quasi-resonant switching. When using quasi-resonant switching, the goal is to reduce the power losses by ensuring soft switching—in other words, switching when the current or voltage is zero. At the same time, the di/dt and du/dt values at the switch are reduced. That increases the reliability of the switch and the circuit as a whole, due to the decrease in the electromagnetic interference.

In practice, quasi-resonant switches are best used in dc converters. Such converters include not only unidirectional and bidirectional semiconductor switches but also reactors and capacitors with small inductance and capacitance (Lee, 1989; Nagy, 2002). In some cases, for example, at high switching frequencies, this function may simply be served by the intrinsic capacitance and inductance of the switches themselves, which has traditionally been regarded as parasitic. Of course, special technology is required in that case.

The use of the switch intrinsic capacitance and inductance in that way permits the formation of the desired switching trajectories, corresponding to soft switching. Essentially, these elements constitute nondissipative circuits for the formation of the switching trajectories. The circuits are nondissipative because they do not contain resistive elements.

8.3.1.1 Zero current switching

Figure 8.15 shows typical unidirectional and bidirectional switches with zero current switching (ZCS). Various controllable switches can be employed here, in particular, transistors with a diode in series or antiparallel. The capacitor C_r and inductance L_r connected to the switch are elements of a resonant circuit for creating current oscillations at frequency $\omega_0 = 1/\sqrt{L_r C_r}$. When a unidirectional switch is turned on (Figure 8.15a and b), the current begins to rise slowly from zero $i_L(0) = i_{VT}(0) = 0$. Then an oscillatory process begins in the circuit $L_r C_r$ and, when current i_L passes through zero during the first half-period, the switch is turned off. In that case, current flows through the switch during the first half-period. Accordingly, in quasi-resonant converters, this is said to be a half-wave operation. When using a bidirectional switch (Figure 8.15c and d), likewise, the current rises smoothly after it has been turned

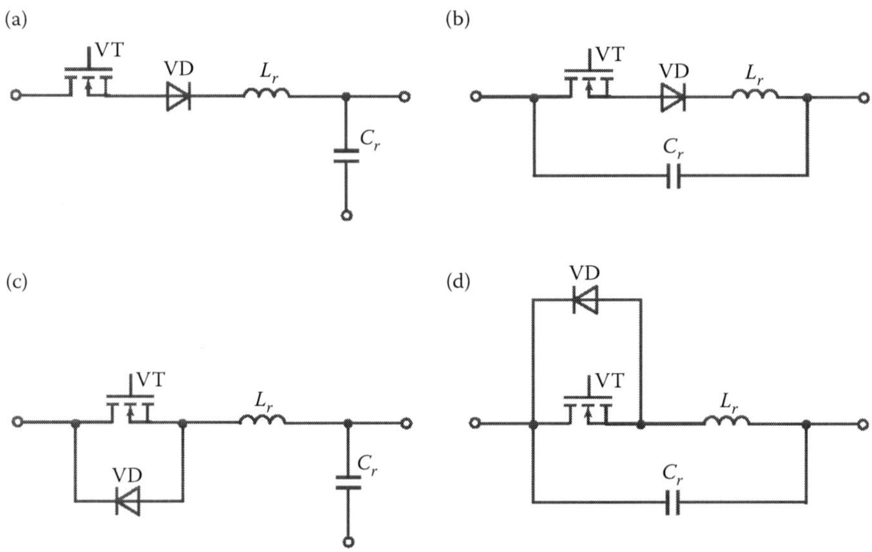

Figure 8.15 (a and b) Half-wave and (c and d) full-wave switch circuits with zero current switching: L_r and C_r are elements of the resonant circuit.

on, and an oscillatory process begins at frequency ω_0. However, this process extends over the whole oscillation period, as the negative current half-wave will pass through antiparallel diode VD. The switch will again be turned off at zero current i_L, but this time at the end of the second half-period. Thus, circuits with full-wave operation are based on bidirectional switches and circuits with half-wave operation on unidirectional switches. In what follows, quasi-resonant converters will be described as half-wave or full-wave circuits, because that better reflects their properties.

8.3.1.2 *Zero voltage switching*

Figure 8.16 shows circuits for quasi-resonant zero voltage switching (ZVS). As for switching at zero current, the elements L_r and C_r create oscillations at the resonant frequency ω_0. It is evident from the topology that the circuits for switching at zero current and switching at zero voltage are dual. Connecting capacitor C_r in parallel with the semiconductor device permits switching at zero voltage, just as the inductance L_r in Figure 8.15 permits switching at zero current. In circuits with a bidirectional switch (Figure 8.16a and b), diode VD shunts the negative half-wave of the oscillations. That results in half-wave operation. When the diode is in series with the switch (Figure 8.16c and d), full-wave operation is observed. The reactive elements allow the semiconductor switches to be turned on and off at zero voltage. Most dc converters are based on switching at zero current or switching at zero voltage. Which

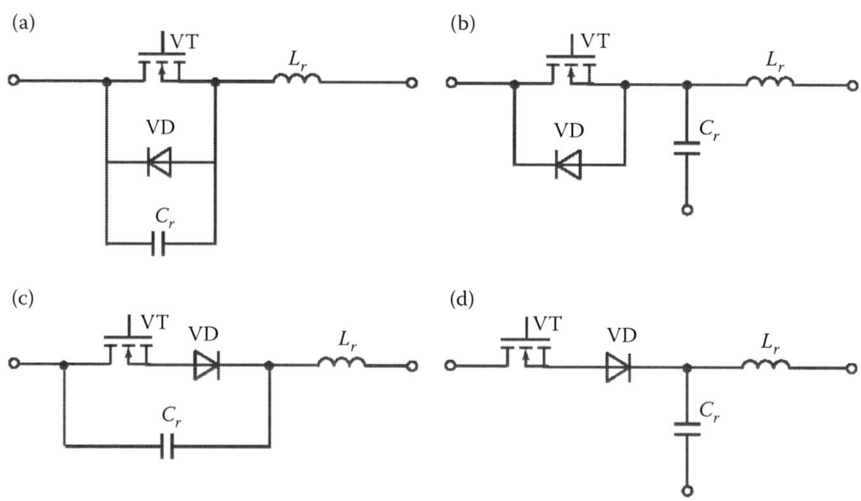

Figure 8.16 (a and b) Half-wave and (c and d) full-wave switch circuits with zero voltage switching.

option is selected will depend on numerous engineering and economic factors, such as the following.

- In switching at zero current, the maximum current is more than double the mean load current.
- In switching at zero voltage, the maximum voltage at the switch is much greater than the converter input voltage.
- In switching at zero current, the intrinsic capacitance of the semiconductor switch produces additional currents when the switch is turned on, with a corresponding increase in the power losses.

8.3.2 *Quasi-resonant dc–dc converters*

8.3.2.1 *ZCS quasi-resonant converters*

Practically, all basic dc–dc converters are capable of being switched off at zero current in quasi-resonant conditions. As an example, we consider the buck converter shown in Figure 8.17a, as well as the version with a half-wave switch in Figure 8.15a. The converter operates in the pulsed mode at frequency $f_s = 1/T_s$. We make the following assumptions:

- Ideal circuit components
- Zero pulsations of the input voltage u_{in} and filter current i_{Lf}
- Steady operation with constant fill factor $\gamma = t_{\text{on}}/T_s$ specified by the sequence of control pulses and the switch

We assume that transistor VT is off when $t < 0$ but is turned on by a control pulse at time $t = 0$. When transistor VT is on, under the given assumptions, the converter can be represented by the equivalent circuits shown in Figure 8.17b, which applies for interval I ($0 < t < t_1$). In this interval, the filter's reactor current I_{Lf} passes through the inverse diode VD, which shunts the output filter and the load. According to the equivalent circuit for interval I,

$$L_r \frac{di_{\text{Lr}}}{dt} = E, \quad i_{\text{Lr}} = \frac{E}{L_{\text{r}}} t, \quad i_{\text{VD}} = I_{\text{Lf}} - i_{\text{Lr}}. \tag{8.15}$$

At time $t = t_1$, current i_{Lr} is equal to current I_{Lf}, and diode VD is switched off. The equivalent circuit of the converter is modified, and the processes that develop correspond to interval II ($t_1 < t < t_2$) in Figure 8.17b. According to the equivalent circuit for interval II,

$$\begin{cases} C_r \dfrac{du_{\text{Cr}}}{dt} = i_{\text{Lr}} - I_{\text{Lf}}, \\ L_r \dfrac{di_{\text{Lr}}}{dt} = E - u_{\text{Cr}}. \end{cases} \tag{8.16}$$

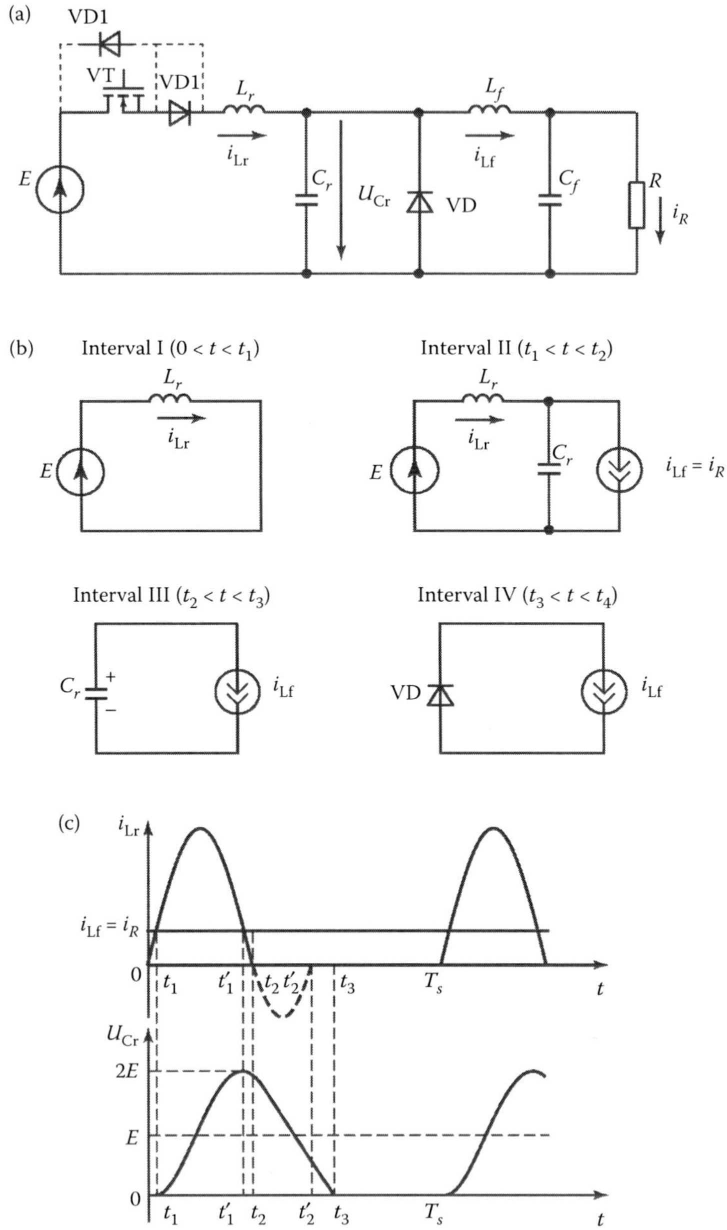

Figure 8.17 (a) A buck converter with zero current switching; (b) equivalent circuits in different operational intervals; and (c) current and voltage waveforms.

Assuming that $t_1 = 0$ and taking into account that $u_{\mathrm{Cr}}(0) = 0$ and $i_{\mathrm{Cr}}(0) = I_{\mathrm{Lf}}$, we obtain the following expressions for interval II from Equation 8.16:

$$\begin{cases} i_{\mathrm{Cr}}(t) = I_{Lf} + \dfrac{E}{\rho}\sin\omega_0 t, \\ u_{\mathrm{Cr}}(t) = E(1 - \cos\omega_0 t), \end{cases} \tag{8.17}$$

where

$$\rho = \sqrt{\frac{L_r}{C_r}} \quad \text{and} \quad \omega_0 = \frac{1}{\sqrt{L_r C_r}}.$$

When $t = t_1'$, the current i_{Lr} becomes less than the load current I_{Lf}. However, in a unidirectional switch, the current i_{Lr} cannot reverse direction. Therefore, it continues to flow through capacitor C_r, and the equivalent circuit for interval II is unchanged.

At time $t = t_2$, i_{Lr} falls to zero, transistor VT is turned off, and interval III begins ($t_2 < t < t_3$). In the corresponding equivalent circuit, the variation in u_{Cr} is linear

$$u_{\mathrm{Cr}} = \frac{1}{C_r}\int I_{\mathrm{Lf}}\,\mathrm{d}t = \frac{I_{\mathrm{Lf}}t}{C_r}. \tag{8.18}$$

Capacitor discharge to zero voltage ends when $t = t_3$. The equivalent circuit again changes, with the onset of interval IV ($t_3 < t < T_s$). In this interval, diode VD at the filter input is turned on and begins to transmit current I_{Lf}. At $t = T_s$, interval IV ends. A pulse is again sent to turn on transistor VT, and operation corresponds to interval I. This sequence repeats periodically (Figure 8.17c).

The use of the full-wave step-down converter shown by the dashed line in Figure 8.17a changes the processes in the system beginning at time $t = t_2$, as the current i_{Lr}, after passing through zero, begins to flow in the opposite direction through the antiparallel diode VD1. As a result, the recharging of capacitor C_r is not linear but oscillatory up to $t = t_2'$, when the second half-wave of current i_{Lr} ends on passing through diode VD1. That corresponds to full-wave operation. Correspondingly, the onset of linear capacitor discharge is shifted from $t = t_2$ to $t = t_2'$. In other words, the bidirectional switch containing diode VD1 is switched off later (at $t = t_2'$). The subsequent operation corresponds to the equivalent circuit for interval III. Full-wave operation of this system improves its properties. First, the load has little influence on the discharge of capacitor C_r. Accordingly, its

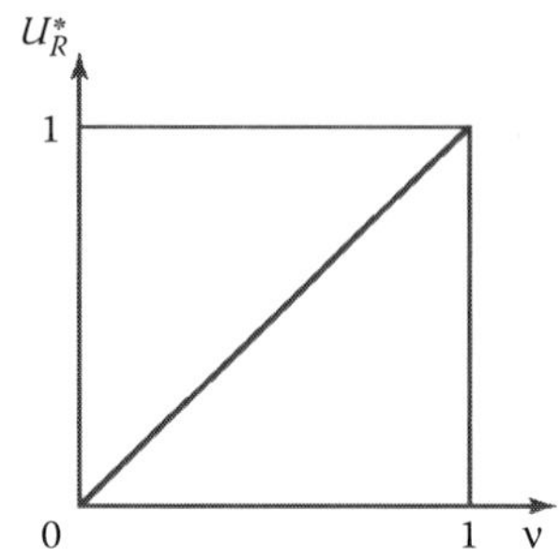

Figure 8.18 Control characteristic of the output voltage for a buck dc–dc converter with zero current switching.

regulated characteristics are practically linear and do not depend greatly on the load. In addition, in a bidirectional switch, some of the energy stored in the reactor L_r is returned to the primary voltage source.

The converter's output voltage may be regulated by adjusting the switching frequency f_s. In zero current switching converters, the controllable switch is on for a fixed period. The variable in the control characteristics is the switching frequency ratio $\nu = \omega_s/\omega_0$, which can be regarded as an analog of the duty ratio γ in pulse-width modulation. The basic factors that perturb the output voltage U_R are the input voltage E and the load R_L. Figure 8.18 shows the control characteristics for a quasi-resonant converter in terms of dimensionless variables $U_R^* = U_R/E$ and $\nu = \omega_s/\omega_0$, with different damping factors $d = R/\rho$. It is evident from Figure 8.18 that when using a bidirectional switch, these factors have no influence on the load.

8.3.2.2 ZVS quasi-resonant converters

Like switching at zero current, switching at zero voltage can be used in all basic dc converters. We will now investigate in more detail the operation of a step-down converter when switching at zero current is replaced by switching at zero voltage, under the same assumptions. We consider the half-wave switch shown in Figure 8.16a. Hence, we may obtain similar variation of the current i_{Lr} in switching at zero current and of the voltage u_{Cr} in switching at zero voltage. Figure 8.19 shows the circuit diagram of the buck converter with ZVS and its equivalent circuits in different operational intervals. At time $t = 0$, transistor VT is switched off, and the equivalent circuit corresponds to interval I ($0 < t < t_1$). In interval I, transistor VT is off, and the capacitor C_r is charged by the load current I_{Lf}. Under the given assumptions, charging is linear, that is,

$$u_{Cr} = \frac{I_{Lf}}{C_r}t. \tag{8.19}$$

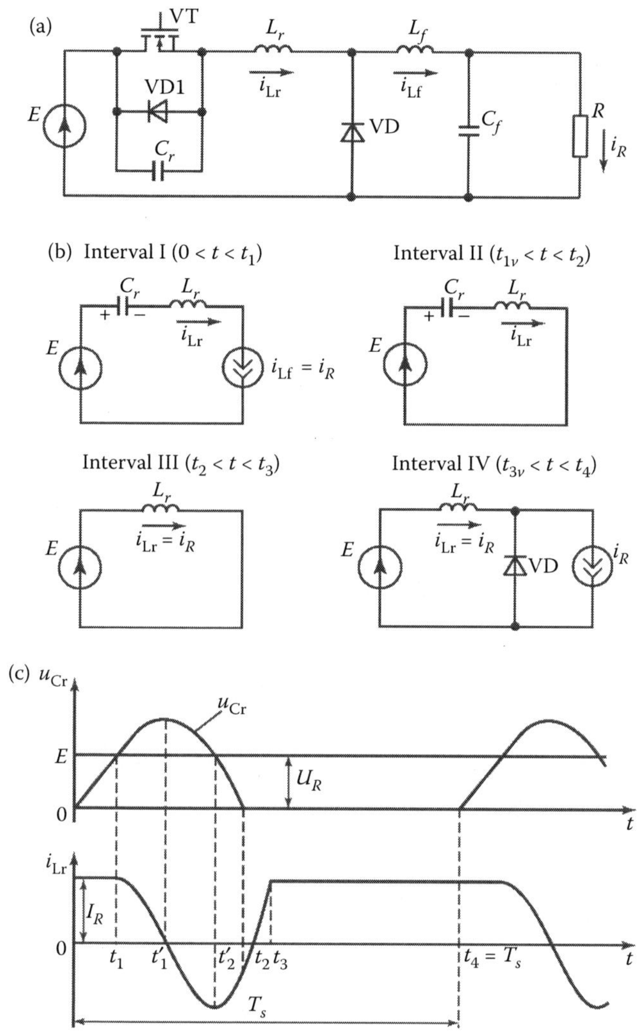

Figure 8.19 (a) A buck converter with zero voltage switching; (b) equivalent circuits in different operational intervals; and (c) current and voltage waveforms.

At $t = t_1$, the voltage at capacitor C_r is equal to the input voltage E. Diode VD_f is turned on and an oscillatory process begins in the circuit containing L_r and C_r. The equivalent circuit now corresponds to Figure 8.19b for interval II ($t_1 < t < t_2$). In this interval, the processes can be described by the relations

$$\left.\begin{aligned} u_{Cr} &= U_{Cr\max}\sin\omega_0 t + E, \\ i_{Lr} &= I_{Lf}\cos\omega_0 t, \end{aligned}\right\} \quad (8.20)$$

where $U_{Crmax} = I_{Lf}\sqrt{L_r/C_r}$.

As a result of the oscillatory process, the capacitor voltage increases to a peak value $E + I_{Lf}\sqrt{L_r/C_r}$, which depends on the load current I_{Lf}, whereas the current i_{Lr} begins to fall to zero. On the assumption of ideal circuit elements, the variation in current i_{Lr} is shifted in phase by $\pi/2$ relative to voltage u_{Lr}, so that the maximum voltage U_{Lrmax} appears when i_{Lr} passes through zero. The load current I_{Lf} must ensure that U_{Crmax} exceeds the input voltage: $I_{Lf}\sqrt{L_r/C_r} > E$. This is necessary in order to ensure that the switch is not turned on at nonzero voltage. After i_{Lr} falls to zero at $t = t_1'$, it reverses direction, and discharge of the capacitor continues. At $t = t_2$, the voltage at capacitor C_r is zero, and the negative component of current i_{Lr} begins to flow through diode VD1, which begins to conduct at $t = t_2$. At $t = t_2$, interval III begins ($t_2 < t < t_3$), with a new equivalent circuit. In this interval, i_{Lr} varies linearly, reaching zero at $t = t_2'$, when diode VD1 is switched off and transistor VT is switched on by a control pulse. Then current i_{Lr} increases linearly to $i_{Lr} = I_{Lf}$ ($t = t_3$). When i_{Lr} reaches the value I_{Lf}, inverse diode VD_f is switched off, as its current becomes equal to zero. A new stage begins, characterized by the equivalent circuit for interval IV ($t_3 < t < t_4$). At $t = t_4$, transistor VT is turned off and a new period begins. The preceding sequence repeats.

The converter output voltage can be regulated by adjusting frequency f_s so as to vary the length of interval IV. Thus, we may contrast switching at zero current, where the converter's output voltage is regulated by adjusting f_s with constant duration of the transistor's on state, and switching at zero voltage, where the duration of the transistor's off state is maintained constant. In the case of converters with half-wave switches, the controllable characteristics depend on the load in switching at zero voltage, just as in switching at zero current. Likewise, as in a converter with switching at zero current, the dependence of the controllable characteristics on the load in a converter with switching at zero voltage can be almost completely eliminated if the half-wave switch is replaced by a full-wave switch (Figure 8.16c and d).

8.3.3 ZVS converters with switch voltage limiting

We now consider converters with switching at zero voltage in the case in which the switch voltage cannot exceed the input voltage. In such converters, at least one arm of the bridge contains bidirectional switches, in parallel with capacitors. The corresponding topology resembles a half-bridge single-phase circuit, which can be regarded as a single-phase module for a bridge inverter. At the same time, the single-arm circuit can be used in dc–dc converters in which switches with a low maximum voltage are preferred. In addition, it permits regulation of the output voltage by pulse-width modulation at constant working frequency. We can note that this circuit corresponds to a controller operating in two quadrants of the output volt–ampere characteristic (Lee, 1989; Kazimierczuk and Charkowski, 1993).

Figure 8.20a shows the power component of such a converter based on transistors VT1 and VT2 and antiparallel diodes VD1 and VD2. Reactor L_r and capacitors $C_{r1} = C_{r2} = C_r/2$ are elements of a circuit with resonant frequency $\omega_0 = \sqrt{L_r C_r} \gg \omega_s$, where ω_s is the switching frequency of the

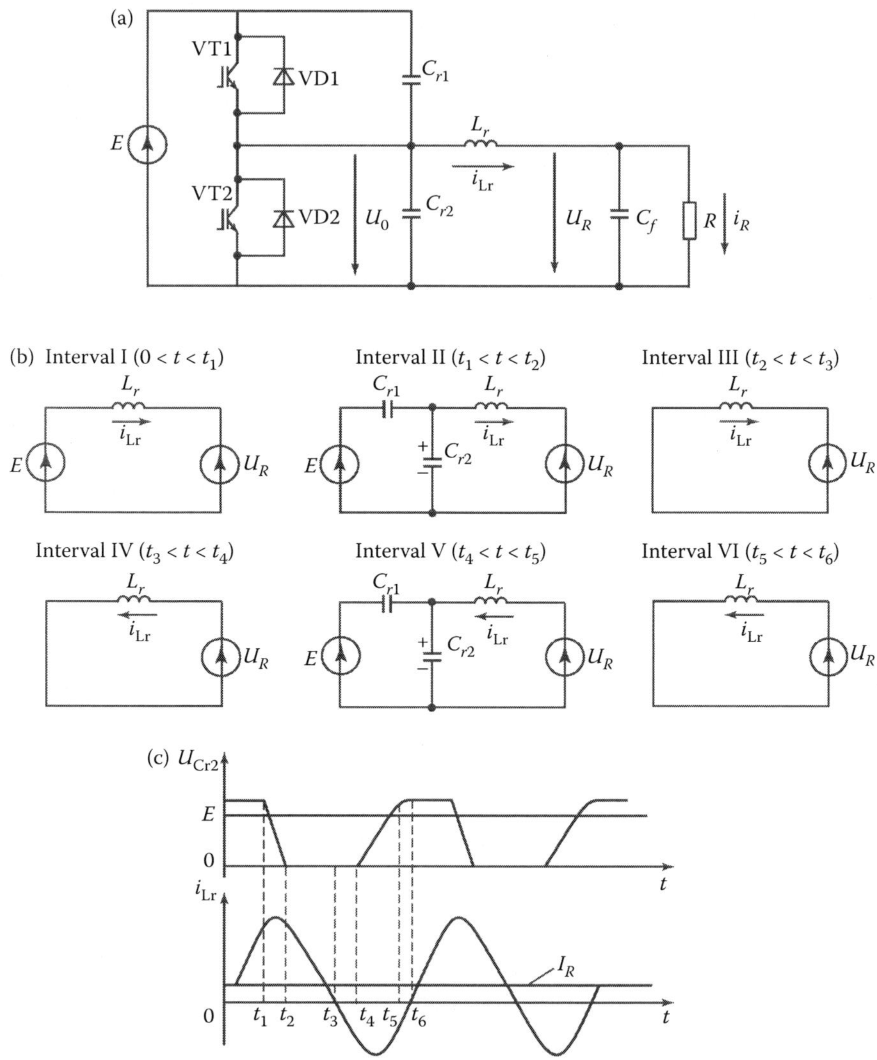

Figure 8.20 (a) A converter with zero voltage switching; (b) equivalent circuits in different operational intervals; and (c) current and voltage waveforms.

transistors. We assume ideal circuit elements and ideally smooth output voltage U_{out} of the filter capacitor C_f. On the basis of the latter assumption, we may regard the load R with filter C_f as a voltage source with mean voltage U_{out}, as we did earlier in formulating equivalent circuits for different operational intervals.

We start to calculate the processes at time $t = 0$, when transistor VT1 is switched on and current i_{Lr} begins to flow. This state corresponds to the equivalent circuit for interval I ($0 < t < t_1$), with linear variation in the current i_{Lr}

$$i_{Lr} = \frac{E}{L_r}t. \tag{8.21}$$

At time $t = t_1$, transistor VT1 is switched off at zero voltage, as the voltage at capacitor C_{r1} is zero. This moment, at which transistors VT1 and VT2 and diodes VD1 and VD2 are turned off, marks the beginning of interval II ($t_1 < t < t_2$). The corresponding equivalent circuit contains a resonant circuit with inductance L_r and capacitance C_r; capacitance C_r consists of capacitors C_{r1} and C_{r2} connected in parallel through the internal impedance of input source E. The transient processes in the circuit are determined not only by the sources E and U_R, but also by the initial conditions $u_{Cr1}(t_2)$ and $u_{Cr2}(t_2)$. The circuit parameters are chosen so that the circuit's characteristic impedance $\rho = \sqrt{L_r/C_r}$ is large, and hence the variation in current i_{Lr} due to oscillatory processes over the relatively short interval t_1–t_2 is slight. In this context, we may approximately assume that the current is constant and determines the transient processes in this interval. As a result, the voltage at capacitors C_{r1} and C_{r2} varies almost linearly. At $t = t_2$, the voltage at capacitor C_{r1} increases to E, whereas that at C_{r2} falls to zero and tends to reverse polarity. However, this is prevented by switching on diode VD2. Interval III begins ($t_2 < t < t_3$). In this interval, i_{Lr} falls linearly to zero under the action of voltage U_R. At $t = t_3$, i_{Lr} reaches zero and begins to change direction. Interval IV begins ($t_3 < t < t_4$). In that interval, current flows through transistor VT2, which is turned on at $t = t_3$ because diode VD2 is turned off in response to the change in direction of current i_{Lr}. Of course, a control pulse to switch on transistor VT2 must be formed at that moment. At $t = t_4$, transistor VT2 is switched off, and interval V begins ($t_4 < t < t_5$). It is similar to interval II, as all the semiconductor switches are off and voltage u_{Cr2} begins to rise from zero to E under the action of the opposing current. At $t = t_5$, diode VD1 is switched on, thereby initiating interval VI; diode VD1 begins to conduct current i_{Lr}, which increases linearly from a negative value to zero at $t = t_6$. The equivalent circuits for intervals III and VI are similar; they differ only in the direction of i_{Lr}. These processes then periodically repeat, running through intervals I–VI.

It follows from Figure 8.20c that, as the resonant processes correspond to short intervals II and V, they may be neglected when $f_r \gg f_s$ and the average output voltage can be assumed to be

$$U_R = \gamma E, \tag{8.22}$$

where

$$\gamma = \frac{\Delta t_{\text{on}}}{T_s} \approx \frac{t_1 + (t_6 - t_5)}{T_s}.$$

According to Equation 8.22, pulse-width modulation can be used to regulate the output voltage.

Zero voltage switching considered here for a dc–dc converter can also be employed for an inverter with an inductive load. In that case, the control algorithm for a bridge with one arm containing parallel capacitors corresponds to the formation of alternating voltage. Depending on the number of phases, we may use a half-bridge inverter or, for example, a three-phase inverter based on three arms.

8.3.4 ZVS inverters with an input resonant circuit

One method of switching at zero voltage in an inverter is to create a voltage with high-frequency ripple on the dc side (Rashid, 1988). This entails the excitement of series or parallel oscillatory circuits by periodic switching so as to produce oscillation of the input voltage, which passes through zero at the resonant frequency of the input circuit. We will consider this approach in more detail for the example of the circuit shown in Figure 8.21a.

We assume that the load at the circuit's output is a current source I_{out}. This source may be represented in the equivalent circuit as a load inductance or the output filter of the inverter phase, which considerably exceeds the input inductance L_r. The oscillatory input circuit contains a reactor (with inductance L_r) and a capacitor C_r. Resistor R represents the total active losses in the system. If we create oscillations in the circuit at the resonant frequency and assume that $R = 0$, the capacitor voltage will vary with frequency ω_0 from zero to E, and the mean value of current I_{Lr} in reactor L_r will be I_{out}. With no losses, this process is undamped (Figure 8.21b) and we can write

$$\begin{cases} i_{\text{Lr}} = E\sqrt{\dfrac{C_r}{L_r}}\sin\omega_0 t + I_{\text{out}}, \\ u_{\text{Cr}} = E(1 - \cos\omega_0 t). \end{cases} \tag{8.23}$$

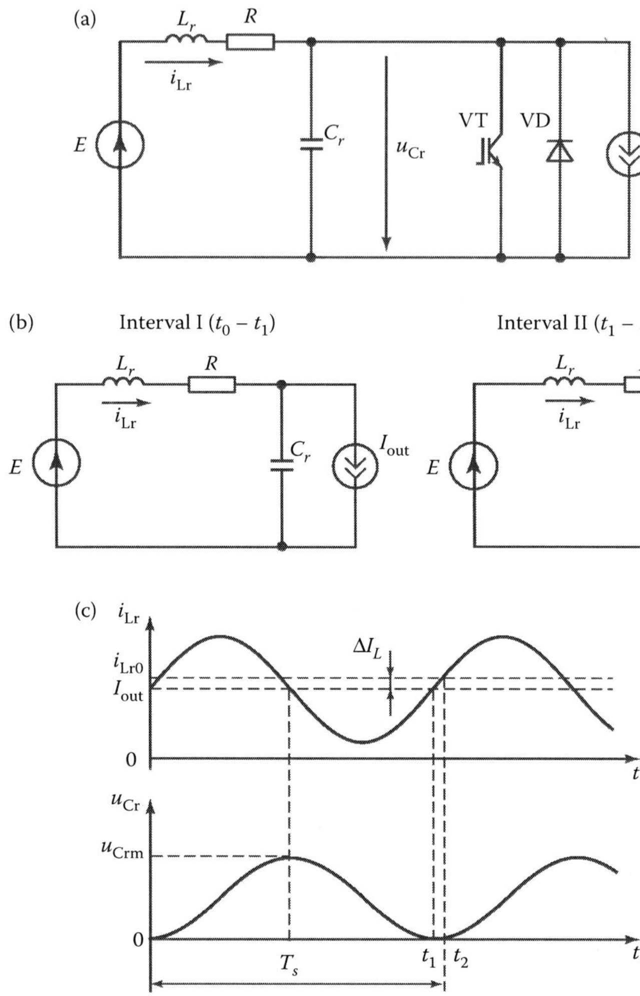

Figure 8.21 (a) An inverter with an input oscillatory circuit; (b) equivalent circuits in different operational intervals; and (c) current and voltage waveforms.

Taking account of the actual losses in the circuit, we see that energy equal to the losses at equivalent resistor R must be introduced in order to sustain the oscillatory process in the system.

In that case, the current can be represented by the approximate expression

$$i_{\mathrm{Cr}}(t) \approx I_{\mathrm{out}} e^{-\delta t}\left(\frac{E}{\omega_r L_r}\sin\omega_r t + \Delta I_L \cos\omega_r t\right), \tag{8.24}$$

where

$$\delta = \frac{R}{2L_r}, \quad \omega_r = \omega_0\sqrt{\frac{1}{L_rC_r} - \frac{R^2}{4L_r^2}}, \quad \Delta I_L = I_{\mathrm{Lr0}} - I_{\mathrm{out}},$$

where I_{Lr0} is the maximum current when the transistor is switched off in order to sustain the oscillations, in light of the power losses R. We calculate ΔI_L so as to permit compensation of the power losses in the resistor R:

$$P_R = \frac{1}{T}\int_0^T i_{\mathrm{Lr}}^2(t)R\mathrm{d}t \approx \frac{\Delta I_L^2 L_r}{2}. \tag{8.25}$$

The oscillatory processes in the circuit (Figure 8.21a) are sustained by periodically switching transistor VT on and off within a near-zero range of voltages at capacitor u_{Cr}. At time $t = t_1$, transistor VT is turned on (Figure 8.21b) and current i_{Lr} in the reactor increases almost linearly under the action of voltage E. At $t = t_2$, the current $i_{\mathrm{Lr}} = I_{\mathrm{Lr0}}$, and transistor VT is switched off (Figure 8.21c). The oscillatory circuit including L_rC_r is restored and oscillation continues. Thus, a quasi-sinusoidal pulsating voltage that varies from zero to almost E appears at the output (Figure 8.21c).

As a result, the inverter input voltage falls periodically to zero. At zero voltage, the switching losses of the inverter components will be a minimum. Obviously, then, the switching strategy should focus on moments when the input voltage passes through zero. That limits the scope for control of the output voltage. Figure 8.22a shows a three-phase inverter bridge circuit with an input resonant link based on the transistor VT0 and L_rC_r circuit. The inverter output voltage is determined by the number of pulsations in each phase when the switches VT1–VT6 are conducting. The switching frequency of transistor VT0 in that case considerably exceeds the frequency of the inverter output voltage (Figure 8.22b). It is evident from Figure 8.22 that this system has the following serious deficiencies:

- The maximum voltage at the inverter switches is greater than the dc input voltage.
- Additional output voltage harmonics appear under the action of the voltage pulsations in the circuit.
- The range of output voltage regulation is limited and discrete.

These problems can be eliminated by modifying the resonant circuit, for example, by using a configuration in which the reactor and the capacitor of the oscillatory circuit are in parallel (Middlebrook, 1978; Nagy, 2002).

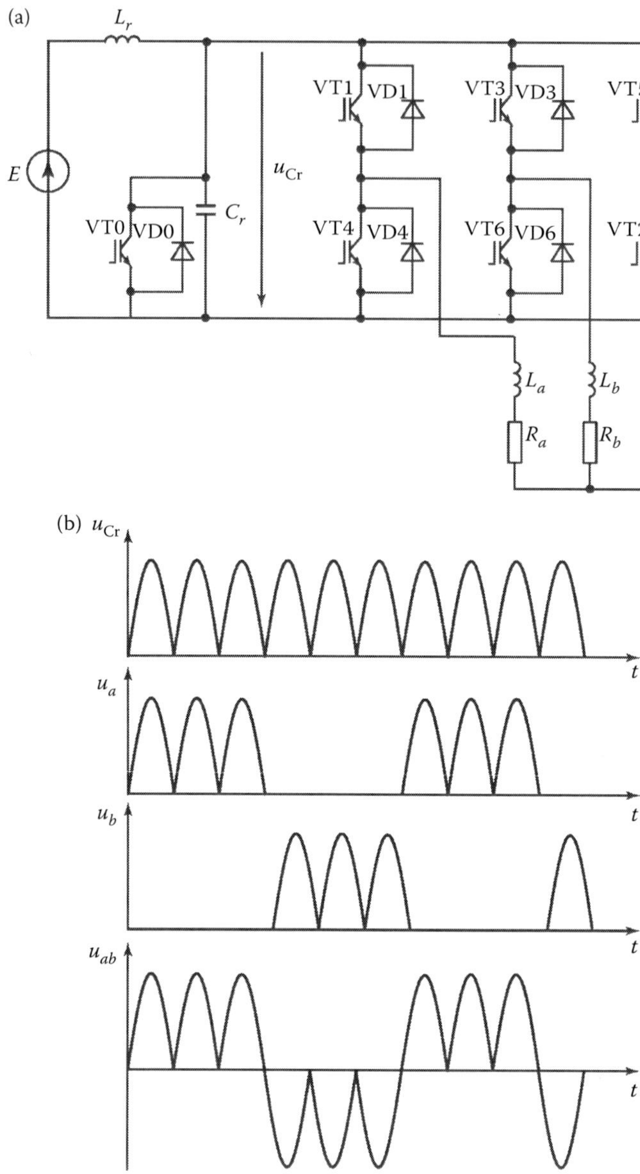

Figure 8.22 (a) An inverter with an input resonant circuit and (b) voltage waveforms.

References

Hui, S.Y. (Ron) and Chung, H.S.H. 2001. Resonant and soft-switching converters. *Power Electronics Handbook*, Rashid, M.H., Ed., pp. 271–305. USA: Academic Press.

Kazimierczuk, M.M. and Charkowski, D. 1993. *Resonant Power Converters*. New York: Wiley-Interscience.

Lee, F.C. 1989. *High-Frequency Resonant, Quasi-Resonant, and Multiresonant Converters*. Blacksburg, VA: Virginia Polytechnic Institute and State University.

Middlebrook, R.D., Cuk, S. 1978. Isolation and multiple output extensions of a new optimum topology switching dc-to-dc converter. IEEE Power Electronics Specialists Conference (PESC '78), Syracuse, New York, June 13–15, p. 256–264.

Nagy, I. 2002. Resonant converters, *The Power Electronics Handbook*, Skvarenina, T.L., Ed., pp. 5-25–5-42. USA: CRC Press.

Rashid, M.H. 1988. *Power Electronics*. USA: Prentice-Hall.

Rozanov, Yu.K. 1987. *Poluprovodnikovye preobrazavateli so zvenom povyshennoi chastity* (*Semiconductor Converters with a High-Frequency Element*). Moscow: Energiya (in Russian).

chapter nine

Multilevel, modular, and multicell converter topologies

9.1 Introduction

In this chapter, we understand a *module* to be a structurally and functionally complete power-electronic device. In modular systems, some additional connections and components of particular type can be used. A *cell* is understood to be a structurally complete device consisting of standard structures. Using these terms, we consider modular design, which permits the connection of functionally different devices, for example, a frequency converter is made by connecting a rectifier and an inverter.

The use of modular and multicell topologies enables to minimize the time and production costs in the design of one-type converters. In addition, this approach is well suited to reducing the higher harmonics of current and voltage and it can also be used for reservation. In general, this approach can be used for the following purposes:

- Increase in the system power with limited parameters of the available components
- Decrease in the design period of new power-electronic devices, with distinctive voltage and current characteristics
- Reservation of devices and their components with the output parameters continuity
- Reduction of higher harmonics in input and output current or voltage
- Matching of the input and output currents and voltages
- Unification and standardization of components

Modular and multicell converters were first used in secondary power sources for autonomous devices, primarily airplanes (Rozanov et al., 2007). This approach is based on the theory of structural–algorithmic synthesis of secondary power sources (Mytsyk, 1989). The same design principles are widely used in the power engineering for the design of dc power systems (Sood, 2001). Various approximations of the harmonic signal were analyzed in Mytsyk (1989). In particular, a signal with N stages was considered (Figure 9.1). Obviously, the distortion in this case depends on two variables: the sampling interval θ_n and the quantization level A_n.

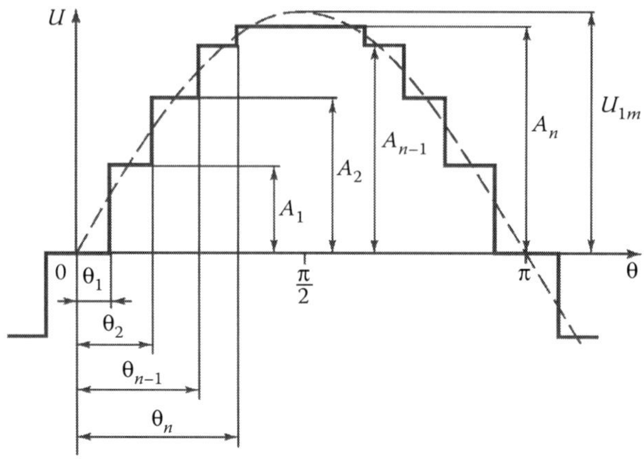

Figure 9.1 Voltage waveform in the case of pulse-amplitude modulation.

Complete minimization of the distortion depends on two variables and entails numerical solution of a system of transcendental equations. The solutions obtained do not lend themselves to technical realization. Therefore, in practice, the relative level is assumed to be a multiple of the minimum level of the first stage. A better method is to combine pulse-amplitude modulation with pulse-width modulation (PWM). In that case, the modulation frequency is minimized, and the number of stages is selected so as to have the best utilization of the components or is determined in accordance with technical and economic optimization.

The most widely used techniques of modular design are the following:

- Parallel connection of ac or dc converters
- Shunt connection of self-commutated inverters
- Multicell topologies of diode–capacitor rectifier circuits of voltage multipliers and voltage dividers
- Multilevel converter topologies
- Cascade converter configurations

9.2 *Parallel connection of rectifiers and dc–dc converters*

Such circuits enable to obtain the following advantages:

- Increase in power
- Reservation of devices
- Improvement of the converter output and input parameters
- Increase in modulation frequency of dc–dc converters

The first and second options basically involve the parallel operation of converters with a dc output (Rozanov, 1992). Note that it is simpler to organize parallel operation for dc–dc converters than for ac–dc converters, as it involves regulating of one or two parameters of the output voltage (or the output current, where necessary). Therefore, we focus here on parallel operation of ac–dc converters. The requirements on the parallel converters will depend on the required function. Thus, for complete reservation (one operational unit and one auxiliary unit), it is sufficient to ensure stable operation of two converters at a common bus without any constraint on the load power distribution between these modules. This follows from the complete reservation principle, when the maximum power of the consumer does not exceed the maximum power capacity of each converter. Depending on the type of consumer, the structure with parallel modules for reservation function can be replaced by a structure in which the output buses of one of the modules operating in cold or hot reserve can be switched at a special command. With partial reservation (e.g., two operational units and one reserve unit) or in a modular circuit with a function of power increase, the power distribution between the parallel converters must not lead to overload of a converter.

In general, there are several types of parallel operation.

1. Operation at common buses with an arbitrary power distribution between the separate converters, provided that the load power does not exceed the rated power of each converter (sometimes known as combined operation rather than parallel operation).
2. Operation at common buses with a distribution of the load power, such that it is proportional to the rated power of each converter. In the case of equal converter power, the load distribution will be balanced.
3. Operation at common buses with an arbitrary load power distribution between the separate converters, except that the load of each converter is limited to the rated overload power (power capability).

Parallel operation of ac converters is associated with additional conditions that will be considered separately, along with the organization of three-phase systems based on single-phase modules.

For dc–dc converters, the simplest form of parallel operation is the connection of converters to common buses through decoupling diodes (Figure 9.2). The diodes primarily protect the output buses from short-circuits within a separate converter. With identical output parameters and precise adjustment of the output voltage control channels, relative load balancing between the converters is obtained. In practice, however, it is difficult to ensure equal load distribution under various operating

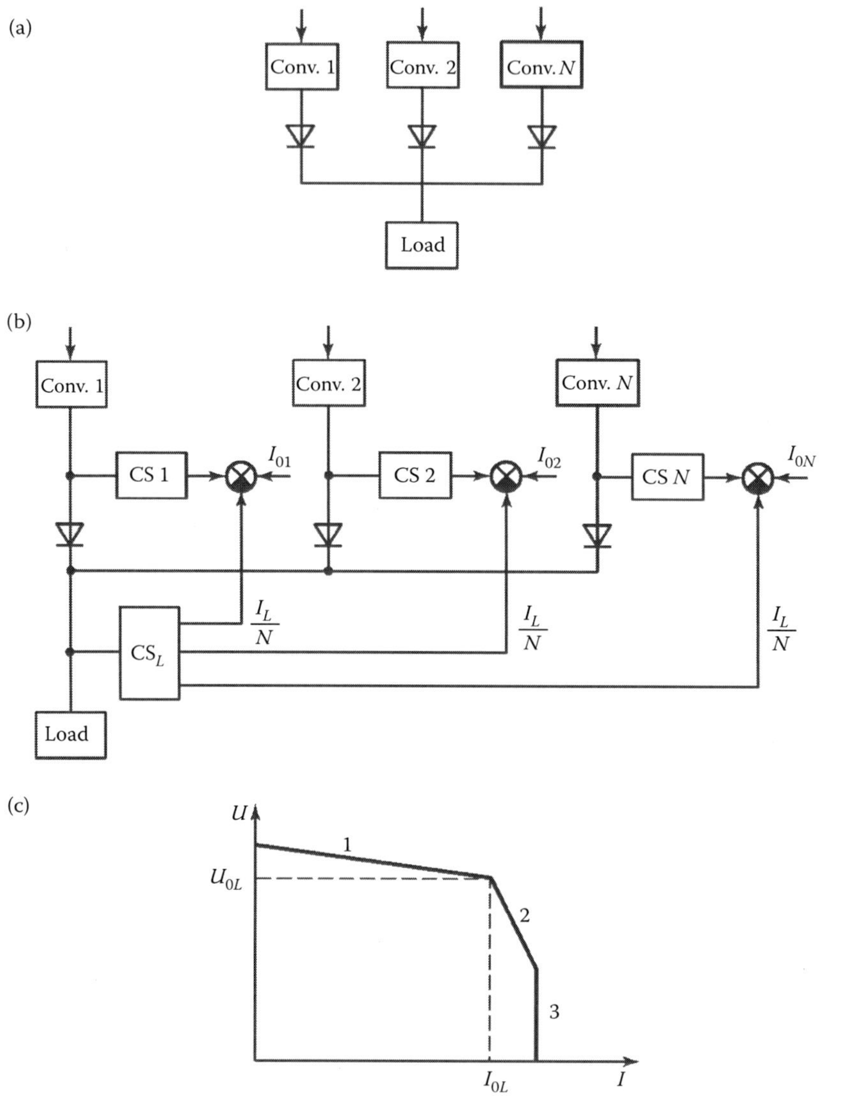

Figure 9.2 Parallel connection of dc–dc converters: (a) circuit with diode decoupling; (b) circuit with load current feedbacks; and (c) output (external) converter characteristic.

conditions, mainly because adjustable converters usually have precise output voltage stabilization. Accordingly, for such operation, we require that

$$I_L \leq I_N, \tag{9.1}$$

where I_L is the load current and I_N is the rated current of converter N (Figure 9.2a).

The second type of parallel operation can be achieved by using feedback between the load and each converter (Figure 9.2b). In that case, the signal from the load current sensor CS_L is split and distributed among the feedback channels and is compared with the current signal of a separate converter. The difference of currents is sent to the output voltage control system. As a result, the load current can be distributed between converters, according to the converter currents. In particular, equal distribution of the load current between the converters is possible ($I_N = I_L/N$). The static accuracy of the distribution will be determined by the current feedback gain. In practice, in such systems, the distribution accuracy is about ±10% when the load of each converter is not more than $0.5I_L$. To ensure stability of parallel operation and to eliminate self-oscillation, we should attend to the design of the frequency-dependent parameters in the current control channels.

The third type of parallel operation is possible for converters whose output characteristic corresponds to the curve shown in Figure 9.2c. Section 1 corresponds to operation with slight output voltage variation. When the rated load is attained, the converter operates with more rapid voltage variation (section 2). In overloads and external short-circuits, the converter operates with stabilization of the output current (section 3) or is switched off. In parallel operation of converters with such a characteristic, the output voltages will be equal, with an accuracy determined by the converter parameters (in practice, within the voltage stabilization zone). The converter structure corresponds to Figure 9.2a. In general case, the load is unbalanced between the converters. However, when the rated load of one of the converters is obtained, its operation will correspond to section 2, and its output voltage will begin to decrease. With a subsequent increase in the load, it is applied to the other converter and so on. This is the most promising approach of parallel operation, as it is not necessary to have additional feedbacks and there are not any structural or circuit constraints on the modular configuration.

The converter begins to operate in a limited current mode (section 2 in Figure 9.2c) when the output current exceeds the specified value and the current limit loop in parallel with the primary voltage feedback is switched on.

In Section 4.2.2, we considered in detail the parallel and series operations of two three-phase bridge converters. In a parallel configuration, the current is doubled, without changing the transformers or semiconductor components; in series circuits, in the same conditions, the voltage is doubled. However, it is more common to improve the spectrum of the input current and output voltage. When connecting two bridges in parallel, a three-step input current waveform is obtained, and the higher harmonics is reduced. In the output voltage, the ripple is reduced, and the frequency of its fundamental component is doubled. Obviously, when three, four, or more rectifiers are connected together, with appropriate changes in the transformer windings configuration, we obtain improved spectrum of the input current and output voltage, which become equivalent to those generated by rectifiers with 18, 24, or more phases. However, it is not usual to employ more than four rectifiers, on account of difficulties in ensuring the specified transformation ratio.

The benefits of parallel operation of converters with a common filter are also evident. In such a circuit (Figure 9.3), each converter operates with a phase shift T/N of the control pulses, where N is the number of cells in

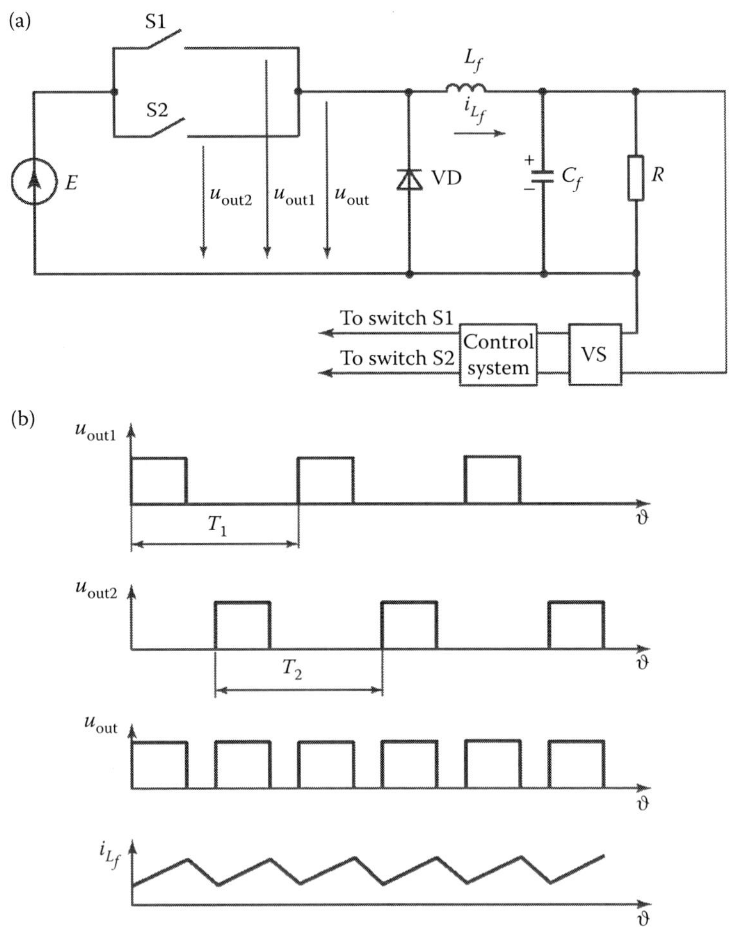

Figure 9.3 Parallel operation of dc–dc converter cells with common filter: (a) circuit diagram and (b) voltage and current waveforms.

parallel, and the modulation frequency and the filter size can be reduced by a factor of N. As a result, the commutation frequency of each switch can be increased to the maximum permissible value on the assumption that the filter is able to eliminate ripple over the expanded frequency range. In that case, the basic converter cells containing the switches will operate in parallel. Other types of converter circuits can also be combined, so as to increase the modulation frequency.

9.3 Parallel connection of inverters

As already noted, it is more difficult to ensure parallel operation of ac modules, because that entails their voltage synchronization (Rozanov et al., 1981; Rozanov, 1992). Figure 9.4a shows a simplified equivalent circuit of two parallel single-phase voltage-source inverter modules, under the assumption of sinusoidal output voltage. That is ensured by means of output filters whose inductance L_f is shown in Figure 9.4a. When the vectors U_{inv1} and U_{inv2} are equal, there is no circulating current between the modules (Figure 9.4b). When the vectors are of the same phase but different amplitude, reactive circulating current I_{cir} appears. For the higher-voltage module, it is equivalent to an inductive load. This current can be calculated as

$$I_{cir} = \frac{\Delta U_{inv}}{2X_{L_f}} \cong \frac{k \cdot \Delta U_d}{2\omega L_f} \tag{9.2}$$

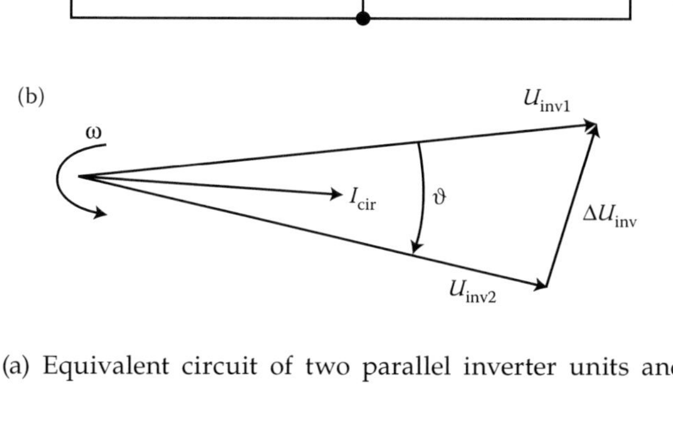

Figure 9.4 (a) Equivalent circuit of two parallel inverter units and (b) vector diagram.

where ΔU_{inv} is the modulus of the vector difference of the RMS values of the first harmonics of the inverter output voltages U_{inv1} and U_{inv2}, ΔU_d the difference in average input voltages, and k the circuit coefficient of the relation between the fundamental output voltage U_{inv} and the average input voltage U_d.

With a phase difference between the vectors U_{inv1} and U_{inv2}, there is an active component of the circulating current between the modules. With a small phase difference, when $U_{inv1} = U_{inv2}$, the circulating current is approximately

$$I_{cir} \approx k \cdot \frac{U_d \, tg\vartheta}{2\omega L_f} \tag{9.3}$$

where ϑ is the phase difference.

Active power is supplied by the module in which voltage U_{inv1} leads, whereas the module with lagging voltage consumes active power (from the buses). This is analogous to the appearance of balancing current in parallel synchronous generators.

Thus, to ensure parallel operation of the inverter modules, we need to regulate the amplitude of the output voltages and ensure that they are synchronous and in phase. Different methods are employed here, depending on the inverter circuit. In modular design, it is expedient to use single-phase modules based on voltage-source inverters.

Various methods can be used to regulate the output voltage in such inverters. In particular, we can regulate the input dc-side voltage by means of a dc voltage controller. In that case, it is expedient to use additional data connection between the modules. To that end, an additional informational bus on the dc side is used in the parallel modules, so as to permit equalization of the inverter output voltages and hence the amplitude of the first harmonic of these voltages.

Synchronization of the inverter modules ensures that the switching functions are in phase. The corresponding algorithm in voltage-source inverters is specified by an accounting and distributed unit, which receives the pulses from the reference oscillator. A special synchronous coupling system is employed for this purpose. On the basis of the standard requirement for uninterruptible power supply systems that any single fault must not disrupt normal operation, the following factors must be taken into account in the synchronization system:

1. The absence of structurally combined centralized devices. However, this does not rule out parametric control.
2. Return of any module to operation (after elimination of the fault) in the system, without reorganization of the links or phase discontinuity.

3. Maintenance of the relations within the system (and equal vector phases) and the frequency in the case of interruption of generation or frequency variation of any reference oscillator and also when any single synchronizing link is broken.

Reference oscillators can be synchronized both by means of a phase-locked loop (PLL) and by direct pulse synchronization. In direct synchronization, a relaxation self-oscillatory circuit is used as the reference generator. In that case, the frequency of the synchronizing signal does not need to exceed the base frequency of the reference oscillator, as it is usually required for such devices. The feature of direct synchronization is the decrease in frequency of the reference generator if the synchronization circuit forms a ring from the output to the input. In that case, the synchronization ring can cover one or more reference oscillators. Thus, in direct synchronization, maintaining constant frequency entails maintaining a closed synchronization ring or an open circuit (i.e., in the corresponding graph figure, maintaining a tree structure of graph). In direct synchronization of a relaxation reference generator, pulses act directly on the switching element of the generator. Therefore, the synchronized generators will be accurate in phase. It only depends on the time delays of the pulsed circuits and switches.

In synchronization by means of a PLL, the reference generator has a smoothly tunable frequency. A control system adjusts this frequency as a function of the sign and magnitude of the phase difference between the synchronizing signal and the output signal of the reference generator. Here, in contrast to direct synchronization, the frequency of the synchronizing signal can be less than the output frequency of the reference generator. The frequency of the PLL reference generator after capture by the synchronization ring will be the same as that in autonomous (unsynchronized) operation. If several PLL generators are closed by the synchronization ring, a frequency becomes close to the average frequency of the operational generators. In structures with a unidirectional system, in both direct synchronization and synchronization by means of a PLL, the whole reference generator system operates at the frequency of the primary generator.

A set of possible synchronization structures arises on account of the contradictory requirements of reliability and structure simplicity, which can be formulated as follows:

a. Minimum number of informational links between the inverter modules
b. Same structure of all the modules
c. Maximum symmetry of the informational links and the dependent operational modes of the modules
d. Minimum complexity of the hardware

Several typical structures of a synchronous link system for in-phase synchronization are shown in Figure 9.5. These structures do not contain centralized devices and hence meet the fundamental reliability requirement. The switches in the signal circuits are assumed to be built into each module. As a centralized device can be regarded a synchronizing bus, its reliability can readily be ensured by structural means. The reliability and simplicity requirements are best met by the following structures: a simple ring for PLL system (Figure 9.5b) and a ring with bypasses (Figure 9.5d) and a simple line (Figure 9.5a) and a line with bypasses for direct synchronization (Figure 9.5c).

In the PLL ring structure, the frequency is close to the average value of the generator base frequencies. Therefore, any variation in the base frequency of a single reference generator will affect the whole system. With a small number of generators, the frequency change can be considerable. To determine the faulty reference generator, we can use a combination of a frequency sensor and a sensor identifying the sign of the generator fault. For example, if the frequency is less than the nominal value and the phase of the reference generator lags that of the corresponding synchronizing signal, the generator base frequency will be too low. However, this method requires relatively large phase variability for the PLL synchronization.

The direct synchronization ring (Figure 9.5b) operates at the highest frequency produced by any of the reference generators. The generator producing that frequency will provide the synchronizing signals that trigger the other generators; it can be regarded as the leading generator (master generator). If for some reason its frequency falls below that of another generator, that generator will become the leading generator, without any disruption of the normal operation of the ring. With an increase in the leading generator frequency, the whole ring begins to operate at that frequency. In that case, the phase lag of the other generators becomes less, and its reliable identification will be difficult in the presence of interference.

It is simpler to prevent frequency variation of a single generator when using a line structure (Figure 9.5c). In this structure, there is always a leading generator that uniquely determines the frequency of all the others. Any frequency fluctuation will be due to frequency deviation of this leading

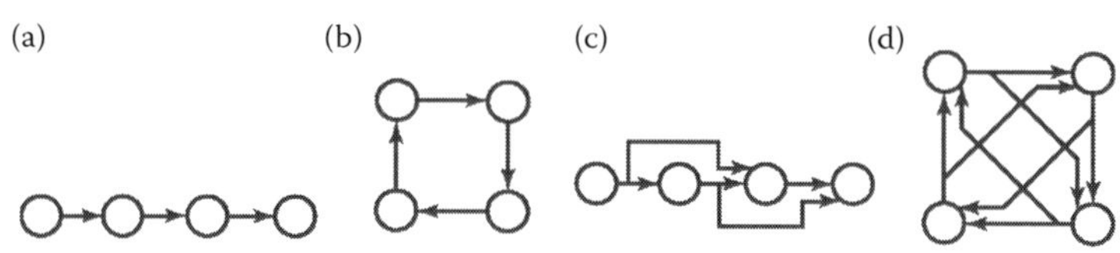

Figure 9.5 Structures of synchronization systems: (a) line; (b) ring; (c) line with bypasses; and (d) ring with bypasses.

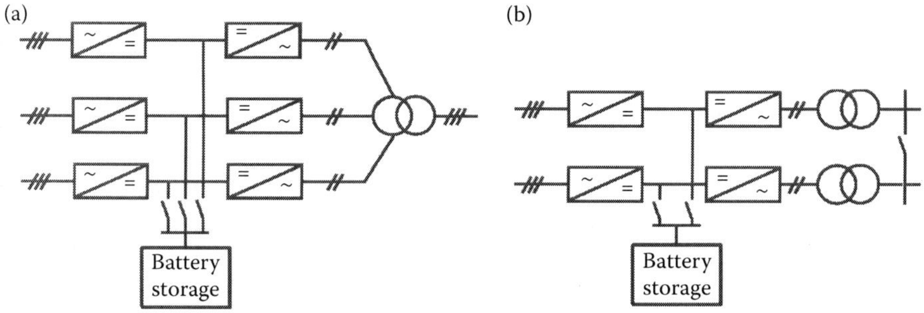

Figure 9.6 Modular topologies of uninterruptible power supplies: (a) a three-phase system of single-phase modules and (b) a duplicate single-phase system.

generator. Frequency deviation of the other generators will not affect system operation as single-phase synchronization will be maintained.

A three-phase power supply system can be assembled from three single-phase inverter modules with the appropriate phases. The output voltage of each module is the phase voltage and is supplied to uncoupled primary windings of a three-phase transformer or a group of single-phase transformers, or else the outputs of the modules are combined directly into a three-phase system, with three or four buses.

Inverter modules can also be connected in order to improve the harmonic composition (spectrum) of the output voltage by pulse-amplitude modulation. In that case, the overall power of the system increases, and reservation of separate modules can be ensured. The output voltage is analogous to the waveform shown in Figure 9.1, except with considerable quantization ratios of the length and height of the voltage steps. Steps with the same output voltage frequency are in parallel. The length of the steps is determined by the frequency divider of a reference generator frequency. Of course, within each step, PWM can be organized. A generalized structure of the inverter modules connection is shown in Figure 9.6.

9.4 Voltage multipliers and voltage dividers based on capacitor–diode cells

The output voltage can be increased or decreased by means of capacitor–diode cells. These are also known as transformer-free circuits (Zinov'ev, 2003). In a conventional transformer circuit, a high-voltage transformer is used to increase the voltage. Systems consisting of low-voltage cells in series are preferable in engineering and economic terms. In voltage dividers, capacitors connected in series and that have the ability to distribute high voltage are employed. Therefore, the transformer will be used for galvanic isolation of the load and consumer circuits and can be eliminated from the circuit whenever possible.

9.4.1 Voltage multipliers

A simplest symmetric voltage doubler circuit is shown in Figure 9.7. During one half-period, the grid is connected through a diode to capacitor C1 and it is charged up to the amplitude of the input voltage. During the negative input voltage half-period, the capacitor C2 is charged. As a result, the load voltage is equal to the double voltage of the secondary windings of transformer *T*. The voltage ripple can be determined by the method developed for rectifiers with a capacitive output filter, taking account of the series connection of capacitors C1 and C2 during discharge. The average output voltage is approximately equal to the double input voltage. Obviously, the ripple level and the average voltage depend on the load value.

In Figure 9.8a, an asymmetric voltage doubler circuit is shown. In the negative half-period ($u_{ab} < 0$), diode VD1 is on, and capacitor C1 is charged up to the voltage amplitude of the secondary transformer winding. During the other half-period, capacitor C2 is charged from two sources in series: capacitor C1 and the secondary winding, through diode VD2. Thus, we obtain approximate doubling of the voltage. Using an asymmetric circuit cell, we can multiply the voltage by a factor of $2N$ for a system consisting of N cells. For example, when $N = 3$, we will have a six-time voltage-multiplying circuit (Figure 9.8b). The voltage ripple strongly depends on the load, which is typical for rectifiers with a capacitive filter.

Using a full-wave rectifier circuit, the capacitors in each cell will be charged alternately. In that case, we obtain a voltage multiplier based on single-phase bridges. This circuit consists of half-wave multipliers based on a six-time voltage-multiplying circuit (Figure 9.8b). In this circuit, the output capacitors are charged in each half-period that corresponds to the operation of bridge circuits.

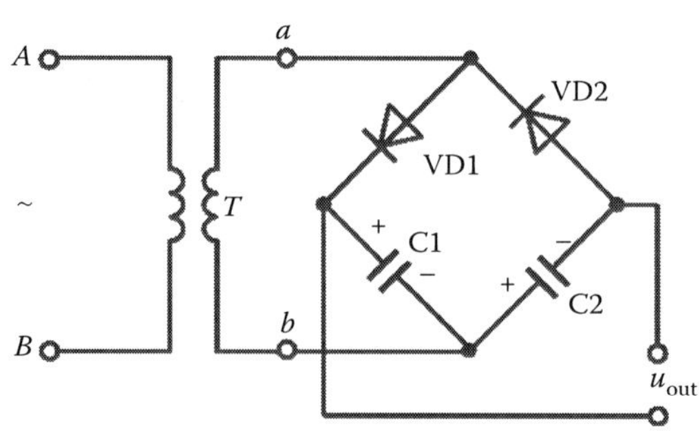

Figure 9.7 A single-phase symmetric voltage doubler.

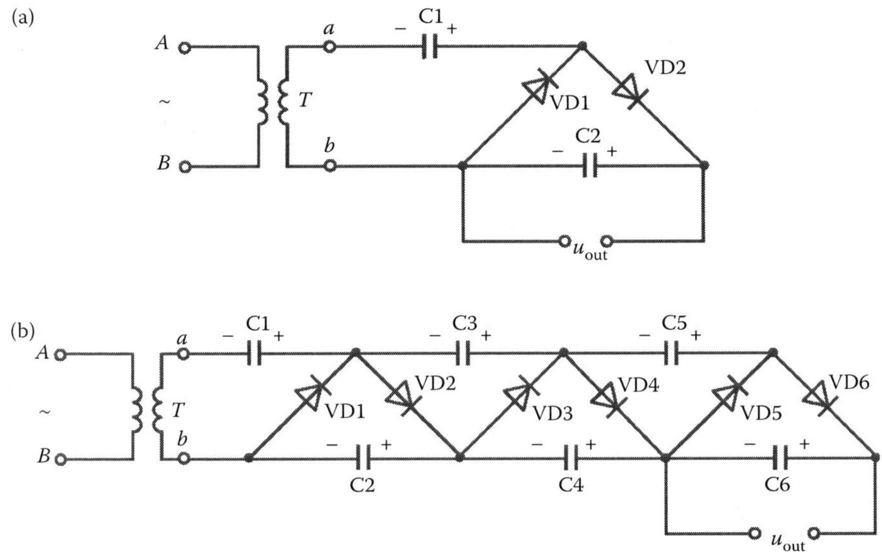

Figure 9.8 Asymmetric voltage multipliers: (a) voltage doubler and (b) six-time voltage-multiplying circuit.

In voltage-multiplying circuits, three-phase voltage can also be used. In that case, we can employ a series connection of capacitors and three-phase bridge diode cells.

9.4.2 *Voltage dividers*

Self-commutated switches are employed in diode–capacitor voltage divider circuits. In one time period, the diode–capacitor cells are connected in series to a high-voltage source. In the next period, these cells are connected in parallel to the load using switches. In Figure 9.9, a simplified voltage divider with self-commutated switches S1 and S2, and S3 and S4 is shown. Both switch pairs operate synchronously, without overlap of the periods in which they are on. When S1 and S2 are on, and S3 and S4 are off, the diode–capacitor cells are in series and capacitors C1, C2, C3, and C4 are charged by the high rectified voltage. The charging current is mainly limited by the internal impedance of the grid and by the resistance of the diodes that are on. When switches S3 and S4 are on, and S1 and S2 are off, the capacitors of all the cells are connected in parallel to the load R_d. The charging current of the output capacitor C_d is determined by the voltage difference between the charged capacitors and capacitor C_d. Obviously, it depends on the load resistance R_d. When the voltages at the cell capacitors and output capacitor are equal, switches S3 and S4 will be

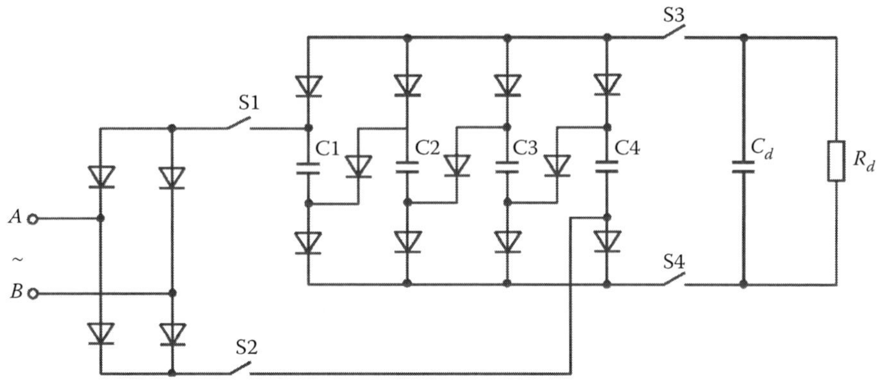

Figure 9.9 A diode–capacitor voltage divider.

turned off, while the input switches will be turned on. By adjusting the duty ratio of switches S1–S4, we can regulate the output voltage.

The main drawbacks of diode–capacitor voltage dividers and voltage multipliers are the low power factor and distortion of the input current, which are typical problems of rectifier circuits with a capacitive filter.

9.5 *Multilevel converter structures*

In order to have high nominal voltage of converters, it is necessary to connect switching devices in series. At the same time, the use of multilevel circuits not only increases the voltage relative to the rated voltage of separate switches, but also improves the harmonic spectrum of currents and voltages. Essentially, pulse-amplitude modulation and PWM can be combined in multilevel converters. At the same time, the switch values of di/dt and du/dt are reduced, and the noise is reduced. In addition, the combination of pulse-amplitude modulation and PWM allows a decrease in the modulation frequency. The term "multilevel" can be used for both inverters and rectifiers. The capabilities of different multilevel topologies are different, and only technical and economic study permits us to select one or another topology. In this section, the basic converter topologies for the example of multilevel inverters are considered (Corzine, 2002). These converters are applied in electric drives for frequency conversion and in the electric power industry for flexible ac transmission lines based on static compensators, active filters, and so on.

Two basic topologies are employed for multilevel converters.

- Series connection of capacitors
- Series connection of single-phase bridge cells

In the first case, diode-clamped circuits and circuits with clamping of capacitors by switches are considered. Also, there are different topologies of cascaded converter circuits.

9.5.1 Diode-clamped circuits

In Figure 9.10, a three-level inverter circuit is presented. The operation of these circuits can be illustrated by considering the equivalent circuits for one arm (e.g., phase a), as shown in Figure 9.11. The three states of the transistors correspond to the following equivalent circuits:

I. The two upper transistors VT11 and VT12 are off, whereas VT41 and VT42 are on.
II. Transistors VT11 and VT42 are off, whereas VT12 and VT41 are on.
III. The two lower transistors VT41 and VT42 are off, whereas VT11 and VT12 are on.

The voltages of capacitors C1 and C2 are equal to half of the input voltage. If we assume only amplitude modulation, we obtain a square wave output voltage in phase a. In case I, the voltage is zero relative to

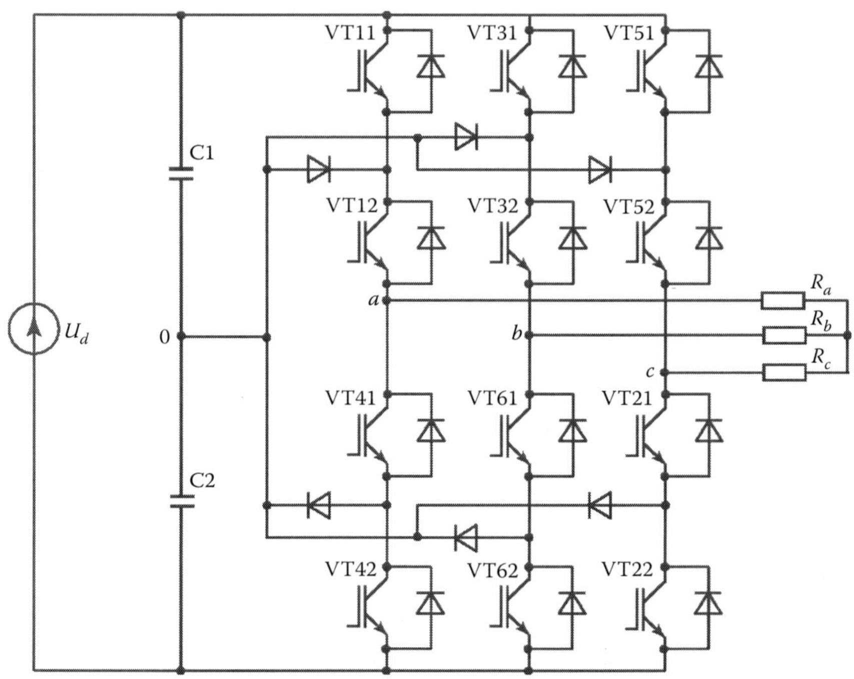

Figure 9.10 A three-phase three-level diode-clamped inverter.

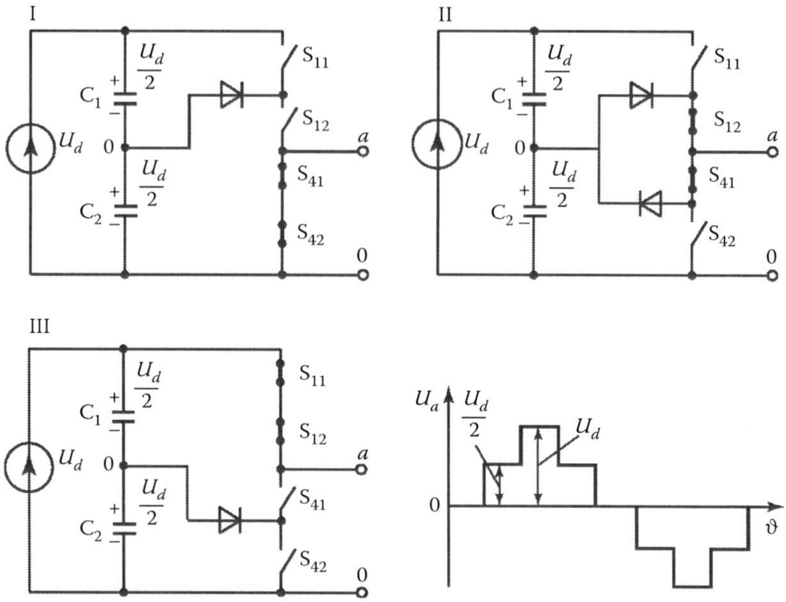

Figure 9.11 Equivalent circuits of one arm and the phase voltage waveform for a three-level inverter.

the bridge lower bus. In case III, it is equal to the input voltage U_d. In case II, the output voltage is equal to the voltage at capacitor C1 or C2. Thus, three voltage levels are ensured: 0, $U_d/2$, and U_d. As the voltage $U_d/2$ is switched, the switching losses are reduced, and the output voltage spectrum is improved. The voltage at the transistors is $U_d/2$, assuming the equal capacitor voltages. However, it is difficult to balance the voltages at the capacitors. This problem is solved by voltage level modulation. Of course, the number of levels can be increased, but that increases the difficulty in voltage balancing.

Figure 9.12 shows the output voltage waveforms for three- and four-level inverters with PWM, which can be realized using different techniques. In particular, space-vector modulation can be used. Obvious drawback of a three-phase diode-clamped inverter is the large number of transistors and the need for clamping diodes. However, the large number of switches ensures the increase in the converter power and the reducing of higher harmonics in the output voltage. As for the diodes, they are simpler and cheaper than the controlled switches. However, an increase in the number of voltage stages leads to an increase in the voltage at the clamping diodes of the first stage. Consequently, they must be connected in series, or else higher-voltage diodes must be employed. Diodes can be eliminated if the controlled switches of the circuit are used for clamping.

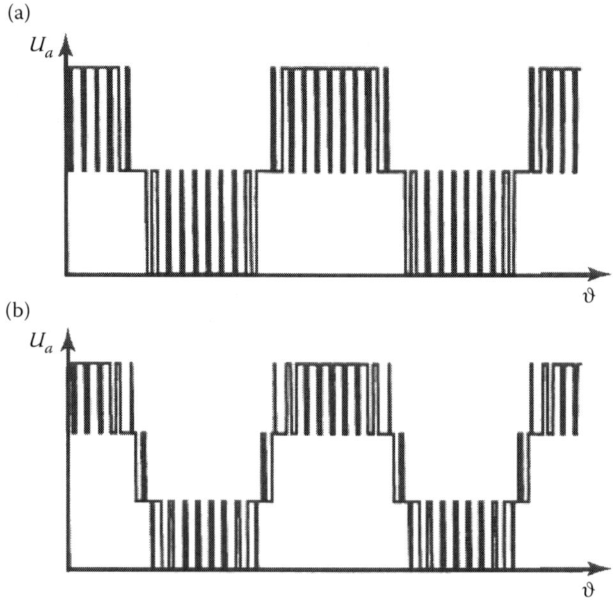

Figure 9.12 Phase voltage waveforms for three-level (a) and four-level (b) PWM converters.

This is a modification of the circuit considered. It is known as a flying capacitor (Corzine, 2002).

9.5.2 Flying-capacitor inverters

In Figure 9.13, an example of a three-level flying-capacitor inverter circuit is shown. The capacitor C is charged to $U_d/2$ and, because it is connected in series with the dc source, ensures a third level in the output voltage. As a result, the upper and lower transistors form a voltage stage of magnitude $U_d/2$, as capacitor C is in series with the phase output. Obviously, it is again difficult to ensure voltage balancing in this case. An increase in the number of capacitors leads to an increase in the number of output voltage levels. For example, for a five-level inverter, the number of capacitors is increased by a factor of 6; the capacitors are connected in series here. The balancing of the voltages is also complicated. In addition, greater capacitance is required than for the other circuits.

9.5.3 Multilevel cascaded converters

Cascaded topology is prospective for multiphase circuits. It is based on addition of separate dc power supply voltages. In these converters,

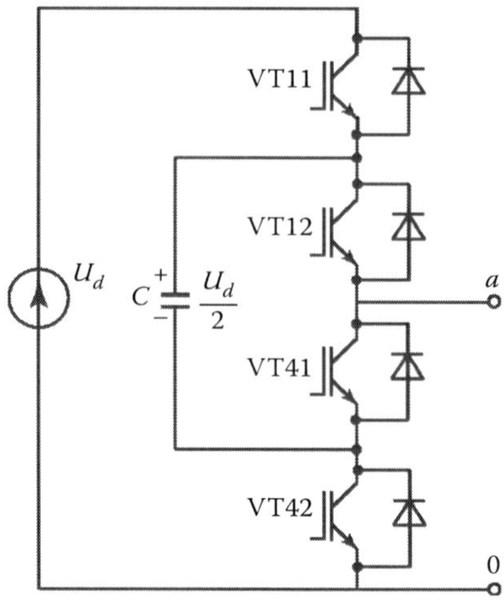

Figure 9.13 Single arm circuit of a three-level flying-capacitor converter.

renewable source elements can be used, for example, a solar battery or a fuel cell, but ordinary rectifiers with input transformers are most commonly employed (Rodriguez et al., 2003). A diode rectifier circuit permits the simple design of such converter cells. It enables to design a multiphase rectifier with stepped input current. In Figure 9.14, one cell of a multilevel multiphase converter is shown. Using three cells in each phase, we can obtain an 18-pulse system on the basis of transformers whose secondary windings have different configurations. Also each cell contains a single-phase voltage-source inverter. Its output voltage varies in sign and magnitude from 0 to U_d. Applying PWM, we obtain a quasi-sinusoidal voltage with low modulation frequency. An increase in the number of stages

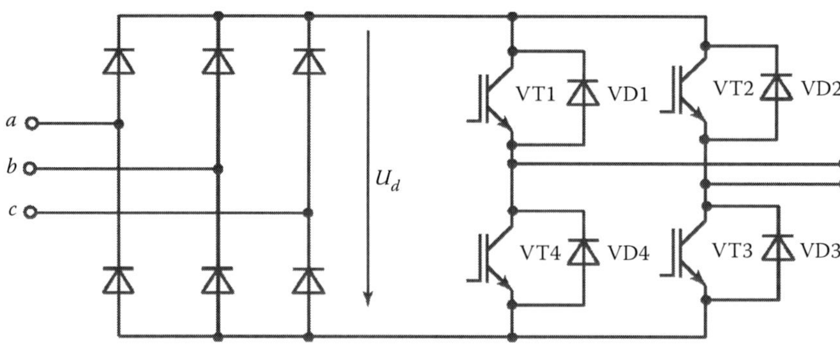

Figure 9.14 Multilevel multiphase converter cell.

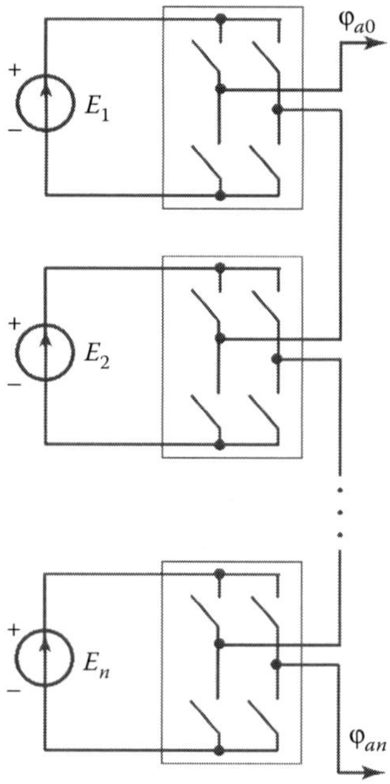

Figure 9.15 Single-phase structure of a multilevel cascaded inverter.

results in a desired approximation of the sinusoidal output voltage on the basis of pulse-amplitude modulation.

Using a cascaded single-phase full-bridge converter, we can obtain a multilevel structure with a required number of levels (Figure 9.15). Each bridge inverter cell generates three-level output voltages. A cascade connection of bridges enables to have multilevel output voltage in each phase. Obviously, the selection of stages and level number is a typical multifactor optimization problem. Its solution depends on the output voltage level, the requirements on the modulation frequency, the application area, the control requirements, and other factors.

References

Corzine, K. 2002. Multilevel converters. *The Power Electronics Handbook*, Skvarenina, T.L., Ed., pp. 6.1–6.21. USA: CRC Press.

Mytsyk, G.S. 1989. Osnovy strukturno-algoritmicheskogo sinteza vtorichnykh istochnikov elektropitaniya (*Principles of Structural–Algorithmic Synthesis of Secondary Power Sources*). Moscow: Energoatomizdat (in Russian).

Rodriguez, J., Moran, L., Pontt, J., Correa, P., and Silva, C. 2003. A high-performance vector control of an 11-level inverter. *IEEE Transactions on Industrial Electronics*, 50(1), 80–85.

Rozanov, Yu.K. 1992. Osnovy silovoi elektroniki (*Fundamentals of Power Electronics*). Moscow: Energoatomizdat (in Russian).

Rozanov, Yu.K., Alferov, N.G., and Mamontov, V.I. 1981. Inverter module for uninterruptible power supplies. *Preobrazovatel'naya Tekhnika*, 1 (in Russian).

Rozanov, Yu.K., Ryabchitskii, M.V., and Kvasnyuk, A.A. 2007. Silovaya elektronika: Uchebnik dlya vuzov (*Power Electronics: A University Textbook*), Rozanov, Yu.K., Ed. Moscow: Izd. MEI (in Russian).

Sood, V.K. 2001. HVDC transmission. *Power Electronics Handbook*, Rashid, M.H., Ed., pp. 575–596. USA: Academic Press.

Zinov'ev, G.S. 2003. Osnovy silovoi elektroniki (*Fundamentals of Power Electronics*). Novosibirsk: Izd. NGTU (in Russian).

chapter ten

Applications of power electronics

10.1 Improvement of the efficiency of power supply

10.1.1 Control of power transmission and power quality

10.1.1.1 Control of ac power flows

Power lines are characterized by distributed parameters, corresponding to active and inductive impedances in series and active conductance and capacitance in parallel. Usually, in overhead ac lines, the inductive impedance exceeds the active impedance, whereas the capacitance exceeds the active conductance. Therefore, to simply the calculations, we consider an idealized line without active losses. In that case, the basic parameter is the wave impedance per unit length. When a matched load is attached to the end of a loss-free line, with wave impedance, all the energy of the incident wave supplied to any line segment is absorbed in the load. At a small load, when its impedance exceeds the wave impedance, the capacitance of the line is overcompensated, and the voltage is increased. With overloads, the line impedance is inductive, with reduction in the voltage. The excess or deficiency of reactive power must be compensated in order to stabilize the voltage in the line. To simplify the analysis, we adopt a point model of the line, consisting of the voltage sources of the two generators (which produce and receive the power), separated by inductive impedance $X = \omega L$ (Figure 10.1). In that case, the active power transmitted is as follows (Burman et al., 2012):

$$P = \frac{U_1 U_2}{X} \sin \delta, \tag{10.1}$$

where δ is the phase shift between the source voltage U_1 and the consumer voltage U_2.

The voltage of the transmitting generator must lead that of the receiving generator. It follows from Equation 10.1 that the maximum transmitted power will correspond to $\delta = \pi/2$. The power flux may be regulated by the following basic means:

- Variation in the impedance X
- Variation in U_1 and U_2
- Variation in δ

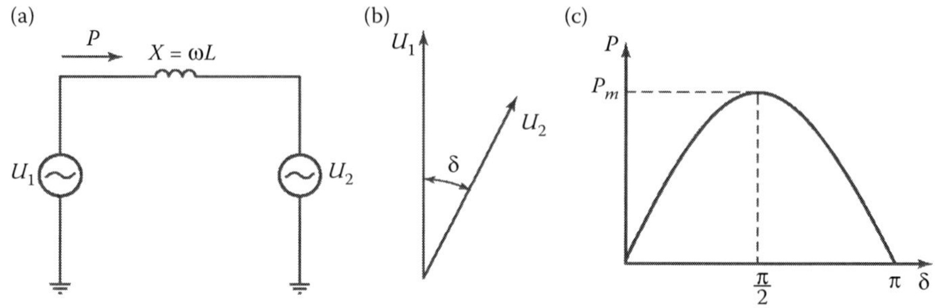

Figure 10.1 Active power transfer in a loss-free line: (a) model of power line; (b) vector diagram; and (c) dependence of the active power on the phase angle δ.

The traditional methods are as follows:

- Compensation of the reactive power by connecting switched reactors or capacitors in parallel with the load
- Compensation by connecting capacitors in series in the line
- Introducing phase shifters, which permit adjustment of δ and hence control of the power flux

10.1.1.2 Reactive-power compensation

The creation of thyristors was the first stage in the development of power electronics, associated with considerable improvement in our ability to control power transmission in ac grids. On the basis of thyristors, high-speed switches and controllers have been created. Figure 10.2a shows a parallel compensation circuit based on capacitors and a thyristor-controlled reactor (Rozanov, 2010).

Antiparallel thyristors permit switching of the reactor, as it is impossible to control thyristors in series with capacitors by natural commutation. In this approach, regulation is based on adjustment of the phase α of the inductive power flux, from zero to the value corresponding to the reactor inductance $Q_L = U^2/\omega L$. In Figure 10.2b, we show the grid voltage and the compensator current with a control angle $\alpha = 2\pi/3$ and also the maximum possible current when $\alpha = \pi/2$.

Combined use of capacitors and a reactor with thyristor control permits smooth regulation of the reactive power, whether it is capacitive or inductive. Note that switches S1–S4 turn on the capacitors at zero voltage. For the thyristor compensator, the range of reactive-power regulation is

$$\frac{U^2\omega C_\Sigma}{U^2/\omega L}, \tag{10.2}$$

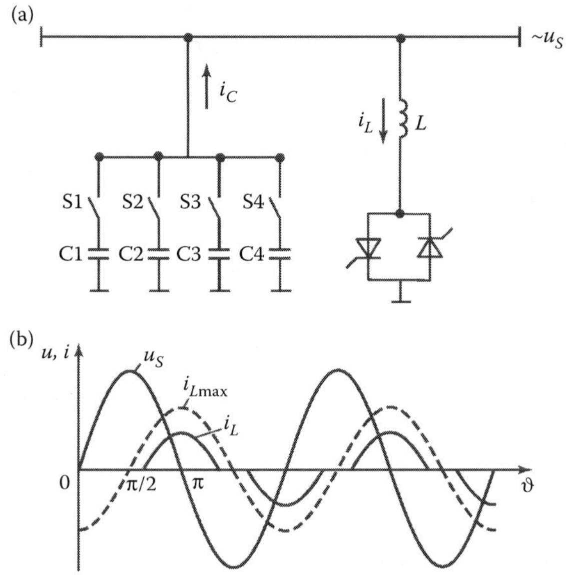

Figure 10.2 Parallel compensation on the basis of capacitors and a thyristor-controlled reactor: (a) circuit and (b) current and voltage waveforms.

where C_Σ and L are the total capacitance in the line and the reactor inductance, respectively, ω the angular frequency of the grid, and U the effective grid voltage.

This device is not only fast, but also reliable. Its main drawbacks are its considerable mass and size (because thyristors designed to control the total compensation power must be employed) and considerable current distortion in the reactor. Various versions of the device are currently in use.

An alternative to series thyristors is the biased reactor. Its benefit is that only low power is required to control the bias current. This is usually a low-voltage, low-current device. It also permits modification of the compensating power by regulating the inductance. However, it is more inertial than the reactor whose current is regulated by antiparallel thyristors.

Two main circuits are used for series capacitor compensation based on thyristors (Figure 10.3). In the first, the capacitive sections are shunted by antiparallel thyristors in series (Figure 10.3a). To eliminate malfunctioning of the semiconducting switches under the action of the capacitors' discharge currents, these devices are switched on when the voltage in the corresponding capacitor section passes through zero. In the second, the capacitors and reactors are connected in series with antiparallel thyristors (Figure 10.3b). The reactors operate analogously to those in the system with parallel compensation. The current is regulated by phase control of

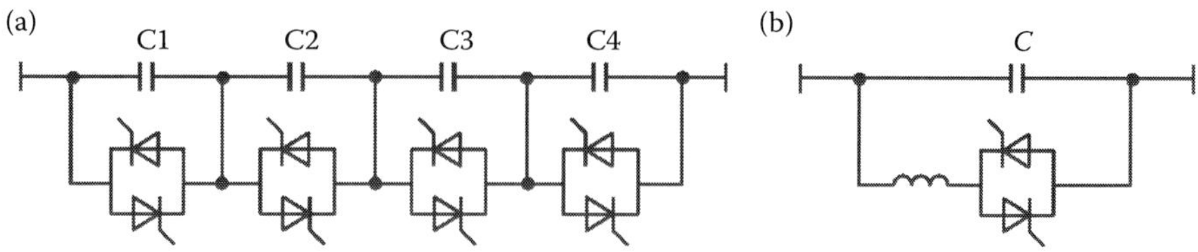

Figure 10.3 Series capacitive compensation: (a) thyristor-switched capacitors and (b) thyristor-controlled circuit.

the thyristors. Depending on the thyristors' control angle, the reactors compensate the series capacitance in the line. In the absence of capacitor compensation, the reactors are switched on to shunt the capacitors and are switched off in total compensation.

10.1.1.3 *Phase shifters*

Traditionally, electromechanical devices with considerable inertia were used to maintain the phase shift between the voltages in power transmission. The speed of phase shifters may be increased on the basis of power electronics. The main application of phase shifters is to regulate δ so as to maintain the power balance in transmission or to increase the stability of transmission systems during transient processes. Electronic switches significantly increase the efficiency of phase shifters.

On the basis of high-power self-commutated devices, such as gate turn-off (GTO) thyristors and power transistors, high-voltage converters can be designed for effective control of power fluxes. On the basis of an ac–dc converter with completely controllable switches, the static compensator has been developed. As interchange of the reactive power is required, electrolytic capacitors are used as temporary energy stores. Such capacitors are included on the dc side of the converter, operating in the voltage-inverter mode, so as to permit reactive-power exchange with the transmission line (Figure 10.4). This compensator is fast and relatively compact. Switching from the consumption of inductive power Q_L to capacitive-power generation Q_C and back takes less than one half-period of the fundamental voltage component. The current distortion due to this compensator is slight, and its performance in transient conditions is excellent.

Due to advances in converters based on self-controlled switches, a new power controller has been developed on the basis solely of electronic power converters. The new device is known as a combined power flow controller. Essentially, the combined power flow controller combines the functions of series and parallel reactive-power compensators and filters. It may be regarded as a universal circuit for the creation of flexible ac lines. Figure 10.5a shows a block diagram of a unified power flow controller connected to a power line. It is based on two ac–dc converters (converter 1

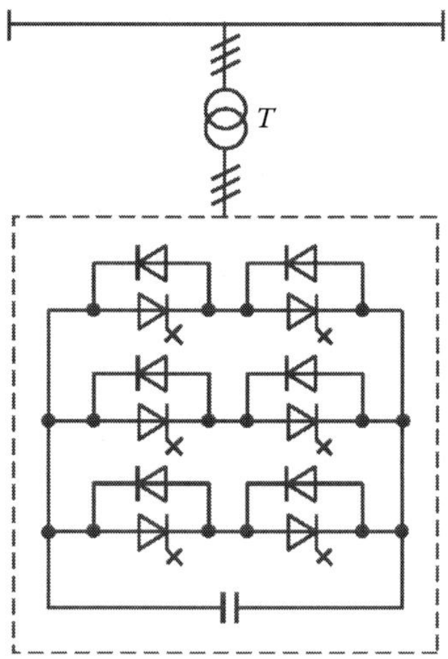

Figure 10.4 A static synchronous compensator.

and converter 2). They are both connected to the same capacitor *C* on the dc side. Converter 1 is connected in parallel to the power line, whereas converter 2 is connected in series with the line, through transformer *T*. As the converters are based on voltage inverters with pulse-width modulation, they may operate in all four quadrants of the complex plane on the ac side. In that case, we may regard converter 1 as a consumer or generator of current whose first harmonic is in any quadrant of the complex plane relative to the voltage vector (Figure 10.5b). Converter 2, whose secondary windings are in series, generates or consumes power on the basis of

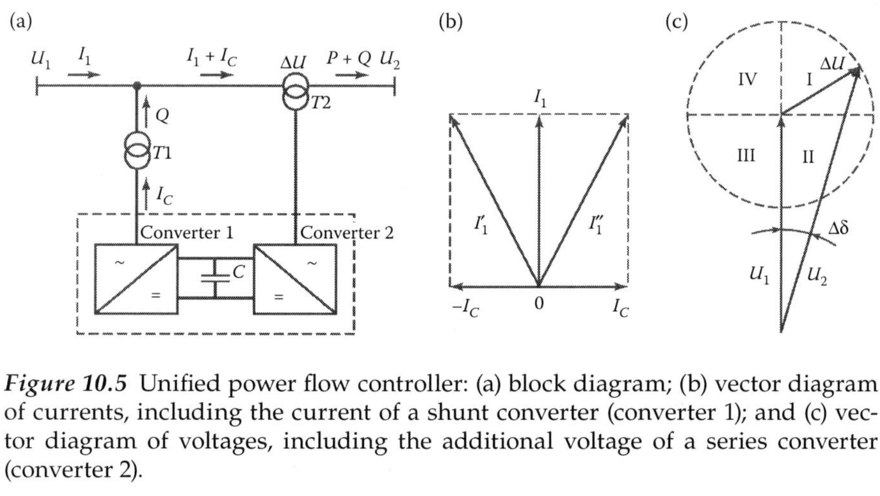

Figure 10.5 Unified power flow controller: (a) block diagram; (b) vector diagram of currents, including the current of a shunt converter (converter 1); and (c) vector diagram of voltages, including the additional voltage of a series converter (converter 2).

the series voltage ΔU, whose first harmonic again can be in any quadrant of the complex plane relative to the current in the transformer winding (Figure 10.5c). Essentially, converter 2 acts as a voltage regulator. It follows from Figure 10.5b and c that the combined power flow controller not only permits reactive-power compensation, but also creates the phase-variable additional voltage ΔU. By adjusting the magnitude and phase of ΔU, we may vary the phase δ between the different points of the line and also control the magnitude and phase of the voltage at transformer T. As result, effective control of the power flux in the line is possible.

10.1.1.4 Power transmission and dc links

Power electronics has significantly improved the efficiency of ac and dc power transmission. Indeed, high-voltage dc power lines are widely used in practice today. The following deficiencies of high-voltage dc power lines have been minimized or eliminated due to power electronics:

- The high cost of the converter equipment
- The high content of higher voltage and current harmonics

At the same time, high-voltage dc power lines have the following benefits in relation to ac transmission:

- No reactive power
- Scalar control of the power flux
- Lower transmission costs over distances greater than 500–800 km.

Figure 10.6 presents the approximate cost–distance graphs for dc and ac transmission (Sood, 2001). We see that the initial cost is considerably less for ac power lines. However, as the distance increases, dc transmission becomes more attractive. The main disadvantage of high-voltage dc

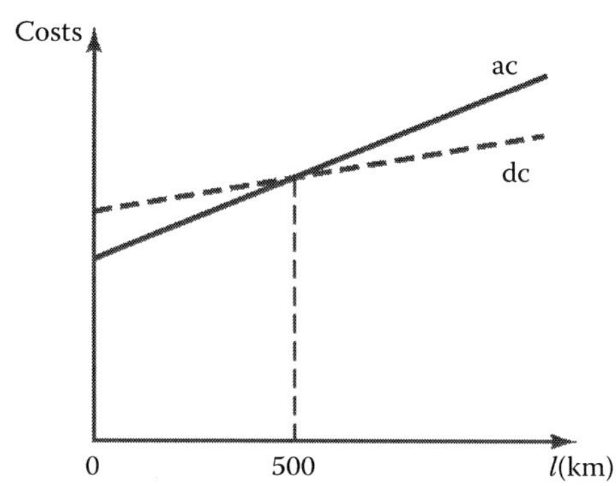

Figure 10.6 Costs of power transmission as a function of the distance.

power lines is that the transformation of direct current is not possible. Therefore, high-voltage dc lines are used for stretches without power delivery to consumers. Between 1987 and 1997, more than 47 high-voltage dc power lines went into operation around the world. The power of many high-voltage dc power lines is as much as 2000 MW, at voltages of 600 kV and above. Note that, in dc transmission, the power losses in the line are 20–30 times less than those in ac transmission.

Potentially, power electronics permits the use of direct current not only in power transmission, but also in other applications, for example, in power generation from alternative sources and in uninterruptible power supply to electric drives. Again, the main obstacle is the impossibility of direct transformation. At the same time, dc–ac conversion costs are significantly greater than the cost of a transformer. However, the situation may change dramatically with the development of faster and more powerful switches based on silicon carbide (SiC). In some countries, the development of power transmission with dc links is underway. When using dc links, one option is to introduce an array of voltage inverters on the dc side. A bank of capacitive batteries is attached to the dc buses. As a result, the actual dc transmission line disappears. This short link is often known as a B2B (back-to-back) system (Figure 10.7). Thus, in this case, there is practically no transmission line due to localization of the converters (Burman et al., 2012).

10.1.1.5 *Power quality control*

Before the creation of completely controllable high-speed power electronic devices, power systems mainly employed thyristor-based converters. Powerful converters of that type were developed for use in electric drives, in electrometallurgy, in information transmission, and elsewhere. The consequence was significant loss of power quality and a decline in the efficiency of power utilization. New electronic devices not only eliminate

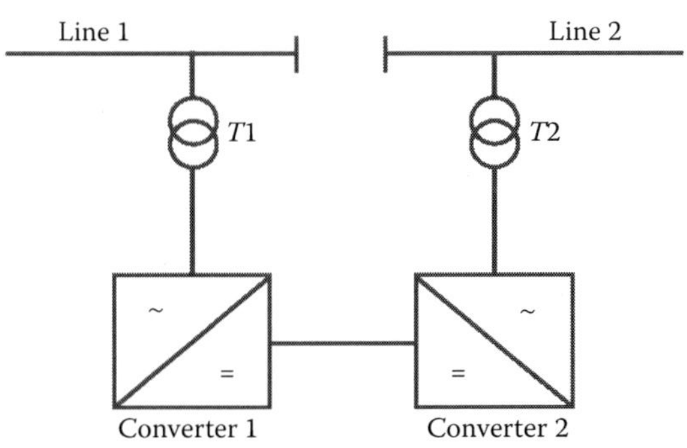

Figure 10.7 Structure of a dc system based on converters between ac lines.

those problems, but also permit the development of equipment that improves power-transmission systems. On the basis of power electronic technology, we can, for example, control the magnitude and sign of the reactive power and hence the power factor cos φ; improve the harmonic composition of the currents and voltages; stabilize the voltage at specified points of the line; and ensure symmetry of the load currents and voltages in three-phase systems. This topic was addressed in more detail in Chapter 7.

10.1.2 Power electronics for renewable energy sources and storages

Our energy problems today are best addressed by the development of renewable energy technologies, including solar, wind, and wave power. Such energy is directly related to the power of the sun, which is practically inexhaustible in comparison with traditional fuels such as coal, oil, and natural gas. New technologies for converting renewable energy to electric power are being developed on the basis of power electronic components. The most important function of such components is to integrate the renewable sources with grid, directly or through intermediate energy stores.

10.1.2.1 Solar cells

Solar energy may be converted into electric power by means of photoelectric converters (solar cells). The development of solar cells depends on the production of silicon and the other materials required, in particular, thin films with a silicon coating (da Rosa, 2009; Sabonnadiere, 2009). Solar cells are based on semiconducting elements consisting of materials that form a p–n junction, resembling a diode. Under the action of photons falling on one of the surface layers of this structure (usually the n-type layer), a photoelectric process converts the energy of the electromagnetic radiation into electric power. Systems based on solar cells are essentially dc sources. Up to 10 kW, ac power is obtained by means of an inverter connected, in the general case, to a single-phase ac grid. In the absence of a grid, and with low-quality requirements on the output current, we may connect the output of the solar cell directly to an inverter, which supplies the consumer. The quality of the output voltage may differ significantly from the standards for ac grid voltage and is determined by special requirements. In the presence of a grid, an additional dc converter is required (Figure 10.8), on account of the extreme instability of the output voltage from the solar cell. In addition, most systems include battery storage, so as to ensure stable voltage at the dc bus (Figure 10.8) (Rozanov and Kriukov, 2008; Kryukov and Baranov, 2012).

Above 10 kW, three-phase power systems have been developed. They usually employ a single centralized inverter capable of operating with

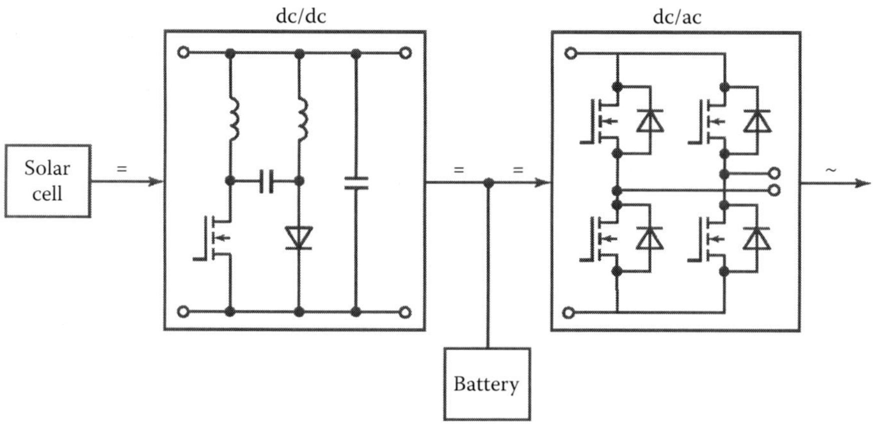

Figure 10.8 Power circuit block diagram for a system with a solar cell.

the grid. At the same time, dc converters are installed at the output of individual solar cells and panels. In these systems, the inverters must be chosen to facilitate integration with the grid and to ensure high quality of the power.

10.1.2.2 Wind turbines

Wind turbines convert the energy of the wind into mechanical energy, which is then transformed to electric power by the rotating turbine blades. These blades essentially form a propeller that initiates turbine operation. As the rotor speed (Alekseev, 2007) varies significantly, a gearbox is installed to regulate the speed of the shaft on which the power generator is mounted. A dc generator is an easy-to-control system for converting the mechanical energy to electric power. However, problems associated with the type, quality, and transmission of the power must be solved in that case. The energy efficiency of wind systems may be significantly increased on the basis of a synchronous generator with power electronic equipment. In that case, the voltage of the synchronous generator is sent to the input of a rectifier, which may be based on uncontrollable switches in some circumstances. Then, the rectified voltage is sent to an inverter (Figure 10.9), which is connected directly to the grid. Either a voltage inverter or a current inverter may be used. Its operation with the grid may be greatly improved by using completely controllable electronic switches. By that means, the inverter may generate specified active and reactive power in the grid. It is also possible to ensure high quality of the current and to supply energy from the grid for battery charging.

Another option is to use an induction generator (IG) with a short-circuited rotor, but that entails regulation of the amplitude and frequency

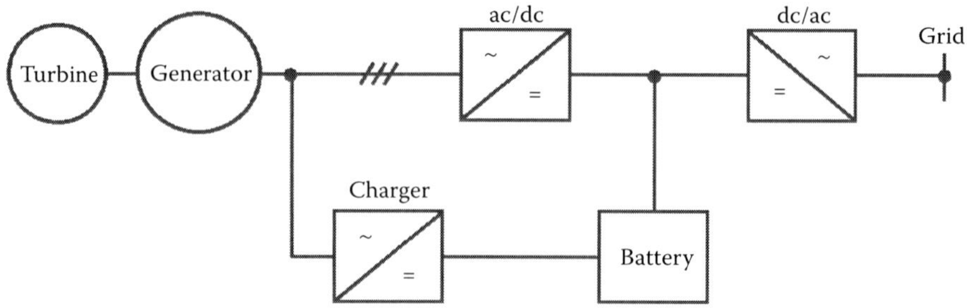

Figure 10.9 Block diagram of a wind power system.

with variation in the rotor speed. Special methods are required here, for example, combined operation with a rectifier–inverter cascade. An inverter based on completely controllable electronic switches must be used in that case; a step-up dc converter may also be needed. However, a more promising approach is to use a source of controllable capacitive power such as a static converter. In that case, the IG can send active energy to the grid and draw from it the required reactive power (Figure 10.10).

A fundamentally new approach involves an asynchronized synchronous generator. Its rotor has two windings, which may effectively be controlled by static converters. In principle, this generator may operate with the grid as its rotor speed varies. Power may be supplied to the rotor windings from a frequency converter in which the voltage is phase-shifted by 90°, and its frequency is the slipping frequency. The resulting magnetic field, which turns at the slipping frequency, ensures synchronous rotation of the exciting field and the generator's stator. The efficiency of the

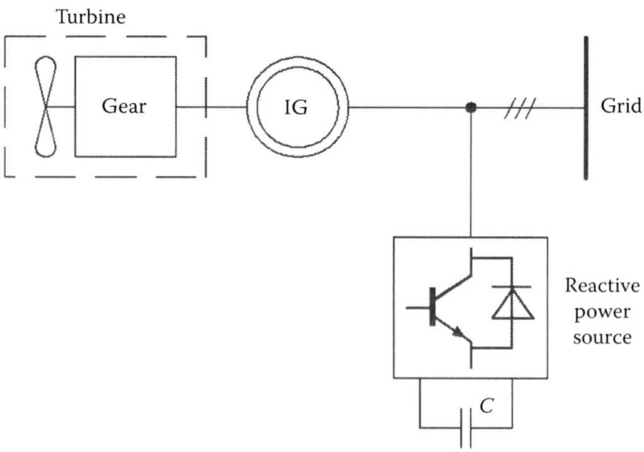

Figure 10.10 Connection of wind turbine to the grid by means of an IG.

asynchronized synchronous generator is especially evident at considerable turbine speeds with a gearbox of limited capabilities. The frequency control of the voltage supplied to the winding ensures highly stable synchronous operation of the asynchronized synchronous generator with the grid.

10.1.2.3 Fuel cells

Traditionally, most power is produced by converting thermal energy into mechanical energy and then into electrical energy. However, the efficiency of thermal machines is not more than 40%. Significantly higher efficiency can be obtained by means of fuel cells (da Rosa, 2009).

In fuel cells, chemical energy is converted into electrical energy. The inverse process is also possible: regeneration (electrolysis) of water, with the liberation of the absorbed heat. In regeneration, the water formed in fuel-cell operation is broken down into the initial elements of the fuel: hydrogen and oxygen. Thus, because the fuel cell permits not only the generation of electric power from fuel, but also the production of hydrogen from water in regeneration, it has a relatively broad spectrum of functional capabilities.

The fuel cell produces direct current. Low-power systems (up to 10 kW) are usually employed with a single-phase grid. The source for the fuel cell is selected so as to ensure the mean rated power, whereas the peak loads must be covered by an energy store such as a battery. A simplified block diagram of the fuel cell connection is shown in Figure 10.11. The basic power components are a dc converter and an inverter. To produce a more compact system, a high-frequency element is generally included. Its output is connected to a transformer T with the required output voltage. In such structures, both dc and ac output channels are possible. To improve the size and mass of the unit, dc converters with galvanic uncoupling may also be employed. In that case, an inverter without an output transformer is connected to the output of the dc converter.

Fuel cells have been successfully used in uninterruptible power sources (UPSs). The fuel cell is connected to a traditional UPS with a battery storage. This significantly increases the possible operating period of the uninterruptible source.

Above 10 kW, fuel cells are used with a three-phase grid. With an increase in power, dc converters with galvanic uncoupling of the output

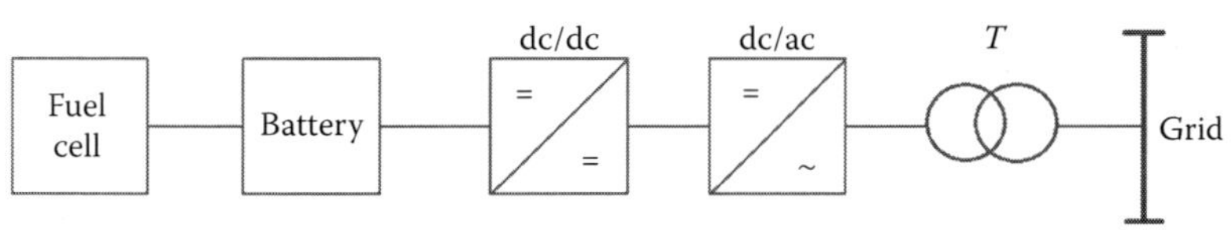

Figure 10.11 Connection of fuel cell to the grid.

buses from the power sources are mainly employed. The inverter directly connected to the grid or load is also based on completely controllable switches of the corresponding power.

Powerful hybrid systems (10 MW or more) operate with output voltages up to 10 kV, based on IGCT and IGBT (these devices were considered in detail in Chapter 2). The microturbine has a generator producing voltage sent to a three-phase rectifier and then to the common dc buses of the system. The output inverter of the system, which connects it to the grid, is a two-level system based on IGCT, corresponding to an input voltage of 6000 V.

Globally, the two main applications of fuel cells today are stationary power sources and automobile drive systems. Stationary power sources currently cover the range from tens of kilowatts to hundreds of megawatts.

10.2 Electric drives

Electric drives are major power consumers. Power electronic components may significantly improve the efficiency of power use and reduce power losses by drawing higher-quality current from the grid and expanding controllability. For example, current and voltage distortion may be almost completed and eliminated over broad ranges; automatic transition from energy consumption to recuperation and back may be ensured; and precise conformity of the operating conditions to the control specifications can be maintained.

10.2.1 Control of dc machines

In dc machines, the conductors in the armature interact with the magnetic field created by the exciting winding on the stator. This produces a torque, which turns the rotor. To maintain the torque as the conductors move in the magnetic field, the direction of the current in the armature winding is reversed. This entails switching of the current from some turns of the winding to others.

The current may be supplied to the exciting winding by various means: independent, series, and mixed. Usually, in vehicle drives, series and mixed excitation is employed. In industrial electric drives, independent excitation is employed. We know that dc machines are reversible: they may operate as motors or as generators. The goal of control is primarily to change the rotor speed. The following types of operation can be considered (Figure 10.12):

- Forward rotation
- Braking in forward rotation
- Reverse rotation
- Braking in reverse rotation

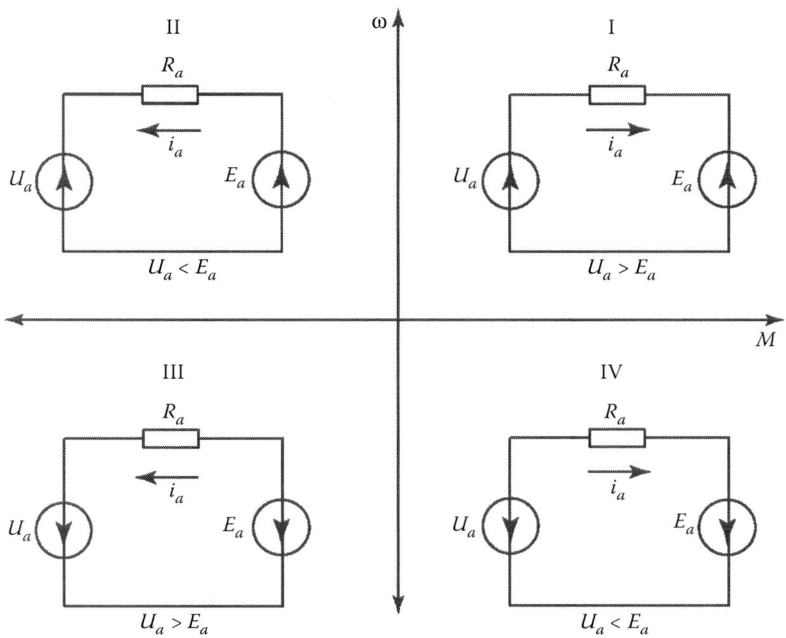

Figure 10.12 Operation modes of dc machines.

In forward rotation, the machine operates at a supply voltage U_a exceeding the emf E_a created in the armature. This corresponds to motor operation in quadrant I. In steady operation, the torque developed at the shaft is equal to the load torque. The machine consumes the armature current. In quadrant II, braking occurs. This is associated with the recuperation of energy to the grid. To accept this energy, the source must have bidirectional conductivity. In this case, the armature emf E_a exceeds the source voltage U_a and the armature current reverses direction, so that it begins to oppose the emf of the source, which consumes the supplied energy. In quadrant III, the rotation of the machine is reversed. It resumes operation in the motor mode, consuming energy from the source. This corresponds to change in polarity of the armature emf at the brushes. However, the direction of the armature current is consistent with the polarity of the source emf, and thus energy is consumed from the source. In quadrant IV, braking begins again, and energy is recuperated from the machine to the source. In this generator mode, the machine rotates in the opposite direction to quadrant II.

At present, the speed or the torque at the shaft is mainly regulated by means of controllable rectifiers, which permit variation in the armature voltage. Rectifiers are also widely used to create the exciting current. However, rectifiers based on completely controllable switches are more efficient. In particular, relatively powerful (1 MW or more) ac–dc converters may be based on power transistors or switchable thyristors of GCT type (see Chapter 2). This is possible for converters with the properties of a current source or a voltage source, on the basis of pulse-width modulation at high modulation frequency. The use of such converters permits an increase in the power factor and a decrease in the distortion of the supply voltage and the current drawn from the grid in all operating conditions, including recuperation.

Systems based on pulsed dc converters may be used to control low- and moderate-power motors. Such converters are effective for components with dc voltage of current sources.

10.2.2 Control of induction motors

Induction motors are commonly used for the conversion of electrical energy into mechanical energy. Accordingly, the control of their speed and torque is of great importance. However, this problem is complicated, in that the rotor speed is controlled by the grid and the motor's starting currents are very large. Therefore, in vehicles, the norm has been to use dc motors, which are much more controllable, even though the use of ac power predominates and ac–dc conversion is required. However, the development of power electronics has changed this situation.

In the induction motor, three-phase alternating current interacts with the current that it induces in a winding or in the short-circuited rotor turns. The speed and torque of an induction motor may be controlled by adjusting the voltage applied to the stator and the rotor slip s. The following control methods are known:

- Adjustment of the voltage at the stator
- Variation in rotor slip by introducing additional impedance or an ac voltage source in the rotor circuit
- Frequency modification of the voltage supplied to the stator
- Simultaneous change in the stator voltage and its frequency

These are known as scalar methods, as they are associated with variation in controllable parameters. A relatively new group of methods has been made possible by the appearance of completely controllable high-speed switches and by the development of pulse-width modulation and digital control. These are known as vector methods. They permit more precise positional control, analogous to the control of dc motors in a servo drive.

10.2.2.1 Scalar control

Control of the stator voltage. Control of the stator voltage by means of an ac thyristor-based voltage regulator is very simple (Figure 10.13). However, it

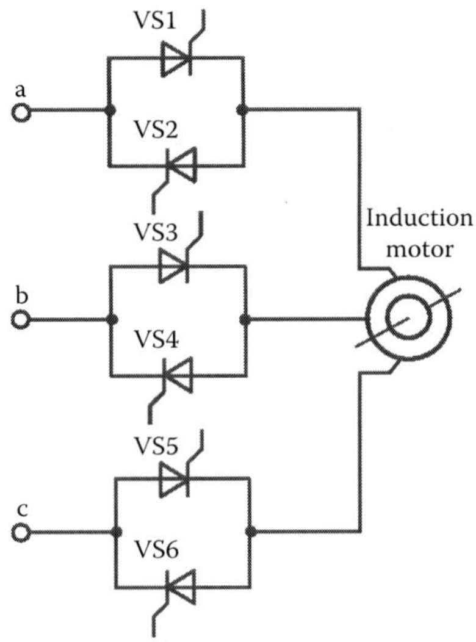

Figure 10.13 A thyristor regulator for an induction motor.

is associated with considerable distortion of the stator voltage and a sharp increase in the higher harmonics, with additional power losses and overall deterioration in motor performance. The content of higher harmonics increases with an increase in the thyristor control angle α. With a larger range of voltage regulation, the change in the voltage distortion is greater.

Control of the rotor slip. With an increase in resistance of the rotor windings, the characteristics change considerably. In particular, the maximum torques are shifted to lower speed. The resistance is usually changed in steps, by means of relay-contact equipment. To eliminate stepwise variation, additional resistance may be introduced in the rotor circuit in pulses, which permits smooth regulation of its equivalent value. This circuit may be based on a completely controllable semiconductor device such as a GTO. The frequency of pulse operation determines (together with the reactor inductance L) the current pulsation in the rotor, which affects the pulsation at the motor shaft.

The energy scattered in the resistor may be replaced by energy supplied to the grid in inversion. To that end, the pulsed regulator and resistor are replaced by a switched cascade.

Regulating the magnitude and frequency of the stator voltage. If the stator voltage is held constant, its frequency may be reduced to increase the flux or increased to decrease the flux.

Such regulation of an induction motor permits its operation with an unsaturated magnetic system; its speed may be adjusted with practically constant motor torque or power. For example, with constant torque M, we obtain $U/f = \text{const}$; for a fan load with $M = k\omega^2$, we obtain $U/f^2 = \text{const}$; and, with constant power P, we obtain $U/\sqrt{f} = \text{const}$. In frequency and voltage regulation, the motor's mechanical characteristics vary in accordance with the control law. In particular, with constant power P at the motor shaft, the speed dependence of the torque is hyperbolic, and the mechanical characteristics correspond to Figure 10.14.

For motors of low and moderate power, frequency control is most often based on frequency converters with a distinct dc element.

Better results are obtained with a frequency converter that may be regarded basically as an inverter with a completely controllable rectifier. In that case, we obtain a two-module frequency converter based on two practically identical ac–dc converters. One converter operates as a completely controllable rectifier with pulse-width modulation; its power factor is close to 1 and there is little distortion of the grid current. In that case, the electric drive resembles a linear active load. The higher harmonics due to voltage modulation are removed by a relatively mild LC filter. This frequency converter may be regarded as ideal, except for its cost.

Direct frequency converters may be used for the control of very powerful motors. When based on thyristors, such converters are characterized by low input power coefficient and limited frequency range.

Frequency converters based on a current inverter are widely used to control induction motors of large and moderate power. In such converters, the rectifier and inverter are separated by a reactor of relatively large inductance, so that they behave as a current source.

10.2.2.2 Vector control

Vector control of induction motors is actively under development at present. These methods are based on resolving the stator-current vector into two orthogonal components. One component determines the magnetic flux, whereas the other corresponds to the motor torque. In this case, we may note the analogy with a dc machine that has independent excitation.

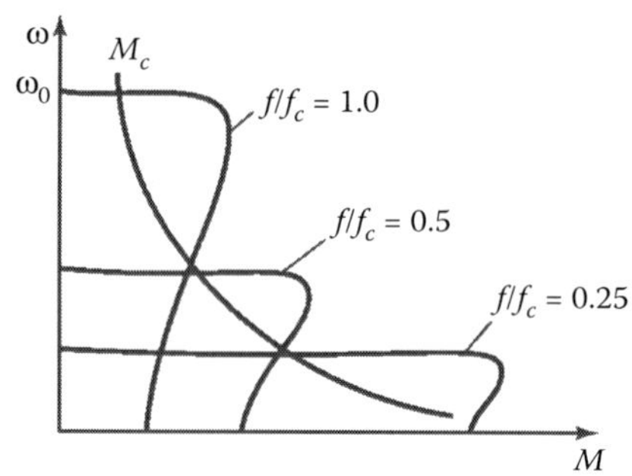

Figure 10.14 Speed–torque characteristics of an electric motor.

The magnetic flux of this machine is determined by the exciting current and the position of the stator poles, whereas the armature current determines the torque at the motor shaft. In turn, different coordinates may be selected in the resolution of the stator-current vector, in particular, coordinates relative to the rotor or stator flux. With different coordinates, the methods will differ in ways that are not fundamental. Essentially, these methods are based on the spatial orientation of the magnetic field; correspondingly, they are sometimes known as field-orientation methods. A detailed classification of vector-control methods may be found in Kazmierkowski et al. (2002). On the basis of the foregoing, induction motors may be controlled by a voltage source with voltage and frequency variation and by means of components such as microcontrollers and sensors.

An example of vector control was presented in Rozanov and Sokolova (2004). This method is based on a modified equivalent circuit of the motor, in which the inductive impedance of the rotor is reduced to zero by selecting the reduction coefficient.

10.2.3 Control of synchronous machines

In synchronous machines, the rotating magnetic field created by the stator winding interacts with the magnetic flux created by the rotor. As the resultant magnetic flux is created by the combined motion of the stator and rotor, the resultant magnetic field in the synchronous machine turns with the same speed as the rotor, in steady conditions. Many different designs of synchronous machines exist. The range has greatly expanded with the development of power electronics. New options include switched motors, brushless motors, and self-synchronizing machines. Rectifiers are widely used in the exciting systems of synchronous machines.

Thus, all types of synchronous machines are characterized by synchronous rotation of the stator's magnetic field and the rotor, in steady conditions. These machines differ not only in design, but also in their applications.

10.2.3.1 Control of synchronous motors with adjustable excitation

Because the speeds of the resultant rotating magnetic field and the rotor are equal, synchronous motors are ideal for applications that require precise control of the speed or torque developed at the shaft. In synchronous machines with nonsalient rotor and stator poles, the air gap is uniformly distributed over the circumference of the stator, and the reactance does not depend on the rotor position. In that case, the angle θ between the armature voltage U_a and the emf E_a induced by the exciting magnetic flux is determined by the total synchronous impedance of the synchronous machine. The value of θ is determined by the load of the synchronous machine; it is known as the load angle. In the generator mode, U_a lags the emf E_a and θ is regarded as positive; in the motor mode, U_a leads E_a and θ is negative. The power factor cos φ characterizes the phase shift between the voltage U applied to the synchronous machine and the current i_a of the armature winding and depends not only on the load, but also on the exciting current i_e. The family of V-shaped curves $I_a = f(I_e)$ (Figure 10.15) can be divided into two regions: overexcitation ($\varphi > 0$) and underexcitation ($\varphi < 0$). As the sign of the angle between the current and the voltage depends on the magnitude of the exciting current, the synchronous generator may be used as a reactive-power compensator with zero active power. In overexcitation, reactive power is generated (capacitive behavior); in underexcitation, reactive power is consumed (inductive behavior).

Semiconductor converters may be used to control synchronous machines, analogous to induction machines. In this respect, the development of power electronics has significantly expanded the potential of electric drives. The control methods are based on variation of the magnitude and frequency of the grid voltage and are implemented by different types of frequency converters. The best option to control the speed and torque of low- and moderate-power synchronous machines is a converter with an explicit dc element based on completely controllable switches. A frequency converter with reversible ac–dc converters and with pulse-width modulation, which behaves as a voltage source, has a mirror structure. Consequently, the system has good performance: the input power coefficient is close to 1, and the input current and output voltage are close to sinusoidal. A deficiency of this approach is high cost, as both components

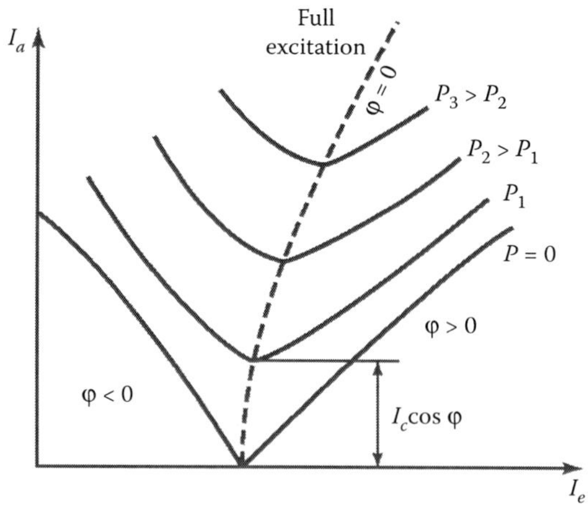

Figure 10.15 Synchronous generator characteristics at different values of the power P and the angle φ between the stator current and voltage.

of the frequency converter are based on completely controllable switches. A common microprocessor controller permits digital control of practically all the functional components. The second component of the frequency converter operates mainly in the inverter mode, ensuring control of the motor speed and torque. However, in braking, the frequency converter operates in the opposite mode: the output component becomes a controllable rectifier, whereas the input component becomes an inverter, recuperating energy to the grid.

For high-power synchronous machines (above 1 MW) operating at low frequency, a thyristor frequency converter with direct coupling is employed. It permits variation in the energy flux in braking and control of the magnitude and frequency of the armature voltage in the synchronous motor. Regulation of the voltage over a wide range leads to considerable impairment of the system's power factor, which must be taken into account in design. Benefits of the traditional frequency converter with direct coupling include a simple control system and the possibility of multiphase configurations for better thyristor utilization and improved harmonic composition of the armature voltage.

10.2.3.2 Control of switched motors

Switched motors are machines in which electronic switches ensure current commutation in the armature. In particular, brush commutation in the armature supply system of dc machines can be replaced by electronic commutation, with a significant improvement in motor life. However, interest in switched motors based on synchronous machines is growing due to the development of power electronics and production technology for permanent magnets. On the basis of power electronics, completely controllable switches may be incorporated in frequency converters with pulse-width modulation; in permanent-magnet production, new materials with magnetic energy have been developed for the rotors, in particular, neodymium–iron–boron alloy (Ni–Fe–B). As a result, it is possible to create completely brushless switched dc machines that match the controllability of dc machines and outperform the induction motors currently in industrial use.

Such switched motors are based on self-synchronization of the rotor with the speed of the armature field. The rotor may be based on a permanent magnet or an exciting winding. The magnitude and frequency of the armature voltage are varied by means of a frequency converter with an explicit inverter component or with direct coupling to the grid. Completely controllable electronic switches are more expensive but permit the creation of high-performance converters. For example, a power factor close to 1 may be maintained over the whole range of the magnitude and frequency of the output voltage, if the rectifier in the converter is based on completely controllable switches. If the frequency converter with

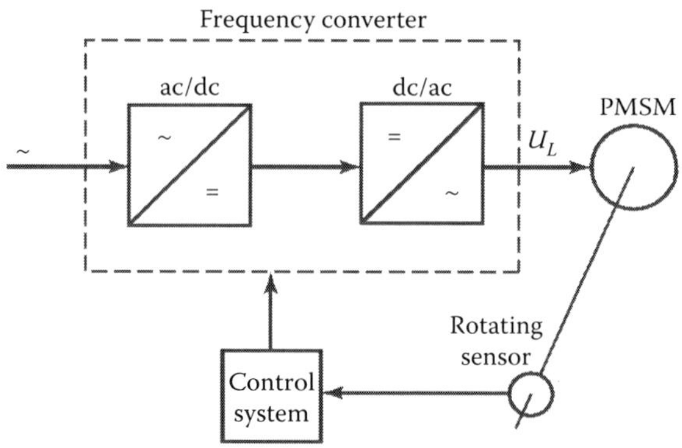

Figure 10.16 Control of the switched motor.

direct coupling is based on completely controllable switches, the power factor will again be close to 1.

Figure 10.16 shows a switched motor with a rotor based on permanent magnets when power is supplied from a frequency converter with a voltage-source inverter (Rahman et al., 2001). In this motor, the rotor position determines when the control pulses for the inverter switches are formed. The rotor turns under the action of the torque produced by the interaction the magnetic fluxes created by the rotor and the armature winding, as their axes approach. In steady conditions, the rotor is synchronous with the inverter's output voltage. The frequency may be regulated by a proportional change in the inverter voltage. Other methods can be used to control the inverter frequency, with self-synchronization of the rotor. For example, for a synchronous machine with a rotor that has an exciting winding, the exciting current may be modified. Motors of this type are characterized by feedback with respect to the rotor position. We can also use a current-source inverter, rather than a voltage-source inverter. That design is cheaper and ensures good motor performance. However, at low speed, the counter-emf in the armature may be insufficient for thyristor commutation. In addition, the dynamic characteristics of the current inverter are worse than those of the voltage inverter, overall.

10.2.3.3 Switched reluctance motors

Researchers are increasingly interested in electric motors whose rotor is based on a laminated stack of electrical-engineering steel in different configurations. These are salient-pole rotors, usually with a toothed outline. In this design, the reluctance between the stator and the rotor will depend on the rotor position. When the exciting winding on the stator is connected to an ac source, a rotating magnetic field is produced. Because of the difference in longitudinal and transverse reluctance, a torque arises and tends to turn the rotor in the direction corresponding to minimum inductive impedance between the stator and the rotor or maximum inductance. Single-phase alternating current can be supplied to the exciting winding. A reactive torque on the rotor appears here. These are sometimes known as reactive motors.

The development of power electronics and microprocessor technology has permitted the creation of high-performance motors on the basis of reactive motors, with electronic switching of the stator winding. The appearance of self-commutated switches permits periodic switching of the stator winding at high frequency. In that case, the pulsed voltage formed by the electronic switches connected to a dc source is supplied to the exciting winding of the motor stator (Figure 10.17b). This is known as a switched reluctance motor (SRM) (Iqbal, 2002). The operation of this motor is based on switching of exciting windings on the stator, which create a rotating magnetic field. The rotor of the SRM tends to occupy a position in which the reluctance between the excited stator pole and the rotor is

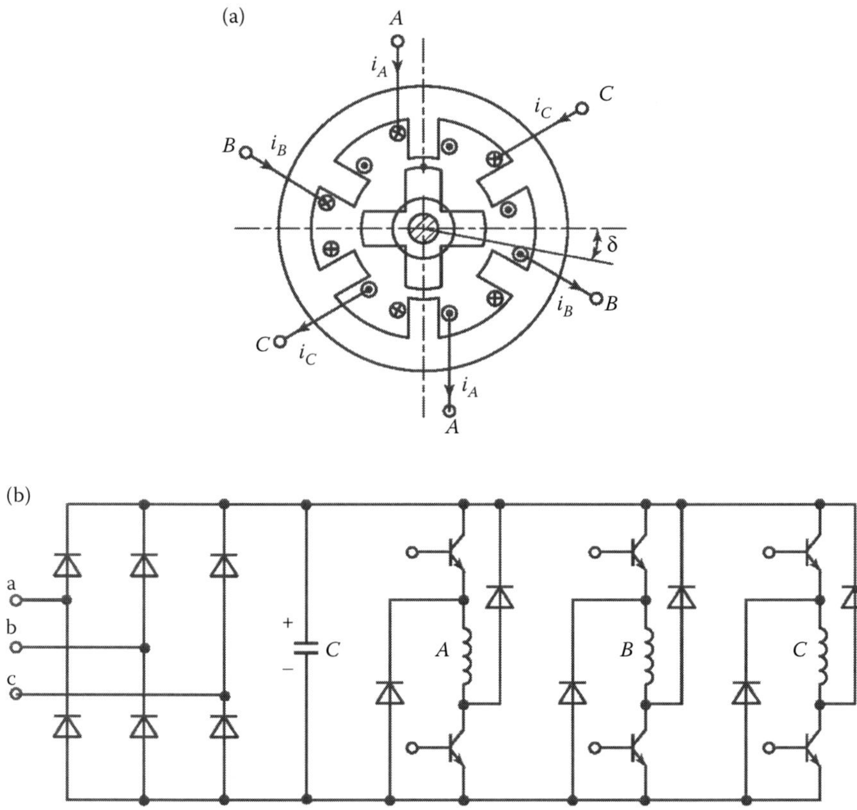

Figure 10.17 Switched induction motor: (a) cross-section and (b) power electronic commutator.

a minimum. That corresponds to the operating principle of the reactive motor. Note that the stator is also usually toothed; hence, this is a salient-pole stator. In Figure 10.17a, as an example, an SRM with a three-phase, six-tooth stator is shown (Rahman et al., 2001). The phase windings are on diametrically opposed teeth. Thus, the magnetic flux in the motor takes a pulsating form due to switching of the windings. In the general case, the magnetic flux is a nonlinear function of the exciting current i_e and the rotor-position angle ϑ. The control pulse responsible for switching the exciting winding is formed by the control system of the electronic commutator on the basis of the signal from the rotor-position sensor.

Different designs can be used for the power electronic commutator. The most common is shown in Figure 10.17b. When the transistors in series with a particular winding are switched on, voltage and hence current will be supplied from the source. As a result, the current in that winding will increase, and the rotor will turn in that direction. Switching on the other winding involves switching off the pair of transistors that are already on and switching off the two transistors connected to that winding. However, switching off the first pair of transistors will turn on the diodes connected to the same winding and the voltage source U_d. As a result, the exciting winding that is being turned off will be demagnetized, with the return of the stored energy to the source.

Various methods can be used to control the motor speed and torque. For instance, the magnitude and frequency of the voltage at the exciting winding may be regulated. A more effective approach is to regulate the rotor position angle, which determines the times ϑ_{on} and ϑ_{off} at which the rotor exciting windings are turned on and off, respectively. In Figure 10.18,

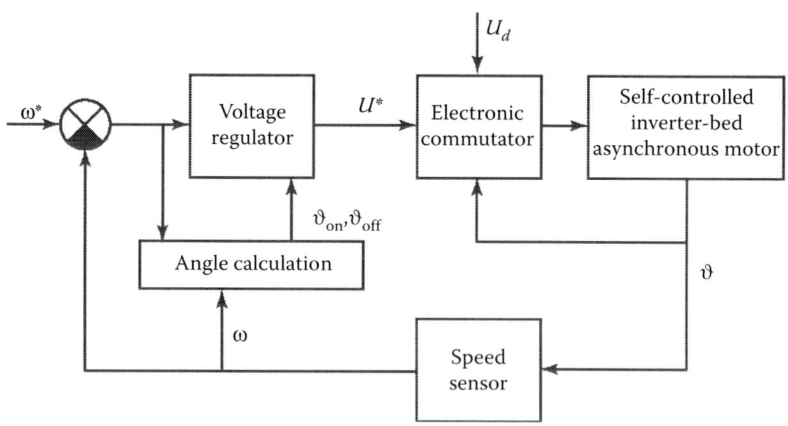

Figure 10.18 Speed control of the SRM with power supply to the windings through an electronic commutator: ϑ is the rotor position angle and ϑ_{on} and ϑ_{off} are the windings switching angles, respectively.

a simplified version of a control system for an SRM is shown. This structure corresponds to control of the voltage U and the angles ϑ_{on} and ϑ_{off}. Tracking of the current in the motor phases is a more effective control method.

10.3 Engineering applications

10.3.1 Lighting

Given that lighting accounts for more than 20% of all power consumption, we see that there is an obvious need to increase the energy efficiency of lighting systems, that is, the ratio of the lamp's light flux to its power consumption. The light output of discharge lamps has significantly increased. Their basic component is the ballast, which determines their efficiency. Traditional ballast has some serious limitations. On the basis of power electronics, the ballast may be replaced by power converters with working frequencies up to 100 kHz, which permits the elimination of undesirable frequencies. Due to the high frequency of the power converter and new circuit designs, converters that are light and compact may be produced. The introduction of such converters in electronic ballast has led to the widespread adoption of discharge lamps in homes and in industry.

The new electronic ballast differs fundamentally from traditional ballast based on chokes and capacitors.

Figure 10.19 shows the generalized structure of standard electronic ballast. It consists of an input filter to eliminate electromagnetic interference, a rectifier, a high-frequency inverter, high-frequency electronic ballast, a control system, and a feedback circuit.

The electromagnetic interference filter consists of interconnected reactors and capacitors within a single framework. The rectifier may be based on diodes or transistors. When transistors are used, control on the basis of pulse-width modulation ensures sinusoidal input current and regulation of the input voltage. The inverter is usually based on resonant circuits that ensure transistor switching at zero current or voltage. This reduces

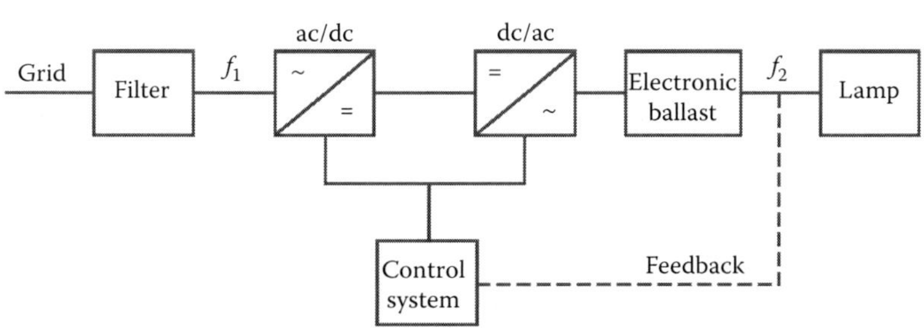

Figure 10.19 Application of electronic ballast based on a frequency converter.

the power losses. At the same time, there are fundamental constraints on the selection of the inverter design. The high-frequency electronic ballast limits the current in the event of short-circuiting of the lamp. This function may be performed both by a low-inductance high-frequency choke and by a small capacitance, determined by the high working frequency of the inverter. The control system of the inverter and rectifier may be based on the same microprocessor. Depending on the parameter being regulated, there may be several feedback channels and sensor locations. For example, the discharge current may be regulated directly on the basis of the lamp circuit current. In addition, the feedback channels can be used for protection in the event of accidents and for diagnostic purposes.

10.3.2 Electrotechnology

Various forms of electrotechnology were developed in the middle of the nineteenth century, in particular, heat treatment, electric welding, and electrophysical and electrochemical machining.

10.3.2.1 Electrical heating

Electrical systems are used for heating and melting of various materials, primarily metals. Resistive, electric-arc, inductive, and other types of heating are employed. Resistive heating is used in electrofurnaces of different types. Such heating depends on thermal energy created by current as it flows directly through a material (direct or contact heating) or through heating elements that are able to transfer the heat to the material (indirect heating). Direct resistive heating is used for metals; indirect heating may be used for any material. In both cases, the heating conditions are regulated by adjusting the current flow. The current may be regulated by various methods, for example, if a transformer is employed, by tapping different points of the winding. It is more effective to use thyristor regulation of the load current. Antiparallel thyristors are used for that purpose. The current is regulated by adjusting the relative length of the periods during which the thyristors are on and off, in terms of the period of the input voltage. With a resistive load, the phase shift of the current and voltage may be neglected, and the regulator may be switched on and off at zero current and voltage, in contrast to phase control of the thyristors. A benefit of this approach is that there are practically no switching losses and no higher current harmonics in the grid. However, a significant disadvantage is the creation of a discrete load, which varies at low frequencies (about 4–10 Hz). Periodic switching of the load leads to low-frequency fluctuation of the voltage. In that case, the spectral composition of the load vibrations hinders filtration by traditional methods. Active filters prove more effective. As the cost of electronic switches declines, active filters are becoming more affordable.

Inductive heating of metals is widely used. In inductive heating, the interaction of the metal with an electromagnetic field generates eddy currents, with consequent loss of active power, which is converted into heat. The power losses depend on factors such as the magnetic field strength, the penetration depth of the eddy currents in the metal, and its shape and electromagnetic characteristics. Frequency converters with a high-frequency inverter are generally used to supply power to induction furnaces. Depending on the circumstances, inverters with a frequency between 50 Hz and 1 MHz can be employed. The penetration depth of the current in the metal is inversely proportional to the square root of the frequency.

Figure 10.20 shows the structure of the frequency converter used for power supply of a standard inductive heater. A system with a resonant inverter is used to create a high-frequency voltage at the converter output. It may be based on a current inverter or a voltage inverter. The resonant circuits in the inverters are formed by connecting a capacitor or a capacitor and a reactor to the load, which in this case is an induction coil containing a metal blank. In a voltage inverter, a capacitor is connected in series with the load. In a current inverter, the resonance loop is formed by connecting a reactor in series and connecting a capacitor and the load in parallel. Frequency regulation permits control of the power in the load, and the capacitance and inductance of the loop determine its resonant frequency. The ratio of the working and resonant frequencies significantly affects the operating conditions and characteristics of the inverter. A current inverter provides better protection of the converter in emergencies. The voltage inverter is lighter and more compact.

By means of resonant inverters, it is simple to organize natural commutation, and so ordinary thyristors may be employed. Natural commutation of the inverter may be associated with limitations on the output frequency regulation. The working frequency of thyristor inverters cannot usually exceed 10 kHz. This limit is determined by the speed of the thyristors. The frequency may be increased in modular systems, by multiplying the frequency of a single module.

The inverter frequency may be significantly increased—by about an order of magnitude, depending on the power—if we use high-speed switches, such as IGBT, which also increase the converter cost. The

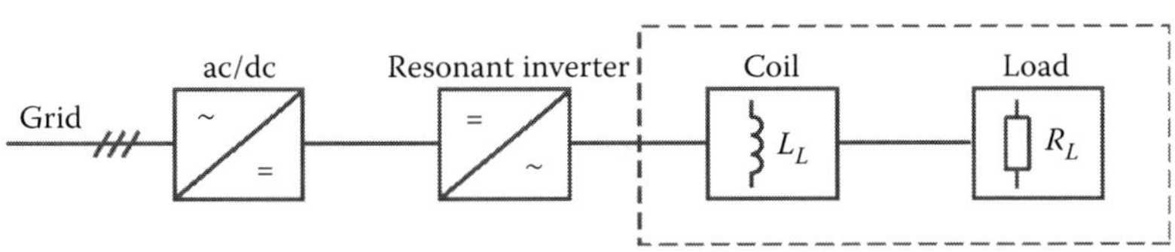

Figure 10.20 Block diagram of a frequency converter with a resonant inverter for power supply of the induction coil and the load.

frequency may be further boosted by using vacuum triodes to excite oscillatory processes in a high-frequency LC circuit.

10.3.2.2 *Electric welding*

Systems with an input transformer are usually employed to supply power to electric welding equipment (Figure 10.21a). For safety purposes, galvanic uncoupling with the grid is essential. In the simplest case, the stability of arc combustion may be ensured by connecting a reactor in series. The volt–ampere characteristic at the output will then decline steeply. An analogous load characteristic may be ensured by means of a special transformer structure that ensures large scattering inductance. To regulate the welding current, various current stabilizers may be introduced at the rectifier output. In particular, dc converters with output-current stabilization may be used for that purpose.

A deficiency of welding equipment with an input transformer is excessive mass and size on account of the transformer, which operates at the low grid frequency. This problem can be eliminated by using a converter with a higher frequency link of a high-frequency voltage-source inverter (Figure 10.21b). The working frequency of such inverters is usually above the audible range. An inverter is used for current regulation within the equipment.

10.3.2.3 *Other uses*

Power electronics can also be used in electrophysical machining. For example, rectifiers may be employed to produce pulse generators used in electrospark machining in electric discharges. Various power sources for electrolysis equipment are used in the production of hydrogen–oxygen mixture in metal welding. Most of these devices are based on rectifiers, autonomous inverters, and thyristor regulators.

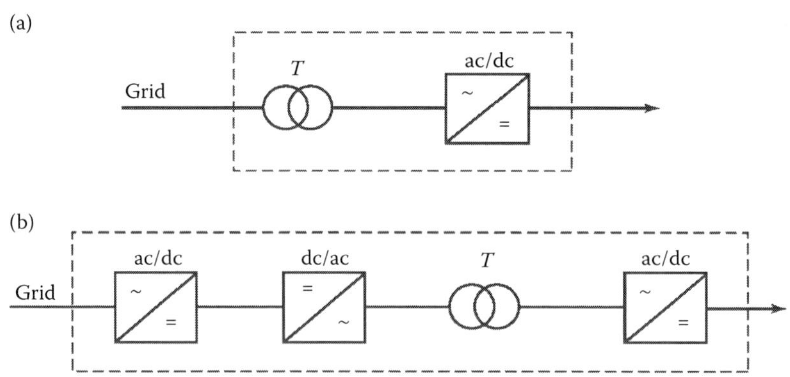

Figure 10.21 A power supply structure of welding equipment: (a) with an input transformer and (b) with a high-frequency converter link.

The use of completely controllable high-speed switches significantly improves the performance of electrophysical machining systems. Microprocessor-based control systems boost the efficiency of power conversion by more precise switch regulation.

Another application of power electronics is in industrial vibrators employed over a broad frequency range. Such systems are used, to name only a few examples, for underwater research in hydroacoustics; for more efficient borehole operation in the oil industry; and for the electromagnetic dosing of powders. Depending on the circumstances, different converters may be employed: from the simplest pulsed thyristor interrupters to frequency converters with controllable output parameters.

10.3.3 Electrical transportation

10.3.3.1 Railway transport

Both dc and ac electrification is used for railroads. The power supply to traction motors in electric trains comes from diesel–electric generators or from external sources (by way of overhead lines or third rails). At present, 1500 and 750 V dc motors predominate. The diesel–electric systems employed use ac or dc generator, as required. External sources take the form of tractional substations containing thyristor-based rectifiers and inverters (ac/dc converters). Note that these substations considerably impair the quality of the power from external ac grids, on account of nonlinear and peak loads. At present, dc grids run at 3 kV. When the locomotive employs ac voltage up to 25 kV, the rectifiers and inverters may be installed directly on the train to the locomotive, which simplifies the traction substations. The ac voltage is supplied to the locomotive step-down transformers and then passes through converters to the traction motor. In the case of a direct dc supply, the locomotive traction motor is controlled by stepped regulation of the armature voltage. Stabilizing resistors permit recuperation of the power, while the direction of motion is changed by reversing the current in the exciting windings. When using ac thyristor converters in the locomotive, phase control of the output voltage is employed. In phase control, the angles at which the thyristors are switched on in rectifier and inverter modes are adjusted. Therefore, this method reduces the power factor of the converter and creates distortion of the current drawn from the grid. To reduce the harmful impact on the ac grid, the locomotives and substations employ passive LC filters and capacitive compensation of the reactive power.

The introduction of ac traction motors began with the development of power electronics. In particular, simple and reliable induction motors were used with frequency converters. The motor speed and torque may be regulated by adjusting the converter voltage and its frequency. Today, manufacturers have developed a number of traction motors based on induction motors and controllable frequency converters with a dc element.

10.3.3.2 Urban transport

This category includes streetcars, trolleybuses, and subways. We will consider the example of streetcars; the characteristics are similar for the other systems. The basic power source is a dc substation with a maximum working voltage of 900 V. Most traction motors used in urban light rail are based on contact–resistor control. At the same time, cyclic operation of the traction motors ensures starting, acceleration, and braking. The periodic modification of such cycles complicates motor operation, shortens motor life, and leads to considerable power losses. These problems may be addressed by introducing semiconductor converters both for control of the traction motors and for providing incidental power for lighting of the cars, door operation, and so on.

10.3.3.3 Automotive applications

The introduction of power electronic components in cars and trucks significantly increases the automation of the vehicles' basic subsystems and expands their functions. The basic source of power in modern vehicles is a switched three-phase synchronous generator with a rotating excitation system. The generator's output voltage is rectified and sent to the dc bus, to which a battery is also connected (a 12 or 24 V battery, depending on the vehicle).

Today, the power of such generators is about 1 kW. From the basic dc bus, power is sent to various subsystems: the ignition, the lights, the air conditioning, the brakes, and so on. The total power consumption by these subsystems is steadily increasing and is expected to reach 1800 kW in the coming decades. The increase in power is associated with expansion of their functions on the basis of power electronic components. We will now briefly consider some examples of their use (Perreault et al., 2001).

In the lighting subsystem, incandescent lamps are being replaced by discharge lamps. This triples the luminous efficiency and extends the working life by a factor of 4–5. Power supply to discharge lamps requires a voltage of 42 V and high-frequency electronic ballast. Accordingly, we need a dc converter to increase the 12–42 V, an inverter based on MOS transistors, and electromagnetic ballast ensuring ignition and maintenance of the discharge in the lamp. The working frequency of the inverter is 96–250 Hz. Prospectively, this configuration can be used for power supply to all discharge lamps, as incandescent lamps are progressively replaced. Pulse-width modulation can be used here to regulate the lamp's luminous flux.

In modern vehicles, a high-frequency ac drive powers the windshield wipers. Controllable high-frequency ac voltage can be created by inverters from the dc voltage at the common bus.

The power source for the motor's electromagnetic valves should also control their operation. To that end, we use inverters regulating the current in the electromagnetic coils of the valve drive.

The air conditioning relies on a switched synchronous motor, controlled by a thyristor-based frequency converter. In that case, the operation of the air conditioner does not depend on the vehicle's engine.

A switched dc drive is used for independent control of the pump drive in the vehicle's hydraulic system.

To reduce the wiring required in modern vehicles, designers have begun to use multiplex control of the subsystems, with transmission of the control signals in code through common conductors, depending on the function and position of the executive devices. This sharply reduces the quantity and weight of the wiring on board the vehicle. Smart relays are used to switch on and off the consumers in multiplex systems. Such relays have built-in amplifiers, as well as safety and monitoring elements. Signals from the microprocessor in the vehicle's control system are sent to the input of the smart relays.

In the future, vehicles may have a dual 14/42 V power system. The 42 V bus will be supplied from a rectifier, in the traditional manner, and will be used for high-power consumers such as the starter and the air conditioner. The 14 V bus will be supplied from a 42/14 V dc converter and will meet the needs of lighting and low-power electronic devices, for instance. In this system, low-voltage batteries may be used to supply the starter. This is possible when using a converter with reversal of the energy flow.

An interesting option is the use of switched reluctance machines in a direct electric drive, rather than the belt drive currently employed. In that case, a power electronic current regulator would be needed. At the same time, the introduction of such a drive might significantly improve the performance of the vehicle's electrical system.

At present, the switch from internal combustion engines to electric drives is beginning. This transition will rely on developments on power electronics, but primarily depends on the identification of efficient power sources and stores. Fuel cells are possible candidates. At the same time, the development of hybrid engines, which combine an internal combustion engine with an electric motor, is well advanced. Hybrid drives effectively address issues of energy conservation and environmental protection.

10.3.3.4 Marine power systems

Electrical systems on ships are characterized by considerable power consumption (as much as tens of megawatts). They also operate autonomously, and power of different quality is required for different subsystems. Shipboard systems use ac and/or dc power. The development of power electronics has permitted the introduction of thyristor rectifiers, whose output voltage is regulated by adjusting the angle at which the thyristors are switched on.

As a rule, the basic power sources on ships supply the propeller drive and associated systems. In addition, many systems receive power from individual diesel–electric generators, which ensure reliable operation of key components. The electrical systems employed are mainly determined by the type of the propeller drive. Figure 10.22a shows a typical power supply system when the propeller drive is based on a dc motor. In this system, the ac generator is part of a diesel–electric unit. Usually, ac frequency of industrial frequency is employed. The ac voltage is sent to thyristor rectifiers, which provide power to the electrical propeller motors. The motor speed is regulated by adjusting the control angle of the thyristor rectifiers.

Frequency converters are used where high-quality power is required. They uncouple the generator buses of the diesel–electric unit and the consumer subsystems. For that purpose, ships are fitted with electromechanical converters consisting of a motor and an industrial-frequency voltage generator mounted on a single shaft. As the reliability of power electronic systems increases, we may expect that such electromechanical converters will be replaced by static converters, with significant improvement in size and mass.

For ships with ac propeller motors, the power system illustrated in Figure 10.22b is used. In this system, industrial-frequency alternating current is converted into voltage of variable frequency, which is supplied to synchronous motors; the frequency converter employed may be based on thyristors with natural commutation if the speed of the electric drive is regulated so as to be less than that at 50 Hz. Progress in power electronics is permitting the creation of static converters based on completely

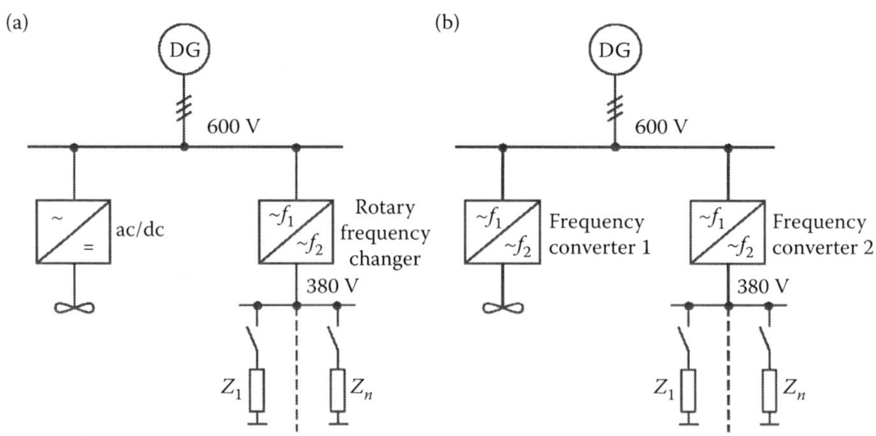

Figure 10.22 Structure of (a) ac and (b) dc propeller drive system.

controllable switches, thereby ensuring satisfactory control of the drive's energy characteristics.

Of course, certain subsystems will require power different from that provided at the common buses. In that case, semiconductor converters may be employed. At the same time, the expediency of centralized power supply from such converters must be verified in the design process. For example, we know that 400 Hz voltage is widely used at radar stations. This frequency is also used for land-based radar stations. For purposes of standardization, 50/400 Hz frequency converters were initially used to supply power to radar stations. However, their general introduction impaired the performance of the other electrical systems. Accordingly, the frequency converters are only used as local power sources for the radar stations. Accordingly, both 50 and 400 Hz voltage is available at radar stations today (Glebov, 1999).

10.3.3.5 Aircraft power systems

In aviation, continuous progress is required in the delivery of electric power to different airplane subsystems. Initially, power was required for ignition of the fuel mixture in the internal combustion engines. Subsequently, the use of power on airplanes significantly expanded. Today, practically, all critically important airplane subsystems depend on electric power; the power required varies from kilowatts to megawatts, depending on the type of airplane.

At first, dc systems were employed. With an increase in the power required and expansion of the necessary applications, designers switched to ac systems. The introduction of ac equipment was facilitated by the following developments:

- The using of brushless ac machines
- Galvanic isolation of the electrical circuits in different subsystems
- Simple voltage matching of the transformers
- The use of simple and reliable rectifiers to supply power to dc components

The development of hybrid electric–hydraulic drives is of great interest. Hydraulic drives are fast and relatively light. In contrast, the operating costs of electric drives are low. These benefits may be combined if hydraulic power is generally employed, but electrical systems are used to control individual devices.

On an airplane, the mass of the power supply system is critical. The mass may be significantly reduced by increasing the magnitude and frequency of the voltage and by multiplexing of the electric circuits. High-frequency voltage may be created by means of intermediate frequency converters. Multiplex systems are based on the transmission of low-power control signals in the coded form within a single channel.

The adoption of these principles relies on electronic information systems and power electronics. In particular, microprocessor systems permit rapid processing of vast quantities of information, and high-speed electronic switches are characterized by low energy consumption and are able to handle large voltages and currents.

Systems with a high-speed ac or dc generator mounted on a turbine shaft can be widely introduced. Such generators, in the range 200–300 V, provide the basic power source for the aircraft. In the case of an ac generator, its voltage is rectified and/or sent to a frequency converter so as to obtain the voltage of stable frequency with variation in the turbine speed (Figure 10.23). To this end, directly coupled frequency converters may be used if the generator frequency considerably exceeds the converter's output frequency. The output voltage of the rectifiers and the ac voltage of stable frequency may be sent to the consumers directly or by way of various converters if adjustment of the voltage is required. For example, an ac voltage of 270 V can be converted to 28 V or to dc voltage for drive control.

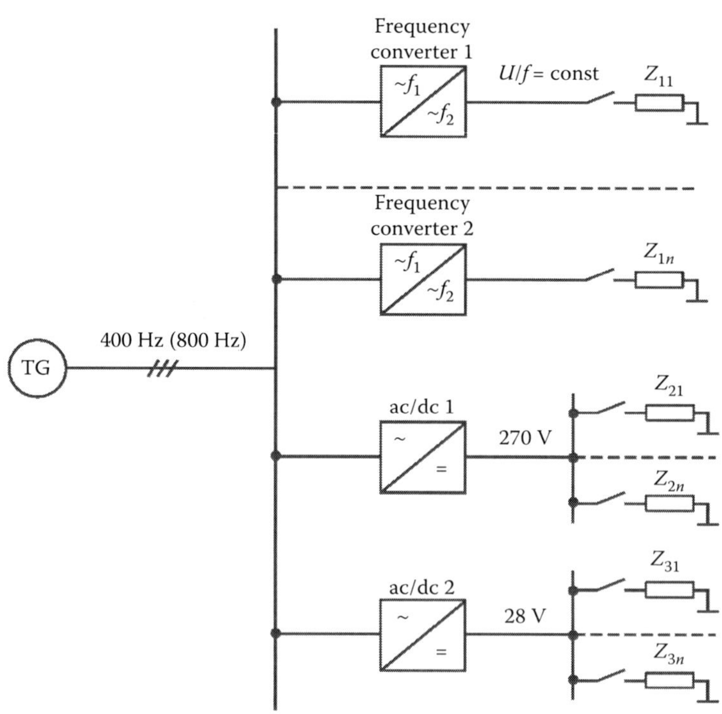

Figure 10.23 Structure of an aircraft power supply system.

10.3.3.6 Rocket power supply

The performance of rockets largely depends on their power supply, which may take various forms. For example, in astronautical rockets, all the electrical systems of the spacecraft must be maintained. Different power sources will be employed on the ground and in motion.

Depending on the power of the electrical systems on the spacecraft and the period for which their operation must be ensured, different onboard energy sources can be employed (Glebov, 1999).

- Batteries and solar cells
- Electrochemical generators
- Isotope sources

The use of these sources requires conversion and regulation of the power produced. As practically all such sources generate dc power, dc converters are widely used on spacecraft. Pulsed converters with a high working frequency (>20 kHz) are used so as to reduce the mass of the input and output passive filters and transformers. To match the voltage of the individual subsystems and obtain ac power, high-frequency inverters and inverters with an output voltage of the required frequency are employed.

Before and during rocket launching, ground-based power sources are important. They provide high-quality power for all the rocket's subsystems prior to the launch and then hand off to onboard sources. The ground-based sources are mainly based on rectifiers with output parameters corresponding to the onboard system in static and dynamic operation. The primary ground-based sources include industrial grids, diesel–electric generators, and dc sources. Depending on the specific circumstances, dc or ac sources producing power of different qualities may be most appropriate. The key consideration is the maintenance of power in the spacecraft until it is launched.

10.3.4 Engineering requirements

10.3.4.1 Basic requirements for power electronic devices

The basic requirements on power electronic components may be divided into two groups:

- Electrical requirements
- Design requirements

Electrical requirements. The electrical requirements are determined by the type of power conversion and the characteristics of the primary power source and the load. Correspondingly, different requirements are imposed on the input and output parameters of the converter.

In the case of an ac source, attention focusses not only on the rated voltage and current, the number of phases, and the frequency but also on the quality of the power, including its static and dynamic stability, the distortion (nonsinusoidal waveform) of the voltage, and the duration and periodicity of dynamic fluctuations. If the power of the primary source is comparable with that of the converter, the requirements on the harmonic composition of the converter input current, the power factor (for the fundamental voltage and current components), and the cyclic operation of the converter must be noted. Besides the requirements on the harmonic composition of the converter's input current, requirements are imposed on the distortion of the source voltage. However, the internal impedance of the source must be noted here, or else characteristics (say, frequency characteristics) permitting the identification of an equivalent electrical component for the primary source (with allowance for the built-in regulator) must be presented. The influence of the nonsinusoidal converter current on the source voltage may be assessed by means of the corresponding mathematical model of the source, taking account of its output parameters. Note that, with long leads between the source and the converter, their impedance must be noted. In some cases, for example, when higher current harmonics are present, the distributed inductance and capacitance of the cables must be noted. Neglecting those parameters may lead to undesirable high-frequency resonance, with greater distortion of the source voltage and may also produce additional voltage surges and dips in dynamic converter operation.

For converters connected to a dc source, its rated voltage, its static and dynamic instability, and also the duration and periodicity of dynamic fluctuations must be noted. The magnitude and frequency of the voltage ripple are also important. When the power of the source is comparable with that of the converter, the dynamic values of the sources internal impedance (with allowance for the built-in regulator and the bidirectional conduction) must be noted. On that basis, the influence of pulsation of the converter's input current on the source voltage and its surges and dips with switching of the converter current (in particular, when the load is applied and removed) may be assessed.

For converters that serve as secondary power sources, the requirements on the output parameters are analogous to those on the input parameters. The only difference is that the input parameters must be regarded as initial data, largely unaffected by the converter, whereas the output parameters are determined by the converter design. In this respect, the converter has a dual nature when it acts as a secondary power source: for the primary source, it is the load; for the consumers, it is the source of power of specific type and quality.

As a rule, the development of converters as power sources is associated with requirements on the range within which the output parameters are controlled. As the relevant standards mandate the required values of the output parameters, regulation allows the specific characteristics of the consumers to be taken into account, as well as the voltage drop in the cables.

For converters supplying power to key systems, the limiting permissible deviation of the output parameters from the standard values is usually noted, in all operating conditions, including emergencies. These requirements provide the basis for the development of the converter internal safety system.

In some cases, for example, when using the converter in special power supply systems, requirements may be imposed on the time required for the converter to accept a load after receiving the command that switches it on. When using converters in UPSs, a requirement is imposed on the duration of the interruption in the power supply at its output buses in all conditions, including malfunctioning of the converter (which must be backed up).

In formulating the requirements on the electrical parameters, the sensitivity of semiconductor devices to instantaneous currents and voltages at the limits of the permitted range must be taken into account. Therefore, increasing the stipulations regarding even brief permissible overloads or voltage surges will entail an increase in the number of semiconductor devices or switching to more powerful devices (i.e., those designed for greater current and voltage). In some cases, for example, for thyristor inverters with capacitive commutation, the rated power of the switching components must be increased in light of possible overloads. Hence, the rated power of the converter must be inflated in order to withstand even brief overloads.

Design requirements. The design requirements on power electronic components are predominantly determined by the operating conditions. They will also be different for an autonomous converter and a converter that is part of an electrical system with specific information and control panels.

Modular design of power electronic components is most common. Therefore, one of the basic design requirements is that the components of the device should be interchangeable.

The design is usually developed on the basis of requirements regarding its mechanical stability. All the possible types of mechanical disturbances (one-time jolts, vibration, etc.) in operation and transportation are noted, along with their magnitudes.

The design must permit monitoring of the electrical parameters in the course of production and operation, by means of built-in and/or external systems. This stipulation is directly related to the requirements on repair and maintenance.

Sealing is required to permit operation in water or at elevated levels of humidity and also in the presence of aggressive agents. Containers may effectively be used for this purpose in converters of moderate power. The coatings on power electronic components must be corrosion-resistant and must also maintain an esthetic appearance. The same considerations apply to the storage conditions, which must also be noted in the design process.

The design must ensure safe operation, in accordance with the relevant regulations. In particular, the design must ensure adequate grounding of all metallic conductors that might be touched by operators or others. In turn, the grounding measures must ensure consistent transient impedance. The impedance between the grounding elements and the accessible parts of the equipment should not exceed specified standard values.

The manufacturing conditions developed for the converter must take account of the comprehensive requirements in the plant standards (in relation to the installation of power components, printed circuits, etc.).

The components must be used in conditions specified by the corresponding standards.

Numerous requirements are determined by the functions of the converter in the electrical system. These requirements apply to the electrical connections between the converter and components such as the primary source, the power consumers, switchgear, control modules, and information-display systems. The specific types of connectors, the type of control signals and information signals, and the requirements on the cables must be noted. For autonomous converters with operator control, the requirements on the monitoring and signaling systems must be specified, in light of standardized metrological requirements.

Ergonomic and esthetic considerations will also be incorporated in the design.

10.3.4.2 Electromagnetic compatibility

Most semiconductor converters rely on periodic switching of nonlinear elements (relays), which results in abrupt current and voltage change in the converters. As a result, electromagnetic radiation is generated over a broad frequency spectrum. These effects are especially pronounced in converters, in which electrical switching often occurs at high frequencies. The electromagnetic radiation produced may interfere with the normal functioning of nearby equipment. Such electromagnetic interference may also disrupt the functioning of the converters. In addition, we need to distinguish between the action of electromagnetic radiation at high frequencies and the distortion of grid current by converters. The distortion of the grid current is assessed in terms of the ratio between the higher harmonics (including the 40th harmonic) and the grid's fundamental component. Current and voltage distortion impairs the quality of electrical energy and is regulated, for example, by the IEEE 519 standard.

We must distinguish between the conductive interference supplied to the grid and the inductive interference transmitted to the surroundings by the electromagnetic field. Conductive interference, including that produced by a radiofrequency electromagnetic field (in the range from 150 kHz to 80 MHz), is measured at frequencies from 3 kHz to 80 MHz. Inductive electromagnetic interference is regulated in the range from 30 to 1000 MHz. The permissible level of the interference is specified in numerous national and international standards in relation to the electromagnetic compatibility required for different components.

Note that electromagnetic interference affects not only engineering systems, but also human beings. Limiting levels of electromagnetic interference in the range from 30 kHz to 300 GHz are established in public-health standards and regulations.

Attention must be paid to electromagnetic compatibility at most stages of the design process. Besides the introduction of devices to suppress electromagnetic interference, the electromagnetic compatibility of a converter will depend on its design, the electrical installation of its individual components, the filtration employed, and the screening of individual components. These issues are complex and demand careful attention.

The designer primary tasks are to distinguish between the electrical circuits in terms of the functions of the components, to minimize the length of the cable required, and to separate the circuits most sensitive to interference from sources of interference by high-frequency filters. In any case, the wiring should not be haphazard and should not be focussed exclusively on reducing production costs. Even in computerized design, the outcome must be scrutinized so as to identify pathways in which interference may penetrate.

In power electronic systems, the following wiring rules must be observed.

- Separation of power circuits from control circuits
- Perpendicularity of the control-system and power-system cables
- Minimum cable length (within the design constraints)
- The combination of three-phase ac circuits and dc circuits in individual pathways with minimum distance between the individual leads

For safety purposes, many components must be connected to a grounding line. However, the use of a common line for different components is undesirable. Note that the grounding line may create resistance to high-frequency interference. As a result, voltage at the frequency of the electromagnetic interference may arise at such resistance, with an undesirable impact on devices that are sensitive to interference. Accordingly, it is expedient to create individual short buses parallel to the grounding line.

Where necessary, point grinding of individual devices may be employed. Note that the length of the line between each device and the common point must be minimized.

In addition to these basic wiring rules, various specific methods are employed for components with particular functions. In particular, leads in a twisted-pair configuration are used to transmit signals from the sensors to the amplifiers of the regulators, the measuring devices, and other highly sensitive components. Transposition of the wires minimizes their induction and compensates the currents induced by interference sources.

Electrostatic, magnetostatic, and electromagnetic screens can be used to protect the leads and components from external electromagnetic fields.

Electrostatic screens are usually made of copper or aluminum foil and surround the interference source. The metallic casing of the screen encloses the wire's electric field, limiting its propagation into the surroundings. However, the effectiveness of the screen depends significantly on the screen's quality and grounding. Magnetostatic screens are made of magnetic materials and reduce the external magnetic field. However, on account of eddy currents, their effectiveness in converters is limited. High-frequency fields may be shielded by means of electromagnetic screens, which reflect the electromagnetic energy. However, magnetostatic and electromagnetic screens are rarely used in converters, on account of their structural complication and the additional active-power losses. The use of such screens may be expedient, for example, in separating magnetic elements and microcircuits.

Electric filters are mainly used to suppress interference in converters. Filters differ in structure, circuit design, and electrical components. In moderate-power converters, L-type filters with pass-through capacitors are most common. The output (input) buses of the converter usually provide the series impedance required for interference suppression; where their inductance is insufficient for this purpose, a series choke is added. It is expedient to surround the filter by a grounded screen. Different types of capacitors may differ in their effectiveness as filters for the suppression of interference at various frequencies. Therefore, it may make sense to employ capacitors of different types in radiofrequency interference filters. Note that radiofrequency filters must be included in converter circuits in the initial stages of the design process. However, their parameters are generally refined experimentally, on account of the innumerable factors that affect the radio interference, such as the wiring configuration, the layout of the functional components, and the grounding system.

10.3.4.3 *Certification of power electronic devices*

The main requirements on power electronic components are outlined in numerous standards. The process of verifying compliance with those standards is known as certification. Certification permits monitoring of product quality, performance, and safety. Therefore, it is an important safeguard of global competitiveness.

Standardization involves certification of products and the quality of their manufacture. At completion, a certificate of compliance is issued for the product or the factory quality-control system.

The International Organization for Standardization maintains satisfactory methodological practices worldwide. For its part, the International Electrotechnical Commission (IEC) operates the most developed certification system for electrical and electronic equipment. National certification systems can be more stringent than the international standards. In Russia, Gosstandart coordinates certification efforts and oversees a network of accredited certification centers and test laboratories that are empowered to verify product quality and manufacturing practices in relation to national and international standards.

In order to eliminate administrative and technical barriers to trade and to improve product safety, Russia has introduced a law on engineering regulation, which outlines the fundamental principles of certification. It lays out technical regulations to improve safety and environmental production. Mandatory and voluntary product certification has been established in accordance with those regulations. Power electronic components are often used in applications in which increased safety is required. In such cases, the products must comply with the technical regulations. The categories of products subject to mandatory and voluntary certification have been determined. In any case, the declared product characteristics must conform to the relevant standards.

As an example, we now briefly consider the requirements on UPSs. Such systems contain rectifiers, inverters, semiconductor regulators, and switches. Therefore, the requirements on UPSs are typical of many power electronic components.

The IEC 62040 standard on UPS consists of three sections. The first contains the basic principles and specifies the electrical and mechanical characteristics of UPSs and their components. The overall test conditions are outlined, along with the basic requirements on the structure and the interface. Requirements on the mechanical strength and electrical strength are considered. The first part consists of two main sections 62040-1.1 and 62040-1.2 for UPS with different operator access. In appendices to the first section, standards regarding test procedures, the types of loads, and ventilation of battery segments are stated, along with recommendations for battery disconnection in transportation.

In the second part 62040-2, the requirements on electromagnetic compatibility are outlined. Measurement conditions are specified. The basic interference stability requirements are formulated. Permissible levels of electromagnetic interference are stated for different frequency ranges.

The third part 62040-3 outlines in detail the technical specifications on UPSs and the corresponding test procedures. Basic information is provided regarding their input and output characteristics. Typical structures and operating conditions for UPSs are presented. Procedures are specified for static and dynamic testing of UPSs with different grid and load conditions. Methods of testing safety and warning circuits are considered. Appendices outline typical faults and types of load.

This brief review of the standard for UPSs illustrates that current standards cover practically all aspects of the functioning and performance of devices, including conditions of transportation and storage. Extremely extensive testing is required to verify compliance of the UPS characteristics with every aspect of the standard. Therefore, manufacturers must reach an agreement with the customer regarding the tests that will be required, in light of the manufacturing and operating conditions. The first priority is safety testing and verification that the requirements in the product technical specifications match the standard requirements.

References

Alekseev, B.A. 2007. Energy and energy problems around the world. *Energetika za Rubezhom*, 5, 31–47 (in Russian).

Burman, A.P., Rozanov, Yu.K., and Shakaryan, Yu. 2012. Upravlenie potokami elektroenergii i povyshenie effektivnosti elektroenergeticheskih system (Control of electric energy and enhancement of power systems efficiency). Moscow: MPEI publishing (in Russian).

da Rosa, A.V. 2009. *Fundamentals of Renewable Energy Processes*, p. 7. USA: Academic Press.

Glebov, I.A. 1999. *Istoriya elektrotekhniki* (*The History of Electrical Engineering*), Glebov, I.A., Ed. Moscow: Izd. MEI (in Russian).

Iqbal, H. 2002. Switched reluctance machines. *The Power Electronics Handbook*, Skvarenina, T.L., Ed., pp. 13-1–13-9. USA: CRC Press.

Kazmierkowski, M.P., Krishnan, R., and Blaabjerg, F. 2002. *Control Power Electronics*. USA: Academic Press.

Kryukov, K.V. and Baranov, N.N. 2012. Extending the functionality of power supply systems with photovoltaic energy converters. *Russian Electr. Eng.*, 83(5), 278–284.

Perreault, D.J., Afridi, K.K., and Khan, I.A. 2001. Automotive applications of power electronics. *Power Electronics Handbook*, Rashid, M.H., Ed., pp. 791–816. USA: Academic Press.

Rahman, M.F., Patterson, D., Cheok, A., and Betts, R. 2001. Motor drives. *Power Electronics Handbook*, Rashid, M.H., Ed., pp. 663–733. USA: Academic Press.

Rozanov, Yu.K. 2010. *Staticheskie kommutatsionnye apparaty i regulyatory peremennogo toka, Elektricheskie i elektronnye apparaty* (*Static Switchgear and ac Controllers: Electrical and Electronic Equipment*), vol. 2. Moscow: Akademiya (in Russian).

Rozanov, Yu.K. and Kriukov, K.V. 2008. Control of the power flow in an energy system based on grid connected with photovoltaic generator. Proceedings of 12th WSEAS International Conference on Circuits, July 2008, Heraklion, Greece.

Rozanov, Yu.K. and Sokolova, E.M. 2004. *Elektronnye ustroistva. Elektromekhanicheskikh system* (*Electronic Devices and Electromechanical Systems*). Moscow: Akademiya (in Russian).

Sabonnadiere, J.-C. 2009. *Renewable Energies Processes*. USA: Academic Press.

Sood, V.K. 2001. HVDC transmission. *Power Electronics Handbook*, Rashid, M.H., Ed., pp. 575–597. USA: Academic Press.

Index

D

J

K

L

M